TEXTBOOKS in MATHEMATICS

Series Editors: Al Boggess and Ken Rosen

PUBLISHED TITLES

ABSTRACT ALGEBRA: AN INQUIRY-BASED APPROACH
Jonathan K. Hodge, Steven Schlicker, and Ted Sundstrom

APPLIED ABSTRACT ALGEBRA WITH MAPLE™ AND MATLAB®, THIRD EDITION
Richard Klima, Neil Sigmon, and Ernest Stitzinger

APPLIED DIFFERENTIAL EQUATIONS: THE PRIMARY COURSE
Vladimir Dobrushkin

COMPUTATIONAL MATHEMATICS: MODELS, METHODS, AND ANALYSIS WITH MATLAB® AND MPI, SECOND EDITION
Robert E. White

DIFFERENTIAL EQUATIONS: THEORY, TECHNIQUE, AND PRACTICE, SECOND EDITION
Steven G. Krantz

DIFFERENTIAL EQUATIONS: THEORY, TECHNIQUE, AND PRACTICE WITH BOUNDARY VALUE PROBLEMS
Steven G. Krantz

DIFFERENTIAL EQUATIONS WITH MATLAB®: EXPLORATION, APPLICATIONS, AND THEORY
Mark A. McKibben and Micah D. Webster

ELEMENTARY NUMBER THEORY
James S. Kraft and Lawrence C. Washington

EXPLORING LINEAR ALGEBRA: LABS AND PROJECTS WITH MATHEMATICA®
Crista Arangala

GRAPHS & DIGRAPHS, SIXTH EDITION
Gary Chartrand, Linda Lesniak, and Ping Zhang

INTRODUCTION TO ABSTRACT ALGEBRA, SECOND EDITION
Jonathan D. H. Smith

INTRODUCTION TO MATHEMATICAL PROOFS: A TRANSITION TO ADVANCED MATHEMATICS, SECOND EDITION
Charles E. Roberts, Jr.

INTRODUCTION TO NUMBER THEORY, SECOND EDITION
Marty Erickson, Anthony Vazzana, and David Garth

PUBLISHED TITLES CONTINUED

LINEAR ALGEBRA, GEOMETRY AND TRANSFORMATION
Bruce Solomon

MATHEMATICAL MODELLING WITH CASE STUDIES: USING MAPLE™ AND MATLAB®, THIRD EDITION
B. Barnes and G. R. Fulford

MATHEMATICS IN GAMES, SPORTS, AND GAMBLING—THE GAMES PEOPLE PLAY, SECOND EDITION
Ronald J. Gould

THE MATHEMATICS OF GAMES: AN INTRODUCTION TO PROBABILITY
David G. Taylor

MEASURE THEORY AND FINE PROPERTIES OF FUNCTIONS, REVISED EDITION
Lawrence C. Evans and Ronald F. Gariepy

NUMERICAL ANALYSIS FOR ENGINEERS: METHODS AND APPLICATIONS, SECOND EDITION
Bilal Ayyub and Richard H. McCuen

ORDINARY DIFFERENTIAL EQUATIONS: AN INTRODUCTION TO THE FUNDAMENTALS
Kenneth B. Howell

RISK ANALYSIS IN ENGINEERING AND ECONOMICS, SECOND EDITION
Bilal M. Ayyub

TRANSFORMATIONAL PLANE GEOMETRY
Ronald N. Umble and Zhigang Han

TEXTBOOKS in MATHEMATICS

ADVANCED LINEAR ALGEBRA

Hugo J. Woerdeman

Drexel University
Philadelphia, Pennsylvania, USA

CRC Press
Taylor & Francis Group
Boca Raton   London   New York

CRC Press is an imprint of the
Taylor & Francis Group an **informa** business

A CHAPMAN & HALL BOOK

First published in paperback 2024

First published 2016
by CRC Press
2385 NW Executive Center Drive, Suite 320, Boca Raton FL 33431

and by CRC Press
4 Park Square, Milton Park, Abingdon, Oxon, OX14 4RN

CRC Press is an imprint of Taylor & Francis Group, LLC

© 2016, 2024 by Taylor & Francis Group, LLC

ISBN: 978-1-032-92140-2 (pbk)
ISBN: 978-1-498-75403-3 (hbk)
ISBN: 978-0-429-08897-1 (ebk)

DOI: 10.1201/b18994

**Visit the Taylor & Francis Web site at
http://www.taylorandfrancis.com**

**and the CRC Press Web site at
http://www.crcpress.com**

To my very supportive family:
Dara, Sloane, Sam, Ruth, and Myra.

Contents

Preface to the Instructor   xi

Preface to the Student   xiii

Acknowledgments   xv

Notation   xvii

List of Figures   xxi

1   Fields and Matrix Algebra   1

    1.1   The field $\mathbb{Z}_3$ 2

    1.2   The field axioms   3

    1.3   Field examples   5

        1.3.1   Complex numbers   7

        1.3.2   The finite field $\mathbb{Z}_p$, with p prime   9

    1.4   Matrix algebra over different fields   11

        1.4.1   Reminders about Cramer's rule and the adjugate matrix.   17

    1.5   Exercises   20

2   Vector Spaces   27

    2.1   Definition of a vector space   27

2.2   Vector spaces of functions   29

    2.2.1   The special case when X is finite   31

2.3   Subspaces and more examples of vector spaces   32

    2.3.1   Vector spaces of polynomials   34

    2.3.2   Vector spaces of matrices   36

2.4   Linear independence, span, and basis   37

2.5   Coordinate systems   45

2.6   Exercises   48

3   Linear Transformations   **55**

3.1   Definition of a linear transformation   55

3.2   Range and kernel of linear transformations   57

3.3   Matrix representations of linear maps   61

3.4   Exercises   65

4   The Jordan Canonical Form   **69**

4.1   The Cayley–Hamilton theorem   69

4.2   Jordan canonical form for nilpotent matrices   71

4.3   An intermezzo about polynomials   75

4.4   The Jordan canonical form   78

4.5   The minimal polynomial   82

4.6   Commuting matrices   84

4.7   Systems of linear differential equations   87

4.8   Functions of matrices   90

4.9   The resolvent   98

4.10 Exercises   100

5 Inner Product and Normed Vector Spaces   **109**

  5.1 Inner products and norms   109

  5.2 Orthogonal and orthonormal sets and bases   119

  5.3 The adjoint of a linear map   122

  5.4 Unitary matrices, QR, and Schur triangularization   125

  5.5 Normal and Hermitian matrices   128

  5.6 Singular value decomposition   132

  5.7 Exercises   137

6 Constructing New Vector Spaces from Given Ones   **147**

  6.1 The Cartesian product   147

  6.2 The quotient space   149

  6.3 The dual space   157

  6.4 Multilinear maps and functionals   166

  6.5 The tensor product   168

  6.6 Anti-symmetric and symmetric tensors   179

  6.7 Exercises   189

7 How to Use Linear Algebra   **195**

  7.1 Matrices you can't write down, but would still like to use . .   196

  7.2 Algorithms based on matrix vector products   198

  7.3 Why use matrices when computing roots of polynomials? . .   203

  7.4 How to find functions with linear algebra?   209

  7.5 How to deal with incomplete matrices   217

  7.6 Solving millennium prize problems with linear algebra   222

    7.6.1 The Riemann hypothesis   223

7.6.2   P vs. NP   225

7.7   How secure is RSA encryption?   229

7.8   Quantum computation and positive maps   232

7.9   Exercises   238

How to Start Your Own Research Project   **247**

Answers to Exercises   **249**

Bibliography   **323**

Index   **325**

Preface to the Instructor

This book is intended for a second linear algebra course. Students are expected to be familiar with (computational) material from a first linear algebra course: matrix multiplication; row reduction; pivots; solving systems of linear equations; checking whether a vector is a linear combination of other vectors; finding eigenvalues and eigenvectors; finding a basis of a nullspace, column space, row space, and eigenspace; computing determinants; and finding inverses. The assumption is that so far they have worked over the real numbers $\mathbb{R}$.

In my view, the **core material** in this book is the following and takes about **24 academic hours**[1] of lectures:

- **Chapter 1**: Introduction to the notion of a general field, with a focus on $\mathbb{Z}_p$ and $\mathbb{C}$, and a refresher of the computational items from the first linear algebra course but now presented over different fields. (4 hours)

- **Chapter 2**: Vector spaces, subspaces, linear independence, span, basis, dimension, coordinate systems. (6 hours)

- **Chapter 3**: Linear transformations, range and kernel, matrix representations. My suggestion is to do a lot of different examples of finding matrix representations. (5 hours)

- **Chapter 4, Sections 4.1, 4.2 and 4.4**: Cayley–Hamilton, presenting the Jordan canonical form (without complete proof) and doing computational examples. (3 hours)

- **Chapter 5**: Inner products, norms, orthogonality, adjoint, QR, normal matrices (including unitary, Hermitian, and positive (semi-)definite), singular value decomposition. (6 hours)

To supplement the core material there are several options:

- **Chapter 4, Sections 4.2–4.5**: Provide the details of the proof of the

---

[1] Academic hour = 50 minutes.

Jordan canonical form and introduce the minimal polynomial (2–3 hours).

- **Chapter 4, Sections 4.6 and 4.7**: These two sections are independent of one another, and each takes about 1 hour. Clearly, how the Jordan canonical form helps in solving differential equations is a classical one for this course. The result on commuting matrices is one that sometimes makes it into my course, but other times does not. (1–2 hours)

- **Chapter 4, Section 4.8 (and 4.9)**: The section "Functions of matrices" provides a way to introduce e^{tA} and discuss the application to systems of linear differential equations in a more conceptual way. Section 4.9 requires knowledge of Cauchy's integral formula and may be somewhat of a stretch for this course. Still, accepting Cauchy's formula, I believe that the corresponding exercises are accessible to the students. (2 hours)

- **Chapter 6, Sections 6.1–6.3**: These three sections are independent of one another. They provide fundamental constructions of new vector spaces from given ones. (1–3 hours)

- **Chapter 6, Sections 6.4–6.6**: Tensor (or Kronecker) products provide a really exciting tool. I especially like how the determinants and permanents show up in anti-symmetric and symmetric tensors, and how for instance, the Cauchy–Binet formula is derived. I would strongly consider including this if I had a semester-long course. (4–5 hours)

- **Chapter 7**: I use the items in this chapter to try to (re)energize the students at the end of a lecture, and ask questions like "What made Google so successful?" (*Response*: Their page rank algorithm). "Does it surprise you when I tell you that to compute roots of a polynomial, one builds a matrix and then computes its eigenvalues?" (*Response*: Yes (hopefully) and isn't the QR algorithm really neat?), "Do you want to win a million bucks?" (*Response*: Solve a millennium prize problem). Of course, there is the option to treat these items in much more detail or assign them as projects (if only I had the time!). (1–7 hours)

I hope that my suggestions are helpful, and that you find this a useful text for your course. I would be very happy to hear from you! I realize that it takes a special effort to provide someone with constructive criticism, so when you take time to do that, I will be especially appreciative.

Preface to the Student

I think that linear algebra is a great subject, and I strongly hope that you (will) agree. It has a strong theoretical side, ample opportunity to explore the subject with computation, and a (continuously growing) number of great applications. With this book, I hope to do justice to all these aspects. I chose to treat the main concepts (vector space and linear transformations) in their full abstraction. Abstraction (taking operations out of their context, and studying them on their own merit) is really **the** strength of mathematics; how else can a theory that started in the 18th and 19th centuries have all these great 21st-century applications (web search engines, data mining, etc.)? In addition, I hope that when you are used to the full abstraction of the theory, it will allow you to think of possibilities of applying the theory in the broadest sense. And, maybe as a more direct benefit, I hope that it will help when you take abstract algebra. Which brings me to my last point. While current curriculum structure has different mathematical subfields neatly separated, this is not reality. Especially when you apply mathematics, you will need to pull from different areas of mathematics. This is why this book does not shy away from occasionally using some calculus, abstract algebra, real analysis and (a little bit of) complex analysis.

Just a note regarding the exercises: I have chosen to include full solutions to almost all exercises. It is up to you how you use these. Of course, if increasingly you rely less on these solutions, the better it is. There are a few exercises (designated as "Honors") for which no solution is included. These are somewhat more challenging. Try them and if you succeed, use them to impress your teacher or yourself!

Preface to the Student

Acknowledgments

In the thirty years that I have taught (starting as a teaching assistant), I have used many textbooks for different versions of a linear algebra course (Linear Algebra I, Linear Algebra II, Graduate Linear Algebra; semester course and quarter course). All these different textbooks have influenced me. In addition, discussions with students and colleagues, sitting in on lectures, reading papers, etc., have all shaped my linear algebra courses. I wish I had a way to thank you all specifically for how you helped me, but I am afraid that is simply impossible. So I hope that a general thank you to all of you who have influenced me, will do: THANK YOU!

This book came about while I was immobile due to an ankle fracture. So, first of all, I would like to thank my wonderful wife Dara and my great kids Sloane, Sam, Ruth, and Myra, for taking care of me during the four months I spent recovering. I am very thankful to my colleagues at Drexel University: Dannis Yang, who used a first version of this text for his course, for providing me with detailed comments on Chapters 1–5; Shari Moskow, R. Andrew Hicks, and Robert Boyer for their feedback on the manuscript, and in Robert Boyer's case for also providing me with one of the figures. In addition, I am grateful to graduate student Charles Burnette for his feedback. I am also very thankful to those at CRC Press who helped me bring this manuscript to publication. Finally, I would like to thank you, the reader, for picking up this book. Without you there would have been no point to produce this. So, MANY THANKS to all of you!

Notation

Here are some often-used notations:

- $\mathbb{N} = \{1, 2, 3, \ldots\}$

- $\mathbb{N}_0 = \{0, 1, 2, \ldots\}$

- $\mathbb{Z}$ = the set of all integers

- $\mathbb{Q}$ = the field of rational numbers

- $\mathbb{R}$ = the field of real numbers

- $\mathbb{R}(t)$ = the field of real rational functions (in t)

- $\mathbb{C}$ = the field of complex numbers

- Re z = real part of z

- Im z = imaginary part of z

- $\bar{z}$ = complex conjugate of z

- $|z|$ = absolute value (modulus) of z

- $\mathbb{Z}_p$ (with p prime) = the finite field $\{0, 1, \ldots, p-1\}$

- $\mathrm{rem}(q|p)$ = remainder of q after division by p

- $\mathbb{F}$ = a generic field

- $\det(A)$ = the determinant of the matrix A

- $\mathrm{tr}(A)$ = the trace of a matrix A (= the sum of its diagonal entries)

- $\mathrm{adj}(A)$ = the adjugate of the matrix A

- $\mathrm{rank}(A)$ = the rank of a matrix A

- $\mathbb{F}[X]$ = the vector space of polynomials in X with coefficients in $\mathbb{F}$

- $\mathbb{F}_n[X]$ = the vector space of polynomials of degree $\leq n$ in X with coefficients in $\mathbb{F}$

- $\mathbb{F}^{n \times m}$ = the vector space of $n \times m$ matrices with entries in $\mathbb{F}$

- $\mathbb{F}^X$ = the vector space of functions acting $X \to \mathbb{F}$

- $H_n = \{A \in \mathbb{C}^{n \times n} : A = A^*\}$, the vector space over $\mathbb{R}$ consisting of all $n \times n$ Hermitian matrices

- $\mathbf{0}$ = the zero vector

- $\dim V$ = the dimension of the vector space V

- $\operatorname{Span}\{\mathbf{v}_1, \ldots, \mathbf{v}_n\}$ = the span of the vectors $\mathbf{v}_1, \ldots, \mathbf{v}_n$

- $\{\mathbf{e}_1, \ldots, \mathbf{e}_n\}$ = the standard basis in $\mathbb{F}^n$

- $[\mathbf{v}]_\mathcal{B}$ = the vector of coordinates of $\mathbf{v}$ relative to the basis $\mathcal{B}$

- $\operatorname{Ker} T$ = the kernel (or nullspace) of a linear map (or matrix) T

- $\operatorname{Ran} T$ = the range of a linear map (or matrix) T

- $T[W] = \{T(\mathbf{w}) : \mathbf{w} \in W\} = \{\mathbf{y} : \text{ there exists } \mathbf{w} \in W \text{ so that } \mathbf{y} = T(\mathbf{w})\} \subseteq \operatorname{Ran} T$

- id_V = the identity map on the vector space V

- $[T]_{\mathcal{C} \leftarrow \mathcal{B}}$ = the matrix representation of T with respect to the bases $\mathcal{B}$ and $\mathcal{C}$

- I_n = the $n \times n$ identity matrix

- $J_k(\lambda)$ = the $k \times k$ Jordan block with eigenvalue λ

- $w_k(A, \lambda) = \dim \operatorname{Ker}(A - \lambda I_n)^k - \dim \operatorname{Ker}(A - \lambda I_n)^{k-1}$; Weyr characteristic of A

- $\oplus_{k=1}^p A_k = \begin{pmatrix} A_1 & 0 & \cdots & 0 \\ 0 & A_2 & \cdots & 0 \\ \vdots & \vdots & \ddots & \vdots \\ 0 & 0 & \cdots & A_p \end{pmatrix}$

- $\operatorname{diag}(d_{ii})_{i=1}^n = \begin{pmatrix} d_{11} & 0 & \cdots & 0 \\ 0 & d_{22} & \cdots & 0 \\ \vdots & \vdots & \ddots & \vdots \\ 0 & 0 & \cdots & d_{nn} \end{pmatrix}$

- $\dotplus$ = direct sum

- $p_A(t)$ = the characteristic polynomial of the matrix A

- $m_A(t)$ = the minimal polynomial of the matrix A

- A^T = the transpose of the matrix A

- A^* = the conjugate transpose of the matrix A

- $T^\star$ = the adjoint of the linear map T

- $\langle \cdot, \cdot \rangle$ = an inner product

- $\| \cdot \|$ = a norm

- $\sigma_j(A)$ = the jth singular value of the matrix A, where $\sigma_1(A) = \|A\|$ is the largest singular value

- $\rho(A) = \max\{|\lambda| : \lambda$ is an eigenvalue of $A\}$ is the spectral radius of A

- $\mathrm{PSD}_n = \{A \in \mathbb{C}^{n \times n} : A$ is positive semidefinite$\} \subseteq H_n$

- $v + W = \{\mathbf{v} + \mathbf{w} : \mathbf{w} \in W\} = \{\mathbf{x} : \mathbf{x} - \mathbf{v} \in W\}$

- $V/W = \{\mathbf{v} + W : \mathbf{v} \in V\}$, the quotient space

- V' = the dual space of V

- $\mathcal{L}(V, W) = \{T : V \to W : T$ is linear$\}$

- $\mathbf{v} \otimes \mathbf{w}$ = the tensor product of $\mathbf{v}$ and $\mathbf{w}$

- $\mathbf{v} \wedge \mathbf{w}$ = the anti-symmetric tensor product of $\mathbf{v}$ and $\mathbf{w}$

- $\mathbf{v} \vee \mathbf{w}$ = the symmetric tensor product of $\mathbf{v}$ and $\mathbf{w}$

- $A[P, Q] = (a_{ij})_{i \in P, j \in Q}$, a submatrix of $A = (a_{ij})_{i,j}$

List of Figures

1.1 The complex number z in the complex plane. 9

7.1 These are the roots of the polynomial $\sum_{k=1}^{10,000} p_k(10,000)x^k$, where $p_k(n)$ is the number of partitions of n in k parts, which is the number of ways n can be written as the sum of k positive integers. 207

7.2 A Meyer wavelet. 210

7.3 Blurring function. 216

7.4 The original image (of size $3000 \times 4000 \times 3$). 217

7.5 The Redheffer matrix of size 500×500. 224

7.6 A sample graph. 225

5.7 The original image (of size $672 \times 524 \times 3$). 299

1

Fields and Matrix Algebra

CONTENTS

1.1   The field $\mathbb{Z}_3$ 2
1.2   The field axioms 3
1.3   Field examples 5
      1.3.1   Complex numbers 7
      1.3.2   The finite field $\mathbb{Z}_p$, with p prime 9
1.4   Matrix algebra over different fields 11
      1.4.1   Reminders about Cramer's rule and the adjugate
              matrix. 17
1.5   Exercises 20

The central notions in linear algebra are *vector spaces* and *linear transformations* that act between vector spaces. We will define these notions in Chapters 2 and 3, respectively. But before we can introduce the general notion of a vector space we need to talk about the notion of a *field*. In your first Linear Algebra course you probably did not worry about fields because it was chosen to only talk about the real numbers $\mathbb{R}$, a field you have been familiar with for a long time. In this chapter we ask you to get used to the general notion of a field, which is a set of mathematical objects on which you can define algebraic operations such as addition, subtraction, multiplication and division with all the rules that also hold for real numbers (commutativity, associativity, distributivity, existence of an additive neutral element, existence of an additive inverse, existence of a multiplicative neutral element, existence of a multiplicative inverse for nonzeros). We start with an example.

1.1   The field $\mathbb{Z}_3$

Let us consider the set $\mathbb{Z}_3 = \{0, 1, 2\}$, and use the following tables to define addition and multiplication:

+	0	1	2
0	0	1	2
1	1	2	0
2	2	0	1

.	0	1	2
0	0	0	0
1	0	1	2
2	0	2	1

So, in other words, $1 + 1 = 2$, $2 + 1 = 0$, $2 \cdot 2 = 1$, $0 \cdot 1 = 0$, etc. In fact, to take the sum of two elements we take the usual sum, and then take the remainder after division by 3. For example, to compute $2 + 2$ we take the remainder of 4 after division by 3, which is 1. Similarly for multiplication.

What you notice in the table is that when you add 0 to any number, it does not change that number (namely, $0 + 0 = 0$, $0 + 1 = 1$, $1 + 0 = 1$, $0 + 2 = 2$, $2 + 0 = 2$). We say that 0 is *the neutral element for addition*. Analogously, 1 is *the neutral element for multiplication*, which means that when we multiply a number in this field by 1, it does not change that number ($0 \cdot 1 = 0$, $1 \cdot 2 = 2$, etc.). Every field has these neutral elements, and they are typically denoted by 0 and 1, although there is no rule that you have to denote them this way.

Another important observation is that in the core part of the addition table

+			
	0	1	2
	1	2	0
	2	0	1

the 0 appears exactly once in every row and column. What this means is that whatever x we choose in $\mathbb{Z}_3 = \{0, 1, 2\}$, we can always find exactly one $y \in \mathbb{Z}_3$ so that

$$x + y = 0.$$

We are going to call y the *additive inverse* of x, and we are going to write $y = -x$. So

$$0 = -0, \quad 2 = -1, \quad 1 = -2.$$

It is important to keep in mind that the equation $y = -x$ is just a shorthand of the equation $x + y = 0$. So, whenever you wonder "what does this $-$ mean?," you have to go back to an equation that only involves $+$ and look at how addition is defined. One of the rules in any field is that *any element of a field has an additive inverse.*

How about multiplicative inverses? For real numbers, any number has a multiplicative inverse **except for** 0. Indeed, no number x satisfies $x \cdot 0 = 1$! In other fields, the same holds true. This means that in looking at the multiplication table for multiplicative inverses, we should only look at the part that does not involve 0:

$$
\begin{array}{c|cc}
\cdot & & \\
\hline
& 1 & 2 \\
& 2 & 1
\end{array}
$$

And here we notice that 1 appears exactly once in each row and column. This means that whenever $x \in \mathbb{Z}_3 \setminus \{0\} = \{1, 2\}$, there exists exactly one y so that

$$x \cdot y = 1.$$

We are going to call y the *multiplicative inverse* of x, and denote this as x^{-1}. Thus

$$1^{-1} = 1, \quad 2^{-1} = 2.$$

In addition to the existence of neutral elements and inverses, the addition and multiplication operations also satisfy commutativity, associativity and distributive laws, so let us next give the full list of axioms that define a field. And after that we will present more examples of fields, both with a finite number of elements (such as the field $\mathbb{Z}_3$ we defined in this subsection) as well as with an infinite number of elements (such as the real numbers $\mathbb{R}$).

1.2 The field axioms

A *field* is a set $\mathbb{F}$ on which two operations

$$+ : \mathbb{F} \times \mathbb{F} \to \mathbb{F}, \quad \cdot : \mathbb{F} \times \mathbb{F} \to \mathbb{F}$$

are defined satisfying the following rules:

1. *Closure of addition:* for all $x, y \in \mathbb{F}$ we have that $x + y \in \mathbb{F}$.

2. *Associativity of addition:* for all $x, y, z \in \mathbb{F}$ we have that $(x + y) + z = x + (y + z)$.

3. *Commutativity of addition*: for all $x, y \in \mathbb{F}$ we have that $x + y = y + x$.

4. *Existence of a neutral element for addition*: there exists a $0 \in \mathbb{F}$ so that $x + 0 = x = 0 + x$ for all $x \in \mathbb{F}$.

5. *Existence of an additive inverse*: for every $x \in \mathbb{F}$ there exists a $y \in \mathbb{F}$ so that $x + y = 0 = y + x$.

6. *Closure of multiplication*: for all $x, y \in \mathbb{F}$ we have that $x \cdot y \in \mathbb{F}$.

7. *Associativity of multiplication*: for all $x, y, z \in \mathbb{F}$ we have that $(x \cdot y) \cdot z = x \cdot (y \cdot z)$.

8. *Commutativity of multiplication*: for all $x, y \in \mathbb{F}$ we have that $x \cdot y = y \cdot x$.

9. *Existence of a neutral element for multiplication*: there exists a $1 \in \mathbb{F} \setminus \{0\}$ so that $x \cdot 1 = x = 1 \cdot x$ for all $x \in \mathbb{F}$.

10. *Existence of a multiplicative inverse for nonzeros*: for every $x \in \mathbb{F} \setminus \{0\}$ there exists a $y \in \mathbb{F}$ so that $x \cdot y = 1 = y \cdot x$.

11. *Distributive law*: for all $x, y, z \in \mathbb{F}$ we have that $x \cdot (y + z) = x \cdot y + x \cdot z$.

    We will denote the additive inverse of x by $-x$, and we will denote the multiplicative inverse of x by x^{-1}.

First notice that any field has at least two elements, namely $0, 1 \in \mathbb{F}$, and part of rule 9 is that $0 \neq 1$. Next, notice that rules 1–5 only involve addition, while rules 6–10 only involve multiplication. The distributive law is the only one that combines both addition and multiplication. In an Abstract Algebra course, one studies various other mathematical notions that involve addition and/or multiplication where only some of the rules above apply.

Some notational shorthands:

- Since addition is associative, we can just write $x + y + z$ instead of $(x + y) + z$ or $x + (y + z)$, because we do not have to worry whether we first add x and y together, and then add z to it, or whether we first add y and z together, and subsequently add x.

- When we are adding several numbers $x_1, \ldots, x_k$ together, we can write this as $x_1 + \cdots + x_k$ or also as $\sum_{j=1}^{k} x_j$. For example, when $k = 5$, we have

$$\sum_{j=1}^{5} x_j = x_1 + x_2 + x_3 + x_4 + x_5 = x_1 + \cdots + x_5.$$

We now also have rules like

$$\sum_{j=1}^{k-1} x_j + x_k = \sum_{j=1}^{k} x_j, \quad \sum_{j=1}^{p} x_j + \sum_{j=p+1}^{q} x_j = \sum_{j=1}^{q} x_j.$$

- While above we use · to denote multiplication, we will often leave the · out. Indeed, instead of writing $x \cdot y$ we will write xy. Occasionally, though, we will write the · just to avoid confusion: for instance, if we want to write 1 times 2, and leave out the · it looks like 1 2. As this looks too much like twelve, we will continue to write $1 \cdot 2$.

- As multiplication is associative, we can just write xyz instead of $(xy)z$ or $x(yz)$.

- When multiplying $x_1, \ldots, x_k$, we write $\prod_{j=1}^{k} x_j$ or $x_1 \cdots x_k$. For instance, when $k = 5$, we have

$$\prod_{j=1}^{5} x_j = x_1 x_2 x_3 x_4 x_5 = x_1 \cdots x_5.$$

We now also have rules like

$$\left(\prod_{j=1}^{k-1} x_j\right) x_k = \prod_{j=1}^{k} x_j, \quad \left(\prod_{j=1}^{p} x_j\right)\left(\prod_{j=p+1}^{q} x_j\right) = \prod_{j=1}^{q} x_j.$$

- We may write x^2 instead of xx, or x^3 instead of xxx, x^{-2} instead of $x^{-1}x^{-1}$, etc. Clearly, when we use a negative exponent we need to insist that $x \neq 0$. Using this convention, we have the familiar rule $x^k x^\ell = x^{k+\ell}$, with the convention that $x^0 = 1$ when $x \neq 0$.

- For the multiplicative inverse we will use both x^{-1} and $\frac{1}{x}$. It is important, though, that we **only** use $\frac{1}{x}$ for certain infinite fields (such as $\mathbb{Q}$, $\mathbb{R}$ and $\mathbb{C}$), as there we are familiar with $\frac{1}{2}$ (half), $\frac{3}{8}$ (three eighths), etc. However, in a finite field such as $\mathbb{Z}_3$ we will always use the notation x^{-1}. So **do not** write $\frac{1}{2}$ when you mean the multiplicative inverse of 2 in $\mathbb{Z}_3$!

1.3 Field examples

In this book we will be using the following fields:

- The real numbers $\mathbb{R}$ with the usual definition of addition and multiplication. As you have already taken a first course in linear algebra, we know that you are familiar with this field.

- The rational numbers $\mathbb{Q}$, which are all numbers of the form $\frac{p}{q}$, where $p \in \mathbb{Z} = \{\ldots, -2, -1, 0, 1, 2, \ldots\}$ and $q \in \mathbb{N} = \{1, 2, 3, \ldots\}$. Again, addition and multiplication are defined as usual. We assume that you are familiar with this field as well. In fact, $\mathbb{Q}$ is a field that is also a subset of the field $\mathbb{R}$, with matching definitions for addition and multiplication. We say that $\mathbb{Q}$ is a *subfield* of $\mathbb{R}$.

- The complex numbers $\mathbb{C}$, which consist of numbers $a + bi$, where $a, b \in \mathbb{R}$ and $i^2 = -1$. We will dedicate the next subsection to this field.

- The finite fields $\mathbb{Z}_p$, where p is a prime number. We already introduced you to $\mathbb{Z}_3$, and later in this section we will see how for any prime number one can define a field $\mathbb{Z}_p$, where addition and multiplication are defined via the usual addition and multiplication of integers followed by taking the remainder after division by p.

- The field $\mathbb{R}(t)$ of rational functions with real coefficients and independent variable t. This field consists of functions $\frac{r(t)}{s(t)}$ where $r(t)$ and $s(t)$ are polynomials in t, with $s(t)$ not being the constant 0 polynomial. For instance,

$$\frac{13t^2 + 5t - 8}{t^8 - 3t^5}, \quad \frac{5t^{10} - 27}{t + 5} \tag{1.1}$$

  are elements of $\mathbb{R}(t)$. Addition and multiplication are defined as usual. We are going to assume that you will be able to work with this field. The only thing that requires some special attention, is to think about the neutral elements. Indeed, the 0 in this field is the constant function 0, where $r(t) \equiv 0$ for all t and $s(t) \equiv 1$ for all t. The 1 in this field is the constant function 1, where $r(t) \equiv 1$ for all t and $s(t) \equiv 1$ for all t. Now sometimes, these elements appear in "hidden" form, for instance,

$$\frac{0}{t + 1} \equiv 0, \quad \frac{t + 5}{t + 5} \equiv 1.$$

In calculus you had to worry that $\frac{t+5}{t+5}$ is not defined at $t = -5$, but in this setting we always automatically get rid of common factors in the numerator and denominator. More formally, $\mathbb{R}(t)$ is defined as the field $\frac{r(t)}{s(t)}$, where $r(t)$ and $s(t) \not\equiv 0$ are polynomials in t that do not have a common factor. If one insists on uniqueness in the representation $\frac{r(t)}{s(t)}$, one can, in addition, require that $s(t)$ is monic which means that the highest power of t has a coefficient 1 (as is the case in (1.1)).

1.3.1 Complex numbers

The complex numbers are defined as

$$\mathbb{C} = \{a + bi \; ; a, b \in \mathbb{R}\},$$

with addition and multiplication defined by

$$(a + bi) + (c + di) := (a + c) + (b + d)i,$$

$$(a + bi)(c + di) := (ac - bd) + (ad + bc)i.$$

Notice that with these rules, we have that $(0 + 1i)(0 + 1i) = -1 + 0i$, or in shorthand $i^2 = -1$. Indeed, this is how to remember the multiplication rule:

$$(a + bi)(c + di) = ac + bdi^2 + (ad + bc)i = ac - bd + (ad + bc)i,$$

where in the last step we used that $i^2 = -1$. It may be obvious, but we should state it clearly anyway: two complex numbers $a + bi$ and $c + di$, with $a, b, c, d \in \mathbb{R}$ are equal if and only if $a = c$ and $b = d$. A typical complex number may be denoted by z or w. When

$$z = a + bi \text{ with } a, b \in \mathbb{R},$$

we say that the *real part* of z equals a and the *imaginary part* of z equals b. The notation for this is,

$$\text{Re } z = a, \quad \text{Im } z = b.$$

It is quite laborious, but in principle elementary, to prove that $\mathbb{C}$ satisfies all the field axioms. In fact, in doing so one needs to use that $\mathbb{R}$ satisfies the field axioms, as addition and multiplication in $\mathbb{C}$ are defined via addition and multiplication in $\mathbb{R}$. As always, it is important to realize what the neutral elements are:

$$0 = 0 + 0i, \quad 1 = 1 + 0i.$$

Another tricky part of this is the multiplicative inverse, for instance,

$$(1 + i)^{-1}, \; (2 - 3i)^{-1}. \tag{1.2}$$

Here it is useful to look at the multiplication

$$(a + bi)(a - bi) = a^2 + b^2 + 0i = a^2 + b^2. \tag{1.3}$$

This means that as soon as a or b is not zero, we have that $(a + bi)(a - bi) = a^2 + b^2$ is a nonzero (actually, positive) real number. From this we can conclude that

$$\frac{1}{a + bi} = (a + bi)^{-1} = \frac{a - bi}{a^2 + b^2} = \frac{a}{a^2 + b^2} - \frac{b}{a^2 + b^2}i.$$

So, getting back to (1.2),

$$\frac{1}{1+i} = \frac{1}{2} - \frac{i}{2}, \quad \frac{1}{2-3i} = \frac{2}{13} + \frac{3i}{13}.$$

Now you should be fully equipped to check all the field axioms for $\mathbb{C}$.

As you notice, the complex number $a - bi$ is a useful "counterpart" of $a + bi$, so that we are going to give it a special name. The *complex conjugate* of $z = a + bi$, $a, b \in \mathbb{R}$, is the complex number $\bar{z} := a - bi$. So, for example,

$$\overline{2 + 3i} = 2 - 3i, \quad \overline{\frac{1}{2} + \frac{6i}{5}} = \frac{1}{2} - \frac{6i}{5}.$$

Thus, we have

$$\operatorname{Re} \bar{z} = \operatorname{Re} z, \quad \operatorname{Im} \bar{z} = -\operatorname{Im} z.$$

Finally, we introduce the *absolute value* or *modulus* of z, via

$$|a + bi| := \sqrt{a^2 + b^2}, \ a, b, \in \mathbb{R}.$$

For example,

$$|1 + 3i| = \sqrt{10}, \quad \left|\frac{1}{2} - \frac{i}{2}\right| = \sqrt{\frac{1}{4} + \frac{1}{4}} = \frac{\sqrt{2}}{2}.$$

Note that we have the rule

$$z\bar{z} = |z|^2,$$

as observed in (1.3), and its consequence

$$\frac{1}{z} = \frac{\bar{z}}{|z|^2}$$

when $z \neq 0$.

A complex number is often depicted as a point in $\mathbb{R}^2$, which we refer to as the *complex plane*. The x-axis is the "real axis" and the y-axis is the "imaginary axis." Indeed, if $z = x + iy$ then we represent z as the point (x, y) as in the following figure.

The distance from the point z to the origin corresponds to $|z| = \sqrt{x^2 + y^2}$. The angle t the point z makes with the positive x-axis is referred to as the *argument* of z. It can be found via

$$\cos t = \frac{\operatorname{Re} z}{|z|}, \quad \sin t = \frac{\operatorname{Im} z}{|z|}.$$

Thus we can write

$$z = |z|(\cos t + i \sin t).$$

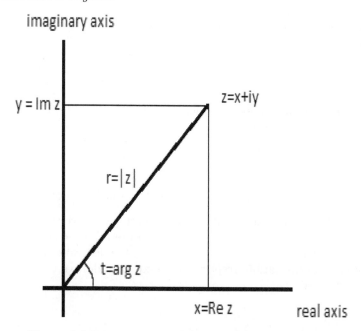

Figure 1.1: The complex number z in the complex plane.

The following notation, due to Euler, is convenient:

$$e^{it} := \cos t + i \sin t.$$

Using the rules for $\cos(t + s)$ and $\sin(t + s)$, one can easily check that

$$e^{it} e^{is} = e^{i(t+s)}.$$

In addition, note that

$$\overline{e^{it}} = e^{-it}.$$

Thus for $z = |z| e^{it} \neq 0$, we have that $z^{-1} = \frac{1}{|z|} e^{-it}$.

1.3.2   The finite field $\mathbb{Z}_p$, with p prime

Addition and multiplication in the field $\mathbb{Z}_p$ are based on the following result you discovered in elementary school when you did long division.

Proposition 1.3.1 *For every $q \in \mathbb{Z}$ and every $p \in \{2, 3, \ldots\}$, there exists unique $a \in \mathbb{Z}$ and $r \in \{0, 1, \ldots, p - 1\}$ so that*

$$q = ap + r.$$

We call r the remainder of q after division by p, and write $r = \text{rem}(q|p)$. For example,

$$\text{rem}(9|2) = 1, \quad \text{rem}(27|5) = 2, \quad \text{rem}(-30|7) = 5, \quad \text{rem}(-19|3) = 2.$$

Let now p be a prime number, and let $\mathbb{Z}_p = \{0, 1, \ldots, p-1\}$. Define the addition and multiplication

$$+ : \mathbb{Z}_p \times \mathbb{Z}_p \to \mathbb{Z}_p, \quad \cdot : \mathbb{Z}_p \times \mathbb{Z}_p \to \mathbb{Z}_p$$

via

$$a + b := \text{rem}(a+b|p), \quad a \cdot b := \text{rem}(ab|p). \tag{1.4}$$

Proposition 1.3.1 guarantees that for any integer q we have that $\text{rem}(q|p) \in \{0, \ldots, p-1\} = \mathbb{Z}_p$, so that the closure rules are clearly satisfied. Also, as expected, 0 and 1 are easily seen to be the neutral elements for addition and multiplication, respectively. Next, the additive inverse $-a$ of a is easily identified via

$$-a = \begin{cases} a & \text{if } a = 0 \\ p - a & \text{if } a \in \{1, \ldots, p-1\}. \end{cases}$$

The trickier part is the multiplicative inverse, and here we are going to use that p is prime. We need to remind you of the following rule for the greatest common divisor $\gcd(a, b)$ of two integers a and b, not both zero.

Proposition 1.3.2 *Let* $a, b \in \mathbb{Z}$ *not both zero. Then there exist* $m, n \in \mathbb{Z}$ *so that*

$$am + bn = \gcd(a, b). \tag{1.5}$$

Equation (1.5) is sometimes referred to as Bezout's identity. To solve Bezout's identity, one applies Euclid's algorithm to find the greatest common divisor (see below), keep track of the division equations, and ultimately put the equations together.

Algorithm 1 Euclid's algorithm

1: **procedure** EUCLID(a, b) ▷ The g.c.d. of a and $b \neq 0$

2:      $r \leftarrow \text{rem}(a|b)$

3:      **while** $r \neq 0$ **do**   ▷ We have the answer if r is 0

4:          $a \leftarrow b$

5:          $b \leftarrow r$

6:          $r \leftarrow \text{rem}(a|b)$

7:      **return** b ▷ The gcd is b

Example 1.3.3 Let $a = 17$ and $b = 5$. Then $2 = \text{rem}(17|5)$, which comes from the equality

$$2 = 17 - 3 \cdot 5. \tag{1.6}$$

Next, we look at the pair 5 and 2, and see that $1 = \text{rem}(5|2)$, which comes from the equality

$$1 = 5 - 2 \cdot 2. \tag{1.7}$$

Next we look at the pair 2 and 1, and see that $0 = \text{rem}(2|1)$. This means that Euclid's algorithm stops and we find that $1 = \gcd(17, 5)$. To next solve Bezout's identity (1.5) with $a = 17$ and $b = 5$, we put (1.7) and (1.6) together, and write

$$1 = 5 - 2 \cdot 2 = 5 - 2(17 - 3 \cdot 5) = -2 \cdot 17 + 7 \cdot 5,$$

and find that with the choices $m = -2$ and $n = 7$ we have solved (1.5).

We have now all we need to be able to prove the following.

Theorem 1.3.4 *Let p be a prime number. Then the set $\mathbb{Z}_p = \{0, 1, \ldots, p - 1\}$ with addition and multiplication defined via (1.4) is a field.*

Proof of existence of a multiplicative inverse. Let $a \in \mathbb{Z}_p$. As p is prime, we have that $\gcd(a, p) = 1$. By Proposition 1.3.2 there exist integers m, n so that $am + pn = 1$. Next we let $r = \text{rem}(m|p)$ and let q be so that $r = m - qp$. We claim that $a^{-1} = r$. Indeed,

$$ar = am - apq = 1 - pn - apq = 1 - p(n + aq).$$

From this we see that

$$1 = \text{rem}(ar|p),$$

and thus in the multiplication defined by (1.4) we have that $a \cdot r = 1$. $\square$

As said, the trickiest part of the proof of Theorem 1.3.4 is the existence of a multiplicative inverse, so the remainder of the proof we leave to the reader.

1.4 Matrix algebra over different fields

All the matrix algebra techniques that you learned in the first Linear Algebra course carry over to any field. Indeed, these algebra techniques were based on elementary algebraic operations, which work exactly the same in another field. In this section we illustrate these techniques by going through several examples with different fields. You will be reminded of matrix multiplication, row reduction, pivots, solving systems of linear equations, checking whether a vector is a linear combination of other vectors, finding a basis of a nullspace, column space, row space, eigenspace, computing determinants, finding inverses, Cramer's rule, etc., but now we do these techniques in other fields.

One notable exception where $\mathbb{R}$ differs from the other fields we are considering, is that $\mathbb{R}$ is an ordered field (that is, $\geq$ defines an order relation on pairs of real numbers, that satisfies $x \geq y \Rightarrow x + z \geq z + y$ and $x, y \geq 0 \Rightarrow xy \geq 0$). So anytime we want to use $\leq$, $<$, $\geq$ or $>$, we will have to make sure we are dealing with real numbers. We will do this when we talk about inner products and related concepts in Chapter 5.

Example 1.4.1 Let $\mathbb{F} = \mathbb{Z}_3$. Compute the product

$$\begin{pmatrix} 1 & 0 & 2 \\ 2 & 2 & 1 \end{pmatrix} \begin{pmatrix} 1 & 2 \\ 2 & 1 \\ 0 & 1 \end{pmatrix}.$$

The product equals

$$\begin{pmatrix} 1 \cdot 1 + 0 \cdot 2 + 2 \cdot 0 & 1 \cdot 2 + 0 \cdot 1 + 2 \cdot 1 \\ 2 \cdot 1 + 2 \cdot 2 + 1 \cdot 0 & 2 \cdot 2 + 2 \cdot 1 + 1 \cdot 1 \end{pmatrix} = \begin{pmatrix} 1 & 1 \\ 0 & 1 \end{pmatrix}.$$

Example 1.4.2 Let $\mathbb{F} = \mathbb{C}$. Compute the product

$$\begin{pmatrix} 1 + i \\ 2 - i \\ i \end{pmatrix} \begin{pmatrix} 1 - i & 2 + i \end{pmatrix}.$$

The product equals

$$\begin{pmatrix} (1+i)(1-i) & (1+i)(2+i) \\ (2-i)(1-i) & (2-i)(2+i) \\ i(1-i) & i(2+i) \end{pmatrix} = \begin{pmatrix} 2 & 1 + 3i \\ 1 - 3i & 5 \\ 1 + i & -1 + 2i \end{pmatrix}.$$

Example 1.4.3 Let $\mathbb{F} = \mathbb{Z}_5$. Put the matrix

$$\begin{pmatrix} 1 & 0 & 2 \\ 2 & 3 & 1 \\ 1 & 4 & 0 \end{pmatrix}$$

in row echelon form. We start with the $(1,1)$ element as our first pivot.

$$\begin{pmatrix} 1 & 0 & 2 \\ 2 & 3 & 1 \\ 1 & 4 & 0 \end{pmatrix} \rightarrow \begin{pmatrix} 1 & 0 & 2 \\ 0 & 3 & (1-4=)2 \\ 0 & 4 & (0-2=)3 \end{pmatrix}.$$

Next, let us multiply the second row with $3^{-1} = 2$, and use the $(2,2)$ entry as our next pivot:

$$\begin{pmatrix} 1 & 0 & 2 \\ 0 & 1 & 4 \\ 0 & 4 & 3 \end{pmatrix} \rightarrow \begin{pmatrix} 1 & 0 & 2 \\ 0 & 1 & 4 \\ 0 & 0 & (3-1=)2 \end{pmatrix},$$

bringing it to row echelon form. After having done this, we can now also easily compute

$$\det \begin{pmatrix} 1 & 0 & 2 \\ 2 & 3 & 1 \\ 1 & 4 & 0 \end{pmatrix} = 3 \det \begin{pmatrix} 1 & 0 & 2 \\ 0 & 1 & 4 \\ 0 & 0 & 2 \end{pmatrix} = 3 \cdot 2 = 1.$$

Alternatively, we can compute the determinant by expanding along (for instance) the first row, giving

$$\det \begin{pmatrix} 1 & 0 & 2 \\ 2 & 3 & 1 \\ 1 & 4 & 0 \end{pmatrix} = 1 \cdot (3 \cdot 0 - 1 \cdot 4) - 0 \cdot (2 \cdot 0 - 1 \cdot 1) + 2 \cdot (2 \cdot 4 - 3 \cdot 1) = 1.$$

Example 1.4.4 Let $\mathbb{F} = \mathbb{Z}_3$. Find the set of all solutions to the system of linear equations

$$\begin{cases} x_1 + 2x_2 & = 0 \\ x_1 + x_2 + x_3 = 1 \end{cases}.$$

We set up the associated augmented system and put it in row reduced echelon form:

$$\left(\begin{array}{ccc|c} 1 & 2 & 0 & 0 \\ 1 & 1 & 1 & 1 \end{array} \right) \rightarrow \left(\begin{array}{ccc|c} 1 & 2 & 0 & 0 \\ 0 & 2 & 1 & 1 \end{array} \right) \rightarrow \left(\begin{array}{ccc|c} 1 & 0 & 2 & 2 \\ 0 & 1 & 2 & 2 \end{array} \right).$$

We find that columns 1 and 2 are pivot columns, and column 3 is not, so x_3 is a free variable, and we get the equalities
$x_1 = 2 - 2x_3 = 2 + x_3, x_2 = 2 - 2x_3 = 2 + x_3$. So we find that all solutions are given by

$$x = \begin{pmatrix} x_1 \\ x_2 \\ x_3 \end{pmatrix} = \begin{pmatrix} 2 \\ 2 \\ 0 \end{pmatrix} + x_3 \begin{pmatrix} 1 \\ 1 \\ 1 \end{pmatrix}, \quad x_3 \in \mathbb{Z}_3.$$

In a typical Linear Algebra I course, systems of linear equations would be over the field of real numbers, and as soon as there was a free variable, one would have infinitely many solutions. This is due to $\mathbb{R}$ being an infinite field. In this example, though, we are dealing with a finite field, and thus when we let x_3 range over all elements of $\mathbb{Z}_3$, we only get a finite number of solutions. This will happen when dealing with any finite field. In this case, all solutions are found by letting $x_3 = 0, 1, 2$, thus we get that

$$\left\{ \begin{pmatrix} 2 \\ 2 \\ 0 \end{pmatrix}, \begin{pmatrix} 0 \\ 0 \\ 1 \end{pmatrix}, \begin{pmatrix} 1 \\ 1 \\ 2 \end{pmatrix} \right\}$$

are all solutions.

Example 1.4.5 Let $\mathbb{F} = \mathbb{C}$. Determine whether $\mathbf{b}$ is a linear combination of $\mathbf{a}_1, \mathbf{a}_2, \mathbf{a}_3$, where

$$\mathbf{a}_1 = \begin{pmatrix} 1+i \\ -1-i \\ 2 \\ 0 \end{pmatrix}, \mathbf{a}_2 = \begin{pmatrix} 0 \\ 2-i \\ -1+2i \\ 3 \end{pmatrix}, \mathbf{a}_3 = \begin{pmatrix} -1+i \\ 3-2i \\ -1+4i \\ 3 \end{pmatrix}, \mathbf{b} = \begin{pmatrix} 2i \\ 2-3i \\ 1 \\ 3+i \end{pmatrix}.$$

We set up the augmented system and put it in echelon form:

$$\begin{pmatrix} 1+i & 0 & -1+i & 2i \\ -1-i & 2-i & 3-2i & 2-3i \\ 2 & -1+2i & -1+4i & 1 \\ 0 & 3 & 3 & 3+i \end{pmatrix} \rightarrow$$

$$\begin{pmatrix} 1 & 0 & i & 1+i \\ -1-i & 2-i & 3-2i & 2-3i \\ 2 & -1+2i & -1+4i & 1 \\ 0 & 3 & 3 & 3+i \end{pmatrix} \rightarrow$$

$$\begin{pmatrix} 1 & 0 & i & 1+i \\ 0 & 2-i & 2-i & 2-i \\ 0 & -1+2i & -1+2i & -1-2i \\ 0 & 3 & 3 & 3+i \end{pmatrix} \rightarrow$$

$$\begin{pmatrix} 1 & 0 & i & 1+i \\ 0 & 1 & 1 & 1 \\ 0 & 1 & 1 & \frac{-3+4i}{5} \\ 0 & 3 & 3 & 3+i \end{pmatrix} \rightarrow \begin{pmatrix} 1 & 0 & i & 1+i \\ 0 & 1 & 1 & 1 \\ 0 & 0 & 0 & 1 \\ 0 & 0 & 0 & 0 \end{pmatrix}.$$

As the augmented column has a pivot, $\mathbf{b}$ is not a linear combination of $\mathbf{a}_1, \mathbf{a}_2, \mathbf{a}_3$.

Example 1.4.6 Let $\mathbb{F} = \mathbb{Z}_5$. Compute the inverse of

$$\begin{pmatrix} 1 & 0 & 2 \\ 2 & 3 & 1 \\ 1 & 4 & 0 \end{pmatrix}.$$

By Example 1.4.3 we know that this matrix is invertible, as every row and column has a pivot (or equivalently, since its determinant is nonzero). Let us compute the inverse:

$$
\left(\begin{array}{ccc|ccc}
1 & 0 & 2 & 1 & 0 & 0 \\
2 & 3 & 1 & 0 & 1 & 0 \\
1 & 4 & 0 & 0 & 0 & 1
\end{array}\right) \rightarrow
\left(\begin{array}{ccc|ccc}
1 & 0 & 2 & 1 & 0 & 0 \\
0 & 3 & (1-4=)2 & (0-2=)3 & 1 & 0 \\
0 & 4 & (0-2=)3 & (0-1=)4 & 0 & 1
\end{array}\right) \rightarrow
$$

$$
\left(\begin{array}{ccc|ccc}
1 & 0 & 2 & 1 & 0 & 0 \\
0 & 1 & 4 & 1 & 2 & 0 \\
0 & 0 & (3-1=)2 & (4-4=)0 & (0-3=)2 & 1
\end{array}\right) \rightarrow
$$

$$
\left(\begin{array}{ccc|ccc}
1 & 0 & 2 & 1 & 0 & 0 \\
0 & 1 & 4 & 1 & 2 & 0 \\
0 & 0 & 1 & 0 & 1 & 3
\end{array}\right) \rightarrow
\left(\begin{array}{ccc|ccc}
1 & 0 & 0 & 1 & 0-2 & 0-1 \\
0 & 1 & 0 & 1 & 2-4 & 0-2 \\
0 & 0 & 1 & 0 & 1 & 3
\end{array}\right),
$$

so the inverse is

$$
\begin{pmatrix}
1 & 3 & 4 \\
1 & 3 & 3 \\
0 & 1 & 3
\end{pmatrix}.
$$

Computing the product

$$
\begin{pmatrix}
1 & 0 & 2 \\
2 & 3 & 1 \\
1 & 4 & 0
\end{pmatrix}
\begin{pmatrix}
1 & 3 & 4 \\
1 & 3 & 3 \\
0 & 1 & 3
\end{pmatrix} =
\begin{pmatrix}
1 & 0 & 0 \\
0 & 1 & 0 \\
0 & 0 & 1
\end{pmatrix},
$$

we see that we computed the inverse correctly.

Example 1.4.7 Let $\mathbb{F} = \mathbb{C}$. Find bases of the column space, row space and null space of the matrix

$$
A = \begin{pmatrix}
i & 1-i & 2-i \\
1+i & -2 & -3+i \\
1-i & 1+2i & 3+3i
\end{pmatrix}.
$$

Let us put A in echelon form:

$$
\begin{pmatrix}
i & 1-i & 2-i \\
1+i & -2 & -3+i \\
1-i & 1+2i & 3+3i
\end{pmatrix} \rightarrow
\begin{pmatrix}
1 & -1-i & -1-2i \\
0 & -2+2i & -4+4i \\
0 & 3+2i & 6+4i
\end{pmatrix} \rightarrow
$$

$$
\begin{pmatrix}
1 & -1-i & -1-2i \\
0 & 1 & 2 \\
0 & 0 & 0
\end{pmatrix}.
$$

There are pivots in columns 1 and 2, and thus we find that

$\left\{\begin{pmatrix} i \\ 1+i \\ 1-i \end{pmatrix}, \begin{pmatrix} 1-i \\ -2 \\ 1+2i \end{pmatrix}\right\}$ is a basis for Col$A$. Next, for a basis of Row$A$ we simply have to pick the nonzero rows of the row echelon form of A, and thus we find that

$$\{(1 \quad -1-i \quad -1-2i), (0 \quad 1 \quad 2)\}$$

is a basis for Row$A$. To find a basis for the null space, we put A in row reduced echelon form:

$$\begin{pmatrix} 1 & -1-i & -1-2i \\ 0 & 1 & 2 \\ 0 & 0 & 0 \end{pmatrix} \rightarrow \begin{pmatrix} 1 & 0 & 1 \\ 0 & 1 & 2 \\ 0 & 0 & 0 \end{pmatrix}.$$

As there is no pivot in column 3, x_3 is a free variable. From

$$\begin{pmatrix} 1 & 0 & 1 \\ 0 & 1 & 2 \\ 0 & 0 & 0 \end{pmatrix} \begin{pmatrix} x_1 \\ x_2 \\ x_3 \end{pmatrix} = \begin{pmatrix} 0 \\ 0 \\ 0 \end{pmatrix}$$

we find $x_1 = -x_3$ and $x_2 = -2x_3$. Thus

$$x = \begin{pmatrix} x_1 \\ x_2 \\ x_3 \end{pmatrix} = x_3 \begin{pmatrix} -1 \\ -2 \\ 1 \end{pmatrix},$$

yielding that $\{ \begin{pmatrix} -1 \\ -2 \\ 1 \end{pmatrix} \}$ is a basis for the null space of A. It is easily checked that

$$\begin{pmatrix} i & 1-i & 2-i \\ 1+i & -2 & -3+i \\ 1-i & 1+2i & 3+3i \end{pmatrix} \begin{pmatrix} -1 \\ -2 \\ 1 \end{pmatrix} = \begin{pmatrix} 0 \\ 0 \\ 0 \end{pmatrix}.$$

Let $A \in \mathbb{F}^{n \times n}$ be a square matrix. Recall that $\lambda \in \mathbb{F}$ is an *eigenvalue* of A, if there exists a nonzero vector $\mathbf{x} \in \mathbb{F}^n$ so that $A\mathbf{x} = \lambda \mathbf{x}$. Such a vector $\mathbf{x} \neq \mathbf{0}$ is called an *eigenvector* of A at the eigenvalue λ. Rewriting $A\mathbf{x} = \lambda \mathbf{x}$ as $(A - \lambda I_n)\mathbf{x} = \mathbf{0}$, one sees that for λ to be an eigenvalue of A, one needs that $A - \lambda I_n$ is singular, and thus $\det(A - \lambda I_n) = 0$. The null space $\mathrm{Ker}(A - \lambda I_n)$ of $A - \lambda I_n$ is called the *eigenspace* of A at λ, and consists of all the eigenvectors of A at λ and the zero vector.

Example 1.4.8 Let $\mathbb{F} = \mathbb{Z}_7$. Find a basis for the eigenspace of
$A = \begin{pmatrix} 4 & 0 & 6 \\ 3 & 0 & 3 \\ 2 & 5 & 5 \end{pmatrix}$ corresponding to the eigenvalue $\lambda = 3$.

We have to find a basis for the null space of $A - 3I = \begin{pmatrix} 1 & 0 & 6 \\ 3 & 4 & 3 \\ 2 & 5 & 2 \end{pmatrix}$, so we put $A - 3I$ in row-reduced echelon form:

$$\begin{pmatrix} 1 & 0 & 6 \\ 3 & 4 & 3 \\ 2 & 5 & 2 \end{pmatrix} \rightarrow \begin{pmatrix} 1 & 0 & 6 \\ 0 & 4 & (3-4=)6 \\ 0 & 5 & (2-5=)4 \end{pmatrix} \rightarrow \begin{pmatrix} 1 & 0 & 6 \\ 0 & 1 & 5 \\ 0 & 0 & (4-4=)0 \end{pmatrix}.$$

We find that x_3 is a free variable, and $x_1 = -6x_3 = x_3, x_2 = -5x_3 = 2x_3$, leading to the basis $\left\{ \begin{pmatrix} 1 \\ 2 \\ 1 \end{pmatrix} \right\}$.

Let us do a check: $A = \begin{pmatrix} 4 & 0 & 6 \\ 3 & 0 & 3 \\ 2 & 5 & 5 \end{pmatrix} \begin{pmatrix} 1 \\ 2 \\ 1 \end{pmatrix} = \begin{pmatrix} 3 \\ 6 \\ 3 \end{pmatrix} = 3 \begin{pmatrix} 1 \\ 2 \\ 1 \end{pmatrix}$, confirming that $\begin{pmatrix} 1 \\ 2 \\ 1 \end{pmatrix}$ is an eigenvector of A corresponding to the eigenvalue $\lambda = 3$.

1.4.1 Reminders about Cramer's rule and the adjugate matrix.

Let $\mathbb{F}$ be a field, and the $n \times n$ matrix $A \in \mathbb{F}^{n \times n}$ and vector $\mathbf{b} \in \mathbb{F}^n$ be given. Let $\mathbf{a}_i$ denote the ith column of A. Now we define

$$A_i(\mathbf{b}) := \begin{pmatrix} \mathbf{a}_1 & \cdots & \mathbf{a}_{i-1} & \mathbf{b} & \mathbf{a}_{i+1} & \cdots & \mathbf{a}_n \end{pmatrix}, \ i = 1, \ldots, n.$$

Thus $A_i(\mathbf{b})$ is the matrix obtained from A by replacing its ith column by $\mathbf{b}$.

We now have the following result.

Theorem 1.4.9 *(Cramer's rule) Let $A \in \mathbb{F}^{n \times n}$ be invertible. For any $\mathbf{b} \in \mathbb{F}^n$, the unique solution $\mathbf{x} = (x_i)_{i=1}^n$ to the equation $A\mathbf{x} = \mathbf{b}$ has entries given by*

$$x_i = \det A_i(\mathbf{b})(\det A)^{-1}, \ i = 1, \ldots, n. \tag{1.8}$$

Proof. We denote the columns of the $n \times n$ identity matrix I by $\mathbf{e}_1, \ldots, \mathbf{e}_n$. Let us compute

$$A \, I_i(\mathbf{x}) = A \begin{pmatrix} \mathbf{e}_1 & \cdots & \mathbf{e}_{i-1} & \mathbf{x} & \mathbf{e}_{i+1} & \cdots & \mathbf{e}_n \end{pmatrix} =$$

$$\begin{pmatrix} A\mathbf{e}_1 & \cdots & A\mathbf{e}_{i-1} & A\mathbf{x} & A\mathbf{e}_{i+1} & \cdots & A\mathbf{e}_n \end{pmatrix} = A_i(\mathbf{b}).$$

But then, using the multiplicativity of the determinant, we get $\det A \det I_i(\mathbf{x}) = \det A_i(\mathbf{b})$. It is easy to see that $\det I_i(\mathbf{x}) = x_i$, and (1.8) follows. $\square$

Example 1.4.10 Let $\mathbb{F} = \mathbb{Z}_3$. Find the solution to the system of linear equations

$$\begin{cases} x_1 + 2x_2 = 0 \\ x_1 + x_2 = 1 \end{cases}.$$

Applying Cramer's rule, we get

$$x_1 = \det \begin{pmatrix} 0 & 2 \\ 1 & 1 \end{pmatrix} (\det \begin{pmatrix} 1 & 2 \\ 1 & 1 \end{pmatrix})^{-1} = 1 \cdot 2^{-1} = 2,$$

$$x_2 = \det \begin{pmatrix} 1 & 0 \\ 1 & 1 \end{pmatrix} (\det \begin{pmatrix} 1 & 2 \\ 1 & 1 \end{pmatrix})^{-1} = 1 \cdot 2^{-1} = 2.$$

Checking the answer $(2 + 2 \cdot 2 = 0, \quad 2 + 2 = 1)$, confirms that the answer is correct.

While Cramer's rule provides a direct formula to solve a system of linear equations (when the coefficient matrix is invertible), in many ways it is much better to solve a system of linear equations via row reduction as the latter requires in general fewer algebraic operations. Cramer's rule can be useful for more theoretical considerations. Here is such an example.

Example 1.4.11 Let $\mathbb{F} = \mathbb{C}$. Consider the matrix vector equation $A\mathbf{x} = \mathbf{b}$ given by

$$\begin{pmatrix} i & 1-i & 2 \\ 1+i & \alpha & 0 \\ 1-i & 1+2i & 3+5i \end{pmatrix} \begin{pmatrix} x_1 \\ x_2 \\ x_3 \end{pmatrix} = \begin{pmatrix} 2 \\ 0 \\ 5i \end{pmatrix}.$$

Find all $\alpha \in \mathbb{C}$ so that A is invertible and x_2 is real.

Applying Cramer's rule, we get

$$x_2 = \det \begin{pmatrix} i & 2 & 2 \\ 1+i & 0 & 0 \\ 1-i & 5i & 3+5i \end{pmatrix} (\det \begin{pmatrix} i & 1-i & 2 \\ 1+i & \alpha & 0 \\ 1-i & 1+2i & 3+5i \end{pmatrix})^{-1}.$$

Expanding along the second row we obtain

$$x_2 = \frac{-(1+i)(2(3+5i) - 2(5i))}{-(1+i)((1-i)(3+5i) - 2(1+2i)) + \alpha(i(3+5i) - 2(1-i))} =$$

$$\frac{-6 - 6i}{-8 - 4i + \alpha(-7 + 5i)}.$$

For $\det A \neq 0$, we need $\alpha \neq \frac{-8-4i}{7-5i} = -\frac{18}{37} - \frac{34i}{37}$. Next, notice that x_2 cannot equal 0, so we may write $-8 - 4i + \alpha(-7 + 5i) = \frac{-6-6i}{x_2}$. Let $t = \frac{1}{x_2}$, and arrive at

$$\alpha = \frac{1}{-7 + 5i}(8 + 4i - (6 + 6i)t) = -\frac{18}{37} - \frac{34i}{37} + (\frac{6}{37} + \frac{36i}{37})t, \ t \in \mathbb{R} \setminus \{0\},$$

as the set of solutions for α.

Given $A = (a_{ij})_{i,j=1}^n \in \mathbb{F}^{n \times n}$. We let $A_{ij} \in \mathbb{F}^{(n-1) \times (n-1)}$ be the matrix obtained from A by removing the ith row and the jth column, and we put

$$C_{ij} = (-1)^{i+j} \det A_{ij}, \ i, j = 1, \ldots, n.$$

The number C_{ij} is called the (i, j)th *cofactor* of A. Given

$$A = \begin{pmatrix} a_{11} & a_{12} & \cdots & a_{1n} \\ a_{21} & a_{22} & \cdots & a_{2n} \\ \vdots & \vdots & & \vdots \\ a_{n1} & a_{n2} & \cdots & a_{nn} \end{pmatrix},$$

the *adjugate* of A is defined by

$$\mathrm{adj}(A) = \begin{pmatrix} C_{11} & C_{21} & \cdots & C_{n1} \\ C_{12} & C_{22} & \cdots & C_{n2} \\ \vdots & \vdots & & \vdots \\ C_{1n} & C_{2n} & \cdots & C_{nn} \end{pmatrix}. \tag{1.9}$$

Thus the (i, j)th entry of $\mathrm{adj}(A)$ is C_{ji} (notice the switch in the indices!).

Example 1.4.12 Let $\mathbb{F} = \mathbb{Z}_5$. Compute the adjugate of

$$A = \begin{pmatrix} 1 & 0 & 2 \\ 2 & 3 & 1 \\ 1 & 4 & 0 \end{pmatrix}.$$

We get

$$\mathrm{adj}(A) = \begin{pmatrix} 3 \cdot 0 - 1 \cdot 4 & -0 \cdot 0 + 2 \cdot 4 & 0 \cdot 1 - 2 \cdot 3 \\ -2 \cdot 0 + 1 \cdot 1 & 1 \cdot 0 - 2 \cdot 1 & -1 \cdot 1 + 2 \cdot 2 \\ 2 \cdot 4 - 3 \cdot 1 & -1 \cdot 4 + 0 \cdot 1 & 1 \cdot 3 - 0 \cdot 2 \end{pmatrix} = \begin{pmatrix} 1 & 3 & 4 \\ 1 & 3 & 3 \\ 0 & 1 & 3 \end{pmatrix}.$$

The usefulness of the adjugate matrix is given by the following result.

Theorem 1.4.13 *Let $A \in \mathbb{F}^{n \times n}$. Then*

$$A \, \mathrm{adj}(A) = (\det A)I_n = \mathrm{adj}(A) \, A. \tag{1.10}$$

In particular, if $\det A \neq 0$, then

$$A^{-1} = (\det A)^{-1} \mathrm{adj}(A). \tag{1.11}$$

Proof. As before, we let $\mathbf{a}_i$ denote the ith column of A. Consider $A_i(\mathbf{a}_j)$, which is the matrix A with the ith column replaced by $\mathbf{a}_j$. Thus, when $i \neq j$ we have that $A_i(\mathbf{a}_j)$ has two identical columns (namely the ith and the jth) and thus $\det A_i(\mathbf{a}_j) = 0$, $i \neq j$. When $i = j$, then $A_i(\mathbf{a}_j) = A$, and thus $\det A_i(\mathbf{a}_j) = \det A$, $i = j$. Computing the (i,j)th entry of the product $\mathrm{adj}(A)\ A$, we get

$$(\mathrm{adj}(A)\ A)_{ij} = \sum_{k=1}^{n} C_{ki} a_{kj} = \det A_i(\mathbf{a}_j) = \begin{cases} \det A & \text{if } i = j \\ 0 & \text{if } i \neq j \end{cases},$$

where we expanded $\det A_i(\mathbf{a}_j)$ along the ith column. This proves the second equality in (1.10). The proof of the first equality in (1.10) is similar. $\square$

Notice that if we apply (1.11) to a 2×2 matrix, we obtain the familiar formula

$$\begin{pmatrix} a & b \\ c & d \end{pmatrix}^{-1} = \frac{1}{ad - bc} \begin{pmatrix} d & -b \\ -c & a \end{pmatrix}.$$

In Example 1.4.3 we have $\det A = 1$, so the adjugate matrix (which we computed in Example 1.4.12) equals in this case the inverse, confirming the computation in Example 1.4.6.

1.5   Exercises

Exercise 1.5.1 The set of integers $\mathbb{Z}$ with the usual addition and multiplication is **not** a field. Which of the field axioms does $\mathbb{Z}$ satisfy, and which one(s) are not satisfied?

Exercise 1.5.2 Write down the addition and multiplication tables for $\mathbb{Z}_2$ and $\mathbb{Z}_5$. How is commutativity reflected in the tables?

Exercise 1.5.3 The addition and multiplication defined in (1.4) also works when p is not prime. Write down the addition and multiplication tables for $\mathbb{Z}_4$. How can you tell from the tables that $\mathbb{Z}_4$ is **not** a field?

Exercise 1.5.4 Solve Bezout's identity for the following choices of a and b:

(i) $a = 25$ and $b = 7$;

(ii) $a = -50$ and $b = 3$.

Exercise 1.5.5 In this exercise we are working in the field $\mathbb{Z}_3$.

(i) $2 + 2 + 2 =$

(ii) $2(2 + 2)^{-1} =$

(iii) Solve for x in $2x + 1 = 2$.

(iv) Find $\det \begin{pmatrix} 1 & 2 \\ 1 & 0 \end{pmatrix}$.

(v) Compute $\begin{pmatrix} 1 & 2 \\ 0 & 2 \end{pmatrix} \begin{pmatrix} 1 & 1 \\ 2 & 1 \end{pmatrix}$.

(vi) Find $\begin{pmatrix} 2 & 0 \\ 1 & 1 \end{pmatrix}^{-1}$.

Exercise 1.5.6 In this exercise we are working in the field $\mathbb{Z}_5$.

(i) $4 + 3 + 2 =$

(ii) $4(1 + 2)^{-1} =$

(iii) Solve for x in $3x + 1 = 3$.

(iv) Find $\det \begin{pmatrix} 4 & 2 \\ 1 & 0 \end{pmatrix}$.

(v) Compute $\begin{pmatrix} 1 & 2 \\ 3 & 4 \end{pmatrix} \begin{pmatrix} 0 & 1 \\ 2 & 1 \end{pmatrix}$.

(vi) Find $\begin{pmatrix} 2 & 2 \\ 4 & 3 \end{pmatrix}^{-1}$.

Exercise 1.5.7 In this exercise we are working in the field $\mathbb{C}$. Make sure you write the final answers in the form $a + bi$, with $a, b \in \mathbb{R}$. For instance, $\frac{1+i}{2-i}$ should not be left as a final answer, but be reworked as

$$\frac{1+i}{2-i} = (\frac{1+i}{2-i})(\frac{2+i}{2+i}) = \frac{2+i+2i+i^2}{2^2+1^2} = \frac{1+3i}{5} = \frac{1}{5} + \frac{3i}{5}.$$

Notice that in order to get rid of i in the denominator, we decided to multiply both numerator and denominator with the complex conjugate of the denominator.

(i) $(1 + 2i)(3 - 4i) - (7 + 8i) =$

(ii) $\frac{1+i}{3+4i} =$

(iii) Solve for x in $(3 + i)x + 6 - 5i = -3 + 2i$.

(iv) Find $\det \begin{pmatrix} 4 + i & 2 - 2i \\ 1 + i & -i \end{pmatrix}$.

(v) Compute $\begin{pmatrix} -1 + i & 2 + 2i \\ -3i & -6 + i \end{pmatrix} \begin{pmatrix} 0 & 1 - i \\ -5 + 4i & 1 - 2i \end{pmatrix}$.

(vi) Find $\begin{pmatrix} 2 + i & 2 - i \\ 4 & 4 \end{pmatrix}^{-1}$.

Exercise 1.5.8 Here the field is $\mathbb{R}(t)$. Find the inverse of the matrix

$$\begin{pmatrix} 2 + 3t & \frac{1}{t^2 + 2t + 1} \\ t + 1 & \frac{3t - 4}{1 + t} \end{pmatrix},$$

if it exists.

Exercise 1.5.9 Let $\mathbb{F} = \mathbb{Z}_3$. Compute the product

$$\begin{pmatrix} 1 & 1 & 0 \\ 2 & 1 & 1 \end{pmatrix} \begin{pmatrix} 1 & 0 & 2 \\ 1 & 2 & 1 \\ 2 & 0 & 1 \end{pmatrix}.$$

Exercise 1.5.10 Let $\mathbb{F} = \mathbb{C}$. Compute the product

$$\begin{pmatrix} 2 - i & 2 + i \\ 2 - i & -10 \end{pmatrix} \begin{pmatrix} 5 + i & 6 - i \\ 1 - i & 2 + i \end{pmatrix}.$$

Exercise 1.5.11 Let $\mathbb{F} = \mathbb{Z}_5$. Put the matrix

$$\begin{pmatrix} 3 & 1 & 4 \\ 2 & 1 & 0 \\ 2 & 2 & 1 \end{pmatrix}$$

in row echelon form, and compute its determinant.

Exercise 1.5.12 Let $\mathbb{F} = \mathbb{Z}_3$. Find the set of all solutions to the system of linear equations

$$\begin{cases} 2x_1 + x_2 & = 1 \\ 2x_1 + 2x_2 + x_3 = 0 \end{cases}.$$

Exercise 1.5.13 Let $\mathbb{F} = \mathbb{C}$. Determine whether $\mathbf{b}$ is a linear combination of $\mathbf{a}_1, \mathbf{a}_2, \mathbf{a}_3$, where

$$\mathbf{a}_1 = \begin{pmatrix} i \\ 1-i \\ 2-i \\ 1 \end{pmatrix}, \mathbf{a}_2 = \begin{pmatrix} 0 \\ 3+i \\ -1+i \\ -3 \end{pmatrix}, \mathbf{a}_3 = \begin{pmatrix} -i \\ 2+2i \\ -3+2i \\ 3 \end{pmatrix}, \mathbf{b} = \begin{pmatrix} 0 \\ 0 \\ 0 \\ 1 \end{pmatrix}.$$

Exercise 1.5.14 Let $\mathbb{F} = \mathbb{Z}_5$. Compute the inverse of

$$\begin{pmatrix} 2 & 3 & 1 \\ 1 & 4 & 1 \\ 1 & 1 & 2 \end{pmatrix}$$

in two different ways (row reduction and by applying (1.11)).

Exercise 1.5.15 Let $\mathbb{F} = \mathbb{C}$. Find bases of the column space, row space and null space of the matrix

$$A = \begin{pmatrix} 1 & 1+i & 2 \\ 1+i & 2i & 3+i \\ 1-i & 2 & 3+5i \end{pmatrix}.$$

Exercise 1.5.16 Let $\mathbb{F} = \mathbb{Z}_7$. Find a basis for the eigenspace of $A = \begin{pmatrix} 3 & 5 & 0 \\ 4 & 6 & 5 \\ 2 & 2 & 4 \end{pmatrix}$ corresponding to the eigenvalue $\lambda = 1$.

Exercise 1.5.17 Let $\mathbb{F} = \mathbb{Z}_3$. Use Cramer's rule to find the solution to the system of linear equations

$$\begin{cases} 2x_1 + 2x_2 = 1 \\ x_1 + 2x_2 = 1 \end{cases}.$$

Exercise 1.5.18 Let $\mathbb{F} = \mathbb{C}$. Consider the matrix vector equation $A\mathbf{x} = \mathbf{b}$ given by

$$\begin{pmatrix} i & 1-i & 2 \\ 1+i & \alpha & 0 \\ 1-i & 1+2i & 3+5i \end{pmatrix} \begin{pmatrix} x_1 \\ x_2 \\ x_3 \end{pmatrix} = \begin{pmatrix} 2 \\ 0 \\ 5i \end{pmatrix}.$$

Determine $\alpha \in \mathbb{C}$ so that A is invertible and $x_1 = x_2$.

Exercise 1.5.19 Let $\mathbb{F} = \mathbb{R}(t)$. Compute the adjugate of

$$A = \begin{pmatrix} \frac{1}{t} & 2+t^2 & 2-t \\ \frac{2}{1+t} & 3t & 1-t \\ 1 & 4+t^2 & 0 \end{pmatrix}.$$

Exercise 1.5.20 Recall that the *trace* of a square matrix is defined to be the sum of its diagonal entries. Thus $\text{tr}[(a_{ij})_{i,j=1}^n] = a_{11} + \cdots + a_{nn} = \sum_{j=1}^n a_{jj}$.

(a) Show that if $A \in \mathbb{F}^{n \times m}$ and $B \in \mathbb{F}^{m \times n}$, then $\text{tr}(AB) = \text{tr}(BA)$.

(b) Show that if $A \in \mathbb{F}^{n \times m}$, $B \in \mathbb{F}^{m \times k}$, and $C \in \mathbb{F}^{k \times n}$, then
$\text{tr}(ABC) = \text{tr}(CAB) = \text{tr}(BCA)$.

(c) Give an example of matrices $A, B, C \in \mathbb{F}^{n \times n}$ so that
$\text{tr}(ABC) \neq \text{tr}(BAC)$.

Exercise 1.5.21 Let $A, B \in \mathbb{F}^{n \times n}$. The *commutator* $[A, B]$ of A and B is defined by $[A, B] := AB - BA$.

(a) Show that $\text{tr}([A, B]) = 0$.

(b) Show that when $n = 2$, we have that $[A, B]^2 = -\det([A, B])I_2$.

(c) Show that if $C \in \mathbb{F}^{n \times n}$ as well, then $\text{tr}(C[A, B]) = \text{tr}([B, C]A)$.

The following two exercises provide a very introductory illustration of how finite fields may be used in coding. To learn more, please look for texts on *linear coding theory*.

Exercise 1.5.22 The 10-digit ISBN number makes use of the field $\mathbb{Z}_{11} = \{0, 1, 2, 3, 4, 5, 6, 7, 8, 9, X\}$ (notice that X is the roman numeral for 10). The first digit(s) present the group. For English-speaking countries, the first digit is a 0 or a 1. The next set of digits represents the publisher. For instance, Princeton University Press has the digits 691. Some of the bigger publishers have a 2-digit publisher code, leaving them more digits for their titles. The next set of digits represent the specific title. Finally, the last digit of the 10-digit ISBN number $a_1 a_2 \ldots a_{10}$ is a check digit, which needs to satisfy the equation

$$a_1 + 2a_2 + 3a_3 + \cdots + 9a_9 + 10a_{10} = 0$$

in $\mathbb{Z}_{11}$. For instance, 0691128898 is the 10-digit ISBN of *Matrix Completions, Moments, and Sums of Hermitian Squares* by Mihály Bakonyi and Hugo J. Woerdeman. Indeed, we have

$$1 \cdot 0 + 2 \cdot 6 + 3 \cdot 9 + 4 \cdot 1 + 5 \cdot 1 + 6 \cdot 2 + 7 \cdot 8 + 8 \cdot 8 + 9 \cdot 9 + 10 \cdot 8 = \text{rem}(341 | 11) = 0.$$

Check that the 10-digit ISBN number 3034806388 has a correct check digit.

Exercise 1.5.23 A not so secure way to convert a secret message is to replace letters by numbers, e.g., AWESOME = 1 23 5 19 15 13 5. Whatever numbers are chosen with the letters, knowing that it corresponds to an English text, one can use general information about English (such as that "E" is the letter that appears most and "Z" is the letter that appears least), to crack the code. What will make cracking the code more challenging is to use a matrix to convert the list of numbers. We are going to work with $\mathbb{Z}_{29}$ and 29 characters, with 0 standing for "space", the numbers 1–26 standing for the letters A–Z, the number 27 standing for "period," and 28 standing for "comma." Thus, for example,

Wow, he said. $\Leftrightarrow$ 23 15 23 28 0 8 5 0 19 1 9 4 27

Next we are going to use 3×3 matrices in $\mathbb{Z}_{29}$ to convert the code as follows. Letting

$$A = \begin{pmatrix} 2 & 1 & 6 \\ 2 & 0 & 10 \\ 11 & 2 & 3 \end{pmatrix},$$

we can take the first three numbers in the sequence, put them in a vector, multiply it by the matrix A, and convert them back to characters:

$$A \begin{pmatrix} 23 \\ 15 \\ 23 \end{pmatrix} = \begin{pmatrix} 25 \\ 15 \\ 4 \end{pmatrix}, \quad 25 \ 15 \ 4 \ \Leftrightarrow \ \text{YOD}.$$

If we do this for the whole sentence, putting the numbers in groups of three, adding spaces (=0) at the end to make sure we have a multiple of three, we have that "Wow, he said. " (notice the two spaces at the end) converts to "YODQTMHZYFMLYYG." In order to decode. one performs the same algorithm with

$$A^{-1} = \begin{pmatrix} 9 & 9 & 10 \\ 17 & 27 & 21 \\ 4 & 7 & 27 \end{pmatrix}.$$

Decode the word "ZWNOWQJJZ."

Exercise 1.5.24 *(Honors)* The field axioms imply several things that one might take for granted but that really require a formal proof. In this exercise, we address the uniqueness of the neutral element and the inverse.

Claim *In a field there is a unique neutral element of addition.*

Proof. Suppose that both 0 and $0'$ satisfy Axiom 4. Thus $0 + x = x = x + 0$ and $0' + x = x = x + 0'$ hold for every x. Then $0 = 0 + 0' = 0'$, proving the uniqueness. $\square$

(i) Prove the uniqueness of the neutral element of multiplication.

(ii) Prove uniqueness of the additive inverse. To do this, one needs to show that if $x + y = 0 = x + z$, it implies that $y = z$. Of course, it is tempting to just remove the x's from the equation $x + y = x + z$ (as you are used to), but the exact purpose of this exercise is to make you aware that these familiar rules need to be reproven by exclusively using the field axioms. So use exclusively the fields axioms to fill in the blanks:

$$y = y + 0 = y + (x + z) = \cdots = \cdots = \cdots = z.$$

Exercise 1.5.25 *(Honors)* Let $\mathbb{F}$ be a field, and $\mathbb{K} \subseteq \mathbb{F}$. Show that $\mathbb{K}$ is a subfield of $\mathbb{F}$ if and only if

(i) $0, 1 \in \mathbb{K}$, and

(ii) $x, y \in \mathbb{K}$ implies $x + y$, xy, $-x$ belong to $\mathbb{K}$, and when $x \neq 0$, x^{-1} also belongs to $\mathbb{K}$.

Exercise 1.5.26 *(Honors)* Let $\mathbb{Q} + \mathbb{Q}\sqrt{2} := \{a + b\sqrt{2} \ : \ a, b \in \mathbb{Q}\}$. So $\mathbb{Q} + \mathbb{Q}\sqrt{2}$ contains elements such as

$$-\frac{5}{6} + \frac{\sqrt{2}}{2} \quad \text{and} \quad \frac{1}{2 + 3\sqrt{2}} = \frac{1}{2 + 3\sqrt{2}} \cdot \frac{2 - 3\sqrt{2}}{2 - 3\sqrt{2}} = -\frac{1}{7} + \frac{3}{14}\sqrt{2}.$$

Show that $\mathbb{Q} + \mathbb{Q}\sqrt{2}$ is a subfield of $\mathbb{R}$.

Exercise 1.5.27 *(Honors)* Let

$$\mathbb{A} = \{z \in \mathbb{C} \ : \ \text{there exist } n \in \mathbb{N} \text{ and } a_0, \ldots, a_n \in \mathbb{Z} \text{ so that } \sum_{k=0}^{n} a_k z^k = 0\}.$$

In other words, $\mathbb{A}$ consists of all roots of polynomials with integer coefficients (also known as *algebraic numbers*). Numbers such as $\sqrt[3]{2} - 5$, $\cos(\frac{\pi}{7})$, and $5 - i\sqrt{3}$ belong to $\mathbb{A}$. The numbers π and e do **not** belong to $\mathbb{A}$ (such numbers are called *transcendental*).

Formulate the statements about polynomials and their roots that would need to be proven to show that $\mathbb{A}$ is closed under addition and multiplication. It turns out that $\mathbb{A}$ is a subfield of $\mathbb{C}$, and you are welcome to look up the proof.

2

Vector Spaces

CONTENTS

2.1   Definition of a vector space 27
2.2   Vector spaces of functions 29
      2.2.1   The special case when X is finite 31
2.3   Subspaces and more examples of vector spaces 32
      2.3.1   Vector spaces of polynomials 34
      2.3.2   Vector spaces of matrices 36
2.4   Linear independence, span, and basis 37
2.5   Coordinate systems 45
2.6   Exercises 48

The foundation for linear algebra is the notion of a vector space over a field. Two operations are important in a vector space (i) addition: any two elements in a vector space can be added together; (ii) multiplication by a scalar: an element in a vector space can be multiplied by a scalar ($=$ an element of the field). Anytime one has mathematical objects where these two operations are well-defined and satisfy some basic properties, one has a vector space. Allowing this generality and developing a theory that just uses these basic rules, leads to results that can be applied in many settings.

2.1   Definition of a vector space

A *vector space* over a field $\mathbb{F}$ is a set V along with two operations

$$+ : V \times V \to V, \quad \cdot : \mathbb{F} \times V \to V$$

satisfying the following rules:

1. *Closure of addition*: for all $\mathbf{u}, \mathbf{v} \in V$ we have that $\mathbf{u} + \mathbf{v} \in V$.

2. *Associativity of addition*: for all $\mathbf{u}, \mathbf{v}, \mathbf{w} \in V$ we have that
   $(\mathbf{u} + \mathbf{v}) + \mathbf{w} = \mathbf{u} + (\mathbf{v} + \mathbf{w})$.

3. *Commutativity of addition*: for all $\mathbf{u}, \mathbf{v} \in V$ we have that $\mathbf{u} + \mathbf{v} = \mathbf{v} + \mathbf{u}$.

4. *Existence of a neutral element for addition*: there exists a $\mathbf{0} \in V$ so that
   $\mathbf{u} + \mathbf{0} = \mathbf{u} = \mathbf{0} + \mathbf{u}$ for all $\mathbf{u} \in V$.

5. *Existence of an additive inverse*: for every $\mathbf{u} \in V$ there exists a $-\mathbf{u} \in V$
   so that $\mathbf{u} + (-\mathbf{u}) = \mathbf{0} = (-\mathbf{u}) + \mathbf{u}$.

6. *Closure of scalar multiplication*: for all $c \in \mathbb{F}$ and $\mathbf{u} \in V$ we have that
   $c\mathbf{u} \in V$.

7. *First distributive law*: for all $c \in \mathbb{F}$ and $\mathbf{u}, \mathbf{v} \in V$ we have that
   $c(\mathbf{u} + \mathbf{v}) = c\mathbf{u} + c\mathbf{v}$.

8. *Second distributive law*: for all $c, d \in \mathbb{F}$ and $\mathbf{u} \in V$ we have that
   $(c + d)\mathbf{u} = c\mathbf{u} + d\mathbf{u}$.

9. *Associativity for scalar multiplication*: for all $c, d \in \mathbb{F}$ and $\mathbf{u} \in V$ we have
   that $c(d\mathbf{u}) = (cd)\mathbf{u}$.

10. *Unit multiplication rule*: for every $\mathbf{u} \in V$ we have that $1\mathbf{u} = \mathbf{u}$.

These axioms imply several rules that seem "obvious," but as all properties in vector spaces have to be traced back to the axioms, we need to reprove these obvious rules. Here are two such examples.

Lemma 2.1.1 *Let V be a vector space over $\mathbb{F}$. Then for all $\mathbf{u} \in V$ we have that*

(i) $0\mathbf{u} = \mathbf{0}$.

(ii) $(-1)\mathbf{u} = -\mathbf{u}$.

Proof. (i) As $0\mathbf{u} \in V$, we have that $0\mathbf{u}$ has an additive inverse; call it $\mathbf{v}$. Then

$$\mathbf{0} = 0\mathbf{u} + \mathbf{v} = (0 + 0)\mathbf{u} + \mathbf{v} = (0\mathbf{u} + 0\mathbf{u}) + \mathbf{v} = 0\mathbf{u} + (0\mathbf{u} + \mathbf{v}) = 0\mathbf{u} + \mathbf{0} = 0\mathbf{u}.$$

For (ii) we observe

$$-\mathbf{u} = \mathbf{0} + (-\mathbf{u}) = 0\mathbf{u} + (-\mathbf{u}) = ((-1) + 1)\mathbf{u} + (-\mathbf{u}) = ((-1)\mathbf{u} + 1\mathbf{u}) + (-\mathbf{u}) =$$

$$(-1)\mathbf{u} + (1\mathbf{u} + (-\mathbf{u})) = (-1)\mathbf{u} + (\mathbf{u} + (-\mathbf{u})) = (-1)\mathbf{u} + \mathbf{0} = (-1)\mathbf{u}.$$

$\square$

2.2 Vector spaces of functions

The set of all functions from a set X to a field $\mathbb{F}$ is denoted by $\mathbb{F}^X$. Thus

$$\mathbb{F}^X := \{f : X \to \mathbb{F} \ : \ f \text{ is a function}\}.$$

When $f, g : X \to \mathbb{F}$ we can define the *sum* of f and g as the function

$$f + g : X \to \mathbb{F}, \quad (f + g)(x) = f(x) + g(x).$$

Thus, by virtue that $\mathbb{F}$ has a well-defined addition, the set $\mathbb{F}^X$ now also has a well-defined addition. It is a fine point, but it is important to recognize that in the equation

$$(f + g)(x) = f(x) + g(x)$$

the first $+$ sign represents addition between functions, while the second $+$ sign represents addition in $\mathbb{F}$, so really the two $+$s are different. We still choose to use the same $+$ sign for both, although technically we could have made them different ($+_{\mathbb{F}^X}$ and $+_{\mathbb{F}}$, say) and written

$$(f +_{\mathbb{F}^X} g)(x) = f(x) +_{\mathbb{F}} g(x).$$

Next, it is also easy to define the *scalar multiplication* on $\mathbb{F}^X$ as follows. Given $c \in \mathbb{F}$ and $f : X \to \mathbb{F}$, we define the function cf via

$$cf : X \to \mathbb{F}, \quad (cf)(x) = c(f(x)).$$

Again, let us make the fine point that there are two different multiplications here, namely the multiplication of a scalar (i.e., an element of $\mathbb{F}$) with a function and the multiplication of two scalars. Again, if we want to highlight this difference, one would write this for instance as

$$(c \cdot_{\mathbb{F}^X} f)(x) = c \cdot_{\mathbb{F}} f(x).$$

We now have the following claim.

Proposition 2.2.1 *The set $\mathbb{F}^X$ with the above definitions of addition and scalar multiplication is a vector space over the field $\mathbb{F}$.*

Checking that all the vector space axioms is not hard. For instance, to check commutativity of addition, we have to show that $f + g = g + f$. This introduces the question: When are two functions equal? The answer to this is:

Two functions $h, k : X \to \mathbb{F}$ are equal if and only if for all $x \in X$:
$h(x) = k(x)$.

Thus, to show that $f + g = g + f$, we simply need to show that for all $x \in X$ we have that $(f + g)(x) = (g + f)(x)$. The proof of this is:

$$(f +_{\mathbb{F}^X} g)(x) = f(x) +_{\mathbb{F}} g(x) = g(x) +_{\mathbb{F}} f(x) = (g +_{\mathbb{F}^X} f)(x) \text{ for all } x \in X,$$

where in the first and third equality we applied the definition of the sum of two functions, while in the middle equality we applied commutativity of addition in $\mathbb{F}$.

Important to realize is what the *neutral element of addition* in $\mathbb{F}^X$ is: it is a function, and when added to another function it should not change the other function. This gives:

the function $\mathbf{0} : X \to \mathbb{F}$ defined via $\mathbf{0}(x) = 0$ for all $x \in X$, is the neutral element in $\mathbb{F}^X$.

Notice that again we have two different mathematical objects: the constant zero function (= the neutral element of addition in $\mathbb{F}^X$) and the neutral element of addition in $\mathbb{F}$. If we want to highlight this difference, one would write for instance:
$$\mathbf{0}_{\mathbb{F}^X}(x) = 0_{\mathbb{F}} \text{ for all } x \in X.$$

For the *inverse element for addition* in $\mathbb{F}^X$, we have similar considerations. Given $f : X \to \mathbb{F}$,

the additive inverse is the function $-f : X \to \mathbb{F}$ defined via
$(-f)(x) = -f(x)$, $x \in X$.

The two minuses are different, which can be highlighted by writing

$$(-_{\mathbb{F}^X} f)(x) = -_{\mathbb{F}} f(x).$$

Now all the ingredients are there to write a complete proof to Proposition 5.1.6. We already showed how to address the commutativity of addition, and as the proofs of the other rules are similar, we will leave them to the reader.

2.2.1   The special case when X is finite

The case when X is a finite set is special in the sense that in this case we can simply write out all the values of the function. For instance, if $X = \{1, \ldots, n\}$, then the function $f : X \to \mathbb{F}$ simply corresponds to choosing elements $f(1), \ldots, f(n) \in \mathbb{F}$. Thus we can identify

$$f : \{1, \ldots n\} \to \mathbb{F} \quad \Leftrightarrow \quad \begin{pmatrix} f(1) \\ \vdots \\ f(n) \end{pmatrix} \in \mathbb{F}^n.$$

When $f, g : \{1, \ldots, n\} \to \mathbb{F}$, the sum function is defined by

$$(f + g)(1) = f(1) + g(1), \ldots, (f + g)(n) = f(n) + g(n),$$

which in the notation above amounts to

$$f + g : \{1, \ldots n\} \to \mathbb{F} \quad \Leftrightarrow \quad \begin{pmatrix} f(1) + g(1) \\ \vdots \\ f(n) + g(n) \end{pmatrix} \in \mathbb{F}^n.$$

This corresponds to the definition of adding elements in $\mathbb{F}^n$:

$$\begin{pmatrix} f(1) \\ \vdots \\ f(n) \end{pmatrix} + \begin{pmatrix} g(1) \\ \vdots \\ g(n) \end{pmatrix} = \begin{pmatrix} f(1) + g(1) \\ \vdots \\ f(n) + g(n) \end{pmatrix}.$$

Similarly, for scalar multiplication we have

$$cf : \{1, \ldots n\} \to \mathbb{F} \quad \Leftrightarrow \quad \begin{pmatrix} cf(1) \\ \vdots \\ cf(n) \end{pmatrix} \in \mathbb{F}^n.$$

This corresponds to scalar multiplication of elements in $\mathbb{F}^n$:

$$c \begin{pmatrix} f(1) \\ \vdots \\ f(n) \end{pmatrix} = \begin{pmatrix} cf(1) \\ \vdots \\ cf(n) \end{pmatrix}.$$

Thus when we deal with function $f : X \to \mathbb{F}$ with X a finite set with n elements, the vector space $\mathbb{F}^X$ corresponds exactly to the vector space $\mathbb{F}^n$. Clearly, when X has n elements, it does not have to equal the set $X = \{1, \ldots, n\}$, however, it will be our default choice. Sometimes, though, it may be convenient to use $X = \{0, \ldots, n-1\}$ instead.

As a final remark in this subsection, we note that typically we write a vector $\mathbf{x} \in \mathbb{F}^n$ as

$$\mathbf{x} = \begin{pmatrix} x_1 \\ \vdots \\ x_n \end{pmatrix},$$

instead of the function notation. Of course, this is just a notational choice. Conceptually, we can still think of this vector as representing a function on a finite set of n elements.

2.3 Subspaces and more examples of vector spaces

Given a vector space V over a field $\mathbb{F}$, and $W \subseteq V$. When $\mathbf{w}, \mathbf{y} \in W$, then as W is a subset of V, we also have $\mathbf{w}, \mathbf{y} \in V$, and thus $\mathbf{w} + \mathbf{y}$ is well-defined. In addition, when $c \in \mathbb{F}$ and $\mathbf{w} \in W \subseteq V$, then $c\mathbf{w}$ is well-defined. Thus we can consider the question whether W with the operations as defined on V, is itself a vector space. If so, we call W a *subspace* of V.

Proposition 2.3.1 *Given a vector space V over a field $\mathbb{F}$, and $W \subseteq V$, then W is a subspace of V if and only if*

(i) $\mathbf{0} \in W$.

(ii) *W is closed under addition: for all $\mathbf{w}, \mathbf{y} \in W$, we have $\mathbf{w} + \mathbf{y} \in W$.*

(iii) *W is closed under scalar multiplication: for all $c \in \mathbb{F}$ and $\mathbf{w} \in W$, we have that $c\mathbf{w} \in W$.*

Proof. If W is a vector space, then (i), (ii) and (iii) are clearly satisfied.

For the converse, we need to check that when W satisfies (i), (ii) and (iii), it satisfies all ten axioms in the definition of a vector space. Clearly properties (i), (ii) and (iii) above take care of axioms 1, 4 and 6 in the definition of a vector space. Axiom 5 follows from (iii) in combination with Lemma 2.1.1(ii). The other properties (associativity, commutativity, distributivity, unit multiplication) are satisfied as they hold for all elements of V, and thus also for elements of W. $\qquad\square$

In Proposition 2.3.1 one may replace (i) by

(i)' $W \neq \emptyset$.

Clearly, if (i) holds then (i)' holds.

For the other direction, note that if $\mathbf{w} \in W$ (existence of such $\mathbf{w}$ is guaranteed by (i)') then by (iii) and Lemma 2.1.1(i), we get that $\mathbf{0} = 0\mathbf{w} \in W$. Thus (i)' and (iii) together imply (i).

Given two subspaces U and W of a vector space V, we introduce

$$U + W := \{\mathbf{v} \in V \ : \ \text{there exist } \mathbf{u} \in U \text{ and } \mathbf{w} \in W \text{ so that } \mathbf{v} = \mathbf{u} + \mathbf{w}\},$$

$$U \cap W := \{\mathbf{v} \in V \ : \ \mathbf{v} \in U \text{ and } \mathbf{v} \in W\}.$$

Proposition 2.3.2 *Given two subspaces U and W of a vector space V over $\mathbb{F}$, then $U + W$ and $U \cap W$ are also subspaces of V.*

Proof. Clearly $\mathbf{0} = \mathbf{0} + \mathbf{0} \in U + W$ as $\mathbf{0} \in U$ and $\mathbf{0} \in W$. Let $\mathbf{v}, \hat{\mathbf{v}} \in U + W$ and $c \in \mathbb{F}$. Then there exist $\mathbf{u}, \hat{\mathbf{u}} \in U$ and $\mathbf{w}, \hat{\mathbf{w}} \in W$ so that $\mathbf{v} = \mathbf{u} + \mathbf{w}$ and $\hat{\mathbf{v}} = \hat{\mathbf{u}} + \hat{\mathbf{w}}$. Then $\mathbf{v} + \hat{\mathbf{v}} = (\mathbf{u} + \mathbf{w}) + (\hat{\mathbf{u}} + \hat{\mathbf{w}}) = (\mathbf{u} + \hat{\mathbf{u}}) + (\mathbf{w} + \hat{\mathbf{w}}) \in U + W$, since $\mathbf{u} + \hat{\mathbf{u}} \in U$ and $\mathbf{w} + \hat{\mathbf{w}} \in W$. Also $c\mathbf{v} = c(\mathbf{u} + \mathbf{w}) = c\mathbf{u} + c\mathbf{w} \in U + W$ as $c\mathbf{u} \in U$ and $c\mathbf{w} \in W$. This proves that $U + W$ is a subspace.

As $\mathbf{0} \in U$ and $\mathbf{0} \in W$, we have that $\mathbf{0} \in U \cap W$. Next, let $\mathbf{v}, \hat{\mathbf{v}} \in U \cap W$ and $c \in \mathbb{F}$. Then $\mathbf{v}, \hat{\mathbf{v}} \in U$, and since U is a subspace, we have $\mathbf{v} + \hat{\mathbf{v}} \in U$. Similarly, $\mathbf{v} + \hat{\mathbf{v}} \in W$. Thus $\mathbf{v} + \hat{\mathbf{v}} \in U \cap W$. Finally, since $\mathbf{v} \in U$ and U is a subspace, $c\mathbf{v} \in U$. Similarly, $c\mathbf{v} \in W$. Thus $c\mathbf{v} \in U \cap W$. $\square$

When $U \cap W = \{\mathbf{0}\}$, then we refer to $U + W$ as a *direct sum* of U and W, and write $U \dotplus W$. More generally, when $U_1, \ldots, U_k$ are subspaces of V, then we define the following

$$+_{j=1}^{k} U_j = U_1 + \cdots + U_k =$$

$$\{\mathbf{v} \in V \ : \ \text{there exist } \mathbf{u}_j \in U_j, j = 1, \ldots, k, \text{ so that } \mathbf{v} = \mathbf{u}_1 + \cdots + \mathbf{u}_k\},$$

$$\cap_{j=1}^{k} U_j = U_1 \cap \ldots \cap U_k = \{\mathbf{v} \in V \ : \ \mathbf{v} \in U_j \text{ for all } j = 1, \ldots, k\}.$$

It is straightforward to prove that $+_{j=1}^{k} U_j$ and $\cap_{j=1}^{k} U_j$ are subspaces of V. We say that $U_1 + \cdots + U_k$ is a *direct sum* if for all $j = 1, \ldots, k$, we have that

$$U_j \cap (U_1 + \cdots + U_{j-1} + U_{j+1} + \cdots + U_k) = \{\mathbf{0}\}.$$

In that case we write $U_1 \dotplus \cdots \dotplus U_k$ or $\dotplus_{j=1}^{k} U_j$.

Proposition 2.3.3 *Consider the direct sum $U_1 \dotplus \cdots \dotplus U_k$, then for every $\mathbf{v} \in U_1 \dotplus \cdots \dotplus U_k$ there exists unique $\mathbf{u}_j \in U_j$, $j = 1, \ldots, k$, so that $\mathbf{v} = \mathbf{u}_1 + \cdots + \mathbf{u}_k$. In particular, if $\mathbf{u}_j \in U_j$, $j = 1, \ldots, k$, are so that $\mathbf{u}_1 + \cdots + \mathbf{u}_k = \mathbf{0}$, then $\mathbf{u}_j = \mathbf{0}$, $j = 1, \ldots, k$.*

Proof. Suppose $\mathbf{v} = \mathbf{u}_1 + \cdots + \mathbf{u}_k = \hat{\mathbf{u}}_1 + \cdots + \hat{\mathbf{u}}_k$, with $\mathbf{u}_j, \hat{\mathbf{u}}_j \in U_j$, $j = 1, \ldots, k$. Then

$$-(\mathbf{u}_j - \hat{\mathbf{u}}_j) = (\mathbf{u}_1 - \hat{\mathbf{u}}_1) + \cdots + (\mathbf{u}_{j-1} - \hat{\mathbf{u}}_{j-1}) + (\mathbf{u}_{j+1} - \hat{\mathbf{u}}_{j+1}) + \cdots + (\mathbf{u}_k - \hat{\mathbf{u}}_k)$$

belongs to both U_j and $U_1 + \cdots + U_{j-1} + U_{j+1} + \cdots + U_k$, and thus to their intersection. As the intersection equals $\{\mathbf{0}\}$, we obtain that $\mathbf{u}_j - \hat{\mathbf{u}}_j = \mathbf{0}$. As $j \in \{1, \ldots, k\}$ was arbitrary, we get $\mathbf{u}_j = \hat{\mathbf{u}}_j$, $j = 1, \ldots, k$, as desired.

When $\mathbf{u}_1 + \cdots + \mathbf{u}_k = \mathbf{0} = \mathbf{0} + \cdots + \mathbf{0}$, then the uniqueness of the representation implies that $\mathbf{u}_j = \mathbf{0}$, $j = 1, \ldots, k$. $\square$

2.3.1   Vector spaces of polynomials

We let $\mathbb{F}[X]$ be the set of all polynomials in X with coefficients in $\mathbb{F}$. Thus a typical element of $\mathbb{F}[X]$ has the form

$$p(X) = \sum_{j=0}^{n} p_j X^j = p_0 X^0 + p_1 X + p_2 X^2 + \cdots + p_n X^n,$$

where $n \in \mathbb{N}$ and $p_0, \ldots, p_n \in \mathbb{F}$. Here X is merely a symbol and so are its powers X^i, with the understanding that $X^i X^j = X^{i+j}$. Often X^0 is omitted, as when we specify X we will have that X^0 is a multiplicative neutral element (as for instance the equality $X^0 X^i = X^i$ suggests).

When we have two polynomials $p(X) = \sum_{j=0}^{n} p_j X^j$ and $q(X) = \sum_{j=0}^{m} q_j X^j$, it is often convenient to have $m = n$. We do this by introducing additional terms with a zero coefficient. For instance, if we want to view

$$p(X) = 1 + X \text{ and } q(X) = 1 + 2X^2 - X^5$$

as having the same number of terms we may view them as

$$p(X) = 1 + X + 0X^2 + 0X^3 + 0X^4 + 0X^5, q(X) = 1 + 0X + 2X^2 + 0X^3 + 0X^4 - X^5.$$

Notice that the term X is really $1X$, and $-X^5$ is $(-1)X^5$.

Two polynomials $p(X) = \sum_{j=0}^{n} p_j X^j$ and $q(X) = \sum_{j=0}^{n} q_j X^j$ are *equal* exactly when all their coefficients are equal: $p_j = q_j$, $j = 0, \ldots, n$.

The sum of two polynomials $p(X) = \sum_{j=0}^{n} p_j X^j$ and $q(X) = \sum_{j=0}^{n} q_j X^j$ is given by

$$(p+q)(X) = \sum_{j=0}^{n} (p_j + q_j) X^j.$$

When $c \in \mathbb{F}$ and $p(X) = \sum_{j=0}^{n} p_j X^j$ are given, we define the polynomial $(cp)(X)$ via

$$(cp)(X) = \sum_{j=0}^{n} (cp_j) X^j.$$

Proposition 2.3.4 *The set $\mathbb{F}[X]$ with the above defined addition and scalar multiplication, is a vector space over $\mathbb{F}$.*

The proof is straightforward, so we will leave it as an exercise. Of course, the zero in $\mathbb{F}[X]$ is the polynomials with all its coefficients equal to 0, and when $p(X) = \sum_{j=0}^{n} p_j X^j$ then $(-p)(X) = \sum_{j=0}^{n} (-p_j) X^j$.

Given two equal polynomial $p(X), q(X) \in \mathbb{F}[X]$, then obviously $p(x) = q(x)$ for all $x \in \mathbb{F}$. However, the converse is not always the case, as the following example shows.

Example 2.3.5 Let $\mathbb{F} = \mathbb{Z}_2$, and $p(X) = 0$ and $q(X) = X - X^2$. Then $p(X)$ and $q(X)$ are different polynomials (e.g., $p_1 = 0 \neq 1 = q_1$), but $p(x) = q(x)$ for all $x \in \mathbb{Z}_2$. Indeed, $p(0) = 0 = q(0)$ and $p(1) = 0 = q(1)$.

We do have the following observation. When $A \in \mathbb{F}^{m \times m}$ (i.e., A is an $m \times m$ matrix with entries in $\mathbb{F}$) and $p(X) = \sum_{j=0}^{n} p_j X^j$ then we define

$$p(A) = p_0 I_m + p_1 A + p_2 A^2 + \cdots + p_n A^n \in \mathbb{F}^{m \times m},$$

where I_m denotes the $m \times m$ identity matrix. For future use, we define $A^0 := I_m$.

Proposition 2.3.6 *Two polynomials $p(X), q(X) \in \mathbb{F}[X]$ are equal if and only if for all $m \in \mathbb{N}$*

$$p(A) = q(A) \text{ for all } A \in \mathbb{F}^{m \times m}. \tag{2.1}$$

Proof. When $p(X) = q(X)$, clearly (2.1) holds for all $m \in \mathbb{N}$.

For the converse, suppose that $p(X) = \sum_{j=0}^{n} p_j X^j$ and $q(X) = \sum_{j=0}^{m} q_j X^j$

satisfy (2.1) for all $m \in \mathbb{N}$. Let J be the $n \times n$ matrix

$$
J_n = \begin{pmatrix}
0 & 1 & 0 & \cdots & 0 \\
0 & 0 & 1 & \cdots & 0 \\
\vdots & \vdots & \ddots & \ddots & \vdots \\
0 & 0 & \cdots & 0 & 1 \\
0 & 0 & \cdots & 0 & 0
\end{pmatrix}.
$$

Then

$$
\begin{pmatrix}
p_0 & p_1 & p_2 & \cdots & p_n \\
0 & p_0 & p_1 & \cdots & p_{n-1} \\
\vdots & \vdots & \ddots & \ddots & \vdots \\
0 & 0 & \cdots & p_0 & p_1 \\
0 & 0 & \cdots & 0 & p_0
\end{pmatrix}
= p(J_n) = q(J_n) =
\begin{pmatrix}
q_0 & q_1 & q_2 & \cdots & q_n \\
0 & q_0 & q_1 & \cdots & q_{n-1} \\
\vdots & \vdots & \ddots & \ddots & \vdots \\
0 & 0 & \cdots & q_0 & q_1 \\
0 & 0 & \cdots & 0 & q_0
\end{pmatrix},
$$

and thus $p_j = q_j$, $j = 0, \ldots n$, follows. $\qquad\square$

For a polynomial $p(X) = \sum_{j=0}^{n} p_j X^j$ with $p_n \neq 0$ we say that its *degree* equals n, and we write $\deg p = n$. It is convenient to assign $-\infty$ as the degree of the zero polynomial (in this way, with the convention that $-\infty + n = -\infty$, we have that the degree of a product of polynomials is the sum of the degrees).

Proposition 2.3.7 *Let* $\mathbb{F}_n[X] := \{ p(X) \in \mathbb{F}[X] : \deg p \leq n \}$, *where* $n \in \{0, 1, \ldots\}$. *Then* $\mathbb{F}_n[X]$ *is a subspace of* $\mathbb{F}[X]$.

Proof. Clearly $0 \in \mathbb{F}_n[X]$. Next, if $\deg p \leq n$, $\deg q \leq n$ and $c \in \mathbb{F}$, then $\deg p + q \leq n$ and $\deg cp \leq n$. Thus, $\mathbb{F}_n[X]$ is closed under addition and scalar multiplication. Apply now Proposition 2.3.1 to conclude that $\mathbb{F}_n[X]$ is a subspace of $\mathbb{F}[X]$. $\qquad\square$

One can also consider polynomials in several variables $X_1, \ldots, X_k$, which can either be commuting variables (so that, for instance, $X_1 X_2$ and $X_2 X_1$ are the same polynomial) or non-commuting variables (so that $X_1 X_2$ and $X_2 X_1$ are different polynomials). We will not pursue this here.

2.3.2   Vector spaces of matrices

Let $\mathbb{F}^{n \times m}$ denote the set of $n \times m$ matrices with entries in $\mathbb{F}$. So a typical element of $\mathbb{F}^{n \times m}$ is

$$
A = (a_{i,j})_{i=1,j=1}^{n \quad m} =
\begin{pmatrix}
a_{11} & \cdots & a_{1m} \\
\vdots & & \vdots \\
a_{n1} & \cdots & a_{nm}
\end{pmatrix}.
$$

Addition and scalar multiplication are defined via

$$(a_{i,j})_{i=1,j=1}^{n\quad m}+(b_{i,j})_{i=1,j=1}^{n\quad m} = (a_{i,j}+b_{i,j})_{i=1,j=1}^{n\quad m}, \quad c(a_{i,j})_{i=1,j=1}^{n\quad m} = (ca_{i,j})_{i=1,j=1}^{n\quad m}.$$

Proposition 2.3.8 *The set $\mathbb{F}^{n\times m}$ with the above definitions of addition and scalar multiplication is a vector space over $\mathbb{F}$.*

When $m = 1$, we have $\mathbb{F}^{n\times 1} = \mathbb{F}^n$. The vector space $\mathbb{F}^{1\times m}$ can be identified with $\mathbb{F}^m$ (by simply turning a row vector into a column vector). In fact, we can identify $\mathbb{F}^{n\times m}$ with $\mathbb{F}^{nm}$, for instance by stacking the columns of a matrix into a large vector. For example, when $n = m = 2$, the identification would be

$$\begin{pmatrix} a & b \\ c & d \end{pmatrix} \leftrightarrow \begin{pmatrix} a \\ c \\ b \\ d \end{pmatrix}.$$

This identification works when we are only interested in the vector space properties of $\mathbb{F}^{n\times m}$. However, if at the same time we are interested in other properties of $n \times m$ matrices (such as, that one can multiply such matrices on the left with a $k \times n$ matrix), one should not make this identification.

2.4 Linear independence, span, and basis

The notion of a basis is a crucial one; it basically singles out few elements in the vector space with which we can reconstruct the whole vector space. For example, the monomials $1, X, X^2, \ldots$ form a basis of the vector space of polynomials. When we start to do certain (namely, linear) operations on elements of a vector space, we will see in the next chapter that it will suffice to know how these operations act on the basis elements. Differentiation is an example: as soon as we know that the derivatives of $1, X, X^2, X^3, \ldots$ are $0, 1, 2X, 3X^2, \ldots$, respectively, it is easy to find the derivative of a polynomial. Before we get to the notion of a basis, we first need to introduce *linear independence* and *span*.

Let V be a vector space over $\mathbb{F}$. A set of vectors $\{\mathbf{v}_1, \ldots, \mathbf{v}_p\}$ in V is said to be *linearly independent* if the vector equation

$$c_1\mathbf{v}_1 + c_2\mathbf{v}_2 + \cdots + c_p\mathbf{v}_p = \mathbf{0}, \tag{2.2}$$

with $c_1, \ldots, c_p \in \mathbb{F}$, only has the solution $c_1 = 0, \ldots, c_p = 0$ (the *trivial solution*). The set $\{\mathbf{v}_1, \ldots, \mathbf{v}_p\}$ is said to be *linearly dependent* if (2.2) has a solution where not all of $c_1, \ldots, c_p$ are zero (a *nontrivial solution*). In such a case, (2.2) with at least one c_i nonzero gives a *linear dependence relation* among $\{\mathbf{v}_1, \ldots, \mathbf{v}_p\}$. An arbitrary set $S \subseteq V$ is said to be linearly independent if every finite subset of S is linearly independent. The set S is linearly dependent, if it is not linearly independent.

Example 2.4.1 Let $V = \mathbb{R}^{\mathbb{R}} = \{f : \mathbb{R} \to \mathbb{R} \ : \ f \text{ is a function}\}$, and consider the finite set of vectors $\{\cos(x), e^x, x^2\}$ in $\mathbb{R}^{\mathbb{R}}$. We claim that this set is linearly independent. For this, consider a linear combination $c_1 \cos(x) + c_2 e^x + c_3 x^2$ and set it equal to the zero function $\mathbf{0}(x)$, which is the neutral element of addition in $\mathbb{R}^{\mathbb{R}}$:

$$c_1 \cos(x) + c_2 e^x + c_3 x^2 = \mathbf{0}(x) = 0 \text{ for all } x \in \mathbb{R}.$$

If we take different values for x we get linear equations for c_1, c_2, c_3. Taking $x = 0$, $x = \frac{\pi}{2}$, $x = -\frac{\pi}{2}$, we get the following three equations:

$$\begin{cases} c_1 + c_2 e & = 0 \\ \quad c_2 e^{\frac{\pi}{2}} + c_3 \frac{\pi^2}{4} = 0 \\ \quad c_2 e^{-\frac{\pi}{2}} + c_3 \frac{\pi^2}{4} = 0. \end{cases}$$

As $\det \begin{pmatrix} 1 & e & 0 \\ 0 & e^{\frac{\pi}{2}} & \frac{\pi^2}{4} \\ 0 & e^{-\frac{\pi}{2}} & \frac{\pi^2}{4} \end{pmatrix} \neq 0$, we get that we must have $c_1 = c_2 = c_3 = 0$.

Thus linear independence of $\{\cos(x), e^x, x^2\}$ follows.

Let us also consider the set of vectors $\{1, \cos(x), \sin(x), \cos^2(x), \sin^2(x)\}$. We claim this set is linearly dependent, as the nontrivial choice $c_1 = 1, c_2 = 0, c_3 = 0, c_4 = -1, c_5 = -1$ gives the linear dependence relation

$$c_1 1 + c_2 \cos(x) + c_3 \sin(x) + c_4 \cos^2(x) + c_5 \sin^2(x) =$$

$$1 - \cos^2(x) - \sin^2(x) = \mathbf{0}(x) = 0 \text{ for all } x \in \mathbb{R}.$$

Example 2.4.2 Let $V = \mathbb{Z}_3^{2 \times 2}$. Let us check whether

$$S = \left\{ \begin{pmatrix} 1 & 0 \\ 2 & 1 \end{pmatrix}, \begin{pmatrix} 1 & 1 \\ 1 & 1 \end{pmatrix}, \begin{pmatrix} 0 & 2 \\ 1 & 1 \end{pmatrix} \right\}$$

is linearly independent or not. Notice that $\begin{pmatrix} 0 & 0 \\ 0 & 0 \end{pmatrix}$ is the neutral element of addition in this vector space. Consider the equation

$$c_1 \begin{pmatrix} 1 & 0 \\ 2 & 1 \end{pmatrix} + c_2 \begin{pmatrix} 1 & 1 \\ 1 & 1 \end{pmatrix} + c_3 \begin{pmatrix} 0 & 2 \\ 1 & 1 \end{pmatrix} = \begin{pmatrix} 0 & 0 \\ 0 & 0 \end{pmatrix}.$$

Rewriting, we get

$$\begin{pmatrix} 1 & 1 & 0 \\ 0 & 1 & 2 \\ 2 & 1 & 1 \\ 1 & 1 & 1 \end{pmatrix} \begin{pmatrix} c_1 \\ c_2 \\ c_3 \end{pmatrix} = \begin{pmatrix} 0 \\ 0 \\ 0 \\ 0 \end{pmatrix}. \tag{2.3}$$

Bringing this 4×3 matrix in row echelon form gives

$$\begin{pmatrix} 1 & 1 & 0 \\ 0 & 1 & 2 \\ 2 & 1 & 1 \\ 1 & 1 & 1 \end{pmatrix} \rightarrow \begin{pmatrix} 1 & 1 & 0 \\ 0 & 1 & 2 \\ 0 & 2 & 1 \\ 0 & 0 & 1 \end{pmatrix} \rightarrow \begin{pmatrix} 1 & 1 & 0 \\ 0 & 1 & 2 \\ 0 & 0 & 1 \\ 0 & 0 & 0 \end{pmatrix}.$$

As there are pivots in all columns, the system (2.3) only has the trivial solution $c_1 = c_2 = c_3 = 0$. Thus S is linearly independent.

Next, consider

$$\hat{S} = \{ \begin{pmatrix} 1 & 0 \\ 2 & 1 \end{pmatrix}, \begin{pmatrix} 1 & 1 \\ 1 & 1 \end{pmatrix}, \begin{pmatrix} 2 & 1 \\ 0 & 2 \end{pmatrix} \}.$$

Following the same reasoning as above we arrive at the system

$$\begin{pmatrix} 1 & 1 & 2 \\ 0 & 1 & 1 \\ 2 & 1 & 0 \\ 1 & 1 & 2 \end{pmatrix} \begin{pmatrix} c_1 \\ c_2 \\ c_3 \end{pmatrix} = \begin{pmatrix} 0 \\ 0 \\ 0 \\ 0 \end{pmatrix}, \tag{2.4}$$

which after row reduction leads to

$$\begin{pmatrix} 1 & 1 & 2 \\ 0 & 1 & 1 \\ 0 & 0 & 0 \\ 0 & 0 & 0 \end{pmatrix} \begin{pmatrix} c_1 \\ c_2 \\ c_3 \end{pmatrix} = \begin{pmatrix} 0 \\ 0 \\ 0 \\ 0 \end{pmatrix}. \tag{2.5}$$

So, c_3 is a free variable. Letting $c_3 = 1$, we get $c_2 = -c_3 = 2$ and $c_1 = -c_2 - 2c_3 = 2$, so we find the linear dependence relation

$$2 \begin{pmatrix} 1 & 0 \\ 2 & 1 \end{pmatrix} + 2 \begin{pmatrix} 1 & 1 \\ 1 & 1 \end{pmatrix} + \begin{pmatrix} 2 & 1 \\ 0 & 2 \end{pmatrix} = \begin{pmatrix} 0 & 0 \\ 0 & 0 \end{pmatrix},$$

and thus $\hat{S}$ is linearly dependent.

Given a set $S \subseteq V$ we define

$$\text{Span } S := \{ c_1 \mathbf{v}_1 + \cdots + c_p \mathbf{v}_p \ : \ p \in \mathbb{N}, c_1, \ldots, c_p \in \mathbb{F}, \mathbf{v}_1, \ldots, \mathbf{v}_p \in S \}.$$

Thus, Span S consists of all linear combinations of a finite set of vectors in S. It is straightforward to check that Span S is a subspace of V. Indeed,

$\mathbf{0} \in \text{Span } S$ as one can choose $p = 1$, $c_1 = 0$, and any $\mathbf{v}_1 \in S$, to get that $\mathbf{0} = 0\mathbf{v}_1 \in \text{Span } S$. Next, the sum of two linear combinations of vectors in S is again a linear combination of vectors of S. Finally, for $c \in \mathbb{F}$ we have that $c \sum_{j=1}^{p} c_j \mathbf{v}_j = \sum_{j=1}^{p} (c c_j) \mathbf{v}_j$ is again a linear combination of elements in S. Thus by Proposition 2.3.1 we have that Span S is a subspace of V.

Example 2.4.3 Let $V = (\mathbb{Z}_5)_3[X] = \{p(X) \in \mathbb{Z}_5[X] : \deg p \le 3\}$. We claim that

$$\text{Span}\{X-1, X^2-2X+1, X^3-3X^2+3X-1\} = \{p(X) \in V : p(1) = 0\} =: W. \tag{2.6}$$

First, observe that if

$$p(X) = c_1(X - 1) + c_2(X^2 - 2X + 1) + c_3(X^3 - 3X^2 + 3X - 1),$$

then $\deg p \le 3$ and $p(1) = 0$, and thus
$\text{Span}\{X - 1, X^2 - 2X + 1, X^3 - 3X^2 + 3X - 1\} \subseteq W$.

To prove the converse inclusion $\supseteq$ in (2.6), let
$p(X) = p_0 + p_1 X + p_2 X^2 + p_3 X^3$ be an arbitrary element of W. The condition $p(1) = 0$ gives that $p_0 + p_1 + p_2 + p_3 = 0$. We need to show that we can write

$$p(X) = c_1(X - 1) + c_2(X^2 - 2X + 1) + c_3(X^3 - 3X^2 + 3X - 1),$$

for some $c_1, c_2, c_3 \in \mathbb{Z}_5$. As two polynomials are equal if and only if all the coefficients are equal, we arrive at the following set of equations

$$\begin{cases} -c_1 + c_2 - c_3 = p_0 \\ c_1 - 2c_2 + 3c_3 = p_1 \\ c_2 - 3c_3 = p_2 \\ c_3 = p_3 \end{cases}.$$

Setting up the corresponding augmented matrix and putting it in row reduced echelon form, we find

$$\left(\begin{array}{ccc|c} 4 & 1 & 4 & p_0 \\ 1 & 3 & 3 & p_1 \\ 0 & 1 & 2 & p_2 \\ 0 & 0 & 1 & p_3 \end{array} \right) \rightarrow \left(\begin{array}{ccc|c} 1 & 0 & 0 & p_1 + 2p_2 + 3p_3 \\ 0 & 1 & 0 & p_2 + 3p_3 \\ 0 & 0 & 1 & p_3 \\ 0 & 0 & 0 & 0 \end{array} \right),$$

where we used that $p_0 + p_1 + p_2 + p_3 = 0$. Thus the system is consistent. We find that

$$p(X) = (p_1 + 2p_2 + 3p_3)(X-1) + (p_2 + 3p_3)(X^2 - 2X + 1) + p_3(X^3 - 3X^2 + 3X - 1).$$

Thus $p(X) \in \text{Span}\{X - 1, X^2 - 2X + 1, X^3 - 3X^2 + 3X - 1\}$, and the proof is complete.

Let W be a vector space. We say that $S \subset W$ is a *basis* for W if the following two conditions are both satisfied:

(i) Span $S = W$.

(ii) S is linearly independent.

If S has a finite number of elements, then for any other basis of W it will have the same number of elements, as the following result shows.

Proposition 2.4.4 *Let* $\mathcal{B} = \{\mathbf{v}_1, \ldots, \mathbf{v}_n\}$ *and* $\mathcal{C} = \{\mathbf{w}_1, \ldots, \mathbf{w}_m\}$ *be bases for the vector space* W*. Then* $n = m$*.*

Proof. Suppose that $n \neq m$. Without loss of generality, we may assume that $n < m$. As $\mathcal{B}$ is a basis, we can express $\mathbf{w}_j$ as a linear combination of elements of $\mathcal{B}$:

$$\mathbf{w}_j = a_{1j}\mathbf{v}_1 + \cdots + a_{nj}\mathbf{v}_n, \ j = 1, \ldots, m.$$

The matrix $A = (a_{ij})_{i=1,j=1}^{n,\ m}$ has more columns than rows (and thus a non-pivot column), so the equation $A\mathbf{c} = \mathbf{0}$ has a nontrivial solution

$$\mathbf{c} = \begin{pmatrix} c_1 \\ \vdots \\ c_m \end{pmatrix} \neq \mathbf{0}.$$ But then it follows that

$$\sum_{j=1}^{m} c_j \mathbf{w}_j = \sum_{j=1}^{m} [c_j \sum_{i=1}^{n} a_{ij} \mathbf{v}_i)] = \sum_{i=1}^{n} (\sum_{j=1}^{m} a_{ij} c_j) \mathbf{v}_i = \sum_{i=1}^{n} 0\mathbf{v}_i = \mathbf{0}.$$

Thus a nontrivial linear combination of elements of $\mathcal{C}$ equals $\mathbf{0}$, and thus $\mathcal{C}$ is linearly dependent. Contradiction. Consequently, we must have $n = m$. □

We can now define the *dimension* of a vector space as:

$$\dim W := \text{ the number of elements in a basis of } W.$$

When no basis with a finite number of elements exists for W, we say $\dim W = \infty$.

Remark 2.4.5 Notice that the proof of Proposition 2.4.4 also shows that in an n-dimensional vector space any set of vectors with more than n elements must be linearly dependent.

Example 2.4.6 Let $W = \{p(X) \in \mathbb{R}_2[X] \ : \ \int_{-1}^{1} p(x)dx = 0\}$. Show that W is a subspace of $\mathbb{R}_2[X]$ and find a basis for W.

Clearly, the zero polynomial $\mathbf{0}(x)$ belongs to W as $\int_{-1}^{1} \mathbf{0}(x)dx = \int_{-1}^{1} 0dx = 0$. Next, when $p(X), q(X) \in W$, then

$$\int_{-1}^{1} (p(x) + q(x))dx = \int_{-1}^{1} p(x)dx + \int_{-1}^{1} q(x)dx = 0 + 0 = 0,$$

so $(p + q)(X) \in W$. Similarly, when $c \in \mathbb{R}$ and $p(X) \in W$, then

$$\int_{-1}^{1} (cp)(x)dx = \int_{-1}^{1} cp(x)dx = c\int_{-1}^{1} p(x)dx = c0 = 0,$$

so $(cp)(X) \in W$. Thus by Proposition 2.3.1, W is a subspace of $\mathbb{R}_2[X]$.

To find a basis, let us take an arbitrary element $p(X) = p_0 + p_1 X + p_2 X^2 \in W$, which means that

$$\int_{-1}^{1} p(x)dx = 2p_0 + \frac{2}{3}p_2 = 0.$$

This yields the linear system

$$\begin{pmatrix} 2 & 0 & \frac{2}{3} \end{pmatrix} \begin{pmatrix} p_0 \\ p_1 \\ p_2 \end{pmatrix} = 0.$$

The coefficient matrix only has a pivot in column 1, so we let p_1 and p_2 be the free variables (as they correspond to the variables corresponding to the 2nd and 3rd column) and observe that $p_0 = -\frac{1}{3}p_2$. Expressing $p(X)$ solely in the free variables we get

$$p(X) = p_0 + p_1 X + p_2 X^2 = p_1 X + p_2(X^2 - \frac{1}{3}).$$

Thus $p(X) \in \mathrm{Span}\{X, X^2 - \frac{1}{3}\}$. As we started with an arbitrary $p(X) \in W$, we now proved that $W \subseteq \mathrm{Span}\{X, X^2 - \frac{1}{3}\}$. As $X \in W$ and $X^2 - \frac{1}{3} \in W$ and W is a subspace, we also have $\mathrm{Span}\{X, X^2 - \frac{1}{3}\} \subseteq W$. Thus $\mathrm{Span}\{X, X^2 - \frac{1}{3}\} = W$.

Next, one easily checks that $\{X, X^2 - \frac{1}{3}\}$ is linearly independent, and thus $\{X, X^2 - \frac{1}{3}\}$ is a basis for W. In particular, $\dim W = 2$.

Example 2.4.7 Let $V = \mathbb{R}^4$,

$$U = \mathrm{Span}\{ \begin{pmatrix} 1 \\ 0 \\ 2 \\ 1 \end{pmatrix}, \begin{pmatrix} 1 \\ 1 \\ 1 \\ 1 \end{pmatrix} \}, W = \mathrm{Span}\{ \begin{pmatrix} 4 \\ 2 \\ 2 \\ 0 \end{pmatrix}, \begin{pmatrix} 2 \\ 0 \\ 2 \\ 0 \end{pmatrix} \}.$$

Find bases for $U \cap W$ and $U + W$.

Vectors in $U \cap W$ are of the form

$$x_1 \begin{pmatrix} 1 \\ 0 \\ 2 \\ 1 \end{pmatrix} + x_2 \begin{pmatrix} 1 \\ 1 \\ 1 \\ 1 \end{pmatrix} = x_3 \begin{pmatrix} 4 \\ 2 \\ 2 \\ 0 \end{pmatrix} + x_4 \begin{pmatrix} 2 \\ 0 \\ 2 \\ 0 \end{pmatrix}. \tag{2.7}$$

Setting up the homogeneous system of linear equations, and subsequently row reducing, we get

$$\begin{pmatrix} 1 & 1 & -4 & -2 \\ 0 & 1 & -2 & 0 \\ 2 & 1 & -2 & -2 \\ 1 & 1 & 0 & 0 \end{pmatrix} \rightarrow \begin{pmatrix} 1 & 1 & -4 & -2 \\ 0 & 1 & -2 & 0 \\ 0 & -1 & 6 & 2 \\ 0 & 0 & 4 & 2 \end{pmatrix} \rightarrow \begin{pmatrix} 1 & 1 & -4 & -2 \\ 0 & 1 & -2 & 0 \\ 0 & 0 & 4 & 2 \\ 0 & 0 & 0 & 0 \end{pmatrix}.$$

This gives that x_4 is free and $x_3 = -\frac{x_4}{2}$. Plugging this into the right-hand side of (2.7) gives

$$-\frac{x_4}{2} \begin{pmatrix} 4 \\ 2 \\ 2 \\ 0 \end{pmatrix} + x_4 \begin{pmatrix} 2 \\ 0 \\ 2 \\ 0 \end{pmatrix} = x_4 \begin{pmatrix} 0 \\ -1 \\ 1 \\ 0 \end{pmatrix}$$

as a typical element of $U \cap W$. So

$$\left\{ \begin{pmatrix} 0 \\ -1 \\ 1 \\ 0 \end{pmatrix} \right\}$$

is a basis for $U \cap W$.

Notice that

$$U + W = \text{Span}\left\{ \begin{pmatrix} 1 \\ 0 \\ 2 \\ 1 \end{pmatrix}, \begin{pmatrix} 1 \\ 1 \\ 1 \\ 1 \end{pmatrix}, \begin{pmatrix} 4 \\ 2 \\ 2 \\ 0 \end{pmatrix}, \begin{pmatrix} 2 \\ 0 \\ 2 \\ 0 \end{pmatrix} \right\}.$$

From the row reductions above, we see that the fourth vector is a linear combination of the first three, while the first three are linearly independent. Thus a basis for $U + W$ is

$$\left\{ \begin{pmatrix} 1 \\ 0 \\ 2 \\ 1 \end{pmatrix}, \begin{pmatrix} 1 \\ 1 \\ 1 \\ 1 \end{pmatrix}, \begin{pmatrix} 4 \\ 2 \\ 2 \\ 0 \end{pmatrix} \right\}.$$

Notice that

$$\dim(U + W) = 3 = 2 + 2 - 1 = \dim U + \dim W - \dim(U \cap W).$$

In Exercise 2.6.15 we will see that this holds in general.

Example 2.4.8 The vectors

$$\mathbf{e}_1 = \begin{pmatrix} 1 \\ 0 \\ \vdots \\ 0 \end{pmatrix}, \mathbf{e}_2 = \begin{pmatrix} 0 \\ 1 \\ \vdots \\ 0 \end{pmatrix}, \dots, \mathbf{e}_n = \begin{pmatrix} 0 \\ 0 \\ \vdots \\ 1 \end{pmatrix}$$

form a basis for $\mathbb{F}^n$. Thus $\dim \mathbb{F}^n = n$. We call this the *standard basis* for $\mathbb{F}^n$.

Example 2.4.9 The previous example shows that $\mathbb{C}^n$ has dimension n. Here it is understood that we view $\mathbb{C}^n$ as a vector space over $\mathbb{C}$, which is the default. However, we may also view $\mathbb{C}^n$ as a vector space over $\mathbb{R}$. In this case, we only allow scalar multiplication with real scalars. In this setting the vectors

$$\mathbf{e}_1 = \begin{pmatrix} 1 \\ 0 \\ \vdots \\ 0 \end{pmatrix}, i\mathbf{e}_1 = \begin{pmatrix} i \\ 0 \\ \vdots \\ 0 \end{pmatrix}, \mathbf{e}_2 = \begin{pmatrix} 0 \\ 1 \\ \vdots \\ 0 \end{pmatrix}, i\mathbf{e}_2 = \begin{pmatrix} 0 \\ i \\ \vdots \\ 0 \end{pmatrix} \dots, \mathbf{e}_n = \begin{pmatrix} 0 \\ 0 \\ \vdots \\ 1 \end{pmatrix}, i\mathbf{e}_n = \begin{pmatrix} 0 \\ 0 \\ \vdots \\ i \end{pmatrix}$$
$$(2.8)$$

are linearly independent. Indeed, if $a_1, b_1, \dots, a_n, b_n \in \mathbb{R}$, then the equality

$$a_1\mathbf{e}_1 + b_1(i\mathbf{e}_1) + \dots + a_n\mathbf{e}_n + b_n(i\mathbf{e}_n) = \mathbf{0},$$

leads to $a_1 + ib_1 = 0, \dots, a_n + ib_n = 0$, which yields $a_1 = b_1 = \dots = a_n = b_n = 0$ (in this last step we used that a_j and b_j are all real). It is easy to check that taking all real linear combinations of the vectors in (2.8) we get all of $\mathbb{C}^n$. Thus the vectors in (2.8) form a basis of $\mathbb{C}^n$ when viewed as a vector space over $\mathbb{R}$, and thus its dimension viewed as a vector space over $\mathbb{R}$ is $2n$. We write this as $\dim_{\mathbb{R}}\mathbb{C}^n = 2n$.

Example 2.4.10 For polynomial spaces we have the following observations.

- The set $\{1, X, X^2, X^3, \dots\}$ is a basis for $\mathbb{F}[X]$. We have $\dim \mathbb{F}[X] = \infty$.

- The set $\{1, X, X^2, \dots, X^n\}$ is a basis for $\mathbb{F}_n[X]$. We have $\dim \mathbb{F}_n[X] = n + 1$.

- The set $\{1, i, X, iX, X^2, iX^2, \ldots, X^n, iX^n\}$ is a basis for $\mathbb{C}_n[X]$ when viewed as a vector space over $\mathbb{R}$. We have $\dim_{\mathbb{R}} \mathbb{C}_n[X] = 2n + 2$.

These bases are all referred to as the "standard" basis for their respective vector space.

We let E_{jk} be the matrix with all zero entries except for the (j, k) entry which equals 1. We expect the size of the matrices E_{jk} to be clear from the context in which we use them.

Example 2.4.11 For matrix spaces we have the following observations.

- The set $\{E_{j,k} : j = 1, \ldots, n, \ k = 1, \ldots, m\}$ is a basis for $\mathbb{F}^{n \times m}$. We have $\dim \mathbb{F}^{n \times m} = nm$.

- The set
$$\{E_{j,k} : j = 1, \ldots, n, \ k = 1, \ldots, m\} \cup \{iE_{j,k} : j = 1, \ldots, n, \ k = 1, \ldots, m\}$$
is a basis for $\mathbb{C}^{n \times m}$ when viewed as a vector space over $\mathbb{R}$. We have $\dim_{\mathbb{R}} \mathbb{C}^{n \times m} = 2nm$.

These bases are all referred to as the standard basis for their respective vector space.

2.5 Coordinate systems

We will see in this section that any n-dimensional vector space over $\mathbb{F}$ "works the same" as $\mathbb{F}^n$, which simplifies the study of such vector spaces tremendously. To make this idea more precise, we have to discuss coordinate systems. We start with the following result.

Theorem 2.5.1 Let $\mathcal{B} = \{\mathbf{v}_1, \ldots, \mathbf{v}_n\}$ be a basis for a vector space V over $\mathbb{F}$. Then for each $\mathbf{v} \in V$ there exists unique $c_1, \ldots, c_n \in \mathbb{F}$ so that

$$\mathbf{v} = c_1 \mathbf{v}_1 + \cdots + c_n \mathbf{v}_n. \tag{2.9}$$

Proof. Let $\mathbf{v} \in V$. As Span $\mathcal{B} = V$, we have that $\mathbf{v} = c_1 \mathbf{v}_1 + \cdots + c_n \mathbf{v}_n$ for some $c_1, \ldots, c_n \in \mathbb{F}$. Suppose that we also have $\mathbf{v} = d_1 \mathbf{v}_1 + \cdots + d_n \mathbf{v}_n$ for some $d_1, \ldots, d_n \in \mathbb{F}$. Then

$$\mathbf{0} = \mathbf{v} - \mathbf{v} = \sum_{j=1}^{n} c_j \mathbf{v}_j - \sum_{j=1}^{n} d_j \mathbf{v}_j = (c_1 - d_1)\mathbf{v}_1 + \cdots + (c_n - d_n)\mathbf{v}_n.$$

As $\{\mathbf{v}_1, \ldots, \mathbf{v}_n\}$ is linearly independent, we must have
$c_1 - d_1 = 0, \ldots, c_n - d_n = 0$. This yields $c_1 = d_1, \ldots, c_n = d_n$, yielding the
uniqueness. $\qquad\qquad\qquad\qquad\qquad\qquad\qquad\qquad\qquad\qquad\qquad\qquad\qquad\qquad\square$

When (2.9) holds, we say that $c_1, \ldots, c_n$ are the *coordinates of* $\mathbf{v}$ *relative to
the basis* $\mathcal{B}$, and we write

$$[\mathbf{v}]_{\mathcal{B}} = \begin{pmatrix} c_1 \\ \vdots \\ c_n \end{pmatrix}.$$

Thus, when $\mathcal{B} = \{\mathbf{v}_1, \ldots, \mathbf{v}_n\}$ we have

$$\mathbf{v} = c_1 \mathbf{v}_1 + \cdots + c_n \mathbf{v}_n \iff [\mathbf{v}]_{\mathcal{B}} = \begin{pmatrix} c_1 \\ \vdots \\ c_n \end{pmatrix}. \qquad (2.10)$$

Clearly, when $\mathbf{v} = c_1 \mathbf{v}_1 + \cdots + c_n \mathbf{v}_n$, $\mathbf{w} = d_1 \mathbf{v}_1 + \cdots + d_n \mathbf{v}_n$, then
$\mathbf{v} + \mathbf{w} = \sum_{j=1}^{n} (c_j + d_j)\mathbf{v}_j$, and thus

$$[\mathbf{v} + \mathbf{w}]_{\mathcal{B}} = \begin{pmatrix} c_1 + d_1 \\ \vdots \\ c_n + d_n \end{pmatrix} = [\mathbf{v}]_{\mathcal{B}} + [\mathbf{w}]_{\mathcal{B}}.$$

Similarly,

$$[\alpha\mathbf{v}]_{\mathcal{B}} = \begin{pmatrix} \alpha c_1 \\ \vdots \\ \alpha c_n \end{pmatrix} = \alpha[\mathbf{v}]_{\mathcal{B}}.$$

Thus adding two vectors in V corresponds to adding their corresponding
coordinate vectors (which are both with respect to the basis $\mathcal{B}$), and
multiplying a vector by a scalar in V corresponds to multiplying the
corresponding coordinate vector by the same scalar. As we will see in the
next chapter, the map $\mathbf{v} \mapsto [\mathbf{v}]_{\mathcal{B}}$ is a bijective linear map (also called an
isomorphism). This map allows one to view an n-dimensional vector space V
over $\mathbb{F}$ as essentially being the vector space $\mathbb{F}^n$.

Example 2.5.2 Let $V = \mathbb{Z}_7^3$ and $\mathcal{B} = \{ \begin{pmatrix} 1 \\ 1 \\ 1 \end{pmatrix}, \begin{pmatrix} 1 \\ 2 \\ 3 \end{pmatrix}, \begin{pmatrix} 1 \\ 3 \\ 6 \end{pmatrix} \}$. Let $\mathbf{v} = \begin{pmatrix} 6 \\ 5 \\ 4 \end{pmatrix}$.

Find $[\mathbf{v}]_{\mathcal{B}}$.

Denoting $[\mathbf{v}]_{\mathcal{B}} = \begin{pmatrix} c_1 \\ c_2 \\ c_3 \end{pmatrix}$ we need to solve for c_1, c_2, c_3 in the vector equation

$$c_1 \begin{pmatrix} 1 \\ 1 \\ 1 \end{pmatrix} + c_2 \begin{pmatrix} 1 \\ 2 \\ 3 \end{pmatrix} + c_3 \begin{pmatrix} 1 \\ 3 \\ 6 \end{pmatrix} = \begin{pmatrix} 6 \\ 5 \\ 4 \end{pmatrix}.$$

Setting up the augmented matrix and row reducing gives

$$\left(\begin{array}{ccc|c} 1 & 1 & 1 & 6 \\ 1 & 2 & 3 & 5 \\ 1 & 3 & 6 & 4 \end{array}\right) \rightarrow \left(\begin{array}{ccc|c} 1 & 1 & 1 & 6 \\ 0 & 1 & 2 & 6 \\ 0 & 2 & 5 & 5 \end{array}\right) \rightarrow$$

$$\left(\begin{array}{ccc|c} 1 & 1 & 1 & 6 \\ 0 & 1 & 2 & 6 \\ 0 & 0 & 1 & 0 \end{array}\right) \rightarrow \left(\begin{array}{ccc|c} 1 & 0 & 0 & 0 \\ 0 & 1 & 0 & 6 \\ 0 & 0 & 1 & 0 \end{array}\right),$$

yielding $c_1 = 0, c_2 = 6, c_3 = 0$. Thus $[\mathbf{v}]_{\mathcal{B}} = \begin{pmatrix} 0 \\ 6 \\ 0 \end{pmatrix}$.

Example 2.5.3 Let $V = \mathbb{C}_3[X]$ and
$\mathcal{B} = \{1, X - 1, X^2 - 2X + 1, X^3 - 3X^2 + 3X - 1\}$. Find $[X^3 + X^2 + X + 1]_{\mathcal{B}}$.

We need to find $c_1, c_2, c_3, c_4 \in \mathbb{C}$ so that

$$c_1 1 + c_2(X - 1) + c_3(X^2 - 2X + 1) + c_4(X^3 - 3X^2 + 3X - 1) = X^3 + X^2 + X + 1.$$

Equating the coefficients of $1, X, X^2, X^3$, setting up the augmented matrix, and row reducing gives

$$\left(\begin{array}{cccc|c} 1 & -1 & 1 & -1 & 1 \\ 0 & 1 & -2 & 3 & 1 \\ 0 & 0 & 1 & -3 & 1 \\ 0 & 0 & 0 & 1 & 1 \end{array}\right) \rightarrow \left(\begin{array}{cccc|c} 1 & -1 & 1 & 0 & 2 \\ 0 & 1 & -2 & 0 & -2 \\ 0 & 0 & 1 & 0 & 4 \\ 0 & 0 & 0 & 1 & 1 \end{array}\right) \rightarrow$$

$$\left(\begin{array}{cccc|c} 1 & -1 & 0 & 0 & -2 \\ 0 & 1 & 0 & 0 & 6 \\ 0 & 0 & 1 & 0 & 4 \\ 0 & 0 & 0 & 1 & 1 \end{array}\right) \rightarrow \left(\begin{array}{cccc|c} 1 & 0 & 0 & 0 & 4 \\ 0 & 1 & 0 & 0 & 6 \\ 0 & 0 & 1 & 0 & 4 \\ 0 & 0 & 0 & 1 & 1 \end{array}\right).$$

Thus we find $[X^3 + X^2 + X + 1]_{\mathcal{B}} = \begin{pmatrix} 4 \\ 6 \\ 4 \\ 1 \end{pmatrix}$.

2.6　Exercises

Exercise 2.6.1 For the proof of Lemma 2.1.1 provide a reason why each equality holds. For instance, the equality $\mathbf{0} = 0u + v$ is due to Axiom 5 in the definition of a vector space and $\mathbf{v}$ being the additive inverse of $0u$.

Exercise 2.6.2 Consider $p(X), q(X) \in \mathbb{F}[X]$ with $\mathbb{F} = \mathbb{R}$ or $\mathbb{F} = \mathbb{C}$. Show that $p(X) = q(X)$ if and only if $p(x) = q(x)$ for all $x \in \mathbb{F}$. (One way to do it is by using derivatives. Indeed, using calculus one can observe that if two polynomials are equal, then so are all their derivatives. Next observe that $p_j = \frac{1}{j!} \frac{d^j p}{dx^j}(0)$.) Where do you use in your proof that $\mathbb{F} = \mathbb{R}$ or $\mathbb{F} = \mathbb{C}$?

Exercise 2.6.3 When the underlying field is $\mathbb{Z}_p$, why does closure under addition automatically imply closure under scalar multiplication?

Exercise 2.6.4 Let $V = \mathbb{R}^{\mathbb{R}}$. For $W \subset V$, show that W is a subspace of V.

(a) $W = \{f : \mathbb{R} \to \mathbb{R} \; : \; f \text{ is continuous}\}$.

(b) $W = \{f : \mathbb{R} \to \mathbb{R} \; : \; f \text{ is differentiable}\}$.

Exercise 2.6.5 For the following choices of $\mathbb{F}$, V and W, determine whether W is a subspace of V over $\mathbb{F}$. In case the answer is yes, provide a basis for W.

(a) Let $\mathbb{F} = \mathbb{R}$ and $V = \mathbb{R}^3$,

$$W = \left\{ \begin{pmatrix} x_1 \\ x_2 \\ x_3 \end{pmatrix} \; : \; x_1, x_2, x_3 \in \mathbb{R}, x_1 - 2x_2 + x_3^2 = 0 \right\}.$$

(b) $\mathbb{F} = \mathbb{C}$ and $V = \mathbb{C}^{3\times3}$,

$$W = \left\{ \begin{pmatrix} a & b & c \\ 0 & a & b \\ 0 & 0 & a \end{pmatrix} \; : \; a, b, c \in \mathbb{C} \right\}.$$

(c) $\mathbb{F} = \mathbb{C}$ and $V = \mathbb{C}^{2\times2}$,

$$W = \left\{ \begin{pmatrix} a & \bar{b} \\ b & c \end{pmatrix} \; : \; a, b, c \in \mathbb{C} \right\}.$$

(d) $\mathbb{F} = \mathbb{R}$, $V = \mathbb{R}_2[X]$ and

$$W = \{p(x) \in V \; : \; \int_0^1 p(x) \cos x dx = 0\}.$$

(e) $\mathbb{F} = \mathbb{R}$, $V = \mathbb{R}_2[X]$ and

$$W = \{p(x) \in V \; : p(1) = p(2)p(3)\}.$$

(f) $\mathbb{F} = \mathbb{C}$, $V = \mathbb{C}^3$, and

$$W = \{ \begin{pmatrix} x_1 \\ x_2 \\ x_3 \end{pmatrix} \in \mathbb{C}^3 \; : \; x_1 - x_2 = x_3 - x_2\}.$$

Exercise 2.6.6 For the following vector spaces (V over $\mathbb{F}$) and vectors, determine whether the vectors are linearly independent or linearly independent.

(a) Let $\mathbb{F} = \mathbb{Z}_5$, $V = \mathbb{Z}_5^4$ and consider the vectors

$$\begin{pmatrix} 3 \\ 0 \\ 2 \\ 1 \end{pmatrix}, \begin{pmatrix} 2 \\ 1 \\ 0 \\ 3 \end{pmatrix}, \begin{pmatrix} 1 \\ 2 \\ 1 \\ 0 \end{pmatrix}.$$

(b) Let $\mathbb{F} = \mathbb{R}$, $V = \{f \mid f : (0, \infty) \to \mathbb{R}$ is a continuous function$\}$, and consider the vectors

$$t, t^2, \frac{1}{t}.$$

(c) Let $\mathbb{F} = \mathbb{Z}_5$, $V = \mathbb{Z}_5^4$ and consider the vectors

$$\begin{pmatrix} 4 \\ 0 \\ 2 \\ 3 \end{pmatrix}, \begin{pmatrix} 2 \\ 1 \\ 0 \\ 3 \end{pmatrix}, \begin{pmatrix} 1 \\ 2 \\ 1 \\ 0 \end{pmatrix}.$$

(d) Let $\mathbb{F} = \mathbb{R}$, $V = \{f \mid f : \mathbb{R} \to \mathbb{R}$ is a continuous function$\}$, and consider the vectors

$$\cos 2x, \sin 2x, \cos^2 x, \sin^2 x.$$

(e) Let $\mathbb{F} = \mathbb{C}$, $V = \mathbb{C}^{2 \times 2}$, and consider the vectors

$$\begin{pmatrix} i & 1 \\ -1 & -i \end{pmatrix}, \begin{pmatrix} 1 & 1 \\ i & -i \end{pmatrix}, \begin{pmatrix} -1 & i \\ -i & 1 \end{pmatrix}.$$

(f) Let $\mathbb{F} = \mathbb{R}$, $V = \mathbb{C}^{2 \times 2}$, and consider the vectors

$$\begin{pmatrix} i & 1 \\ -1 & -i \end{pmatrix}, \begin{pmatrix} 1 & 1 \\ i & -i \end{pmatrix}, \begin{pmatrix} -1 & i \\ -i & 1 \end{pmatrix}.$$

(g) Let $\mathbb{F} = \mathbb{Z}_5$, $V = \mathbb{F}^{3 \times 2}$, and consider the vectors

$$\begin{pmatrix} 3 & 4 \\ 1 & 0 \\ 1 & 0 \end{pmatrix}, \begin{pmatrix} 1 & 1 \\ 4 & 2 \\ 1 & 2 \end{pmatrix}, \begin{pmatrix} 1 & 2 \\ 3 & 1 \\ 1 & 2 \end{pmatrix}.$$

(h) Let $\mathbb{F} = \mathbb{R}$, $V = \{f \mid f : \mathbb{R} \to \mathbb{R}$ is a continuous function$\}$, and consider the vectors

$$1, e^t, e^{2t}.$$

Exercise 2.6.7 Let $\mathbf{v}_1, \mathbf{v}_2, \mathbf{v}_3$ be linearly independent vectors in a vector space V.

(a) For which k are $k\mathbf{v}_1 + \mathbf{v}_2, k\mathbf{v}_2 - \mathbf{v}_3, \mathbf{v}_3 + \mathbf{v}_1$ linearly independent?

(b) Show that if $\mathbf{v}$ is in the span of $\mathbf{v}_1, \mathbf{v}_2$ and in the span of $\mathbf{v}_2 + \mathbf{v}_3, \mathbf{v}_2 - \mathbf{v}_3$, then $\mathbf{v}$ is a multiple of $\mathbf{v}_2$.

Exercise 2.6.8 (a) Show that if the set $\{\mathbf{v}_1, \ldots, \mathbf{v}_k\}$ is linearly independent, and $\mathbf{v}_{k+1}$ is not in $\mathrm{Span}\{\mathbf{v}_1, \ldots, \mathbf{v}_k\}$, then the set $\{\mathbf{v}_1, \ldots, \mathbf{v}_k, \mathbf{v}_{k+1}\}$ is linearly independent.

(b) Let W be a subspace of an n-dimensional vector space V, and let $\{\mathbf{v}_1, \ldots, \mathbf{v}_p\}$ be a basis for W. Show that there exist vectors $\mathbf{v}_{p+1}, \ldots, \mathbf{v}_n \in V$ so that $\{\mathbf{v}_1, \ldots, \mathbf{v}_p, \mathbf{v}_{p+1}, \ldots, \mathbf{v}_n\}$ is a basis for V.

(Hint: once $\mathbf{v}_1, \ldots, \mathbf{v}_k$ are found and $k < n$, observe that one can choose $\mathbf{v}_{k+1} \in V \setminus (\mathrm{Span}\{\mathbf{v}_1, \ldots, \mathbf{v}_k\})$. Argue that this process stops when $k = n$, and that at that point a basis for V is found.)

Exercise 2.6.9 Let $V = \mathbb{R}_2[X]$ and

$$W = \{p \in V \ : \ p(2) = 0\}.$$

(a) Show that W is a subspace of V.

(b) Find a basis for W.

Exercise 2.6.10 For the following choices of subspaces U and W in V, find bases for $U + W$ and $U \cap W$.

(a) $V = \mathbb{R}_5[X]$, $U = \text{Span}\{X + 1, X^2 - 1\}$, $W = \{p(X) \ : \ p(2) = 0\}$.

(b) $V = \mathbb{Z}_5^4$,

$$U = \text{Span}\left\{ \begin{pmatrix} 3 \\ 0 \\ 2 \\ 1 \end{pmatrix}, \begin{pmatrix} 2 \\ 1 \\ 0 \\ 0 \end{pmatrix} \right\}, \ W = \text{Span}\left\{ \begin{pmatrix} 1 \\ 2 \\ 1 \\ 0 \end{pmatrix}, \begin{pmatrix} 4 \\ 4 \\ 1 \\ 1 \end{pmatrix} \right\}.$$

Exercise 2.6.11 Let $\{\mathbf{v}_1, \mathbf{v}_2, \mathbf{v}_3, \mathbf{v}_4, \mathbf{v}_5\}$ be linearly independent vectors in a vector space V. Determine whether the following sets are linearly dependent or linearly independent.

(a) $\{\mathbf{v}_1 + \mathbf{v}_2 + \mathbf{v}_3 + \mathbf{v}_4, \mathbf{v}_1 - \mathbf{v}_2 + \mathbf{v}_3 - \mathbf{v}_4, \mathbf{v}_1 - \mathbf{v}_2 - \mathbf{v}_3 - \mathbf{v}_4\}$.

(b) $\{\mathbf{v}_1 + \mathbf{v}_2, \mathbf{v}_2 + \mathbf{v}_3, \mathbf{v}_3 + \mathbf{v}_4, \mathbf{v}_4 + \mathbf{v}_5, \mathbf{v}_5 + \mathbf{v}_2\}$.

(c) $\{\mathbf{v}_1 + \mathbf{v}_3, \mathbf{v}_4 - \mathbf{v}_2, \mathbf{v}_5 + \mathbf{v}_1, \mathbf{v}_4 - \mathbf{v}_2, \mathbf{v}_5 + \mathbf{v}_3, \mathbf{v}_1 + \mathbf{v}_2\}$.

When you did this exercise, did you make any assumptions on the underlying field?

Exercise 2.6.12

Let $\{\mathbf{v}_1, \mathbf{v}_2, \mathbf{v}_3, \mathbf{v}_4\}$ be a basis for a vector space V over $\mathbb{Z}_3$. Determine whether the following are also bases for V.

(a) $\{\mathbf{v}_1 + \mathbf{v}_2 + \mathbf{v}_3 + \mathbf{v}_4, \mathbf{v}_1 - \mathbf{v}_2 + \mathbf{v}_3 - \mathbf{v}_4, \mathbf{v}_1 - \mathbf{v}_2 - \mathbf{v}_3 - \mathbf{v}_4\}$.

(b) $\{\mathbf{v}_1, \mathbf{v}_2 + \mathbf{v}_3 + \mathbf{v}_4, \mathbf{v}_1 - \mathbf{v}_2 + \mathbf{v}_3 - \mathbf{v}_4, \mathbf{v}_1 - \mathbf{v}_2 - \mathbf{v}_3 - \mathbf{v}_4\}$.

(c) $\{\mathbf{v}_1 + \mathbf{v}_2 + \mathbf{v}_3 + \mathbf{v}_4, \mathbf{v}_1 - \mathbf{v}_2 + \mathbf{v}_3 - \mathbf{v}_4, \mathbf{v}_1 - \mathbf{v}_2 - \mathbf{v}_3 - \mathbf{v}_4, \mathbf{v}_2 + \mathbf{v}_4, \mathbf{v}_1 + \mathbf{v}_3\}$.

Exercise 2.6.13

For the following choices of vector spaces V over the field $\mathbb{F}$, bases $\mathcal{B}$ and vectors $\mathbf{v}$, determine $[\mathbf{v}]_\mathcal{B}$.

(a) Let $\mathbb{F} = \mathbb{Z}_5$, $V = \mathbb{Z}_5^4$,

$$\mathcal{B} = \left\{ \begin{pmatrix} 3 \\ 0 \\ 2 \\ 1 \end{pmatrix}, \begin{pmatrix} 2 \\ 1 \\ 0 \\ 0 \end{pmatrix}, \begin{pmatrix} 1 \\ 2 \\ 1 \\ 0 \end{pmatrix}, \begin{pmatrix} 0 \\ 2 \\ 1 \\ 0 \end{pmatrix} \right\}, \ \mathbf{v} = \begin{pmatrix} 1 \\ 3 \\ 2 \\ 2 \end{pmatrix}.$$

(b) Let $\mathbb{F} = \mathbb{R}$, $\mathcal{B} = \{t, t^2, \frac{1}{t}\}$, $V = \text{Span}\mathcal{B}$ and $\mathbf{v} = \frac{t^3 + 3t^2 + 5}{t}$.

(c) Let $\mathbb{F} = \mathbb{C}$, $V = \mathbb{C}^{2 \times 2}$,

$$\mathcal{B} = \{\begin{pmatrix} 0 & 1 \\ -1 & -i \end{pmatrix}, \begin{pmatrix} 1 & 1 \\ i & -i \end{pmatrix}, \begin{pmatrix} i & 0 \\ -1 & -i \end{pmatrix}, \begin{pmatrix} i & 1 \\ -1 & -i \end{pmatrix}\}, \mathbf{v} = \begin{pmatrix} -2+i & 3-2i \\ -5-i & 10 \end{pmatrix}.$$

(d) Let $\mathbb{F} = \mathbb{R}$, $V = \mathbb{C}^{2 \times 2}$, and consider the vectors

$$\mathcal{B} = \{E_{11}, E_{12}, E_{21}, E_{22}, iE_{11}, iE_{12}, iE_{21}, iE_{22}\}, \mathbf{v} = \begin{pmatrix} -1 & i \\ -i & 1 \end{pmatrix}.$$

(e) Let $\mathbb{F} = \mathbb{Z}_5$, $V = \text{Span}\mathcal{B}$,

$$\mathcal{B} = \left\{ \begin{pmatrix} 3 & 4 \\ 1 & 0 \\ 1 & 0 \end{pmatrix}, \begin{pmatrix} 1 & 1 \\ 4 & 2 \\ 1 & 2 \end{pmatrix}, \begin{pmatrix} 1 & 2 \\ 3 & 3 \\ 3 & 0 \end{pmatrix} \right\}, \mathbf{v} = \begin{pmatrix} 0 & 2 \\ 3 & 0 \\ 0 & 2 \end{pmatrix}.$$

Exercise 2.6.14 Given a matrix $A = (a_{jk})_{j=1,k=1}^{n \quad m} \in \mathbb{C}^{n \times m}$, we define $A^* = (\overline{a_{kj}})_{j=1,k=1}^{m \quad n} \in \mathbb{C}^{m \times n}$. For instance,

$$\begin{pmatrix} 1+2i & 3+4i & 5+6i \\ 7+8i & 9+10i & 11+12i \end{pmatrix}^* = \begin{pmatrix} 1-2i & 7-8i \\ 3-4i & 9-10i \\ 5-6i & 11-12i \end{pmatrix}.$$

We call a matrix $A \in \mathbb{C}^{n \times n}$ *Hermitian* if $A^* = A$. For instance, $\begin{pmatrix} 2 & 1-3i \\ 1+3i & 5 \end{pmatrix}$ is Hermitian. Let $H_n \subseteq \mathbb{C}^{n \times n}$ be the set of all $n \times n$ Hermitian matrices.

(a) Show that H_n is **not** a vector space over $\mathbb{C}$.

(b) Show that H_n is a vector space over $\mathbb{R}$. Determine $\dim_{\mathbb{R}} H_n$.

(Hint: Do it first for 2×2 matrices.)

Exercise 2.6.15 (a) Show that for finite-dimensional subspaces U and W of V we have that $\dim(U + W) = \dim U + \dim W - \dim(U \cap W)$.

(Hint: Start with a basis $\{\mathbf{v}_1, \ldots, \mathbf{v}_p\}$ for $U \cap W$. Next, find $\mathbf{u}_1, \ldots, \mathbf{u}_k$ so that $\{\mathbf{v}_1, \ldots, \mathbf{v}_p, \mathbf{u}_1, \ldots, \mathbf{u}_k\}$ is a basis for U. Similarly, find $\mathbf{w}_1, \ldots, \mathbf{w}_l$ so that $\{\mathbf{v}_1, \ldots, \mathbf{v}_p, \mathbf{w}_1, \ldots, \mathbf{w}_l\}$ is a basis for W. Finally, argue that $\{\mathbf{v}_1, \ldots, \mathbf{v}_p, \mathbf{u}_1, \ldots, \mathbf{u}_k, \mathbf{w}_1, \ldots, \mathbf{w}_l\}$ is a basis for $U + W$.)

(b) Show that for a direct sum $U_1 \dot{+} \cdots \dot{+} U_k$ of finite-dimensional subspaces $U_1, \ldots, U_k$, we have that

$$\dim(U_1 \dot{+} \cdots \dot{+} U_k) = \dim U_1 + \cdots + \dim U_k.$$

Exercise 2.6.16 *(Honors)* Let
$P_n = \{(p_i)_{i=1}^n \in \mathbb{R}^n \; : \; p_i > 0, i = 1, \ldots, n, \text{ and } \sum_{i=1}^n p_i = 1\}$. Define the operations

$$\oplus : P_n \times P_n \to P_n, \quad \circ : \mathbb{R} \times P_n \to P_n,$$

via

$$(p_i)_{i=1}^n \oplus (q_i)_{i=1}^n := \frac{1}{\sum_{j=1}^n p_j q_j} (p_i q_i)_{i=1}^n,$$

and

$$c \circ (p_i)_{i=1}^n := \frac{1}{\sum_{j=1}^n p_j^c} (p_i^c)_{i=1}^n.$$

Show that P_n with the operations $\oplus$ and $\circ$ is a vector space over $\mathbb{R}$. Why is P_n **not** a subspace of $\mathbb{R}^n$?

Hint: observe that $\begin{pmatrix} \frac{1}{n} \\ \vdots \\ \frac{1}{n} \end{pmatrix}$ is the neutral element for $\oplus$.

This exercise is based the paper [A. Sgarro, An informational divergence geometry for stochastic matrices. *Calcolo* **15** (1978), no. 1, 41–49.] Thanks are due to Valerie Girardin for making the author aware of the example.

Exercise 2.6.17 *(Honors)* Let V be a vector space over $\mathbb{F}$ and $W \subseteq V$ a subspace. Define the relation

$$\mathbf{v} \sim \hat{\mathbf{v}} \;\Leftrightarrow\; \mathbf{v} - \hat{\mathbf{v}} \in W.$$

(a) Show that $\sim$ is an equivalence relation.

Let

$$\mathbf{v} + W := \{\hat{\mathbf{v}} \; : \; \mathbf{v} \sim \hat{\mathbf{v}}\}$$

denote the equivalence class of $\mathbf{v} \in V$, and let

$$V/W := \{\mathbf{v} + W \; : \; \mathbf{v} \in V\}$$

denote the set of equivalence classes. Define addition and scalar multiplication on V/W via

$$(\mathbf{v} + W) + (\hat{\mathbf{v}} + W) := (\mathbf{v} + \hat{\mathbf{v}}) + W \;, \quad c(\mathbf{v} + W) := (c\mathbf{v}) + W.$$

(b) Show that addition on V/W is well-defined. (It needs to be shown that if $\mathbf{v} + W = \mathbf{w} + W$ and $\hat{\mathbf{v}} + W = \hat{\mathbf{w}} + W$, then $(\mathbf{v} + \hat{\mathbf{v}}) + W := (\mathbf{w} + \hat{\mathbf{w}}) + W$ as the sum of two equivalence classes should be independent on the particular representatives chosen.)

(c) Show that scalar multiplication on V/W is well-defined.

(d) Show that V/W is a vector space.

(e) Let $V = \mathbb{R}^2$ and $W = \text{Span}\{\begin{pmatrix} 1 \\ 1 \end{pmatrix}\}$. Explain that V/W consists of all lines parallel to W, and explain how the addition and scalar multiplication are defined on these parallel lines.

3

Linear Transformations

CONTENTS

3.1   Definition of a linear transformation 55
3.2   Range and kernel of linear transformations 57
3.3   Matrix representations of linear maps 61
3.4   Exercises 65

Now that we have introduced vector spaces, we can move on to the next main object in linear algebra: linear transformations. These are functions between vector spaces that behave nicely with respect to the two fundamental operations on a vector space: addition and scalar multiplication. Differentiation and taking integrals are two important examples of linear transformation. Regarding the nice behavior, note for example that if we take the derivative of the sum of two functions, it is the same as if we would take the derivative of each and then the sum. Let us start with the precise definition.

3.1   Definition of a linear transformation

Let V and W be vector spaces over the same field $\mathbb{F}$. A function $T : V \to W$ is called *linear* if

(i) $T(\mathbf{u} + \mathbf{v}) = T(\mathbf{u}) + T(\mathbf{v})$ for all $\mathbf{u}, \mathbf{v} \in V$, and

(ii) $T(c\mathbf{u}) = cT(\mathbf{u})$ for all $c \in \mathbb{F}$ and all $\mathbf{u} \in V$. In this case, we say that T is a *linear transformation* or a *linear map*.

When T is linear, we must have that $T(\mathbf{0}) = \mathbf{0}$. Indeed, by using (ii) we have $T(\mathbf{0}) = T(0 \cdot \mathbf{0}) = 0T(\mathbf{0}) = \mathbf{0}$, where in the first and last step we used Lemma 2.1.1.

Example 3.1.1 Let $T : \mathbb{Z}_3^2 \to \mathbb{Z}_3^3$ be defined by

$$T \begin{pmatrix} x_1 \\ x_2 \end{pmatrix} = \begin{pmatrix} 2x_1 + x_2 \\ x_1 + x_2 \\ x_2 \end{pmatrix}.$$

Then

$$T(\begin{pmatrix} x_1 \\ x_2 \end{pmatrix} + \begin{pmatrix} y_1 \\ y_2 \end{pmatrix}) = T \begin{pmatrix} x_1 + y_1 \\ x_2 + y_2 \end{pmatrix} = \begin{pmatrix} 2(x_1 + y_1) + x_2 + y_2 \\ x_1 + y_1 + x_2 + y_2 \\ x_2 + y_2 \end{pmatrix} =$$

$$\begin{pmatrix} 2x_1 + x_2 \\ x_1 + x_2 \\ x_2 \end{pmatrix} + \begin{pmatrix} 2y_1 + y_2 \\ y_1 + y_2 \\ y_2 \end{pmatrix} = T \begin{pmatrix} x_1 \\ x_2 \end{pmatrix} + T \begin{pmatrix} x_1 \\ x_2 \end{pmatrix},$$

and

$$T(c \begin{pmatrix} x_1 \\ x_2 \end{pmatrix}) = T \begin{pmatrix} cx_1 \\ cx_2 \end{pmatrix} = \begin{pmatrix} 2cx_1 + cx_2 \\ cx_1 + cx_2 \\ cx_2 \end{pmatrix} = c \begin{pmatrix} 2x_1 + x_2 \\ x_1 + x_2 \\ x_2 \end{pmatrix} = cT \begin{pmatrix} x_1 \\ x_2 \end{pmatrix}.$$

Thus T is linear.

Example 3.1.2 Let $T : \mathbb{C}^3 \to \mathbb{C}^2$ be defined by

$$T \begin{pmatrix} x_1 \\ x_2 \\ x_3 \end{pmatrix} = \begin{pmatrix} x_1 x_2 \\ x_1 + x_2 + x_3 \end{pmatrix}.$$

Then

$$T \begin{pmatrix} 1 \\ 1 \\ 1 \end{pmatrix} = \begin{pmatrix} 1 \\ 3 \end{pmatrix}, \ T \begin{pmatrix} 2 \\ 2 \\ 2 \end{pmatrix} = \begin{pmatrix} 4 \\ 6 \end{pmatrix} \neq 2 \begin{pmatrix} 1 \\ 3 \end{pmatrix} = 2T \begin{pmatrix} 1 \\ 1 \\ 1 \end{pmatrix},$$

thus T fails to satisfy (ii) above. Thus T is not linear.

Notice that in order to show that a function is not linear, one only needs to provide one example where the above rules (i) or (ii) are not satisfied.

The linear map in Example 3.1.1 can be written in the form

$$T \begin{pmatrix} x_1 \\ x_2 \end{pmatrix} = \begin{pmatrix} 2 & 1 \\ 1 & 1 \\ 0 & 1 \end{pmatrix} \begin{pmatrix} x_1 \\ x_2 \end{pmatrix}.$$

We have the following general result, from which linearity in Example 3.1.1 directly follows.

Proposition 3.1.3 *Let $A \in \mathbb{F}^{n \times m}$ and define $T : \mathbb{F}^m \to \mathbb{F}^n$ via $T(\mathbf{x}) := A\mathbf{x}$. Then T is linear.*

Proof. This follows directly from rules on matrix vector multiplication: $A(\mathbf{x} + \mathbf{y}) = A\mathbf{x} + A\mathbf{y}$ and $A(c\mathbf{x}) = cA\mathbf{x}$. □

3.2 Range and kernel of linear transformations

With a linear transformation there are two subspaces associated with it: the range (which lies in the co-domain) and the kernel (which lies in the domain). These subspaces provide us with crucial information about the linear transformation. We start with discussing the range.

Let $T : V \to W$ be a linear map. Define the *range* of T by

$$\text{Ran } T := \{\mathbf{w} \in W \; : \; \text{there exists a } \mathbf{v} \in V \text{ so that } T(\mathbf{v}) = \mathbf{w}\}.$$

Proposition 3.2.1 *Let $T : V \to W$ be a linear map. Then $\text{Ran } T$ is a subspace of W. Moreover, if $\{\mathbf{v}_1, \ldots, \mathbf{v}_p\}$ is a basis for V, then $\text{Ran } T = \text{Span}\{T(\mathbf{v}_1), \ldots, T(\mathbf{v}_p)\}$. In particular $\dim \text{Ran } T \leq \dim V$.*

Proof. First observe that $T(\mathbf{0}) = \mathbf{0}$ gives that $\mathbf{0} \in \text{Ran } T$. Next, let $\mathbf{w}$, $\hat{\mathbf{w}} \in \text{Ran } T$ and $c \in \mathbb{F}$. Then there exist $\mathbf{v}, \hat{\mathbf{v}} \in V$ so that $T(\mathbf{v}) = \mathbf{w}$ and $T(\hat{\mathbf{v}}) = \hat{\mathbf{w}}$. Then $\mathbf{w} + \hat{\mathbf{w}} = T(\mathbf{v} + \hat{\mathbf{v}}) \in \text{Ran } T$ and $c\mathbf{w} = T(c\mathbf{v}) \in \text{Ran } T$. Thus, by Proposition 2.3.1, $\text{Ran } T$ is a subspace of W.

Clearly, $T(\mathbf{v}_1), \ldots, T(\mathbf{v}_p) \in \text{Ran } T$, and since $\text{Ran } T$ is a subspace we have that $\text{Span}\{T(\mathbf{v}_1), \ldots, T(\mathbf{v}_p)\} \subseteq \text{Ran } T$. For the converse inclusion, let $\mathbf{w} \in \text{Ran } T$. Then there exists a $\mathbf{v} \in V$ so that $T(\mathbf{v}) = \mathbf{w}$. As $\{\mathbf{v}_1, \ldots, \mathbf{v}_p\}$ is a basis for V, there exist $c_1, \ldots, c_p \in \mathbb{F}$ so that $\mathbf{v} = c_1\mathbf{v}_1 + \cdots + c_p\mathbf{v}_p$. Then

$$\mathbf{w} = T(\mathbf{v}) = T\left(\sum_{j=1}^{p} c_j \mathbf{v}_j\right) = \sum_{j=1}^{p} c_j T(\mathbf{v}_j) \in \text{Span}\{T(\mathbf{v}_1), \ldots, T(\mathbf{v}_p)\}.$$

Thus $\text{Ran } T \subseteq \text{Span}\{T(\mathbf{v}_1), \ldots, T(\mathbf{v}_p)\}$. We have shown both inclusions, and consequently $\text{Ran } T = \text{Span}\{T(\mathbf{v}_1), \ldots, T(\mathbf{v}_p)\}$ follows. □

We say that $T : V \to W$ is *onto* (or *surjective*) if $\text{Ran } T = W$. Equivalently, T is onto if and only if for every $\mathbf{w} \in W$ there exists a $\mathbf{v} \in V$ so that $T(\mathbf{v}) = \mathbf{w}$.

Example 3.1.1 continued. As the standard basis $\{\mathbf{e}_1, \mathbf{e}_2\}$ is a basis for $\mathbb{Z}_3^2$, we have that

$$\text{Ran } T = \text{Span}\{T(\mathbf{e}_1), T(\mathbf{e}_2)\} = \begin{pmatrix} 2 \\ 1 \\ 0 \end{pmatrix}, \begin{pmatrix} 1 \\ 1 \\ 1 \end{pmatrix}\}.$$

In fact, as these two vectors are linearly independent, they form a basis for Ran T. The map T is not onto as $\dim \text{Ran } T = 2$, while $\dim W = \dim \mathbb{Z}_3^3 = 3$, and thus Ran $T \neq \mathbb{Z}_3^3$.

Define the *kernel* of T by

$$\text{Ker } T := \{\mathbf{v} \in V \ : \ T(\mathbf{v}) = \mathbf{0}\}.$$

Proposition 3.2.2 *Let $T : V \to W$ be a linear map. Then* Ker T *is subspace of V.*

Proof. First observe that $T(\mathbf{0}) = \mathbf{0}$ gives that $\mathbf{0} \in \text{Ker } T$. Next, let $\mathbf{v}$, $\hat{\mathbf{v}} \in \text{Ker } T$ and $c \in \mathbb{F}$. Then $T(\mathbf{v} + \hat{\mathbf{v}}) = T(\mathbf{v}) + T(\hat{\mathbf{v}}) = \mathbf{0} + \mathbf{0} = \mathbf{0}$ and $T(c\mathbf{v}) = cT(\mathbf{v}) = c\mathbf{0} = \mathbf{0}$, so $\mathbf{v} + \hat{\mathbf{v}}, c\mathbf{v} \in \text{Ker } T$. Thus, by Proposition 2.3.1, Ker T is a subspace of V. $\qquad\square$

We say that $T : V \to W$ is *one-to-one* (or *injective*) if $T(\mathbf{v}) = T(\mathbf{w})$ only holds when $\mathbf{v} = \mathbf{w}$. We have the following way to check for injectivity of a linear map.

Lemma 3.2.3 *The linear map T is one-to-one if and only if* Ker $T = \{\mathbf{0}\}$.

Proof. Suppose that T is one-to-one, and $\mathbf{v} \in \text{Ker } T$. Then $T(\mathbf{v}) = \mathbf{0} = T(\mathbf{0})$, where in the last step we used that T is linear. Since T is one-to-one, $T(\mathbf{v}) = T(\mathbf{0})$ implies that $\mathbf{v} = \mathbf{0}$. Thus Ker $T = \{\mathbf{0}\}$.

Next, suppose that Ker $T = \{\mathbf{0}\}$, and let $T(\mathbf{v}) = T(\mathbf{w})$. Then, using linearity we get $\mathbf{0} = T(\mathbf{v}) - T(\mathbf{w}) = T(\mathbf{v} - \mathbf{w})$, implying that $\mathbf{v} - \mathbf{w} \in \text{Ker } T = \{\mathbf{0}\}$, and thus $\mathbf{v} - \mathbf{w} = \mathbf{0}$. Thus $\mathbf{v} = \mathbf{w}$, and we can conclude that T is one-to-one. $\square$

Example 3.2.4 Let $V = \mathbb{R}_3[X]$, $W = \mathbb{R}^2$, and

$$T(p(X)) = \begin{pmatrix} p(1) \\ \int_0^2 p(x)dx \end{pmatrix}.$$

Let $p(X) = a + bX + cX^2 + dX^3 \in \operatorname{Ker} T$, then

$$\mathbf{0} = T(p(X)) = \begin{pmatrix} a + b + c + d \\ 2a + 2b + \frac{8}{3}c + 4d \end{pmatrix} = \begin{pmatrix} 1 & 1 & 1 & 1 \\ 2 & 2 & \frac{8}{3} & 4 \end{pmatrix} \begin{pmatrix} a \\ b \\ c \\ d \end{pmatrix}.$$

Row reducing

$$\begin{pmatrix} 1 & 1 & 1 & 1 \\ 2 & 2 & \frac{8}{3} & 4 \end{pmatrix} \to \begin{pmatrix} 1 & 1 & 1 & 1 \\ 0 & 0 & \frac{2}{3} & 2 \end{pmatrix}, \tag{3.1}$$

gives that b and d are free variables and $c = -3d$, $a = -b - c - d = -b + 2d$. Thus

$$p(X) = b(-1 + X) + d(2 - 3X^2 + X^3).$$

We get that $\operatorname{Ker} T = \operatorname{Span}\{-1 + X, 2 - 3X^2 + X^3\}$. In fact, the two polynomials are a basis for $\operatorname{Ker} T$.

As $\{1, X, X^2, X^3\}$ is a basis for $\mathbb{R}_3[X]$, we get that

$$\operatorname{Ran} T = \operatorname{Span}\{T(1), T(X), T(X^2), T(X^3)\} =$$

$$\operatorname{Span}\{\begin{pmatrix} 1 \\ 2 \end{pmatrix}, \begin{pmatrix} 1 \\ 2 \end{pmatrix}, \begin{pmatrix} 1 \\ \frac{8}{3} \end{pmatrix}, \begin{pmatrix} 1 \\ 4 \end{pmatrix}\} = \operatorname{Span}\{\begin{pmatrix} 1 \\ 2 \end{pmatrix}, \begin{pmatrix} 1 \\ \frac{8}{3} \end{pmatrix}\}.$$

In the last step, we reduced the set of vectors to a basis for $\operatorname{Ran} T$ by just keeping the columns corresponding to pivot columns in (3.1).

Notice that

$$\dim \operatorname{Ker} T + \dim \operatorname{Ran} T = 2 + 2 = 4 = \dim \mathbb{R}_3[X].$$

As the next result shows, this is not a coincidence.

Theorem 3.2.5 *Let $T : V \to W$ be linear, and suppose that $\dim V < \infty$. Then*

$$\dim \operatorname{Ker} T + \dim \operatorname{Ran} T = \dim V. \tag{3.2}$$

Proof. Let $\{\mathbf{v}_1, \ldots, \mathbf{v}_p\}$ be a basis for $\operatorname{Ker} T (\subseteq V)$, and $\{\mathbf{w}_1, \ldots, \mathbf{w}_q\}$ a basis for $\operatorname{Ran} T$ (notice that by Proposition 3.2.1 it follows that $\operatorname{Ran} T$ is finite dimensional as V is finite dimensional). Let $\mathbf{x}_1, \ldots, \mathbf{x}_q \in V$ be so that $T(\mathbf{x}_j) = \mathbf{w}_j$, $j = 1, \ldots, q$. We claim that $\mathcal{B} = \{\mathbf{v}_1, \ldots, \mathbf{v}_p, \mathbf{x}_1, \ldots, \mathbf{x}_q\}$ is a basis for V, which then yields that $\dim V = p + q = \dim \operatorname{Ker} T + \dim \operatorname{Ran} T$.

Let $\mathbf{v} \in V$. Then $T(\mathbf{v}) \in \operatorname{Ran} T$, and thus there exist $b_1, \ldots, b_q$ so that $T(\mathbf{v}) = \sum_{j=1}^{q} b_j \mathbf{w}_j$. Then

$$T(\mathbf{v} - \sum_{j=1}^{q} b_j \mathbf{x}_j) = T(\mathbf{v}) - \sum_{j=1}^{q} b_j \mathbf{w}_j = \mathbf{0}.$$

Thus $\mathbf{v} - \sum_{j=1}^{q} b_j \mathbf{x}_j \in \text{Ker } T$. Therefore, there exist $a_1, \ldots, a_p \in \mathbb{F}$ so that $\mathbf{v} - \sum_{j=1}^{q} b_j \mathbf{x}_j = \sum_{j=1}^{p} a_j \mathbf{v}_j$. Consequently, $\mathbf{v} = \sum_{j=1}^{p} a_j \mathbf{v}_j + \sum_{j=1}^{q} b_j \mathbf{x}_j \in \text{Span } \mathcal{B}$. This proves that $V = \text{Span } \mathcal{B}$.

It remains to show that $\mathcal{B}$, is linearly independent, so assume $\sum_{j=1}^{p} a_j \mathbf{v}_j + \sum_{j=1}^{q} b_j \mathbf{x}_j = \mathbf{0}$. Then

$$\mathbf{0} = T\left(\sum_{j=1}^{p} a_j \mathbf{v}_j + \sum_{j=1}^{q} b_j \mathbf{x}_j\right) = \sum_{j=1}^{p} a_j T(\mathbf{v}_j) + \sum_{j=1}^{q} b_j T(\mathbf{x}_j) = \sum_{j=1}^{q} b_j \mathbf{w}_j,$$

where we use that $\mathbf{v}_j \in \text{Ker } T$, $j = 1, \ldots, p$. As $\{\mathbf{w}_1, \ldots, \mathbf{w}_q\}$ is linearly independent, we now get that $b_1 = \cdots = b_q = 0$. But then we obtain that $\sum_{j=1}^{p} a_j \mathbf{v}_j = \mathbf{0}$, and as $\{\mathbf{v}_1, \ldots, \mathbf{v}_p\}$ is linearly independent, we get $a_1 = \cdots = a_p = 0$. Thus $\sum_{j=1}^{p} a_j \mathbf{v}_j + \sum_{j=1}^{q} b_j \mathbf{x}_j = \mathbf{0}$ implies $a_1 = \cdots = a_p = b_1 = \cdots = b_q = 0$, showing the linear independence of $\mathcal{B}$. $\square$

We say that T is *bijective* if T is both onto and one-to-one. We let $id_V : V \to V$ denote the *identity mapping*, that is $id_V(\mathbf{v}) = \mathbf{v}$, $\mathbf{v} \in V$.

Proposition 3.2.6 *Let $T : V \to W$ be bijective. Then T has an inverse T^{-1}. That is, $T^{-1} : W \to V$ exists so that $T \circ T^{-1} = id_W$ and $T^{-1} \circ T = id_V$. Moreover, T^{-1} is linear. Conversely, if T has an inverse, then T is bijective.*

Proof. Let $\mathbf{w} \in W$. As T is onto, there exists a $\mathbf{v} \in V$ so that $T(\mathbf{v}) = \mathbf{w}$, and as T is one-to-one, this $\mathbf{v}$ is unique. Define $T^{-1}(\mathbf{w}) := \mathbf{v}$, making $T^{-1} : W \to V$ well-defined. It is straightforward to check that $T(T^{-1}(\mathbf{w})) = \mathbf{w}$ for all $\mathbf{w} \in W$, and $T^{-1}(T(\mathbf{v})) = \mathbf{v}$ for all $\mathbf{v} \in V$.

Next suppose $T^{-1}(\mathbf{w}) = \mathbf{v}$ and $T^{-1}(\hat{\mathbf{w}}) = \hat{\mathbf{v}}$. This means that $T(\mathbf{v}) = \mathbf{w}$ and $T(\hat{\mathbf{v}}) = \hat{\mathbf{w}}$. Thus $T(\mathbf{v} + \hat{\mathbf{v}}) = \mathbf{w} + \hat{\mathbf{w}}$. But then, by definition, $T^{-1}(\mathbf{w} + \hat{\mathbf{w}}) = \mathbf{v} + \hat{\mathbf{v}}$ and, consequently, $T^{-1}(\mathbf{w} + \hat{\mathbf{w}}) = T^{-1}(\mathbf{w}) + T^{-1}(\hat{\mathbf{w}})$. Similarly, one proves $T^{-1}(c\mathbf{w}) = cT^{-1}(\mathbf{w})$. Thus T^{-1} is linear.

Next, suppose that T has an inverse T^{-1}. Let $\mathbf{w} \in W$. Put $\mathbf{v} = T^{-1}(\mathbf{w})$. Then $T(\mathbf{v}) = \mathbf{w}$, and thus $\mathbf{w} \in \text{Ran } T$. This shows that T is onto. Finally, suppose that $T(\mathbf{v}) = T(\hat{\mathbf{v}})$. Applying T^{-1} on both sides, gives $\mathbf{v} = T^{-1}(T(\mathbf{v})) = T^{-1}(T(\hat{\mathbf{v}})) = \hat{\mathbf{v}}$, showing that T is one-to-one. $\square$

A bijective linear map T is also called an *isomorphism*. We call two vector spaces V and W *isomorphic* if there exists an isomorphism $T : V \to W$. When two vector spaces are isomorphic they essentially have the same vector space properties. Indeed, whatever vector space properties V has, are carried over by T to W, and whatever vector space properties W has, are carried over by T^{-1} to V. As the following results shows, any n-dimensional vector space over the field $\mathbb{F}$ is isomorphic to $\mathbb{F}^n$.

Theorem 3.2.7 *Let V be a n-dimensional vector space over $\mathbb{F}$. Let $\mathcal{B} = \{\mathbf{v}_1, \ldots, \mathbf{v}_n\}$ be a basis for V. Then the map $T : V \to \mathbb{F}^n$ defined via $T(\mathbf{v}) = [\mathbf{v}]_\mathcal{B}$ is an isomorphism. In particular, V and $\mathbb{F}^n$ are isomorphic.*

Proof. In Section 2.5 we have already seen that T is linear. Next suppose that $T(\mathbf{v}) = \mathbf{0}$. Thus $[\mathbf{v}]_\mathcal{B} = \begin{pmatrix} 0 \\ \vdots \\ 0 \end{pmatrix}$, which means that $\mathbf{v} = \sum_{j=1}^n 0\mathbf{v}_j = \mathbf{0}$.

Thus Ker $T = \{\mathbf{0}\}$, giving that T is one-to-one. Next, let $\begin{pmatrix} c_1 \\ \vdots \\ c_n \end{pmatrix} \in \mathbb{F}^n$. Put

$\mathbf{v} = \sum_{j=1}^n c_j\mathbf{v}_j$. Then $T(\mathbf{v}) = [\mathbf{v}]_\mathcal{B} = \begin{pmatrix} c_1 \\ \vdots \\ c_n \end{pmatrix} \in$ Ran T. This shows that

Ran $T = V$, and thus T is onto. $\qquad\square$

The following example illustrates this result.

Example 3.2.8 Let $T : \mathbb{F}_{n-1}[X] \to \mathbb{F}^n$ be defined by

$$T(a_0 + a_1 X + \cdots + a_{n-1}X^{n-1}) := \begin{pmatrix} a_0 \\ a_1 \\ \vdots \\ a_{n-1} \end{pmatrix}.$$

It is easy to see that T is an isomorphism, and thus $\mathbb{F}_{n-1}[X]$ and $\mathbb{F}^n$ are isomorphic. The underlying basis $\mathcal{B}$ here is the standard basis $\{1, X, \ldots, X^{n-1}\}$.

3.3 Matrix representations of linear maps

The following results show that any linear map between finite-dimensional spaces allows a matrix representation with respect to chosen bases. The significance of this result is that one can study linear maps between finite-dimensional spaces by studying matrices.

Theorem 3.3.1 *Given vector spaces V and W over $\mathbb{F}$, with bases $\mathcal{B} = \{\mathbf{v}_1, \ldots, \mathbf{v}_n\}$ and $\mathcal{C} = \{\mathbf{w}_1, \ldots, \mathbf{w}_m\}$, respectively. Let $T : V \to W$. Represent $T(\mathbf{v}_j)$ with respect to the basis $\mathcal{C}$:*

$$T(\mathbf{v}_j) = a_{1j}\mathbf{w}_1 + \cdots + a_{mj}\mathbf{w}_m \;\Leftrightarrow\; [T(\mathbf{v}_j)]_{\mathcal{C}} = \begin{pmatrix} a_{1j} \\ \vdots \\ a_{mj} \end{pmatrix}, j = 1, \ldots, n. \quad (3.3)$$

Introduce the matrix $[T]_{\mathcal{C} \leftarrow \mathcal{B}} = (a_{ij})_{i=1,j=1}^{m \quad n}$. Then we have that

$$T(\mathbf{v}) = \mathbf{w} \;\Leftrightarrow\; [\mathbf{w}]_{\mathcal{C}} = [T]_{\mathcal{C} \leftarrow \mathcal{B}}[\mathbf{v}]_{\mathcal{B}}. \quad (3.4)$$

Conversely, if $A = (a_{ij})_{i=1,j=1}^{m \quad n} \in \mathbb{F}^{m \times n}$ is given, then defining $T : V \to W$ via (3.3) and extending by linearity via $T(\sum_{j=1}^{n} c_j \mathbf{v}_j) := \sum_{j=1}^{n} c_j T(\mathbf{v}_j)$, yields a linear map $T : V \to W$ with matrix representation $[T]_{\mathcal{C} \leftarrow \mathcal{B}} = A$.

Proof. The proof follows directly from the following observation. If

$$\mathbf{v} = c_1\mathbf{v}_1 + \cdots + c_n\mathbf{v}_n \;\Leftrightarrow\; [\mathbf{v}]_{\mathcal{B}} = \begin{pmatrix} c_1 \\ \vdots \\ c_n \end{pmatrix},$$

then

$$\mathbf{w} = T(\mathbf{v}) = \sum_{j=1}^{n} c_j T(\mathbf{v}_j) =$$

$$\sum_{j=1}^{n} c_j (\sum_{k=1}^{m} a_{kj}\mathbf{w}_k) \;\Leftrightarrow\; [\mathbf{w}]_{\mathcal{C}} = \begin{pmatrix} \sum_{j=1}^{n} a_{1j}c_j \\ \vdots \\ \sum_{j=1}^{n} a_{mj}c_j \end{pmatrix} = (a_{ij})_{i=1,j=1}^{m \quad n} \begin{pmatrix} c_1 \\ \vdots \\ c_n \end{pmatrix}.$$

$\square$

Example 3.3.2 Let $V = \mathbb{C}^{2 \times 2}$ and $\mathbb{F} = \mathbb{C}$. Let $\mathcal{B}$ be the standard basis $\{E_{11}, E_{12}, E_{21}, E_{22}\}$. Define $T : V \to V$ via

$$T(A) = \begin{pmatrix} 1 & 2 \\ 3 & 4 \end{pmatrix} A \begin{pmatrix} i & 3i \\ 5i & 7i \end{pmatrix}.$$

Find the matrix representation $[T]_{\mathcal{B} \leftarrow \mathcal{B}}$.

Compute

$$T(E_{11}) = \begin{pmatrix} i & 3i \\ 3i & 9i \end{pmatrix} = iE_{11} + 3iE_{12} + 3iE21 + 9iE_{22},$$

$$T(E_{12}) = \begin{pmatrix} 5i & 7i \\ 15i & 21i \end{pmatrix} = 5iE_{11} + 7iE_{12} + 15iE21 + 21iE_{22},$$

$$T(E_{21}) = \begin{pmatrix} 2i & 6i \\ 4i & 12i \end{pmatrix} = 2iE_{11} + 6iE_{12} + 4iE21 + 12iE_{22},$$

$$T(E_{22}) = \begin{pmatrix} 10i & 14i \\ 20i & 28i \end{pmatrix} = 10iE_{11} + 14iE_{12} + 20iE21 + 28iE_{22}.$$

This gives that

$$[T]_{\mathcal{B}\leftarrow\mathcal{B}} = \begin{pmatrix} i & 5i & 2i & 10i \\ 3i & 7i & 6i & 14i \\ 3i & 15i & 4i & 20i \\ 9i & 21i & 12i & 28i \end{pmatrix}.$$

Example 3.3.3 Let $V = \mathbb{Z}_5^3$ and

$$\mathcal{B} = \{ \begin{pmatrix} 3 \\ 0 \\ 1 \end{pmatrix}, \begin{pmatrix} 2 \\ 3 \\ 4 \end{pmatrix}, \begin{pmatrix} 1 \\ 4 \\ 1 \end{pmatrix} \}, \quad \mathcal{C} = \{ \begin{pmatrix} 0 \\ 2 \\ 4 \end{pmatrix}, \begin{pmatrix} 1 \\ 0 \\ 3 \end{pmatrix}, \begin{pmatrix} 2 \\ 1 \\ 0 \end{pmatrix} \}.$$

Find the matrix representation $[id_V]_{\mathcal{C}\leftarrow\mathcal{B}}$.

Compute

$$id_V \begin{pmatrix} 3 \\ 0 \\ 1 \end{pmatrix} = \begin{pmatrix} 3 \\ 0 \\ 1 \end{pmatrix} = 2 \begin{pmatrix} 0 \\ 2 \\ 4 \end{pmatrix} + 1 \begin{pmatrix} 1 \\ 0 \\ 3 \end{pmatrix} + 1 \begin{pmatrix} 2 \\ 1 \\ 0 \end{pmatrix},$$

$$id_V \begin{pmatrix} 2 \\ 3 \\ 4 \end{pmatrix} = \begin{pmatrix} 2 \\ 3 \\ 4 \end{pmatrix} = 1 \begin{pmatrix} 0 \\ 2 \\ 4 \end{pmatrix} + 0 \begin{pmatrix} 1 \\ 0 \\ 3 \end{pmatrix} + 1 \begin{pmatrix} 2 \\ 1 \\ 0 \end{pmatrix},$$

$$id_V \begin{pmatrix} 1 \\ 4 \\ 1 \end{pmatrix} = \begin{pmatrix} 1 \\ 4 \\ 1 \end{pmatrix} = 2 \begin{pmatrix} 0 \\ 2 \\ 4 \end{pmatrix} + 1 \begin{pmatrix} 1 \\ 0 \\ 3 \end{pmatrix} + 0 \begin{pmatrix} 2 \\ 1 \\ 0 \end{pmatrix}.$$

This gives that

$$[id_V]_{\mathcal{C}\leftarrow\mathcal{B}} = \begin{pmatrix} 2 & 1 & 2 \\ 1 & 0 & 1 \\ 1 & 1 & 0 \end{pmatrix}.$$

The next result shows that composition of linear maps corresponds to matrix multiplication of the matrix representation, when the bases match. Please be reminded that the composition is defined via $(S \circ T)(\mathbf{x}) = S(T(\mathbf{x}))$.

Theorem 3.3.4 Let $T : V \to W$ and $S : W \to X$ be linear maps between finite-dimensional vector spaces over $\mathbb{F}$, and let $\mathcal{B}, \mathcal{C}$, and $\mathcal{D}$ be bases for V, W, and X, respectively. Then

$$[S \circ T]_{\mathcal{D}\leftarrow\mathcal{B}} = [S]_{\mathcal{D}\leftarrow\mathcal{C}}[T]_{\mathcal{C}\leftarrow\mathcal{B}}. \tag{3.5}$$

Proof. Denoting

$$\mathcal{B} = \{\mathbf{v}_1, \ldots, \mathbf{v}_n\}, \mathcal{C} = \{\mathbf{w}_1, \ldots, \mathbf{w}_m\}, \mathcal{D} = \{\mathbf{x}_1, \ldots, \mathbf{x}_p\},$$

$$[S \circ T]_{\mathcal{D} \leftarrow \mathcal{B}} = (c_{ij})_{i=1,j=1}^{p \quad n}, \ [S]_{\mathcal{D} \leftarrow \mathcal{C}} = (b_{ij})_{i=1,j=1}^{p \quad m}, \ [T]_{\mathcal{C} \leftarrow \mathcal{B}} = (a_{ij})_{i=1,j=1}^{m \quad n}.$$

We thus have that

$$T(\mathbf{v}_j) = \sum_{i=1}^{m} a_{ij} \mathbf{w}_i, j = 1, \ldots, n, \quad S(\mathbf{w}_k) = \sum_{l=1}^{p} b_{lk} \mathbf{x}_l, k = 1, \ldots, m.$$

Then

$$(S \circ T)(\mathbf{v}_j) = S(T(\mathbf{v}_j)) = S(\sum_{i=1}^{m} a_{ij} \mathbf{w}_j) = \sum_{i=1}^{m} a_{ij} S(\mathbf{w}_i) =$$

$$\sum_{i=1}^{m} [a_{ij} \sum_{l=1}^{p} b_{li} \mathbf{x}_l] = \sum_{l=1}^{p} (\sum_{i=1}^{m} b_{li} a_{ij}) \mathbf{x}_l, j = 1, \ldots, n.$$

Thus we get that $c_{lj} = \sum_{i=1}^{m} b_{li} a_{ij}, l = 1, \ldots, p, \ j = 1, \ldots, n$, which corresponds exactly to (3.5). $\qquad\square$

Corollary 3.3.5 *Let V be a n-dimensional vector space over $\mathbb{F}$ with bases $\mathcal{B}$ and $\mathcal{C}$. Then*

$$[id_V]_{\mathcal{B} \leftarrow \mathcal{C}}^{-1} = [id_V]_{\mathcal{C} \leftarrow \mathcal{B}}. \tag{3.6}$$

Proof. Clearly, $id_V \circ id_V = id_V$. In addition, it is easy to see that $[id_V]_{\mathcal{B} \leftarrow \mathcal{B}} = I_n = [id_V]_{\mathcal{C} \leftarrow \mathcal{C}}$. Then from Theorem 3.3.4 we get that

$$[id_V]_{\mathcal{B} \leftarrow \mathcal{C}}[id_V]_{\mathcal{C} \leftarrow \mathcal{B}} = [id_V]_{\mathcal{B} \leftarrow \mathcal{B}} = I_n.$$

As the matrices involved are all square, we can now conclude that (3.6) holds. $\qquad\square$

Example 3.3.3 continued.

$$[id_V]_{\mathcal{B} \leftarrow \mathcal{C}} = \begin{pmatrix} 2 & 1 & 2 \\ 1 & 0 & 1 \\ 1 & 1 & 0 \end{pmatrix}^{-1} = \begin{pmatrix} 4 & 2 & 1 \\ 1 & 3 & 0 \\ 1 & 4 & 4 \end{pmatrix}.$$

Let us check:

$$id_V \begin{pmatrix} 0 \\ 2 \\ 4 \end{pmatrix} = \begin{pmatrix} 0 \\ 2 \\ 4 \end{pmatrix} = 4 \begin{pmatrix} 3 \\ 0 \\ 1 \end{pmatrix} + 1 \begin{pmatrix} 2 \\ 3 \\ 4 \end{pmatrix} + 1 \begin{pmatrix} 1 \\ 4 \\ 1 \end{pmatrix},$$

$$id_V \begin{pmatrix} 1 \\ 0 \\ 3 \end{pmatrix} = \begin{pmatrix} 1 \\ 0 \\ 3 \end{pmatrix} = 2 \begin{pmatrix} 3 \\ 0 \\ 1 \end{pmatrix} + 3 \begin{pmatrix} 2 \\ 3 \\ 4 \end{pmatrix} + 4 \begin{pmatrix} 1 \\ 4 \\ 1 \end{pmatrix},$$

$$idv\begin{pmatrix}2\\1\\0\end{pmatrix} = \begin{pmatrix}2\\1\\0\end{pmatrix} = 1\begin{pmatrix}3\\0\\1\end{pmatrix} + 0\begin{pmatrix}2\\3\\4\end{pmatrix} + 4\begin{pmatrix}1\\4\\1\end{pmatrix},$$

confirming that our calculations were correct.

In the next corollary, we present an important special case where we change bases in a vector space, and express a linear map with respect to the new basis. Recall that two $n \times n$ matrices A and B are called *similar* if there exists an invertible $n \times n$ matrix P so that

$$A = PBP^{-1}.$$

We have the following corollary.

Corollary 3.3.6 *Let $T : V \to V$ and let $\mathcal{B}$ and $\mathcal{C}$ be two bases in the n-dimensional vector space V. Then*

$$[T]_{\mathcal{B} \leftarrow \mathcal{B}} = [id_V]_{\mathcal{B} \leftarrow \mathcal{C}}[T]_{\mathcal{C} \leftarrow \mathcal{C}}[id_V]_{\mathcal{C} \leftarrow \mathcal{B}} = [id_V]_{\mathcal{C} \leftarrow \mathcal{B}}^{-1}[T]_{\mathcal{C} \leftarrow \mathcal{C}}[id_V]_{\mathcal{C} \leftarrow \mathcal{B}}. \qquad (3.7)$$

In particular, $[T]_{\mathcal{B} \leftarrow \mathcal{B}}$ and $[T]_{\mathcal{C} \leftarrow \mathcal{C}}$ are similar.

In the next chapter we will find bases of generalized eigenvectors of a linear T, making the corresponding matrix representation of a particular simple form (the Jordan canonical form). In the case of a basis of eigenvectors, the matrix representation is diagonal.

3.4 Exercises

Exercise 3.4.1 Let $T : V \to W$ and $S : W \to X$ be linear maps. Show that the composition $S \circ T : V \to X$ is also linear.

Exercise 3.4.2 For the following choices of V, W and $T : V \to W$, determine whether T is linear or not.

(a) $V = \mathbb{R}^3$, $W = \mathbb{R}^4$,

$$T\begin{pmatrix}x_1\\x_2\\x_3\end{pmatrix} = \begin{pmatrix}x_1 - 5x_3\\7x_2 + 5\\3x_1 - 6x_2\\8x_3\end{pmatrix}.$$

(b) $V = \mathbb{Z}_5^3$, $W = \mathbb{Z}_5^2$,

$$T\begin{pmatrix} x_1 \\ x_2 \\ x_3 \end{pmatrix} = \begin{pmatrix} x_1 - 2x_3 \\ 3x_2 x_3 \end{pmatrix}.$$

(c) $V = W = \mathbb{C}^{2\times 2}$ (over $\mathbb{F} = \mathbb{C}$), $T(A) = A - A^T$.

(d) $V = W = \mathbb{C}^{2\times 2}$ (over $\mathbb{F} = \mathbb{C}$), $T(A) = A - A^*$.

(e) $V = W = \mathbb{C}^{2\times 2}$ (over $\mathbb{F} = \mathbb{R}$), $T(A) = A - A^*$.

(f) $V = \{f : \mathbb{R} \to \mathbb{R} \ : \ f \text{ is differentiable}\}$, $W = \mathbb{R}^{\mathbb{R}}$,

$$(T(f))(x) = f'(x)(x^2 + 5).$$

(g) $V = \{f : \mathbb{R} \to \mathbb{R} \ : \ f \text{ is continuous}\}$, $W = \mathbb{R}$,

$$T(f) = \int_{-5}^{10} f(x)\,dx.$$

Exercise 3.4.3 Show that if $T : V \to W$ is linear and the set $\{T(\mathbf{v}_1), \ldots, T(\mathbf{v}_k)\}$ is linearly independent, then the set $\{\mathbf{v}_1, \ldots, \mathbf{v}_k\}$ is linearly independent.

Exercise 3.4.4 Show that if $T : V \to W$ is linear and onto, and $\{\mathbf{v}_1 \ldots, \mathbf{v}_k\}$ is a basis for V, then the set $\{T(\mathbf{v}_1), \ldots, T(\mathbf{v}_k)\}$ spans W. When is $\{T(\mathbf{v}_1), \ldots, T(\mathbf{v}_k)\}$ a basis for W?

Exercise 3.4.5 Let $T : V \to W$ be linear, and let $U \subseteq V$ be a subspace of V. Define

$$T[U] := \{\mathbf{w} \in W \ ; \ \text{there exists } \mathbf{u} \in U \text{ so that } \mathbf{w} = T(\mathbf{u})\}.$$

Observe that $T[V] = \operatorname{Ran} T$.

(a) Show that $T[U]$ is a subspace of W.

(b) Assuming $\dim U < \infty$, show that $\dim T[U] \leq \dim U$.

(c) If $\hat{U}$ is another subspace of V, is it always true that $T[U + \hat{U}] = T[U] + T[\hat{U}]$? If so, provide a proof. If not, provide a counterexample.

(d) If $\hat{U}$ is another subspace of V, is it always true that $T[U \cap \hat{U}] = T[U] \cap T[\hat{U}]$? If so, provide a proof. If not, provide a counterexample.

Exercise 3.4.6 Let $\mathbf{v}_1, \mathbf{v}_2, \mathbf{v}_3, \mathbf{v}_4$ be a basis for a vector space V.

(a) Let $T : V \to V$ be given by $T(\mathbf{v}_i) = \mathbf{v}_{i+1}, i = 1, 2, 3$, and $T(\mathbf{v}_4) = \mathbf{v}_1$. Determine the matrix representation of T with respect to the basis $\{\mathbf{v}_1, \mathbf{v}_2, \mathbf{v}_3, \mathbf{v}_4\}$.

(b) If the matrix representation of a linear map $S : V \to V$ with respect to the $\{\mathbf{v}_1, \mathbf{v}_2, \mathbf{v}_3, \mathbf{v}_4\}$ is given by

$$\begin{pmatrix} 1 & 0 & 1 & 1 \\ 0 & 2 & 0 & 2 \\ 1 & 2 & 1 & 3 \\ -1 & 0 & -1 & -1 \end{pmatrix},$$

determine $S(\mathbf{v}_1 - \mathbf{v}_4)$.

(c) Determine bases for Ran S and Ker S.

Exercise 3.4.7 Consider the linear map $T : \mathbb{R}_2[X] \to \mathbb{R}^2$ given by $T(p(X)) = \begin{pmatrix} p(1) \\ p(3) \end{pmatrix}$.

(a) Find a basis for the kernel of T.

(b) Find a basis for the range of T.

Exercise 3.4.8 Let $T : V \to W$ with $V = \mathbb{Z}_5^4$ and $W = \mathbb{Z}_5^{2 \times 2}$ be defined by

$$T\left(\begin{pmatrix} a \\ b \\ c \\ d \end{pmatrix}\right) = \begin{pmatrix} a+b & b+c \\ c+d & d+a \end{pmatrix}.$$

(a) Find a basis for the kernel of T.

(b) Find a basis for the range of T.

Exercise 3.4.9 For the following $T : V \to W$ with bases $\mathcal{B}$ and $\mathcal{C}$, respectively, determine the matrix representation for T with respect to the bases $\mathcal{B}$ and $\mathcal{C}$. In addition, find bases for the range and kernel of T.

(a) $\mathcal{B} = \mathcal{C} = \{\sin t, \cos t, \sin 2t, \cos 2t\}$, $V = W = \text{Span } \mathcal{B}$, and $T = \frac{d^2}{dt^2} + \frac{d}{dt}$.

(b) $\mathcal{B} = \{1, t, t^2, t^3\}, \mathcal{C} = \{\begin{pmatrix} 1 \\ 0 \end{pmatrix}, \begin{pmatrix} 1 \\ -1 \end{pmatrix}\}, V = \mathbb{C}_3[X], \text{ and } W = \mathbb{C}^2, \text{ and}$

$$T(p) = \begin{pmatrix} p(3) \\ p(5) \end{pmatrix}.$$

(c) $\mathcal{B} = \mathcal{C} = \{e^t \cos t, e^t \sin t, e^{3t}, te^{3t}\}, V = W = \text{Span } \mathcal{B}, \text{ and } T = \frac{d}{dt}.$

(d) $\mathcal{B} = \{1, t, t^2\}, \mathcal{C} = \{\begin{pmatrix} 1 \\ 1 \end{pmatrix}, \begin{pmatrix} 1 \\ 0 \end{pmatrix}\}, V = \mathbb{C}_2[X], \text{ and } W = \mathbb{C}^2, \text{ and}$

$$T(p) = \begin{pmatrix} \int_0^1 p(t)dt \\ p(1) \end{pmatrix}.$$

Exercise 3.4.10 Let $V = \mathbb{C}^{n \times n}$. Define $L : V \to V$ via $L(A) = \frac{1}{2}(A + A^T)$.

(a) Let

$$\mathcal{B} = \{\begin{pmatrix} 1 & 0 \\ 0 & 0 \end{pmatrix}, \begin{pmatrix} 0 & 1 \\ 0 & 0 \end{pmatrix}, \begin{pmatrix} 0 & 0 \\ 1 & 0 \end{pmatrix}, \begin{pmatrix} 0 & 0 \\ 0 & 1 \end{pmatrix}\}.$$

Determine the matrix representation of L with respect to the basis $\mathcal{B}$.

(b) Determine the dimensions of the subspaces

$$W = \{A \in V \; : \; L(A) = A\}, \text{ and}$$
$$\text{Ker } L = \{A \in V \; : \; L(A) = 0\}.$$

(c) Determine the eigenvalues of L.

Exercise 3.4.11 Let $\mathcal{B} = \{1, t, \ldots, t^n\}, \mathcal{C} = \{1, t, \ldots, t^{n+1}\}, V = \text{Span } \mathcal{B}$ and $W = \text{Span } \mathcal{C}$. Define $A : V \to W$ via

$$Af(t) := (2t^2 - 3t + 4)f'(t),$$

where f' is the derivative of f.

(a) Find the matrix representation of A with respect to the bases $\mathcal{B}$ and $\mathcal{C}$.

(b) Find bases for Ran A and Ker A.

Exercise 3.4.12 *(Honors)* Let V and W be vector spaces. Let $\mathcal{L}(V, W)$ be the set of all linear maps acting $V \to W$:

$$\mathcal{L}(V, W) = \{T : V \to W \; : \; T \text{ is linear}\}.$$

Notice that $\mathcal{L}(V, W) \subseteq W^V$, and as addition and scalar multiplication are defined in W, one may define addition and scalar multiplication on W^V as is done in vector spaces of functions. Show that $\mathcal{L}(V, W)$ is a subspace of W^V. What is the dimension of $\mathcal{L}(V, W)$ when $\dim V = n$ and $\dim W = m$?

4

The Jordan Canonical Form

CONTENTS

4.1   The Cayley–Hamilton theorem 69
4.2   Jordan canonical form for nilpotent matrices 71
4.3   An intermezzo about polynomials 75
4.4   The Jordan canonical form 78
4.5   The minimal polynomial 82
4.6   Commuting matrices 84
4.7   Systems of linear differential equations 87
4.8   Functions of matrices 90
4.9   The resolvent 98
4.10   Exercises 100

The main result in this chapter allows us to write a square matrix A as $A = SJS^{-1}$, where J is a particularly simple matrix (in some cases a diagonal matrix). In light of the results in Section 3.3, this means that for a linear transformation on a finite-dimensional vector space we can find a simple matrix representation J (called the Jordan canonical form). This is helpful when one wants to work with this linear transformation. For example, we will see how the Jordan canonical form is helpful in solving a system of linear differential equations.

4.1   The Cayley–Hamilton theorem

It will take a few sections before we get to the general Jordan canonical form. First we need to develop the following polynomial identity for a matrix.

Let $A \in \mathbb{F}^{n \times n}$. We define the *characteristic polynomial* $p_A(\lambda)$ of A to be the degree n polynomial

$$p_A(\lambda) := \det(\lambda I_n - A).$$

Note that $p_A(\lambda)$ has the form

$$p_A(\lambda) = \lambda^n + a_{n-1}\lambda^{n-1} + \cdots + a_1\lambda + a_0, \tag{4.1}$$

where $a_{n-1}, \ldots, a_0 \in \mathbb{F}$. When the leading coefficient in a polynomial is 1, we call the polynomial *monic*. Thus the characteristic polynomial of A is monic. We have the following result.

Theorem 4.1.1 *(Cayley–Hamilton) Let $A \in \mathbb{F}^{n \times n}$ with characteristic polynomial $p_A(\lambda)$ as in (4.1). Then*

$$p_A(A) = A^n + a_{n-1}A^{n-1} + \cdots + a_1A + a_0I_n = 0. \tag{4.2}$$

With the convention $A^0 = I_n$ and $a_n = 1$, we can write (4.2) also as $p_A(A) = \sum_{j=0}^{n} a_j A^j = 0$.

Example 4.1.2 Let $A = \begin{pmatrix} 1 & 2 \\ 3 & 4 \end{pmatrix}$. Then
$p_A(\lambda) = (\lambda - 1)(\lambda - 4) - (-2)(-3) = \lambda^2 - 5\lambda - 2$. Let us check (4.1) for this matrix:

$$\begin{pmatrix} 1 & 2 \\ 3 & 4 \end{pmatrix}^2 - 5\begin{pmatrix} 1 & 2 \\ 3 & 4 \end{pmatrix} - 2\begin{pmatrix} 1 & 0 \\ 0 & 1 \end{pmatrix} = \begin{pmatrix} 7 - 5 - 2 & 10 - 10 - 0 \\ 15 - 15 - 0 & 22 - 20 - 2 \end{pmatrix} = \begin{pmatrix} 0 & 0 \\ 0 & 0 \end{pmatrix}.$$

In the proof of Theorem 4.1.1 we use matrices in which the entries are polynomials in λ, such as for instance

$$\begin{pmatrix} \lambda^2 - 6\lambda + 1 & 2\lambda - 10 \\ 3\lambda^2 + 5\lambda - 7 & -\lambda^2 + 4\lambda - 25 \end{pmatrix}. \tag{4.3}$$

We will rewrite such polynomials in the form $\sum_{j=0}^{n} \lambda^j A_j$, with A_j constant matrices (i.e., A_j does not depend on λ). For (4.3) it looks like

$$\lambda^2 \begin{pmatrix} 1 & 0 \\ 3 & -1 \end{pmatrix} + \lambda \begin{pmatrix} -6 & 2 \\ 5 & 4 \end{pmatrix} + \begin{pmatrix} 1 & -10 \\ -7 & -25 \end{pmatrix}.$$

Proof of Theorem 4.1.1. Applying Theorem 1.4.13 to the matrix $\lambda I_n - A$, we get that

$$(\lambda I_n - A)\, \text{adj}(\lambda I_n - A) = p_A(\lambda)I_n. \tag{4.4}$$

It is easy to see that $\text{adj}(\lambda I_n - A)$ is of the form

$$\text{adj}(\lambda I_n - A) = \lambda^{n-1}I_n + \lambda^{n-2}A_{n-2} + \cdots + \lambda A_1 + A_0,$$

with $A_j \in \mathbb{F}^{n \times n}$ constant matrices. Using the notation (4.1) and equating the coefficients of λ^j, $j = 0, \ldots, n$, on both sides of (4.4) we get

$$-A + A_{n-2} = a_{n-1}I_n, \quad -AA_{n-2} + A_{n-3} = a_{n-2}I_n, \quad \ldots,$$

$$-AA_1 + A_0 = a_1 I_n, \quad -AA_0 = a_0 I_n.$$

But then $p_A(A)$ equals

$$\sum_{j=0}^{n} a_j A^j = A^n + A^{n-1}(-A + A_{n-2}) + A^{n-2}(-AA_{n-2} + A_{n-3}) +$$

$$\cdots + A(-AA_1 + A_0) - AA_0 = 0.$$

$\square$

4.2   Jordan canonical form for nilpotent matrices

We will see that the Cayley–Hamilton theorem (Theorem 4.1.1) plays a crucial role in obtaining the Jordan canonical of a matrix. In this section we focus on the case when $p_A(\lambda) = \lambda^n$. Thus $A^n = 0$. A matrix with this property is called *nilpotent*.

Given a matrix A, we introduce the following quantities:

$$w_k(A, \lambda) = \dim \, \mathrm{Ker}(A - \lambda I_n)^k - \dim \, \mathrm{Ker}(A - \lambda I_n)^{k-1}, k = 1, \ldots, n. \quad (4.5)$$

Here $(A - \lambda I_n)^0 = I_n$, so $w_1(A, \lambda) = \dim \, \mathrm{Ker}(A - \lambda I_n)$. The numbers $w_k(A, \lambda)$ are collectively called the *Weyr characteristics* of A. The spaces $\mathrm{Ker}(A - \lambda I_n)^k$ are called *generalized eigenspaces* of A at λ.

We also introduce the *Jordan block* $J_k(\lambda)$ of size k at λ, as being the $k \times k$ upper triangular matrix

$$J_k(\lambda) = \begin{pmatrix} \lambda & 1 & 0 & \cdots & 0 \\ 0 & \lambda & 1 & \cdots & 0 \\ \vdots & & \ddots & \ddots & \vdots \\ 0 & 0 & \cdots & \lambda & 1 \\ 0 & 0 & \cdots & 0 & \lambda \end{pmatrix} \quad (4.6)$$

We write $\oplus_{k=1}^{p} A_k$ for the block diagonal matrix

$$\oplus_{k=1}^{p} A_k = \begin{pmatrix} A_1 & 0 & \cdots & 0 \\ 0 & A_2 & \cdots & 0 \\ \vdots & \vdots & \ddots & \vdots \\ 0 & 0 & \cdots & A_p \end{pmatrix}.$$

When we have a block diagonal matrix with p copies of the same matrix B, we write $\oplus_{k=1}^{p} B$.

Theorem 4.2.1 *Let $A \in \mathbb{F}^{n \times n}$ be so that $A^n = 0$. Let $w_j = w_j(A, 0)$, $j = 1, \ldots, n+1$. Note that $w_{n+1} = 0$. Then A is similar to the matrix*

$$J = \oplus_{k=1}^{n} (\oplus_{j=1}^{w_k - w_{k+1}} J_k(0)).$$

Thus J is a block diagonal matrix with Jordan blocks, where for $k = 1, \ldots, n$ the Jordan block $J_k(0)$ appears exactly $w_k - w_{k+1}$ times.

Example 4.2.2 Let

$$A = \begin{pmatrix} 0 & 1 & 0 & -1 & 1 & -1 \\ 0 & 1 & 1 & -2 & 2 & -2 \\ 0 & 1 & 0 & -1 & 2 & -2 \\ 0 & 1 & 0 & -1 & 2 & -2 \\ 0 & 1 & 0 & -1 & 1 & -1 \\ 0 & 1 & 0 & -1 & 1 & -1 \end{pmatrix}.$$

Then one finds that

$$\dim \operatorname{Ker} A = 3, \ \dim \operatorname{Ker} A^2 = 5, \text{ and } \dim \operatorname{Ker} A^j = 6, j = 3, 4, 5, 6.$$

Thus $w_1 = 3, w_2 = 2, w_3 = 1, w_4 = w_5 = w_6 = 0$. Theorem 4.2.1 now states that A is similar to the matrix

$$J = \begin{pmatrix} 0 & & & & & \\ & 0 & 1 & & & \\ & 0 & 0 & & & \\ & & & 0 & 1 & 0 \\ & & & 0 & 0 & 1 \\ & & & 0 & 0 & 0 \end{pmatrix}, \tag{4.7}$$

where the empty entries are zeros.

Proof of Theorem 4.2.1. Put $s_k = w_k - w_{k+1}$. Choose linearly independent vectors $\mathbf{x}_{n1}, \ldots, \mathbf{x}_{ns_n}$ so that

$$\operatorname{Span}\{\mathbf{x}_{n1}, \ldots, \mathbf{x}_{n,s_n}\} \dotplus \operatorname{Ker} A^{n-1} = \operatorname{Ker} A^n (= \mathbb{F}^n).$$

Next, for $j = n - 1, \ldots, 1$, choose linearly independent vectors $\mathbf{x}_{j,1}, \ldots, \mathbf{x}_{j,s_j}$ so that

$$\mathrm{Span}\{\mathbf{x}_{j,1}, \ldots, \mathbf{x}_{j,s_j}\} \dot{+} \left(\dot{+}_{k=j+1}^{n} \mathrm{Span}\{A^{k-j}\mathbf{x}_{k,1}, \ldots, A^{k-j}\mathbf{x}_{k,s_k}\} \right)$$

$$\dot{+} \mathrm{Ker} A^{j-1} = \mathrm{Ker} A^j. \tag{4.8}$$

We claim that the set of vectors

$$\mathcal{B} = \cup_{k=1}^{n} \left(\cup_{j=1}^{s_k} \{A^{k-1}\mathbf{x}_{k,j}, \ldots, A\mathbf{x}_{k,j}, \mathbf{x}_{k,j} \right) \tag{4.9}$$

is a basis of $\mathbb{F}^n$, and that

$$[A]_{\mathcal{B} \leftarrow \mathcal{B}} = J.$$

First we observe that the number of elements in $\mathcal{B}$ equals

$$\sum_{k=1}^{n} k s_k = w_1 - w_2 + 2(w_2 - w_3) + 3(w_3 - w_4) + \cdots + n(w_n - 0) = \sum_{k=1}^{n} w_k =$$

$$\sum_{k=1}^{n} (\dim \mathrm{Ker} A^k - \dim \mathrm{Ker} A^{k-1}) = \dim \mathrm{Ker} A^n - \dim \mathrm{Ker} I_n = n,$$

where in the last step we used that $A^n = 0$. Thus it remains to prove that $\mathcal{B}$ is linearly independent. For this purpose, let $c_{k,j}^{(l)} \in \mathbb{F}$ be so that

$$\sum_{k=1}^{n} \sum_{j=1}^{s_k} \sum_{l=0}^{k-1} c_{k,j}^{(l)} A^l \mathbf{x}_{k,j} = 0. \tag{4.10}$$

When we multiply (4.10) on the left with A^{n-1} and use that $A^k \mathbf{x}_{k,j} = 0$, we get that

$$A^{n-1} \left(\sum_{j=1}^{s_n} c_{k,j}^{(0)} \mathbf{x}_{n,j} \right) = \mathbf{0}.$$

Then

$$\sum_{j=1}^{s_n} c_{n,j}^{(0)} \mathbf{x}_{n,j} \in (\mathrm{Span}\{\mathbf{x}_{n1}, \ldots, \mathbf{x}_{n,s_n}\}) \cap \mathrm{Ker} A^{n-1} = \{\mathbf{0}\},$$

and thus

$$\sum_{j=1}^{s_n} c_{n,j}^{(0)} \mathbf{x}_{n,j} = \mathbf{0}.$$

As $\{\mathbf{x}_{n1}, \ldots, \mathbf{x}_{n,s_n}\}$ is linearly independent, we get that $c_{n,j}^{(0)} = 0$, $j = 1, \ldots, s_n$. If $n = 1$, we are done. If $n \geq 2$, we multiply (4.10) with A^{n-2} on the left, to obtain

$$A^{n-2} \left(\sum_{j=1}^{s_{n-1}} c_{n-1,j}^{(0)} \mathbf{x}_{n-1,j} \right) + A^{n-1} \left(\sum_{j=1}^{s_n} c_{n,j}^{(1)} \mathbf{x}_{n,j} \right) = \mathbf{0}.$$

Then

$$\sum_{j=1}^{s_{n-1}} c_{n-1,j}^{(0)} \mathbf{x}_{n-1,j} + A \sum_{j=1}^{s_n} c_{n,j}^{(1)} \mathbf{x}_{n,j} \in$$

$$[\mathrm{Span}\{\mathbf{x}_{n-1,1}, \ldots, \mathbf{x}_{n-1,s_{n-1}}\} \dotplus \mathrm{Span}\{A\mathbf{x}_{n,1}, \ldots, A\mathbf{x}_{n,s_n}\}] \cap \mathrm{Ker}A^{n-2}.$$

By (4.8) this intersection equals $\{\mathbf{0}\}$, and thus

$$\sum_{j=1}^{s_{n-1}} c_{n-1,j}^{(0)} \mathbf{x}_{n-1,j} + A \sum_{j=1}^{s_n} c_{n,j}^{(1)} \mathbf{x}_{n,j} = \mathbf{0}.$$

Next, using Proposition 2.3.3, we get that

$$\sum_{j=1}^{s_{n-1}} c_{n-1,j}^{(0)} \mathbf{x}_{n-1,j} = \mathbf{0}, A \sum_{j=1}^{s_n} c_{n,j}^{(1)} \mathbf{x}_{n,j} = \mathbf{0}.$$

Since $\{\mathbf{x}_{n-1,1}, \ldots, \mathbf{x}_{n-1,s_{n-1}}\}$ is linearly independent, we get $c_{n-1,j}^{(0)} = 0$, $j = 1, \ldots, s_{n-1}$. In addition, as $\mathrm{Ker}A \subseteq \mathrm{Ker}A^{n-1}$ we get that

$$\sum_{j=1}^{s_n} c_{n,j}^{(1)} \mathbf{x}_{n,j} \in \mathrm{Span}\{\mathbf{x}_{n1}, \ldots, \mathbf{x}_{n,s_n}\} \cap \mathrm{Ker}A^{n-1} = \{\mathbf{0}\}.$$

and using linear independence of $\{\mathbf{x}_{n1}, \ldots, \mathbf{x}_{n,s_n}\}$, we obtain $c_{n,j}^{(1)} = 0$, $j = 1, \ldots, s_n$. If $n = 2$, we are done. If $n \geq 3$, we continue by multiplying (4.10) with A^{n-3} on the left and argue in a similar manner as above. Ultimately, we arrive at $c_{k,j}^{(l)} = 0$ for all k, j, and l, showing that $\mathcal{B}$ is linearly independent, and thus a basis for $\mathbb{F}^n$.

To show that $[A]_{\mathcal{B} \leftarrow \mathcal{B}} = J$, notice that if we apply A to an element of $\mathcal{B}$, two possibilities occur: we either get $\mathbf{0}$ or we get another element of the basis $\mathcal{B}$. Indeed, taking the element $A^{k-1}\mathbf{x}_{k,j} \in \mathcal{B}$ and applying A to it, we get (since $\mathbf{x}_{k,j} \in \mathrm{Ker}A^k$) that

$$A(A^{k-1}\mathbf{x}_{k,j}) = A^k \mathbf{x}_{k,j} = \mathbf{0},$$

and thus the corresponding column in $[A]_{\mathcal{B} \leftarrow \mathcal{B}}$ consists of only zeros. If we apply A to any other element $A^{l-1}\mathbf{x}_{k,j}$, $l < k$, of $\mathcal{B}$, we get

$$A(A^{l-1}\mathbf{x}_{k,j}) = A^l \mathbf{x}_{k,l} \in \mathcal{B},$$

and as $A^l \mathbf{x}_{k,l}$ precedes $A^{l-1}\mathbf{x}_{k,j}$ in $\mathcal{B}$, we get exactly a 1 in the entry above the diagonal in the column of $[A]_{\mathcal{B} \leftarrow \mathcal{B}}$ corresponding to $A^{l-1}\mathbf{x}_{k,j}$, and zeros elsewhere in this column. This shows that $[A]_{\mathcal{B} \leftarrow \mathcal{B}} = J$, completing the proof. $\qquad \square$

Example 4.2.2 continued. We compute

$$A^2 = \begin{pmatrix} 0 & 0 & 1 & -1 & 0 & 0 \\ 0 & 0 & 1 & -1 & 0 & 0 \\ 0 & 0 & 1 & -1 & 0 & 0 \\ 0 & 0 & 1 & -1 & 0 & 0 \\ 0 & 0 & 1 & -1 & 0 & 0 \\ 0 & 0 & 1 & -1 & 0 & 0 \end{pmatrix}, \quad A^j = 0, j \geq 3.$$

Letting $\mathbf{e}_j$, $j = 1, \ldots, 6$, denote the standard basis elements of $\mathbb{F}^6$, we find

$$\mathrm{Ker}A^j = \mathbb{F}^6, j \geq 3, \quad \mathrm{Ker}A^2 = \mathrm{Span}\{\mathbf{e}_1, \mathbf{e}_2, \mathbf{e}_3 + \mathbf{e}_4, \mathbf{e}_5, \mathbf{e}_6\},$$

$$\mathrm{Ker}A = \mathrm{Span}\{\mathbf{e}_1, \mathbf{e}_2 + \mathbf{e}_3 + \mathbf{e}_4, \mathbf{e}_5 + \mathbf{e}_6\}.$$

We can now choose $\mathbf{x}_{3,1} = \mathbf{e}_3$. Next, we need to choose $\mathbf{x}_{2,1}$ so that

$$\mathrm{Span}\{\mathbf{x}_{2,1}\} \dotplus \mathrm{Span}\{A\mathbf{x}_{3,1}(= \mathbf{e}_2)\} \dotplus \mathrm{Ker}A = \mathrm{Ker}A^2.$$

Take for instance $\mathbf{x}_{2,1} = \mathbf{e}_5$. Finally, we need to choose $\mathbf{x}_{1,1}$ so that

$$\mathrm{Span}\{\mathbf{x}_{1,1}\} \dotplus \mathrm{Span}\{A^2\mathbf{x}_{3,1}, A\mathbf{x}_{2,1}\} \dotplus \mathrm{Ker}I_6 = \mathrm{Ker}A.$$

One can for instance choose $\mathbf{x}_{1,1} = \mathbf{e}_1$. We now get that

$$\mathcal{B} = \{\mathbf{x}_{1,1}, A\mathbf{x}_{2,1}, \mathbf{x}_{2,1}, A^2\mathbf{x}_{3,1}, A\mathbf{x}_{3,1}, \mathbf{x}_{3,1}\} = \{\mathbf{e}_1, A\mathbf{e}_5, \mathbf{e}_5, A^2\mathbf{e}_3, A\mathbf{e}_3, \mathbf{e}_3\}.$$

Letting $P = [id_{\mathbb{F}^6}]_{\mathcal{E} \leftarrow \mathcal{B}}$, we get that the columns of P are exactly the vectors in $\mathcal{B}$ (with coordinates with respect to the standard basis $\mathcal{E}$), and thus

$$P = \begin{pmatrix} 1 & 1 & 0 & 1 & 0 & 0 \\ 0 & 2 & 0 & 1 & 1 & 0 \\ 0 & 2 & 0 & 1 & 0 & 1 \\ 0 & 2 & 0 & 1 & 0 & 0 \\ 0 & 1 & 1 & 1 & 0 & 0 \\ 0 & 1 & 0 & 1 & 0 & 0 \end{pmatrix}.$$

Then we find indeed that $P^{-1}AP = J$ with J as in (4.7). Writing this equality as $AP = PJ$, it is easy to verify this by hand.

4.3   An intermezzo about polynomials

Given two polynomials $f(X), g(X) \in \mathbb{F}[X]$, we say that $f(X)$ *divides* $g(X)$ (notation: $f(X)|g(X)$) if there exists an $h(X) \in \mathbb{F}[X]$ so that

$f(X)h(X) = g(X)$. Clearly, if $f(X)|g(X)$ and $g(X)$ is not the zero polynomial, then $\deg f \leq \deg g$. We say that $f(X)$ is a *common divisor* of $g(X)$ and $h(X)$ if $f(X)$ divides both $g(X)$ and $h(X)$. We call a nonzero polynomial $f(X)$ a *greatest common divisor* of the nonzero polynomials $g(X)$ and $h(X)$ if $f(X)$ is a common divisor of $g(X)$ and $h(X)$ and among all common divisors $f(X)$ has the highest possible degree.

Analogous to the results on integers as presented in Subsection 1.3.2 we have the following result for polynomials. We will not provide proofs for these results.

Proposition 4.3.1 *For every pair of nonzero polynomials* $g(X), h(X) \in \mathbb{F}[X]$, *there exists unique* $q(X), r(X) \in \mathbb{F}[X]$ *so that*

$$g(X) = h(X)q(X) + r(X) \text{ and } \deg r < \deg h.$$

We call $r(X)$ the remainder of $g(X)$ after division by $h(X)$. One can find $q(X)$ and $r(X)$ via long division. We present an example.

Example 4.3.2 Let $g(X) = X^3 + X^2 - 1$ and $h(X) = X - 1$. Then we perform the long division

$$
\begin{array}{r}
X^2 + 2X + 2 \\
\hline
X - 1) X^3 + X^2 - 1 \\
-X^3 + X^2 \\
\hline
2X^2 \\
-2X^2 + 2X \\
\hline
2X - 1 \\
-2X + 2 \\
\hline
1
\end{array}
$$

resulting in $q(X) = X^2 + 2X + 2$ and $r(X) = 1$.

Proposition 4.3.3 *For every pair of nonzero polynomials* $g(X), h(X) \in \mathbb{F}[X]$, *the greatest common divisor is unique up to multiplication with a nonzero element of* $\mathbb{F}$. *Consequently, every pair of nonzero polynomials* $g(X), h(X) \in \mathbb{F}[X]$ *has a unique monic greatest common divisor.*

We denote the unique monic greatest common divisor of $g(X)$ and $h(X)$ by $\gcd(g(X), h(X))$. We say that $g(X)$ and $h(X)$ are *coprime* if $\gcd(g(X), h(X)) = 1$. In this setting we now also have a Bezout equation result.

Proposition 4.3.4 *For every pair of nonzero polynomials* $g(X), h(X) \in \mathbb{F}[X]$, *there exists* $a(X), b(X) \in \mathbb{F}[X]$ *so that*

$$a(X)g(X) + b(X)h(X) = \gcd(g(X), h(X)). \tag{4.11}$$

In particular, if $g(X)$ *and* $h(X)$ *are coprime, then there exists* $a(X), b(X) \in \mathbb{F}[X]$ *so that*

$$a(X)g(X) + b(X)h(X) = 1. \tag{4.12}$$

As in Subsection 1.3.2, to solve Bezout's identity (4.11), one applies Euclid's algorithm to find the greatest common divisor, keep track of the division equations, and ultimately put the equations together.

Example 4.3.5 Let us solve (4.11) for $g(X) = X^4 - 2X^3 + 2X^2 - 2X + 1$ and $h(X) = X^3 + X^2 - X - 1$, both in $\mathbb{R}[X]$. We perform Euclid's algorithm:

$$X^4 - 2X^3 + 2X^2 - 2X + 1 = (X^3 + X^2 - X - 1)(X - 3) + (6X^2 - 4X - 2)$$

$$X^3 + X^2 - X - 1 = (6X^2 - 4X - 2)(\frac{1}{6}X + \frac{5}{18}) + (\frac{4}{9}X - \frac{4}{9})$$

$$6X^2 - 4X - 2 = (\frac{4}{9}X - \frac{4}{9})(\frac{27}{2}X + \frac{9}{2}) + 0.$$

$$\tag{4.13}$$

So we find that $\frac{4}{9}X - \frac{4}{9}$ is a greatest common divisor. Making it monic, we get $\gcd(g(X), h(X)) = X - 1$. Using the above equations, we get

$$X - 1 = \frac{9}{4}[X^3 + X^2 - X - 1 - (6X^2 - 4X - 2)(\frac{1}{6}X + \frac{5}{18})] =$$

$$\frac{9}{4}[X^3 + X^2 - X - 1 - [X^4 - 2X^3 + 2X^2 - 2X - 1 - (X^3 + X^2 - X - 1)(X-3)](\frac{1}{6}X + \frac{5}{18})].$$

Thus we find

$$a(X) = -\frac{9}{4}(\frac{1}{6}X + \frac{5}{18}) = -\frac{3}{8}X - \frac{5}{8},$$

and

$$b(X) = \frac{9}{4}[1 + (X-3)(\frac{1}{6}X + \frac{5}{18})] = \frac{3}{8}X^2 - \frac{1}{2}X + \frac{3}{8}.$$

Given nonzero polynomials $g_1(X), \ldots, g_k(X) \in \mathbb{F}[X]$, we call $f(X) \in \mathbb{F}[X]$ a common divisor of $g_1(X), \ldots, g_k(X)$ if $f(X)|g_j(X)$, $j = 1, \ldots, k$. A common divisor of $g_1(X), \ldots, g_k(X)$ is called a *greatest common divisor* of $g_1(X), \ldots, g_k(X)$ if among all common divisors of $g_1(X), \ldots, g_k(X)$ it has the highest possible degree. Analogous to the case $k = 2$, we have the following.

Proposition 4.3.6 *For every k nonzero polynomials*
$g_1(X), \ldots, g_k(X) \in \mathbb{F}[X]$, *the greatest common divisor is unique up to multiplication with a nonzero element of* $\mathbb{F}$. *Consequently, every pair of nonzero polynomials* $g_1(X), \ldots, g_k(X) \in \mathbb{F}[X]$ *has a unique monic greatest common divisor (notation:* $\gcd(g_1(X), \ldots, g_k(X))$). *Moreover, there exists* $a_1(X), \ldots, a_k(X) \in \mathbb{F}[X]$ *so that*

$$a_1(X)g_1(X) + \cdots + a_k(X)g_k(X) = \gcd(g_1(X), \ldots, g_k(X)). \qquad (4.14)$$

The above result follows easily from the $k = 2$ case after first observing that

$$\gcd(g_1(X), \ldots, g_k(X)) = \gcd(g_1(X), \gcd(g_2(X) \ldots, g_k(X))).$$

4.4   The Jordan canonical form

We now come to the main result of this chapter.

Theorem 4.4.1 *(Jordan canonical form) Let $A \in \mathbb{F}^{n \times n}$ and suppose we may write $p_A(\lambda) = (\lambda - \lambda_1)^{n_1} \cdots (\lambda - \lambda_m)^{n_m}$, where $\lambda_1, \ldots, \lambda_m \in \mathbb{F}$ are the different roots of $p_A(\lambda)$. Then A is similar to the matrix*

$$J = \begin{pmatrix} J(\lambda_1) & 0 & \cdots & 0 \\ 0 & J(\lambda_2) & \cdots & 0 \\ \vdots & \vdots & \ddots & \vdots \\ 0 & 0 & \cdots & J(\lambda_m) \end{pmatrix},$$

where $J(\lambda_j)$ is the $n_j \times n_j$ matrix

$$J(\lambda_j) = \oplus_{k=1}^{n_j} \left(\oplus_{l=1}^{w_k(A,\lambda_j) - w_{k+1}(A,\lambda_j)} J_k(\lambda_j) \right), \ j = 1, \ldots, m.$$

Here $w_k(A, \lambda)$ is defined in (4.5).

Remark 4.4.2 A field $\mathbb{F}$ is called *algebraically closed* if every polynomial $p(X) \in \mathbb{F}[X]$ with $\deg p \geq 1$ has a root in $\mathbb{F}$. If a field is algebraically closed, one can factor any monic polynomial $p(\lambda)$ of degree ≥ 1 as $p(\lambda) = (\lambda - \lambda_1)^{n_1} \cdots (\lambda - \lambda_m)^{n_m}$ with $\lambda_1, \ldots, \lambda_m \in \mathbb{F}$. Thus for an algebraically closed field it is not necessary to assume in Theorem 4.4.1 that $p_A(\lambda)$ factors in this way, as it automatically does. By the fundamental theorem of algebra $\mathbb{C}$ is an algebraically closed field. The fields $\mathbb{Z}_p$ and $\mathbb{R}$ are not algebraically closed: $1 + X(X-1)(X-2) \cdots (X - (p-1))$ does not have any roots in $\mathbb{Z}_p$, while $X^2 + 1$ does not have any real roots.

Our first step in the proof of Theorem 4.4.1 is the following.

Proposition 4.4.3 *Let $A \in \mathbb{F}^{n \times n}$ and suppose*
$p_A(\lambda) = (\lambda - \lambda_1)^{n_1} \cdots (\lambda - \lambda_m)^{n_m}$, *where $\lambda_1, \ldots, \lambda_m \in \mathbb{F}$ are different. Then*

$$\mathbb{F}^n = \operatorname{Ker}(A - \lambda_1 I_n)^{n_1} \dotplus \cdots \dotplus \operatorname{Ker}(A - \lambda_m I_n)^{n_m}. \tag{4.15}$$

Proof. Let $g_j(\lambda) = p_A(\lambda)/(\lambda - \lambda_j)^{n_j} \in \mathbb{F}[\lambda]$, $j = 1, \ldots, m$. Then $\gcd(g_1(\lambda), \ldots, g_m(\lambda)) = 1$, thus by Proposition 4.3.6 there exist $a_1(\lambda), \ldots, a_m(\lambda) \in \mathbb{F}[\lambda]$ so that

$$a_1(\lambda)g_1(\lambda) + \cdots + a_m(\lambda)g_m(\lambda) = 1.$$

But then,

$$a_1(A)g_1(A) + \cdots + a_m(A)g_m(A) = I_n. \tag{4.16}$$

Let now $\mathbf{v} \in \mathbb{F}^n$ be arbitrary, and put $\mathbf{v}_j = a_j(A)g_j(A)\mathbf{v}$, $j = 1, \ldots, m$. Then, due to (4.16) we have that $\mathbf{v} = \mathbf{v}_1 + \cdots + \mathbf{v}_m$. Moreover, $(A - \lambda_j I_n)^{n_j} \mathbf{v}_j = a_j(A)p_A(A)\mathbf{v}_j = \mathbf{0}$, due to Theorem 4.1.1. Thus $\mathbf{v}_j \in \operatorname{Ker}(A - \lambda_j I_n)^{n_j}$, $j = 1, \ldots, m$, and thus

$$\mathbf{v} = \mathbf{v}_1 + \cdots + \mathbf{v}_m \in \operatorname{Ker}(A - \lambda_1 I_n)^{n_1} + \cdots + \operatorname{Ker}(A - \lambda_m I_n)^{n_m},$$

proving the inclusion $\subseteq$ in (4.15). The other inclusion $\supseteq$ is trivial, so equality in (4.15) holds.

It remains to show that the right-hand side of (4.15) is a direct sum. We show that the first $+$ is a direct sum, as this is notationwise the most convenient. The argument is that same for all the others. Thus we let

$$\mathbf{v} \in \operatorname{Ker}(A - \lambda_1 I_n)^{n_1} \cap [\operatorname{Ker}(A - \lambda_2 I_n)^{n_2} + \cdots + \operatorname{Ker}(A - \lambda_m I_n)^{n_m}].$$

We need to show that $\mathbf{v} = \mathbf{0}$. Using that $(\lambda - \lambda_1)^{n_1}$ and $g_1(\lambda)$ are coprime, we have by Proposition 4.3.3 that there exist $a(\lambda), b(\lambda) \in \mathbb{F}[\lambda]$ so that

$$a(\lambda)(\lambda - \lambda_1)^{n_1} + b(\lambda)g_1(\lambda) = 1.$$

Thus

$$a(A)(A - \lambda_1 I_n)^{n_1} + b(A)g_1(A) = I_n. \tag{4.17}$$

Next, observe that $\mathbf{v} \in \operatorname{Ker}(A - \lambda_1 I_n)^{n_1}$ gives that $(A - \lambda_1 I_n)^{n_1}\mathbf{v} = \mathbf{0}$, and that

$$\mathbf{v} \in \operatorname{Ker}(A - \lambda_2 I_n)^{n_2} + \cdots + \operatorname{Ker}(A - \lambda_m I_n)^{n_m}$$

implies that $g_1(A)\mathbf{v} = \mathbf{0}$. But then, using (4.17), we get that

$$\mathbf{v} = a(A)(A - \lambda_1 I_n)^{n_1}\mathbf{v} + b(A)g_1(A)\mathbf{v} = \mathbf{0} + \mathbf{0} = \mathbf{0},$$

showing that

$$\operatorname{Ker}(A - \lambda_1 I_n)^{n_1} \cap [\operatorname{Ker}(A - \lambda_2 I_n)^{n_2} + \cdots + \operatorname{Ker}(A - \lambda_m I_n)^{n_m}] = \{\mathbf{0}\},$$

as desired. $\square$

Lemma 4.4.4 *Let $A \in \mathbb{F}^{n \times n}$, $\lambda \in \mathbb{F}$, and $s \in \mathbb{N}$. Put $W = \text{Ker}(A - \lambda I_n)^s$. Then*

$$A[W] \subseteq W.$$

Let $B : W \to W$ be defined by $B\mathbf{w} = A\mathbf{w}$. Then B is a linear map, and $(B - \lambda \, id_W)^s = 0$. Moreover, λ is the only eigenvalue of B.

When W is a subspace satisfying $A[W] \subseteq W$, we say that W is an *invariant subspace* of A. We denote the linear map B in Lemma 4.4.4 by $A|_W$ and call it the *restriction of A* to the invariant subspace W.

Proof of Lemma 4.4.4. Let $\mathbf{w} \in W = \text{Ker}(A - \lambda I_n)^s$, thus $(A - \lambda I_n)^s\mathbf{w} = \mathbf{0}$. But then $(A - \lambda I_n)^s A\mathbf{w} = A(A - \lambda I_n)^s\mathbf{w} = \mathbf{0}$, and thus $A\mathbf{w} \in W$. Clearly, B is linear. Notice that for any $\mathbf{w} \in W$, we have that

$$(B - \lambda \, id_W)^s\mathbf{w} = (A - \lambda I_n)^s\mathbf{w} = \mathbf{0},$$

due to $\mathbf{w} \in \text{Ker}(A - \lambda I_n)^s$. This shows that $(B - \lambda \, id_W)^s = 0$. Finally, let μ be an eigenvalue of B, with eigenvector $\mathbf{v}(\neq 0)$, say. Then $(B - \lambda \, id_{W_j})\mathbf{v} = (\mu - \lambda)\mathbf{v}$, and thus $\mathbf{0} = (B - \lambda \, id_W)^s\mathbf{v} = (\mu - \lambda)^s\mathbf{v}$. As $\mathbf{v} \neq 0$, this implies that $\mu = \lambda$. $\qquad\square$

Proof of Theorem 4.4.1. Let $W_j = \text{Ker}(A - \lambda_j I_n)^{n_j}$, $j = 1, \ldots, m$. First note that by Proposition 4.4.3 and Lemma 4.4.4 we have that $A = \oplus_{j=1}^m A|_{W_j}$, and thus

$$p_A(\lambda) = \det(\lambda I_n - A) = \prod_{j=1}^m \det(\lambda \, id_{W_j} - A|_{W_j}) = \prod_{j=1}^m \det(\lambda - \lambda_j)^{\dim W_j}.$$

We now obtain that $\dim W_j = n_j$, $j = 1, \ldots, m$.

Next, by Lemma 4.4.4 we have that $(A|_{W_j} - \lambda_j \, id_{W_j})^{n_j}$ is nilpotent, and thus by Theorem 4.2.1 there is a basis $\mathcal{B}_j$ for W_j, so that

$$[(A - \lambda_j \, id_{W_j})|_{W_j}]_{\mathcal{B}_j \leftarrow \mathcal{B}_j}$$

is in Jordan form as described in Theorem 4.2.1. But then, using that $[id_{W_j}]_{\mathcal{B}_j \leftarrow \mathcal{B}_j} = I_{n_j}$, we get that

$$[A|_{W_j}]_{\mathcal{B}_j \leftarrow \mathcal{B}_j} = \lambda_j I_{n_j} + [(A - \lambda_j \, id_{W_j})|_{W_j}]_{\mathcal{B}_j \leftarrow \mathcal{B}_j} = J(\lambda_j).$$

Letting now $\mathcal{B} = \cup_{j=1}^m \mathcal{B}_j$, we get by Proposition 4.4.3 that $\mathcal{B}$ is a basis for $\mathbb{F}^n$. Moreover,

$$[A]_{\mathcal{B} \leftarrow \mathcal{B}} = \oplus_{j=1}^m [A|_{W_j}]_{\mathcal{B}_j \leftarrow \mathcal{B}_j} = \oplus_{j=1}^m J(\lambda_j) = J,$$

proving the result. $\qquad\square$

Example 4.4.5 Let $A = \begin{pmatrix} 2 & 2 & 3 \\ 1 & 3 & 3 \\ -1 & -2 & -2 \end{pmatrix}$. Computing the characteristic polynomial p_A of A we find $p_A(\lambda) = (\lambda - 1)^3$. Thus 1 is the only eigenvalue of A. Computing the eigenspace at $\lambda = 1$, we row reduce

$$A - I = \begin{pmatrix} 1 & 2 & 3 \\ 1 & 2 & 3 \\ -1 & -2 & -3 \end{pmatrix} \to \begin{pmatrix} 1 & 2 & 3 \\ 0 & 0 & 0 \\ 0 & 0 & 0 \end{pmatrix}.$$

Thus

$$\text{Ker } (A - I) = \text{Span}\left\{ \begin{pmatrix} -2 \\ 1 \\ 0 \end{pmatrix}, \begin{pmatrix} -3 \\ 0 \\ 1 \end{pmatrix} \right\}.$$

One finds that $(A - I)^2 = 0$, and thus $w_1(A, 1) = 2$, $w_j(A, 1) = 3$, $j \geq 2$. Thus A has one Jordan block of size 1 and one of size 2, giving that A is similar to

$$J = \begin{pmatrix} 1 & & \\ & 1 & 1 \\ & 0 & 1 \end{pmatrix}.$$

For the basis $\mathcal{B} = \{\mathbf{b}_1, \mathbf{b}_2, \mathbf{3}\}$, we choose $\mathbf{b}_3$ so that

$$\text{Ker } (A - I) \dot{+} \text{Span}\{\mathbf{b}_3\} = \text{Ker } (A - I)^2 = \mathbb{C}^3.$$

Choose, for instance, $\mathbf{b}_3 = \mathbf{e}_1$. Then $\mathbf{b}_2 = (A - I)\mathbf{e}_1 = \begin{pmatrix} 1 \\ 1 \\ -1 \end{pmatrix}$. Next we choose $\mathbf{b}_1$ so that

$$\text{Span}\{\mathbf{b}_1\} \dot{+} \text{Span}\{\mathbf{b}_2\} = \text{Ker } (A - I).$$

For instance $\mathbf{b}_1 = \begin{pmatrix} -2 \\ 1 \\ 0 \end{pmatrix}$. Letting

$$P = [id_{\mathbb{C}^3}]_{\mathcal{E} \leftarrow \mathcal{B}} = \begin{pmatrix} -2 & 1 & 1 \\ 1 & 1 & 0 \\ 0 & -1 & 0 \end{pmatrix},$$

we indeed find that $P^{-1}AP = J$.

4.5   The minimal polynomial

As we have seen in Theorem 4.1.1, the characteristic polynomial $p_A(t)$ of a matrix A has the property that $p_A(A) = 0$. There are many other monic polynomials $p(t)$ that also satisfy $p(A) = 0$. Of particular interest is the one of lowest possible degree. This so-called "minimal polynomial" of A captures some essential features of the Jordan canonical form of the matrix A.

Given $A \in \mathbb{F}^{n \times n}$ we define its *minimal polynomial* $m_A(t)$ to be the lowest-degree monic polynomial so that $m_A(A) = 0$.

Example 4.5.1 Let

$$A = \begin{pmatrix} 1 & 0 & 0 \\ 0 & 1 & 0 \\ 0 & 0 & 2 \end{pmatrix}.$$

Then $m_A(t) = (t-1)(t-2)$. Indeed,

$$m_A(A) = (A - I_3)(A - 2I_3) = 0,$$

and any monic degree-1 polynomial has the form $t - \lambda$, but $A - \lambda I_3 \neq 0$ for all λ.

Proposition 4.5.2 *Every $A \in \mathbb{F}^n$ has a unique minimal polynomial $m_A(t)$, and every eigenvalue of A is a root of $m_A(t)$. Moreover, if $p(A) = 0$, then $m_A(t)$ divides $p(t)$. In particular, $m_A(t)$ divides $p_A(t)$.*

Proof. As $p_A(A) = 0$, there certainly exists a degree-n polynomial satisfying $p(A) = 0$, and thus there exists also a nonzero polynomial of lowest degree which can always be made monic by multiplying by a nonzero element of $\mathbb{F}$. Next suppose that $m_1(t)$ and $m_2(t)$ are both monic polynomials of lowest possible degree k so that $m_1(A) = 0 = m_2(A)$. Then by Proposition 4.3.1 there exists $q(t)$ and $r(t)$ with $\deg r < k$ so that

$$m_1(t) = q(t)m_2(t) + r(t).$$

Note that $r(A) = m_1(A) - q(A)m_2(A) = 0$. If $r(t)$ is not the zero polynomial, then after multiplying by a nonzero constant $r(t)$ will be a monic polynomial of degree $< k$ so that $r(A) = 0$. This contradicts $m_1(t)$ and $m_2(t)$ being minimal polynomials for A. Thus $r(t)$ is the zero polynomial, and thus $m_1(t) = q(t)m_2(t)$. Since $\deg m_1 = \deg m_2 = k$ and m_1

and m_2 are both monic, we must have that $q(t) \equiv 1$, and thus $m_1(t) = m_2(t)$. This proves uniqueness.

Let λ be an eigenvalue with corresponding eigenvector $\mathbf{v}(\neq 0)$. Thus $A\mathbf{v} = \lambda\mathbf{v}$. Then $\mathbf{0} = m_A(A)\mathbf{v} = m_A(\lambda)\mathbf{v}$, and since $\mathbf{v} \neq 0$ it follows that $m_A(\lambda) = 0$. Thus λ is a root of $m_A(t)$.

Finally, let $p(t)$ be so that $p(A) = 0$. If $p(t) \equiv 0$, then clearly $m_A(t)$ divides $p(t)$. If $p(t)$ is not the zero polynomial, apply Proposition 4.3.1 providing the existence of $q(t)$ and $r(t)$ with $\deg r < \deg m_A$ so that

$$p_A(t) = q(t)m_A(t) + r(t).$$

As in the previous paragraph, $r(t)$ not being the zero polynomial contradicts that $m_A(t)$ is the minimal polynomial. Thus $r(t) \equiv 0$, yielding that $m_A(t)$ divides $p(t)$. As $p_A(A)$ by Theorem 4.1.1 we get in particular that $m_A(t)$ divides $p_A(t)$. □

Theorem 4.5.3 *Let $A \in \mathbb{F}^{n \times n}$ and suppose* $p_A(t) = (t - \lambda_1)^{n_1} \cdots (t - \lambda_m)^{n_m}$, *where $\lambda_1, \ldots, \lambda_m \in \mathbb{F}$ are different. Then*

$$m_A(t) = (t - \lambda_1)^{k_1} \cdots (t - \lambda_m)^{k_m}, \qquad (4.18)$$

where k_j is the size of the largest Jordan block at λ_j, $j = 1, \ldots, m$. Equivalently, k_j is the largest index k so that $w_{k-1}(A, \lambda_j) \neq w_k(A, \lambda_j)$.

Proof. It is easy to see that the minimal polynomial for $J_k(\lambda)$ is $(t - \lambda)^k$. As $m_A(t)$ divides $p_A(t)$ we must have that $m_A(t)$ is of the form (4.18) for some $k_j \leq n_j$, $j = 1, \ldots, m$. Observing that $A = PJP^{-1}$ implies $m(A) = Pm(J)P^{-1}$ for any polynomial $m(t)$, it is easy to see by inspection that k_j must correspond exactly to the size of the largest Jordan block corresponding to λ_j. □

Example 4.2.2 continued. The minimal polynomial for A is $m_A(t) = t^3$ as 0 is the only eigenvalue of A and the largest Jordan block associated with it is of size 3.

Example 4.5.4 Let $A \in \mathbb{Z}_5^{4 \times 4}$ satisfy $A^3 - 4A^3 + 2I_6 = 0$. What are the possible Jordan canonical forms of A?

Let $p(t) = t^3 - 4t^2 - 2 = (t-1)^2(t-2)$. Then $p(A) = 0$. Since $m_A(t)$ divides $p(t)$, there are 5 possibilities:
$m_A(t) = t - 1$, $m_A(t) = (t-1)^2$, $m_A(t) = t - 2$, $m_A(t) = (t-1)(t-2)$, or

$m_A(t) = (t-1)^2(t-2)$. Possibilities for the Jordan canonical form are:

$$J = \begin{pmatrix} 1 & 0 & 0 & 0 \\ 0 & 1 & 0 & 0 \\ 0 & 0 & 1 & 0 \\ 0 & 0 & 0 & 1 \end{pmatrix}, \begin{pmatrix} 1 & 1 & 0 & 0 \\ 0 & 1 & 0 & 0 \\ 0 & 0 & 1 & 0 \\ 0 & 0 & 0 & 1 \end{pmatrix}, \begin{pmatrix} 1 & 1 & 0 & 0 \\ 0 & 1 & 0 & 0 \\ 0 & 0 & 1 & 1 \\ 0 & 0 & 0 & 1 \end{pmatrix},$$

$$\begin{pmatrix} 2 & 0 & 0 & 0 \\ 0 & 2 & 0 & 0 \\ 0 & 0 & 2 & 0 \\ 0 & 0 & 0 & 2 \end{pmatrix}, \begin{pmatrix} 1 & 0 & 0 & 0 \\ 0 & 1 & 0 & 0 \\ 0 & 0 & 1 & 0 \\ 0 & 0 & 0 & 2 \end{pmatrix}, \begin{pmatrix} 1 & 0 & 0 & 0 \\ 0 & 1 & 0 & 0 \\ 0 & 0 & 2 & 0 \\ 0 & 0 & 0 & 2 \end{pmatrix},$$

$$\begin{pmatrix} 1 & 0 & 0 & 0 \\ 0 & 2 & 0 & 0 \\ 0 & 0 & 2 & 0 \\ 0 & 0 & 0 & 2 \end{pmatrix}, \begin{pmatrix} 1 & 1 & 0 & 0 \\ 0 & 1 & 0 & 0 \\ 0 & 0 & 2 & 0 \\ 0 & 0 & 0 & 2 \end{pmatrix}, \begin{pmatrix} 1 & 1 & 0 & 0 \\ 0 & 1 & 0 & 0 \\ 0 & 0 & 1 & 0 \\ 0 & 0 & 0 & 2 \end{pmatrix}.$$

We say that a matrix A is *diagonalizable* if its Jordan canonical form is a diagonal matrix. In other words, a matrix is diagonalizable if and only if all its Jordan blocks are of size 1.

Corollary 4.5.5 *A matrix A is diagonalizable if and only if its minimal polynomial $m_A(t)$ has only roots of multiplicity 1.*

Proof. Follows directly from Theorem 4.5.3 as a matrix is diagonalizable if and only if the largest Jordan block for each eigenvalue is 1. □

4.6   Commuting matrices

One learns early on when dealing with matrices that in general they do not commute (indeed, in general $AB \neq BA$). Sometimes, though, one does encounter commuting matrices; for example, if they are matrix representations of taking partial derivatives with respect to different variables on a vector space of "nice" functions. It is of interest to relate such commuting matrices to one another. We focus on the case when one of the matrices is nonderogatory.

We call a matrix *nonderogatory* if the matrix only has a single Jordan block associated with each eigenvalue. The following results is easily proven.

Proposition 4.6.1 *Let $A \in \mathbb{F}^{n \times n}$. The following are equivalent.*

(i) *A is nonderogatory.*

(ii) *$w_1(A, \lambda) = \dim \operatorname{Ker}(A - \lambda I_n) = 1$ for every eigenvalue λ of A.*

(iii) *$m_A(t) = p_A(t)$.*

The main result of this section is the following. We say that matrices A and B *commute* if $AB = BA$.

Theorem 4.6.2 *Let $A \in \mathbb{F}^{n \times n}$ be nonderogatory with $p_A(\lambda) = \prod_{i=1}^{m}(\lambda - \lambda_i)^{n_i}$ with $\lambda_1, \ldots, \lambda_m \in \mathbb{F}$ all different. Then $B \in \mathbb{F}^{n \times n}$ commutes with A if and only if there exists a polynomial $p(X) \in \mathbb{F}[X]$ so that $B = p(A)$. In that case, one can always choose $p(X)$ to have degree $\leq n - 1$.*

When A is not nonderogatory, there is no guarantee that commuting matrices have to be of the form $p(A)$, as the following example shows.

Example 4.6.3 Let $\mathbb{F} = \mathbb{R}$, $A = \begin{pmatrix} 1 & 0 \\ 0 & 1 \end{pmatrix}$, and $B = \begin{pmatrix} 1 & 2 \\ 0 & 3 \end{pmatrix}$. Clearly $AB = BA$. If $p(X)$ is some polynomial, then $p(A) = \begin{pmatrix} p(1) & 0 \\ 0 & p(1) \end{pmatrix}$, which never equals B.

We will need the following result.

Lemma 4.6.4 *Let $\lambda \neq \mu$, $C = J_n(\lambda) \in \mathbb{F}^{n \times n}$, $D = J_m(\mu) \in \mathbb{F}^{m \times m}$, and $Y \in \mathbb{F}^{n \times m}$. Suppose that $CY = YD$. Then $Y = 0$.*

Proof. We first show that

$$C^k Y = Y D^k \quad \text{for all } k \in \{0, 1, 2, \ldots\}. \tag{4.19}$$

For $k = 0$ this is trivial, while for $k = 1$ it is an assumption of this lemma. Next, $C^2 Y = C(CY) = C(YD) = (CY)D = (YD)D = YD^2$. Proceeding by induction, assume that $C^k Y = YD^k$ holds for some $k \in \{2, 3 \ldots\}$. Then $C^{k+1}Y = CC^k Y = CYD^k = YDD^k = YD^{k+1}$. This proves (4.19). By taking linear combinations, we get that for all polynomials $p(t)$ we have $p(C)Y = Yp(D)$. Let now $p(t) = (t - \lambda)^n$. Then $p(C) = 0$, while $p(D)$ is upper triangular with $(\mu - \lambda)^n \neq 0$ on the main diagonal. Thus $p(D)$ is invertible. So from $p(C)Y = Yp(D)$, we get that $0 = Yp(D)$, and since $p(D)$ is invertible, we get $Y = 0(p(D))^{-1} = 0$. □

Proof of Theorem 4.6.2. When $B = p(A)$, then clearly A and B commute. Thus the main part concerns the converse statement. Thus, suppose that A is as in the statement, and let B commute with A.

We first consider the case when $A = J_n(0)$. Let $B = (b_{ij})_{i,j=1}^n$. Then $AB = (b_{i+1,j})_{i,j=1}^n$, where we let $b_{n+1,j} = 0$ for all j. Furthermore, $BA = (b_{i,j-1})_{i,j=1}^n$, where we let $b_{i,0} = 0$ for all i. Equating AB and BA, we therefore obtain

$$b_{i+1,j} = b_{i,j-1}, \ i,j = 1,\ldots,n, \ \text{where } b_{i,0} = 0 = b_{n+1,j}. \tag{4.20}$$

Set now $b_k = b_{i,j}$, whenever $j - i = k$ and $i \in \{1,\ldots,n+1\}$ and $j \in \{0,\ldots,n\}$. This is well-defined due to (4.20). Then we see that $b_k = 0$ when $k < 0$, and that B is the upper-triangular Toeplitz matrix

$$B = \begin{pmatrix} b_0 & b_1 & b_2 & \cdots & b_{n-1} \\ 0 & b_0 & b_1 & \cdots & b_{n-2} \\ \vdots & \vdots & \ddots & \ddots & \vdots \\ 0 & 0 & \cdots & b_0 & b_1 \\ 0 & 0 & \cdots & 0 & b_0 \end{pmatrix}.$$

If we put $p(X) = b_0 + b_1 X + \cdots + b_{n-1} X^{n-1}$, we get $B = p(A)$.

Next, if $A = J_n(\lambda)$, then we have that $AB = BA$ if and only if $(A - \lambda I_n)B = B(A - \lambda I_n)$. By the previous paragraph, we have that $B = p(A - \lambda I_n)$ for some polynomial $p(X)$. But then, $B = q(A)$, where $q(X) = p(X - \lambda)$. Notice that $\deg q = \deg p$, so we are done in this case as well.

Next, let $A = \oplus_{j=1}^m J_{n_j}(\lambda_j)$ with $\lambda_1,\ldots,\lambda_m$ different, and decompose $B = (B_{ij})_{i,j=1}^m$ where B_{ij} has size $n_i \times n_j$. The equation $AB = BA$ leads now to the equalities

$$J_{n_i}(\lambda_i)B_{ij} = B_{ij}J_{n_j}(\lambda_j), \ i,j = 1,\ldots,m. \tag{4.21}$$

When $i \neq j$, we get by Lemma 4.6.4 that $B_{ij} = 0$, and for $j = i$ we get that B_{ii} is upper-triangular Toeplitz. Define

$$q_j(t) = \frac{p_A(t)}{(t - \lambda_j)^{n_j}}.$$

Then $q_j(t)$ is a polynomial of degree $n - n_j$, and $q_j(J_{n_i}(\lambda_i)) = 0$, $i \neq j$. Also, observe that $q_j(J_{n_j}(\lambda_j))$ is an upper-triangular invertible Toeplitz matrix, and thus $(q_j(J_{n_j}(\lambda_j)))^{-1}B_{jj}$ is upper-triangular Toeplitz. But then there exists a polynomial $r_j(t)$ of degree $\leq n_j - 1$ so that $r(J_{n_j}(\lambda_j)) = (q_j(J_{n_j}(\lambda_j)))^{-1} B_{jj}$. It is now straightforward to check that the polynomial

$$p(t) = q_1(t)r_1(t) + \cdots + q_m(t)r_m(t)$$

satisfies $p(A) = B$.

Finally, we consider $A = P(\oplus_{j=1}^m J_{n_j}(\lambda_j))P^{-1}$. Then $AB = BA$ implies that $\hat{B} = P^{-1}BP$ commutes with $\oplus_{j=1}^m J_{n_j}(\lambda_j)$. The polynomial from the previous paragraph, now establishes $p(\oplus_{j=1}^m J_{n_j}(\lambda_j)) = \hat{B}$. But then $p(A) = B$ also holds. $\qquad\square$

4.7 Systems of linear differential equations

The Jordan canonical form is useful for solving systems of linear differential equations. We set $\mathbb{F} = \mathbb{C}$, as we are dealing with differentiable functions. A system of linear differential equations has the form

$$\begin{cases} x_1'(t) = a_{11}x_1(t) + \cdots + a_{1n}x_n(t), & x_1(0) = c_1, \\ \quad\vdots & \quad\vdots \\ x_n'(t) = a_{n1}x_1(t) + \cdots + a_{nn}x_n(t), & x_n(0) = c_n, \end{cases}$$

which in shorthand we can write as

$$\begin{cases} x'(t) = Ax(t) \\ x(0) = c. \end{cases}$$

If $A = J_n(0)$ (and, for later convenience, changing x to z), the system is

$$\begin{cases} z_1'(t) = z_2(t), & z_1(0) = c_1, \\ \quad\vdots & \quad\vdots \\ z_{n-1}'(t) = z_n(t), & z_{n-1}(0) = c_{n-1}, \\ z_n'(t) = 0, & z_n(0) = c_n. \end{cases}$$

Solving from the bottom up, one easily sees that the solution is

$$z_n(t) = c_n, \, z_{n-1}(t) = c_{n-1} + c_n t, \, z_{n-2} = c_{n-2} + c_{n-1}t + \frac{c_n}{2!}t^2,$$

$$\ldots, z_1(t) = \sum_{k=1}^n \frac{c_k}{(k-1)!}t^{k-1}.$$

Next, if $A = J_n(\lambda) = \lambda I_n + J_n(0)$, then with $z(t) = \begin{pmatrix} z_1(t) \\ \vdots \\ z_n(t) \end{pmatrix}$ as above, it is

straightforward to see that $y(t) = e^{\lambda t} z(t)$ solves

$$\begin{cases} y'(t) = J_n(\lambda)y(t), \\ y(0) = c. \end{cases}$$

Clearly, $y(0) = z(0) = c$. Furthermore,

$$y'(t) = \lambda e^{\lambda t} z(t) + e^{\lambda t} z'(t) = \lambda y(t) + e^{\lambda t} J_n(0)z(t) = \lambda y(t) + J_n(0)y(t) = J_n(\lambda)y(t).$$

To solve the general system

$$\begin{cases} x'(t) = Ax(t) \\ x(0) = c, \end{cases} \tag{4.22}$$

one writes A in Jordan canonical form $A = PJP^{-1}$. If we now put $y(t) = P^{-1}x(t)$, then we get that $y(t)$ satisfies

$$\begin{cases} y'(t) = Jy(t) \\ y(0) = P^{-1}c. \end{cases}$$

With $J = \oplus_{j=1}^m J_{n_j}(\lambda_j)$, this system converts to m systems treated in the previous paragraph. We can subsequently solve these m systems, leading to a solution $y(t)$. Then, in the end, $x(t) = Py(t)$ solves the system (4.22). We will illustrate this in an example below.

We have the following observation.

Theorem 4.7.1 *Consider the system 4.22, where A is similar to*
$\oplus_{j=1}^m J_{n_j}(\lambda_j)$. *Then the solution* $x(t) = \begin{pmatrix} x_1(t) \\ \vdots \\ x_n(t) \end{pmatrix}$ *consists of functions* $x_j(t)$
that are linear combinations of the functions

$$e^{\lambda_j t}, t e^{\lambda_j t}, \ldots, t^{n_j - 1} e^{\lambda_j t}, \ j = 1, \ldots, m.$$

Example 4.7.2 Consider the system

$$\begin{cases} x_1'(t) = 5x_1(t) + 4x_2(t) + 2x_3(t) + x_4(t), & x_1(0) = 0, \\ x_2'(t) = x_2(t) - x_3(t) - x_4(t), & x_2(0) = 1, \\ x_3'(t) = -x_1(t) - x_2(t) + 3x_3(t), & x_3(0) = 1, \\ x_4'(t) = x_1(t) + x_2(t) - x_3(t) + 2x_4(t), & x_4(0) = 0. \end{cases}$$

We find that $A = P^{-1}JP$, where

$$P = \begin{pmatrix} -1 & 1 & 1 & 1 \\ 1 & -1 & 0 & 0 \\ 0 & 0 & -1 & 0 \\ 0 & 1 & 1 & 0 \end{pmatrix}, \quad J = \begin{pmatrix} 1 & 0 & 0 & 0 \\ 0 & 2 & 0 & 0 \\ 0 & 0 & 4 & 1 \\ 0 & 0 & 0 & 4 \end{pmatrix}.$$

Thus

$$\begin{cases} y'(t) = Jy(t) \\ y(0) = Pc = \begin{pmatrix} 2 & -1 & -1 & 2 \end{pmatrix}^T, \end{cases}$$

has the solution

$$\begin{pmatrix} y_1(t) \\ y_2(t) \\ y_3(t) \\ y_4(t) \end{pmatrix} = \begin{pmatrix} 2e^t \\ -e^{2t} \\ -e^{4t} + 2te^{4t} \\ 2e^{4t} \end{pmatrix}.$$

And thus

$$x(t) = P^{-1}y(t) = \begin{pmatrix} -e^{2t} + e^{4t} + 2te^{4t} \\ e^{4t} + 2te^{4t} \\ e^{4t} - 2te^{4t} \\ 2e^t - e^{2t} - e^{4t} + 2te^{4t} \end{pmatrix}.$$

A higher-order linear differential equation can be converted to a system of first-order differential equations, as in the following example.

Example 4.7.3 Consider the third-order differential equation

$$f^{(3)}(t) - 5f^{(2)}(t) + 8f'(t) - 4f(t) = 0, \quad f^{(2)}(0) = 3, f'(0) = 2, f(0) = 1.$$

If we let $x_1(t) = f(t), x_2(t) = f'(t), x_3(t) = f^{(2)}(t)$, we get the system

$$\begin{pmatrix} x_1'(t) \\ x_2'(t) \\ x_3'(t) \end{pmatrix} = \begin{pmatrix} 0 & 1 & 0 \\ 0 & 0 & 1 \\ 4 & -8 & 5 \end{pmatrix} \begin{pmatrix} x_1(t) \\ x_2(t) \\ x_3(t) \end{pmatrix}, \quad \begin{pmatrix} x_1(0) \\ x_2(0) \\ x_3(0) \end{pmatrix} = \begin{pmatrix} 1 \\ 1 \\ 3 \end{pmatrix}.$$

For the eigenvalues of the coefficient matrix, we find 1,2,2, and we find that there is a Jordan block of size 2 at the eigenvalue 2. Thus the solution is a linear combination of e^t, e^{2t}, and te^{et}. Letting $f(t) = c_1e^t + c_2e^{2t} + c_3te^{2t}$, and plugging in the initial conditions $f^{(2)}(0) = 3$, $f'(0) = 2$, $f(0) = 1$, we get the equations

$$\begin{cases} c_1 + c_2 & = 1, \\ c_1 + 2c_2 + c_3 = 2, \\ c_1 + 4c_2 + 4c_3 = 3. \end{cases}$$

Solving, we obtain $c_1 = -1, c_2 = 2, c_3 = -1$, yielding the solution $f(t) = -e^t + 2e^{2t} - te^{2t}$.

4.8 Functions of matrices

We have already used many times that for a polynomial $p(t) = \sum_{j=1}^m p_j t^j$ and a square matrix A, the matrix $p(A)$ is well-defined, simply by setting $p(A) = \sum_{j=1}^m p_j A^j$. Can we also define in a sensible way $f(A)$, where F is some other function, such as for instance $f(t) = e^t$, $f(t) = \sin t$, etc.? For this, let us start with the case when A is a Jordan block $A = J_k(\lambda)$. We first observe that in this case

$$
A^2 = \begin{pmatrix} \lambda^2 & 2\lambda & 1 & & & \\ & \lambda^2 & 2\lambda & 1 & & \\ & & \ddots & \ddots & \ddots & \\ & & & \ddots & \ddots & \\ & & & & \lambda^2 & 2\lambda \\ & & & & & \lambda^2 \end{pmatrix}, \quad A^3 = \begin{pmatrix} \lambda^3 & 3\lambda^2 & 3\lambda & 1 & & \\ & \lambda^3 & 3\lambda^2 & 3\lambda & 1 & \\ & & \ddots & \ddots & \ddots & \\ & & & \ddots & \ddots & \\ & & & & \lambda^3 & 3\lambda^2 \\ & & & & & \lambda^3 \end{pmatrix},
$$

$$
A^{-1} = \begin{pmatrix} \frac{1}{\lambda} & -\frac{1}{\lambda^2} & \frac{1}{\lambda^3} & \cdots & \cdots & \frac{(-1)^{k-1}}{\lambda^k} \\ & \frac{1}{\lambda} & -\frac{1}{\lambda^2} & \frac{1}{\lambda^3} & \cdots & \frac{(-1)^{k-2}}{\lambda^{k-1}} \\ & & \ddots & \ddots & & \vdots \\ & & & \ddots & \ddots & \vdots \\ & & & & \frac{1}{\lambda} & -\frac{1}{\lambda^2} \\ & & & & & \frac{1}{\lambda} \end{pmatrix}, \quad \lambda \neq 0.
$$

In all cases, it has the form

$$
f(J_k(\lambda)) = \begin{pmatrix} f(\lambda) & \frac{f'(\lambda)}{1!} & \frac{f''(\lambda)}{2!} & \cdots & \cdots & \frac{f^{(k-1)}(\lambda)}{(k-1)!} \\ & f(\lambda) & \frac{f'(\lambda)}{1!} & \frac{f''(\lambda)}{2!} & \cdots & \frac{f^{(k-2)}(\lambda)}{(k-2)!} \\ & & \ddots & \ddots & & \vdots \\ & & & \ddots & \ddots & \vdots \\ & & & & f(\lambda) & \frac{f'(\lambda)}{1!} \\ & & & & & f(\lambda) \end{pmatrix}. \tag{4.23}
$$

This observation leads to the following definition. Let $A \in \mathbb{C}^{n \times n}$ have minimal polynomial $m_A(t) = \prod_{j=1}^m (t - \lambda_j)^{k_j}$, and let f be a complex-valued function on a domain in $\mathbb{C}$ so that $f(\lambda_j), f'(\lambda_j), \ldots, f^{(k_j-1)}(\lambda_j), j = 1, \ldots, m$, are well-defined. If A is given in

Jordan canonical form $A = SJS^{-1}$, with

$$
J = \begin{pmatrix} J(\lambda_1) & 0 & \cdots & 0 \\ 0 & J(\lambda_2) & \cdots & 0 \\ \vdots & \vdots & \ddots & \vdots \\ 0 & 0 & \cdots & J(\lambda_m) \end{pmatrix},
$$

where $J(\lambda_j)$ is the $n_j \times n_j$ matrix

$$
J(\lambda_j) = \oplus_{k=1}^{n_j} \left(\oplus_{l=1}^{w_k(A,\lambda_j)-w_{k+1}(A,\lambda_j)} J_k(\lambda_j) \right), \quad j = 1, \ldots, m,
$$

we define

$$
f(A) := Sf(J)S^{-1}, f(J) := \begin{pmatrix} f(J(\lambda_1)) & 0 & \cdots & 0 \\ 0 & f(J(\lambda_2)) & \cdots & 0 \\ \vdots & \vdots & \ddots & \vdots \\ 0 & 0 & \cdots & f(J(\lambda_m)) \end{pmatrix} \quad (4.24)
$$

and

$$
f(J(\lambda_j)) := \oplus_{k=1}^{n_j} \left(\oplus_{l=1}^{w_k(A,\lambda_j)-w_{k+1}(A,\lambda_j)} f(J_k(\lambda_j)) \right), \quad j = 1, \ldots, m,
$$

with $f(J_k(\lambda_j))$ given via (4.23).

Let us do an example.

Example 4.8.1 Let $A = \begin{pmatrix} 2 & 2 & 3 \\ 1 & 3 & 3 \\ -1 & -2 & -2 \end{pmatrix}$. In Example 4.4.5 we calculated that $A = SJS^{-1}$, where

$$
J = \begin{pmatrix} 1 & & \\ & 1 & 1 \\ & & 1 \end{pmatrix}, S = \begin{pmatrix} -2 & 1 & 1 \\ 1 & 1 & 0 \\ 0 & -1 & 0 \end{pmatrix}.
$$

If $f(t) = e^{wt}$, we find that $f(1) = e^w$ and $f'(1) = we^w$, and thus

$$
f(A) = \begin{pmatrix} -2 & 1 & 1 \\ 1 & 1 & 0 \\ 0 & -1 & 0 \end{pmatrix} \begin{pmatrix} e^w & & \\ & e^w & we^w \\ & & e^w \end{pmatrix} \begin{pmatrix} -2 & 1 & 1 \\ 1 & 1 & 0 \\ 0 & -1 & 0 \end{pmatrix}^{-1}.
$$

Notice that we need to check that $f(A)$ is well-defined. Indeed, we need to check that if $A = SJS^{-1} = \tilde{S}J\tilde{S}^{-1}$, then $Sf(J)S^{-1} = \tilde{S}f(J)\tilde{S}^{-1}$ (where we used that J is unique up to permutation of its blocks, so we do not have to worry about different J's). In other words, if we let $T = \tilde{S}^{-1}S$, we need to check that

$$TJ = JT \quad \text{implies} \quad Tf(J) = f(J)T. \tag{4.25}$$

Using the techniques in Section 4.6, this is fairly straightforward to check, and we will leave this for the reader.

Remark 4.8.2 It should be noticed that with $m_A(t) = \prod_{j=1}^{m}(t - \lambda_j)^{k_j}$ and with functions f and g so that

$$f^{(r)}(\lambda_j) = g^{(r)}(\lambda_j), \ r = 0, \ldots, k_j - 1, \ j = 1, \ldots, m, \tag{4.26}$$

we have that $f(A) = g(A)$. Thus, as an alternative way of defining $f(A)$, one can construct a polynomial g satisfying (4.26) and define $f(A)$ via $f(A) := g(A)$. In this way, one avoids having to use the Jordan canonical form in the definition of $f(A)$, which may be preferable in some cases.

When $h(t) = f(t)g(t)$ we expect that $h(A) = f(A)g(A)$. This is indeed true, but it is something that we need to prove. For this we need to remind ourselves of the product rule for differentiation:

$$h(t) = f(t)g(t) \quad \text{implies that} \quad h'(t) = f(t)g'(t) + f'(t)g(t).$$

Taking a second and third derivative we obtain

$$h''(t) = f(t)g''(t) + 2f'(t)g'(t) + f''(t)g(t),$$

$$h^{(3)}(t) = f(t)g^{(3)}(t) + 3f'(t)g''(t) + 3f''(t)g'(t) + f^{(3)}(t)g(t).$$

In general, we obtain that the kth derivative of h is given by

$$h^{(k)}(t) = \sum_{r=0}^{k} \binom{k}{r} f^{(r)}(t)g^{(k-r)}(t), \tag{4.27}$$

which is referred to as the *Leibniz rule*. We will use the Leibniz rule in the following proof.

Theorem 4.8.3 *Let $A \in \mathbb{C}^{n \times n}$ with minimal polynomial $m_A(t) = \prod_{j=1}^{m}(t - \lambda_j)^{k_j}$ and let f and g be functions so that $f(\lambda_j), f'(\lambda_j), \ldots, f^{(k_j-1)}(\lambda_j), g(\lambda_j), g'(\lambda_j), \ldots, g^{(k_j-1)}(\lambda_j), j = 1, \ldots, m$, are well-defined. Put $k(t) = f(t) + g(t)$ and $h(t) = f(t)g(t)$. Then*

$$k(A) = f(A) + g(A) \quad \text{and} \quad h(A) = f(A)g(A).$$

Proof. We will show the equation $h(A) = f(A)g(A)$. The equation $k(A) = f(A) + g(A)$ can be proven in a similar manner (and is actually easier to prove, as $k^{(j)}(\lambda) = f^{(j)}(\lambda) + g^{(j)}(\lambda)$ for all j).

First, let $A = J_k(\lambda)$. Then

$$f(A)g(A) = \begin{pmatrix} f(\lambda) & \cdots & \frac{f^{(k-1)}(\lambda)}{(k-1)!} \\ & \ddots & \vdots \\ & & f(\lambda) \end{pmatrix} \begin{pmatrix} g(\lambda) & \cdots & \frac{g^{(k-1)}(\lambda)}{(k-1)!} \\ & \ddots & \vdots \\ & & g(\lambda) \end{pmatrix} =$$

$$\begin{pmatrix} f(\lambda)g(\lambda) & \cdots & \sum_{j=0}^{k-1} \frac{f^{(j)}(\lambda)}{j!} \frac{g^{(k-j-1)}(\lambda)}{(k-j-1)!} \\ & \ddots & \vdots \\ & & f(\lambda)g(\lambda) \end{pmatrix} = \begin{pmatrix} h(\lambda) & \cdots & \frac{h^{(k-1)}(\lambda)}{(k-1)!} \\ & \ddots & \vdots \\ & & h(\lambda) \end{pmatrix} = h(A),$$

where we used that Leibniz's rule yields

$$\sum_{j=0}^{k-1} \frac{f^{(j)}(\lambda)}{j!} \frac{g^{(k-j-1)}(\lambda)}{(k-j-1)!} = \frac{1}{(k-1)!} \sum_{j=0}^{k-1} \binom{k-1}{j} f^{(j)}(\lambda)g^{(k-j-1)}(\lambda) = \frac{h^{(k-1)}(\lambda)}{(k-1)!}.$$

As the rule works for a Jordan block, it will also work for a direct sum of Jordan blocks. Finally, when $A = SJS^{-1}$, we get that $f(A)g(A) = Sf(J)S^{-1}Sg(J)S^{-1} = S[f(J)g(J)]S^{-1} = Sh(J)S^{-1} = h(A)$. $\square$

Observe that the matrix in (4.23) can be written as

$$f(\lambda)J_k(0)^0 + \frac{f'(\lambda)}{1!}J_k(0)^1 + \frac{f''(\lambda)}{2!}J_k(0)^2 + \cdots + \frac{f^{(k-1)}(\lambda)}{(k-1)!}J_k(0)^{k-1},$$

where $J_k(0)^0 = I_k$. Applying this to each summand in (4.24), we arrive at the following theorem.

Theorem 4.8.4 *Let $A \in \mathbb{C}^{n \times n}$ with minimal polynomial $m_A(t) = \prod_{j=1}^{m}(t - \lambda_j)^{k_j}$. Then there exist matrices*

$$P_{jk}, \ j = 1, \ldots, m, k = 0, \ldots, k_j - 1,$$

so that for every complex-valued function f so that $f(\lambda_j)$, $f'(\lambda_j)$, ..., $f^{(k_j-1)}(\lambda_j)$, $j = 1, \ldots, m$, are well-defined, we have that

$$f(A) = \sum_{j=1}^{m} \sum_{k=0}^{k_j-1} f^{(k)}(\lambda_j)P_{jk}. \tag{4.28}$$

Moreover, these matrices P_{jk} satisfy

(i) $P_{j0}^2 = P_{j0}$,

(ii) $P_{jk}P_{rs} = 0$, $j \neq r$,

(iii) $P_{jk}P_{js} = \binom{k+s}{k} P_{j,k+s}$, *and*

(iv) $\sum_{j=1}^{m} P_{j0} = I_n$.

Here $P_{jk} = 0$ *when* $k \geq k_j - 1$.

Proof. Let A be given in Jordan canonical form $A = SJS^{-1}$, with

$$J = \begin{pmatrix} J(\lambda_1) & 0 & \cdots & 0 \\ 0 & J(\lambda_2) & \cdots & 0 \\ \vdots & \vdots & \ddots & \vdots \\ 0 & 0 & \cdots & J(\lambda_m) \end{pmatrix},$$

where $J(\lambda_j)$ is the $n_j \times n_j$ matrix

$$J(\lambda_j) = \oplus_{k=1}^{n_j} \left(\oplus_{l=1}^{w_k(A,\lambda_j)-w_{k+1}(A,\lambda_j)} J_k(\lambda_j) \right), \quad j = 1, \ldots, m.$$

We define

$$P_{j0} := S \begin{pmatrix} 0 & & & & & & \\ & \ddots & & & & & \\ & & 0 & & & & \\ & & & I_{n_j} & & & \\ & & & & 0 & & \\ & & & & & \ddots & \\ & & & & & & 0 \end{pmatrix} S^{-1},$$

$$P_{jk} := \frac{1}{k!} S \begin{pmatrix} 0 & & & & & & \\ & \ddots & & & & & \\ & & 0 & & & & \\ & & & J_j^k & & & \\ & & & & 0 & & \\ & & & & & \ddots & \\ & & & & & & 0 \end{pmatrix} S^{-1}, \qquad (4.29)$$

where

$$J_j = \oplus_{k=1}^{n_j} \left(\oplus_{l=1}^{w_k(A,\lambda_j)-w_{k+1}(A,\lambda_j)} J_k(0) \right), \quad j = 1, \ldots, m.$$

Notice that $J_j^s = 0$ when $s \geq k_j$. Equality (4.28) now follows directly from (4.24). Moreover, from the definitions (4.29) it is easy to check that (i)–(iv) hold. $\qquad \square$

Let us compute decomposition (4.28) for an example.

Example 4.8.5 Let

$$
A = \begin{pmatrix}
4 & 0 & 0 & 0 & 2 & 0 \\
-2 & 3 & -1 & 0 & -2 & -1 \\
0 & -2 & 6 & 0 & 2 & 2 \\
2 & -4 & 8 & 2 & 6 & 6 \\
-2 & 3 & -5 & 0 & -2 & -3 \\
2 & -3 & 5 & 0 & 4 & 5
\end{pmatrix}.
$$

Then $A = SJS^{-1}$, where

$$
S = \begin{pmatrix}
1 & 0 & 0 & -1 & 0 & 1 \\
1 & 0 & 1 & 1 & -1 & 0 \\
1 & 0 & 0 & 1 & 0 & 0 \\
-1 & 1 & 0 & 1 & 1 & 0 \\
-1 & 0 & 0 & 0 & -1 & 0 \\
0 & 0 & 1 & 0 & 1 & 0
\end{pmatrix}, J = \begin{pmatrix}
2 & 0 & 0 & 0 & 0 & 0 \\
0 & 2 & 2 & 0 & 0 & 0 \\
0 & 0 & 2 & 0 & 0 & 0 \\
0 & 0 & 0 & 4 & 2 & 0 \\
0 & 0 & 0 & 0 & 4 & 2 \\
0 & 0 & 0 & 0 & 0 & 4
\end{pmatrix}.
$$

Thus $\lambda_2 = 1$ and $\lambda_2 = 4$,

$$
J_1 = \begin{pmatrix}
0 & 0 & 0 \\
0 & 0 & 1 \\
0 & 0 & 0
\end{pmatrix}, J_2 = \begin{pmatrix}
0 & 1 & 0 \\
0 & 0 & 1 \\
0 & 0 & 0
\end{pmatrix}.
$$

Now

$$
P_{10} = S \begin{pmatrix} I_3 & 0 \\ 0 & 0 \end{pmatrix} S^{-1} = \begin{pmatrix}
0 & \frac{1}{2} & -\frac{1}{2} & 0 & -1 & -\frac{1}{2} \\
0 & 1 & -1 & 0 & -1 & 0 \\
0 & \frac{1}{2} & -\frac{1}{2} & 0 & -1 & -\frac{1}{2} \\
0 & 1 & -2 & 1 & -1 & -1 \\
0 & -\frac{1}{2} & \frac{1}{2} & 0 & 1 & \frac{1}{2} \\
0 & \frac{1}{2} & -\frac{1}{2} & 0 & 0 & \frac{1}{2}
\end{pmatrix},
$$

$$
P_{11} = S \begin{pmatrix} J_1 & 0 \\ 0 & 0 \end{pmatrix} S^{-1} = \begin{pmatrix}
0 & 0 & 0 & 0 & 0 & 0 \\
0 & 0 & 0 & 0 & 0 & 0 \\
0 & 0 & 0 & 0 & 0 & 0 \\
0 & \frac{1}{2} & -\frac{1}{2} & 0 & 0 & \frac{1}{2} \\
0 & 0 & 0 & 0 & 0 & 0 \\
0 & 0 & 0 & 0 & 0 & 0
\end{pmatrix}.
$$

We leave the other computations as an exercise.

We will next see that the formalism introduced above provides a useful tool in the setting of systems of differential equations. We first need that if $B(t) = (b_{ij})_{i=1,j=1}^{n,\ m}$ is a matrix whose entries are functions in t, then we define

$$
\frac{d}{dt}B(t) = B'(t) := (b'_{ij})_{i=1,j=1}^{n,\ m} = (\frac{d}{dt}b_{ij})_{i=1,j=1}^{n,\ m}.
$$

Thus the derivative of a matrix function is simply defined by taking the derivative in each entry. For instance

$$\frac{d}{dt}\begin{pmatrix} t^2 & \cos t \\ e^{5t} & 7 \end{pmatrix} = \begin{pmatrix} 2t & -\sin t \\ 5e^{5t} & 0 \end{pmatrix}.$$

As taking the derivative is a linear operation, we have that

$$\frac{d}{dt}(SB(t)W) = S(\frac{d}{dt}(B(t))W, \tag{4.30}$$

where S and W are matrices of appropriate size. Indeed, looking at the (r, s) entry of this product, we have that

$$\frac{d}{dt}(\sum_i \sum_j s_{ri} b_{ij}(t) w_{js}) = \sum_i \sum_j s_{ri}(\frac{d}{dt} b_{ij}(t)) w_{js}.$$

The following proposition now shows that the equality $\frac{d}{dt} e^{at} = a e^{at}$ generalizes to the case when a is replaced by a matrix A.

Proposition 4.8.6 *Given* $A \in \mathbb{C}^{n \times n}$. *Then*

$$\frac{d}{dt} e^{tA} = A e^{tA} = e^{tA} A. \tag{4.31}$$

Proof. We first show (4.31) for a Jordan block $A = J_k(\lambda)$. If $A = J_k(\lambda)$ and $t \neq 0$, we have that

$$tA = \begin{pmatrix} t\lambda & t & & & \\ & t\lambda & t & & \\ & & \ddots & \ddots & \\ & & & t\lambda & t \\ & & & & t\lambda \end{pmatrix} = \begin{pmatrix} 1 & & & & \\ & \frac{1}{t} & & & \\ & & \ddots & & \\ & & & \frac{1}{t^{k-2}} & \\ & & & & \frac{1}{t^{k-1}} \end{pmatrix} \times$$

$$\begin{pmatrix} t\lambda & 1 & & & \\ & t\lambda & 1 & & \\ & & \ddots & \ddots & \\ & & & t\lambda & 1 \\ & & & & t\lambda \end{pmatrix} \begin{pmatrix} 1 & & & & \\ & t & & & \\ & & \ddots & & \\ & & & t^{k-2} & \\ & & & & t^{k-1} \end{pmatrix},$$

bringing A in the $SJ_k(t\lambda)S^{-1}$ format. Then with $f(x) = e^x$ we get $f(tA) = Sf(J_k(t\lambda))S^{-1}$, yielding

$$e^{tA} = \begin{pmatrix} 1 & & & & \\ & \frac{1}{t} & & & \\ & & \ddots & & \\ & & & \frac{1}{t^{k-2}} & \\ & & & & \frac{1}{t^{k-1}} \end{pmatrix} \begin{pmatrix} e^{t\lambda} & \frac{e^{t\lambda}}{1!} & \cdots & \cdots & \frac{e^{t\lambda}}{(k-1)!} \\ & e^{t\lambda} & \frac{e^{t\lambda}}{1!} & \cdots & \frac{e^{t\lambda}}{(k-2)!} \\ & & \ddots & \ddots & \vdots \\ & & & e^{t\lambda} & \frac{e^{t\lambda}}{1!} \\ & & & & e^{t\lambda} \end{pmatrix} \times$$

$$\begin{pmatrix} 1 & & & & \\ & t & & & \\ & & \ddots & & \\ & & & t^{k-2} & \\ & & & & t^{k-1} \end{pmatrix}.$$

Thus we find

$$e^{tA} = \begin{pmatrix} e^{t\lambda} & \frac{te^{t\lambda}}{1!} & \cdots & \cdots & \frac{t^{k-1}e^{t\lambda}}{(k-1)!} \\ & e^{t\lambda} & \frac{te^{t\lambda}}{1!} & \cdots & \frac{t^{k-2}e^{t\lambda}}{(k-2)!} \\ & & \ddots & \ddots & \vdots \\ & & & e^{t\lambda} & \frac{te^{t\lambda}}{1!} \\ & & & & e^{t\lambda} \end{pmatrix},$$

which also holds when $t = 0$. As $\frac{d}{dt}\left(\frac{t^j e^{t\lambda}}{j!}\right) = \frac{t^{j-1}e^{t\lambda}}{(j-1)!} + \frac{\lambda t^j e^{t\lambda}}{j!}$, $j \geq 1$, one easily sees that that (4.31) holds for $A = J_k(\lambda)$).

Next, one needs to observe that (4.31) holds for A a direct sum of Jordan blocks. Finally, using (4.30), one obtains that (4.31) holds when $A = SJS^{-1}$, thus proving the statement for general A.

We can now write down the solution of a system of differential equations very efficiently as follows.

Corollary 4.8.7 *The system of differential equations*

$$\begin{cases} x'(t) = Ax(t) \\ x(0) = c. \end{cases}$$

has the solution

$$x(t) = e^{tA}x_0.$$

Proof. With $x(t) = e^{tA}x_0$, we have $x(0) = e^0 x_0 = Ix_0 = x_0$, and $\frac{d}{dt}x(t) = \frac{d}{dt}e^{tA}x_0 = Ae^{tA}x_0 = Ax(t)$. $\qquad\square$

Using these techniques, we can now also handle non-homogenous systems of differential equations of the form

$$x'(t) = Ax(t) + B(t). \tag{4.32}$$

Indeed, if we set $x(t) = e^{tA}f(t)$, for some differentiable function $f(t)$, then using the product rule we obtain

$$x'(t) = Ae^{tA}f(t) + e^{tA}f'(t) = Ax(t) + e^{tA}f'(t).$$

If $x(t)$ is a solution to (4.32), then we need $B(t) = e^{tA}f'(t)$, yielding

$f'(t) = e^{-tA}B(t)$, and thus

$$f(t) = \int_{t_0}^{t} s^{-sA}B(s)ds + K,$$

where K is some constant vector. Let us illustrate how this works in an example.

Example 4.8.8 Consider the system

$$\begin{cases} x_1'(t) = -x_2(t), \\ x_2'(t) = x_1(t) + t. \end{cases}$$

Then

$$A = \begin{pmatrix} 0 & -1 \\ 1 & 0 \end{pmatrix}, B(t) = \begin{pmatrix} 0 \\ t \end{pmatrix},$$

and

$$e^{-sA} = \begin{pmatrix} \cos s & \sin s \\ -\sin s & \cos s \end{pmatrix}.$$

Thus

$$f(t) := \int_0^t e^{-sA}B(s)ds + K = \int_0^t \begin{pmatrix} \cos s & \sin s \\ -\sin s & \cos s \end{pmatrix} \begin{pmatrix} 0 \\ s \end{pmatrix} + \begin{pmatrix} K_1 \\ K_2 \end{pmatrix} =$$

$$= \int_0^t \begin{pmatrix} s\sin s \\ s\cos s \end{pmatrix} ds + \begin{pmatrix} K_1 \\ K_2 \end{pmatrix} = \begin{pmatrix} \sin t - t\cos t + K_1 \\ \cos t + t\sin t - 1 + K_2 \end{pmatrix}.$$

We now find

$$\begin{pmatrix} x_1(t) \\ x_2(t) \end{pmatrix} = \begin{pmatrix} \cos t & -\sin t \\ \sin t & \cos t \end{pmatrix} \begin{pmatrix} \sin t - t\cos t + K_1 \\ \cos t + t\sin t - 1 + K_2 \end{pmatrix} =$$

$$\begin{pmatrix} -t + K_1\cos t + (1 - K_2)\sin t \\ 1 - (1 - K_2)\cos t + K_1\sin t \end{pmatrix}.$$

4.9 The resolvent

One matrix function that is of particular interest is the resolvent. The *resolvent* of a matrix $A \in \mathbb{C}^{n \times n}$ is the function

$$R(\lambda) := (\lambda I_n - A)^{-1},$$

which is well-defined on $\mathbb{C} \setminus \sigma(A)$, where
$\sigma(A) = \{z \in \mathbb{C} : z \text{ is an eigenvalue of } A\}$ is the *spectrum* of A. We have the following observation.

Proposition 4.9.1 *Let* $A \in \mathbb{C}^{n \times n}$ *with minimal polynomial*
$m_A(t) = \prod_{j=1}^{m}(t - \lambda_j)^{k_j}$, *and let* P_{jk}, $j = 1, \ldots, m, k = 0, \ldots, k_j - 1$, *be as in Theorem 4.8.4. Then*

$$R(\lambda) = (\lambda I_n - A)^{-1} = \sum_{j=1}^{m} \sum_{k=0}^{n_j-1} \frac{k!}{(\lambda - \lambda_j)^{k+1}} P_{jk}. \qquad (4.33)$$

Proof. Fix $\lambda \in \mathbb{C} \setminus \sigma(A)$, and define $g(z) = \frac{1}{\lambda - z}$, which is well-defined and k times differentiable for every $k \in \mathbb{N}$ on the domain $\mathbb{C} \setminus \{\lambda\}$. Notice that $g(A) = (\lambda I_n - A)^{-1} = R(\lambda)$. Also observe that

$$g'(t) = \frac{1}{(\lambda - t)^2}, g''(t) = \frac{2}{(\lambda - t)^3}, \ldots, g^{(k)}(t) = \frac{k!}{(\lambda - t)^{k+1}}.$$

Thus, by Theorem 4.8.4,

$$R(\lambda) = g(A) = \sum_{j=1}^{m} \sum_{k=0}^{n_j-1} g^{(k)}(t) P_{jk} = \sum_{j=1}^{m} \sum_{k=0}^{n_j-1} \frac{k!}{(\lambda - \lambda_j)^{k+1}} P_{jk}.$$

$\square$

If we make use of a fundamental complex analysis result, Cauchy's integral formula, we can develop an integral formula for $f(A)$ that is used, for instance, in analyzing differential operators. Let us start by stating Cauchy's result. A function f of a complex variable is called *analytic* on an open set $D \subseteq \mathbb{C}$ if f is continuously differentiable at every point $z \in D$. If f is analytic on a domain D bounded by a contour γ and continuous on the closure $\overline{D}$, then *Cauchy's integral formula* states that

$$f^{(j)}(\lambda_0) = \frac{j!}{2\pi i} \int_{\gamma} \frac{f(z)}{(z - \lambda_0)^{j+1}} dz, \text{ for all } \lambda_0 \in D \text{ and } j = 0, 1, \ldots . \qquad (4.34)$$

Applying this to Proposition 4.9.1 we obtain the following result.

Theorem 4.9.2 *Let* $A \in \mathbb{C}^{n \times n}$ *with spectrum* $\sigma(A) = \{\lambda_1, \ldots, \lambda_m\}$. *Let* D *be a domain bounded by the contour* γ, *and assume that* $\sigma(A) \subset D$. *For functions* f *analytic on* D *and continuous on the closure* $\overline{D}$, *we have that*

$$f(A) = \frac{1}{2\pi i} \int_{\gamma} f(\lambda)(\lambda I - A)^{-1} d\lambda = \frac{1}{2\pi i} \int_{\gamma} f(\lambda) R(\lambda) d\lambda. \qquad (4.35)$$

Proof. Follows directly from combining Proposition 4.9.1 with Cauchy's integral formula (4.34) and equation (4.28). $\square$

By choosing particular functions for f we can retrieve the matrices P_{jk} from Theorem 4.8.4.

Theorem 4.9.3 *Let $A \in \mathbb{C}^{n \times n}$ with minimal polynomial*
$m_A(t) = \prod_{j=1}^{m}(t - \lambda_j)^{k_j}$. *Let γ_j be a contour that contains λ_j in its interior, but none of the other eigenvalues of A are in the interior of or on γ_j. Then the matrices P_{jk} as defined in Theorem 4.8.4 can be found via*

$$P_{jk} = \frac{1}{2\pi i} \int_{\gamma_j} (\lambda - \lambda_j)^k (\lambda I - A)^{-1} d\lambda = \frac{1}{2\pi i} \int_{\gamma_k} (\lambda - \lambda_j)^k R(\lambda) d\lambda, \quad (4.36)$$

where $j = 1, \ldots, m, k = 0, \ldots, k_j - 1$.

4.10   Exercises

Exercise 4.10.1 Let $\mathbb{F} = \mathbb{Z}_3$. Check the Cayley–Hamilton theorem on the matrix

$$A = \begin{pmatrix} 1 & 0 & 2 \\ 2 & 1 & 0 \\ 2 & 2 & 2 \end{pmatrix}.$$

Exercise 4.10.2 For the following matrices A (and B) determine its Jordan canonical form J and a similarity matrix P, so that $P^{-1}AP = J$.

(a)

$$A = \begin{pmatrix} -1 & 1 & 0 & 0 \\ -1 & 0 & 1 & 0 \\ -1 & 0 & 0 & 1 \\ -1 & 0 & 0 & 1 \end{pmatrix}.$$

This matrix is nilpotent.

(b)

$$A = \begin{pmatrix} 10 & -1 & 1 & -4 & -6 \\ 9 & -1 & 1 & -3 & -6 \\ 4 & -1 & 1 & -3 & -1 \\ 9 & -1 & 1 & -4 & -5 \\ 10 & -1 & 1 & -4 & -6 \end{pmatrix}.$$

This matrix is nilpotent.

(c)

$$A = \begin{pmatrix} 0 & 1 & 0 \\ -1 & 0 & 0 \\ 1 & 1 & 1 \end{pmatrix}.$$

(d)

$$A = \begin{pmatrix} 2 & 0 & -1 & 1 \\ 0 & 1 & 0 & 0 \\ 1 & 0 & 0 & 0 \\ 0 & 0 & 0 & 1 \end{pmatrix}.$$

(e)

$$B = \begin{pmatrix} 1 & -5 & 0 & -3 \\ 1 & 1 & -1 & 0 \\ 0 & -3 & 1 & -2 \\ -2 & 0 & 2 & 1 \end{pmatrix}.$$

(Hint: 1 is an eigenvalue.)

(f) For the matrix B, compute B^{100}, by using the decomposition $B = PJP^{-1}$.

Exercise 4.10.3 Let

$$A = \begin{pmatrix} 3 & 1 & 0 & 0 & 0 & 0 & 0 \\ 0 & 3 & 1 & 0 & 0 & 0 & 0 \\ 0 & 0 & 3 & 0 & 0 & 0 & 0 \\ 0 & 0 & 0 & 3 & 1 & 0 & 0 \\ 0 & 0 & 0 & 0 & 3 & 0 & 0 \\ 0 & 0 & 0 & 0 & 0 & 3 & 1 \\ 0 & 0 & 0 & 0 & 0 & 0 & 3 \end{pmatrix}.$$

Determine bases for the following spaces:

(a) $\mathrm{Ker}(3I - A)$.

(b) $\mathrm{Ker}(3I - A)^2$.

(c) $\mathrm{Ker}(3I - A)^3$.

Exercise 4.10.4 Let M and N be 6×6 matrices over $\mathbb{C}$, both having minimal polynomial x^3.

(a) Prove that M and N are similar if and only if they have the same rank.

(b) Give a counterexample to show that the statement is false if 6 is replaced by 7.

(c) Compute the minimal and characteristic polynomials of the following matrix. Is it diagonalizable?

$$\begin{pmatrix} 5 & -2 & 0 & 0 \\ 6 & -2 & 0 & 0 \\ 0 & 0 & 0 & 6 \\ 0 & 0 & 1 & -1 \end{pmatrix}$$

Exercise 4.10.5 (a) Let A be a 7×7 matrix of rank 4 and with minimal polynomial equal to $q_A(\lambda) = \lambda^2(\lambda + 1)$. Give all possible Jordan canonical forms of A.

(b) Let $A \in \mathbb{C}^n$. Show that if there exists a vector $\mathbf{v}$ so that $\mathbf{v}, A\mathbf{v}, \ldots, A^{n-1}\mathbf{v}$ are linearly independent, then the characteristic polynomial of A equals the minimal polynomial of A. (Hint: use the basis $\mathcal{B} = \{\mathbf{v}, A\mathbf{v}, \ldots, A^{n-1}\mathbf{v}\}$.)

Exercise 4.10.6 Let $A \in \mathbb{F}^{n \times n}$ and A^T denote its transpose. Show that $w_k(A, \lambda) = w_k(A^T, \lambda)$, for all $\lambda \in \mathbb{F}$ and $k \in \mathbb{N}$. Conclude that A and A^T have the same Jordan canonical form, and are therefore similar.

Exercise 4.10.7 Let $A \in \mathbb{C}^{4 \times 4}$ matrix satisfying $A^2 = -I$.

(a) Determine the possible eigenvalues of A.

(b) Determine the possible Jordan structures of A.

Exercise 4.10.8 Let $p(x) = (x - 2)^2(x - 3)^2$. Determine a matrix A for which $p(A) = 0$ and for which $q(A) \neq 0$ for all nonzero polynomials q of degree ≤ 3. Explain why $q(A) \neq 0$ for such q.

Exercise 4.10.9 Let $m_A(t) = (t - 1)^2(t - 2)(t - 3)$ be the minimal polynomial of $A \in M_6$.

(a) What possible Jordan forms can A have?

(b) If it is known that $\text{rank}(A - I) = 3$, what possible Jordan forms can A have?

Exercise 4.10.10 Let A be a 4×4 matrix satisfying $A^2 = -A$.

(a) Determine the possible eigenvalues of A.

(b) Determine the possible Jordan structures of A (Hint: notice that $(A + I)A = 0$.)

Exercise 4.10.11 Let $A \in \mathbb{C}^{n \times n}$. For the following, answer True or False. Provide an explanation.

(a) If $\det(A) = 0$, then 0 is an eigenvalue of A.

(b) If $A^2 = 0$, then the rank of A is at most $\frac{n}{2}$.

(c) There exists a matrix A with minimal polynomial $m_A(t) = (t-1)(t-2)$ and characteristic polynomial $p_A(t) = t^{n-2}(t-1)(t-2)$ (here $n > 2$).

(d) If all eigenvalues of A are 1, then $A = I_n$ (=the $n \times n$ identity matrix).

Exercise 4.10.12 Show that if A is similar to B, then tr $A = $ tr B.

Exercise 4.10.13 Let P be a matrix so that $P^2 = P$.

(a) Show that P only has eigenvalues 0 or 1.

(b) Show that rank $P = $ trace P. (Hint: determine the possible Jordan canonical form of P.)

Exercise 4.10.14 Let $A = PJP^{-1}$. Show that Ran $A = P[\text{Ran } J]$ and Ker $A = P[\text{Ker } J]$. In addition, dim Ran $A = $ dim Ran J and dim Ker $A = $ dim Ker J.

Exercise 4.10.15 Show that matrices A and B are similar if and only if they have the same Jordan canonical form.

Exercise 4.10.16 Show that if A and B are square matrices of the same size, with A invertible, then AB and BA have the same Jordan canonical form.

Exercise 4.10.17 Let $A \in \mathbb{F}^{n \times m}$ and $B \in \mathbb{F}^{m \times n}$. Observe that

$$\begin{pmatrix} I_n & -A \\ 0 & I_m \end{pmatrix} \begin{pmatrix} AB & 0 \\ B & 0_m \end{pmatrix} \begin{pmatrix} I_n & A \\ 0 & I_m \end{pmatrix} = \begin{pmatrix} 0_n & 0 \\ B & BA \end{pmatrix}.$$

(a) Show that the Weyr characteristics at $\lambda \neq 0$ of AB and BA satisfy

$$w_k(AB, \lambda) = w_k(BA, \lambda), \quad k \in \mathbb{N}.$$

(b) Show that $\lambda \neq 0$ is an eigenvalue of AB if and only if it is an eigenvalue of BA, and that AB and BA have the same Jordan structure at λ.

(c) Provide an example of matrices A and B so that AB and BA have different Jordan structures at 0.

Exercise 4.10.18 Let $A, B \in \mathbb{C}^{n \times n}$ be such that $(AB)^n = 0$. Prove that $(BA)^n = 0$.

Exercise 4.10.19 (a) Let $A \in \mathbb{R}^{8 \times 8}$ with characteristic polynomial
$p(x) = (x + 3)^4 (x^2 + 1)^2$ and minimal polynomial
$m(x) = (x + 3)^2 (x^2 + 1)$. What are the possible Jordan canonical form(s) for A (up to permutation of Jordan blocks)?

(b) Suppose that $A \in \mathbb{C}^{n \times n}$ satisfies $A^k \neq 0$ and $A^{k+1} = 0$. Prove that there exists $\mathbf{x} \in \mathbb{C}^n$ such that $\{\mathbf{x}, A\mathbf{x}, \ldots, A^k \mathbf{x}\}$ is linearly independent.

(c) Let $A, B \in \mathbb{C}^{n \times n}$ be such that $A^2 - 2AB + B^2 = 0$. Prove that every eigenvalue of B is an eigenvalue of A, and conversely that every eigenvalue of A is an eigenvalue of B.

Exercise 4.10.20 (a) Prove Proposition 4.6.1.

(b) Let $A = \begin{pmatrix} 0 & 0 & 0 & \cdots & -a_0 \\ 1 & 0 & 0 & \cdots & -a_1 \\ 0 & 1 & 0 & \cdots & -a_2 \\ \vdots & \vdots & \ddots & \ddots & \vdots \\ 0 & \cdots & 1 & 0 & -a_{n-2} \\ 0 & \cdots & 0 & 1 & -a_{n-1} \end{pmatrix}$. Show that

$$p_A(t) = t^n + a_{n-1} t^{n-1} + \cdots + a_1 t + a_0 = m_A(t).$$

This matrix is called the *companion matrix* of the polynomial $p(t) = p_A(t)$. Thus a companion matrix is nonderogatory.

Exercise 4.10.21 For the following pairs of matrices A and B, find a polynomial $p(t)$ so that $p(A) = B$, or show that it is impossible.

(a) $A = \begin{pmatrix} 1 & 1 & 0 \\ 0 & 1 & 1 \\ 0 & 0 & 1 \end{pmatrix}, B = \begin{pmatrix} 1 & 2 & 3 \\ 0 & 2 & 3 \\ 0 & 0 & 3 \end{pmatrix}$.

(b) $A = \begin{pmatrix} 1 & 1 & 0 \\ 0 & 1 & 1 \\ 0 & 0 & 1 \end{pmatrix}, B = \begin{pmatrix} 1 & 2 & 3 \\ 0 & 1 & 2 \\ 0 & 0 & 1 \end{pmatrix}$.

Exercise 4.10.22 Solve the system of differential equations

$$x'(t) = Ax(t), \quad x(0) = \begin{pmatrix} 1 \\ -1 \\ 0 \end{pmatrix},$$

where

$$A = \begin{pmatrix} 1 & -1 & 1 \\ 0 & 1 & -1 \\ 0 & 1 & 0 \end{pmatrix} \begin{pmatrix} 2 & 1 & 0 \\ 0 & 2 & 1 \\ 0 & 0 & 2 \end{pmatrix} \begin{pmatrix} 1 & -1 & 1 \\ 0 & 1 & -1 \\ 0 & 1 & 0 \end{pmatrix}^{-1}.$$

Exercise 4.10.23 Solve the following systems of linear differential equations:

(a)

$$\begin{cases} x_1'(t) = 3x_1(t) - x_2(t), & x_1(0) = 1, \\ x_2'(t) = x_1(t) + x_2(t), & x_2(0) = 2. \end{cases}$$

(b)

$$\begin{cases} x_1'(t) = 3x_1(t) + x_2(t) + x_3(t), & x_1(0) = 1, \\ x_2'(t) = 2x_1(t) + 4x_2(t) + 2x_3(t), & x_2(0) = -1, \\ x_3'(t) = -x_1(t) - x_2(t) + x_3(t), & x_3(0) = 1. \end{cases}$$

(c)

$$\begin{cases} x_1'(t) = -x_2(t), & x_1(0) = 1, \\ x_2'(t) = x_1(t), & x_2(0) = 2. \end{cases}$$

(d)

$$x''(t) - 6x'(t) + 9x(t) = 0, \quad x(0) = 2, x'(0) = 1.$$

(e)

$$x''(t) - 4x'(t) + 4x(t) = 0, \quad x(0) = 6, x'(0) = -1.$$

Exercise 4.10.24 For the following matrices, we determined their Jordan canonical form in Exercise 4.10.2.

(a) Compute $\cos A$ for

$$A = \begin{pmatrix} -1 & 1 & 0 & 0 \\ -1 & 0 & 1 & 0 \\ -1 & 0 & 0 & 1 \\ -1 & 0 & 0 & 1 \end{pmatrix}.$$

(b) Compute A^{24} for

$$A = \begin{pmatrix} 0 & 1 & 0 \\ -1 & 0 & 0 \\ 1 & 1 & 1 \end{pmatrix}.$$

(c) Compute e^A for

$$A = \begin{pmatrix} 2 & 0 & -1 & 1 \\ 0 & 1 & 0 & 0 \\ 1 & 0 & 0 & 0 \\ 0 & 0 & 0 & 1 \end{pmatrix}.$$

Exercise 4.10.25 (a) Find matrices $A, B \in \mathbb{C}^{n \times n}$ so that $e^A e^B \neq e^{A+B}$.

(b) When $AB = BA$, then $e^A e^B = e^{A+B}$. Prove this statement when A is nonderogatory.

Exercise 4.10.26 Compute the matrices P_{20}, P_{21}, P_{22} from Example 4.8.5.

Exercise 4.10.27 (a) Show that if $A = A^*$, then e^A is positive definite.

(b) If e^A is positive definite, is A necessarily Hermitian?

(c) What can you say about e^A when A is skew-Hermitian?

Exercise 4.10.28 Let $A = \begin{pmatrix} \frac{\pi}{2} & 1 & -1 \\ 0 & \frac{\pi}{2} & -\frac{\pi}{4} \\ 0 & 0 & \frac{\pi}{4} \end{pmatrix}$.

(a) Compute $\cos A$ and $\sin A$.

(b) Check that $(\cos A)^2 + (\sin A)^2 = I$.

Exercise 4.10.29 Show that for $A \in \mathbb{C}^{4 \times 4}$, one has that

$$\sin 2A = 2 \sin A \cos A.$$

Exercise 4.10.30 Solve the inhomogeneous system of differential equations

$$\begin{cases} x_1'(t) = x_1(t) + 2x_2(t) + e^{-2t}, \\ x_2'(t) = 4x_1(t) - x_2(t). \end{cases}$$

Exercise 4.10.31 With the notation of Section 4.9 show that

$$I = \frac{1}{2\pi i} \int_\gamma R(\lambda)d\lambda, \quad A = \frac{1}{2\pi i} \int_\gamma \lambda R(\lambda)d\lambda.$$

Exercise 4.10.32 Show that the resolvent satisfies

(a) $\frac{R(\lambda) - R(\mu)}{\lambda - \mu} = -R(\lambda)R(\mu)$.

(b) $\frac{dR(\lambda)}{d\lambda} = -R(\lambda)^2$.

(c) $\frac{d^j R(\lambda)}{d\lambda^j} = (-1)^j j! R(\lambda)^{j+1}$.

Exercise 4.10.33 With the notation of Theorem 4.9.3, show that

$$\lambda_j P_{j0} + P_{j1} = A P_{j0} = \frac{1}{2\pi i} \int_{\gamma_k} \lambda R(\lambda) d\lambda.$$

Exercise 4.10.34 *(Honors)* In this exercise we develop the real Jordan canonical form. Let $A \in \mathbb{R}^{n \times n}$.

(a) Show that if $\lambda \in \mathbb{C}$ is an eigenvalue of A, then so is $\bar{\lambda}$.

(b) Show that if $A\mathbf{x} = \lambda\mathbf{x}$ with $\lambda = a + ib \in \mathbb{C} \setminus \mathbb{R}$, then $A\bar{\mathbf{x}} = \bar{\lambda}\bar{\mathbf{x}}$,

$$A \operatorname{Re}\mathbf{x} = a \operatorname{Re}\mathbf{x} - b \operatorname{Im}\mathbf{x} \text{ and } A \operatorname{Im}\mathbf{x} = b \operatorname{Re}\mathbf{x} + a \operatorname{Im}\mathbf{x}.$$

Here $\bar{\mathbf{x}}$ is the vector obtained from $\mathbf{x}$ by taking the complex conjugate of each entry, $\operatorname{Re}\mathbf{x}$ is the vector obtained from $\mathbf{x}$ by taking the real part of each entry, $\operatorname{Im}\mathbf{x}$ is the vector obtained from $\mathbf{x}$ by taking the imaginary part of each entry.

(c) Show that for all $\lambda \in \mathbb{C}$, we have that $w_k(A, \lambda) = w_k(A, \bar{\lambda})$, $k \in \mathbb{N}$.

(d) Show that $J_k(\lambda) \oplus J_k(\bar{\lambda})$, where $\lambda = a + ib$, is similar to the $2k \times 2k$ matrix

$$K_k(a, b) := \begin{pmatrix} C(a,b) & I_2 & 0 & \cdots & 0 \\ 0 & C(a,b) & I_2 & \cdots & 0 \\ \vdots & & \ddots & \ddots & \vdots \\ 0 & 0 & \cdots & C(a,b) & I_2 \\ 0 & 0 & \cdots & 0 & C(a,b) \end{pmatrix},$$

where $C(a, b) = \begin{pmatrix} a & -b \\ b & a \end{pmatrix}$.

(e) Show that if $A \in \mathbb{R}^{n \times n}$, then there exists a real invertible matrix S and a matrix K so that $A = SKS^{-1}$, where K is a block diagonal matrix with blocks $J_k(\lambda)$, $\lambda \in \mathbb{R}$, and blocks $K_k(a, b)$ on the diagonal.

(Hint: First find the Jordan canonical form of A over $\mathbb{C}$, where for complex eigenvalues the (generalized) eigenvectors $\mathbf{x}$ and $\bar{\mathbf{x}}$ are paired up. Then use the similarity in (d) to simultaneously convert P to a real matrix S and J to the matrix K.)

(f) Show that for systems of real differential equations with real initial conditions, the solutions are combinations of functions $t^k e^{\lambda t}$, $k \in \mathbb{N}_0$, $\lambda \in \mathbb{R}$, and $t^k e^{\alpha t} \cos(\beta t), t^k e^{\alpha t} \sin(\beta t)$, $k \in \mathbb{N}_0$, $\alpha, \beta \in \mathbb{R}$.

Exercise 4.10.35 *(Honors)* Show that the function $f : \mathbb{C}^{2\times 2} \times \mathbb{C}^{2\times 2} \to \mathbb{C}^5$ defined by

$$f(A, B) = (\operatorname{tr}A, \det A, \operatorname{tr}B, \det B, \operatorname{tr}(AB))$$

is surjective. What happens for other fields? (Hint: Notice that a 2×2 matrix A has a single eigenvalue if and only if $(\operatorname{tr}A)^2 = 4\det A$. To show that (a, b, c, d, e) lies in the range of f, first consider the case when $a^2 \neq 4b$, so that A has two different eigenvalues.)

This exercise is based on a result that can be found in Section 1.2 of the book by L. Le Bruyn, entitled *Noncommutative geometry and Cayley-smooth orders*, Volume 290 of Pure and Applied Mathematics, Chapman & Hall/CRC, Boca Raton, FL. Thanks are due to Paul Muhly for making the author aware of this result.

5

Inner Product and Normed Vector Spaces

CONTENTS

5.1   Inner products and norms 109
5.2   Orthogonal and orthonormal sets and bases 119
5.3   The adjoint of a linear map 122
5.4   Unitary matrices, QR, and Schur triangularization 125
5.5   Normal and Hermitian matrices 128
5.6   Singular value decomposition 132
5.7   Exercises 137

Vector spaces may have additional structure. For instance, there may be a natural notion of length of a vector and/or angle between vectors. The properties of length and angle will be formally captured in the notions of *norm* and *inner product*. These notions require us to restrict ourselves to vector spaces over $\mathbb{R}$ or $\mathbb{C}$. Indeed, a length is always nonnegative and thus we will need the inequalities $\leq, \geq, <, >$ (with properties like $x, y \geq 0 \Rightarrow xy \geq 0$ and $x \geq y \Rightarrow x + z \geq y + z$).

5.1   Inner products and norms

Let $\mathbb{F}$ be $\mathbb{R}$ or $\mathbb{C}$. We will write most results for the choice $\mathbb{F} = \mathbb{C}$. To interpret these results for the choice $\mathbb{F} = \mathbb{R}$, one simply ignores the complex conjugates that are part of the definitions.

Let V be a vector space over $\mathbb{F}$. A function

$$\langle \cdot, \cdot \rangle : V \times V \to \mathbb{F}$$

is called a *Hermitian form* if

(i) $\langle \mathbf{x} + \mathbf{y}, \mathbf{z} \rangle = \langle \mathbf{x}, \mathbf{z} \rangle + \langle \mathbf{y}, \mathbf{z} \rangle$ for all $\mathbf{x}, \mathbf{y}, \mathbf{z} \in V$.

(ii) $\langle a\mathbf{x}, \mathbf{y} \rangle = a\langle \mathbf{x}, \mathbf{y} \rangle$ for all $\mathbf{x}, \mathbf{y} \in V$ and all $a \in \mathbb{F}$.

(iii) $\langle \mathbf{x}, \mathbf{y} \rangle = \overline{\langle \mathbf{y}, \mathbf{x} \rangle}$, for all $\mathbf{x}, \mathbf{y} \in V$.

Notice that (iii) implies that $\langle \mathbf{x}, \mathbf{x} \rangle \in \mathbb{R}$ for all $\mathbf{x} \in V$. In addition, (ii) implies that $\langle \mathbf{0}, \mathbf{y} \rangle = 0$ for all $\mathbf{y} \in V$. Also, (ii) and (iii) imply that $\langle \mathbf{x}, a\mathbf{y} \rangle = \overline{a}\langle \mathbf{x}, \mathbf{y} \rangle$ for all $\mathbf{x}, \mathbf{y} \in V$ and all $a \in \mathbb{F}$. Finally, (i) and (iii) imply that $\langle \mathbf{x}, \mathbf{y} + \mathbf{z} \rangle = \langle \mathbf{x}, \mathbf{y} \rangle + \langle \mathbf{x}, \mathbf{z} \rangle$ for all $\mathbf{x}, \mathbf{y}, \mathbf{z} \in V$. Also,

The Hermitian form $\langle \cdot, \cdot \rangle$ is called an *inner product* if in addition

(iv) $\langle \mathbf{x}, \mathbf{x} \rangle > 0$ for all $\mathbf{0} \neq \mathbf{x} \in V$.

If V has an inner product $\langle \cdot, \cdot \rangle$ (or sometimes we say "V is endowed with the inner product $\langle \cdot, \cdot \rangle$"), then we call the pair $(V, \langle \cdot, \cdot \rangle)$ an *inner product space*. At times we do not explicitly mention the inner product $\langle \cdot, \cdot \rangle$, and we refer to V as an inner product space. In the latter case it is implicitly understood what the underlying inner product is, and typically it would be one of the standard inner products which we will encounter below.

Example 5.1.1 On $\mathbb{F}^n$ define

$$\langle \begin{pmatrix} x_1 \\ \vdots \\ x_n \end{pmatrix}, \begin{pmatrix} y_1 \\ \vdots \\ y_n \end{pmatrix} \rangle = x_1\overline{y}_1 + \cdots + x_n\overline{y}_n,$$

or in shorthand notation $\langle \mathbf{x}, \mathbf{y} \rangle = \mathbf{y}^*\mathbf{x}$, where $\mathbf{y}^* = \begin{pmatrix} \overline{y}_1 & \cdots & \overline{y}_n \end{pmatrix}$. Properties (i)–(iv) are easily checked. This is the *standard inner product* or *Euclidean inner product* on $\mathbb{F}^n$, where $\mathbb{F} = \mathbb{R}$ or $\mathbb{C}$.

Example 5.1.2 On $\mathbb{F}^2$ define

$$\langle \begin{pmatrix} x_1 \\ x_2 \end{pmatrix}, \begin{pmatrix} y_1 \\ y_2 \end{pmatrix} \rangle = 2x_1\overline{y}_1 + x_1\overline{y}_2 + x_2\overline{y}_1 + 3x_2\overline{y}_2.$$

Properties (i)–(iii) are easily checked. For (iv) observe that $\langle \mathbf{x}, \mathbf{x} \rangle = |x_1|^2 + |x_1 + x_2|^2 + 2|x_2|^2$, so as soon as $x_1 \neq 0$ or $x_2 \neq 0$, we have that $\langle \mathbf{x}, \mathbf{x} \rangle > 0$.

Example 5.1.3 On $\mathbb{F}^2$ define

$$\langle \begin{pmatrix} x_1 \\ x_2 \end{pmatrix}, \begin{pmatrix} y_1 \\ y_2 \end{pmatrix} \rangle = x_1\overline{y}_1 + 2x_1\overline{y}_2 + 2x_2\overline{y}_1 + x_2\overline{y}_2.$$

Properties (i)–(iii) are easily checked, so it is a Hermitian form. In order to check (iv) observe that $\langle \mathbf{x}, \mathbf{x} \rangle = -|x_1|^2 + 2|x_1 + x_2|^2 - |x_2|^2$. So for instance

$$\left\langle \begin{pmatrix} 1 \\ -1 \end{pmatrix}, \begin{pmatrix} 1 \\ -1 \end{pmatrix} \right\rangle = -1 + 0 - 1 = -2,$$

so $\langle \cdot, \cdot \rangle$ is not an inner product.

Example 5.1.4 Let $V = \{f : [0,1] \to \mathbb{F} \ : \ f \text{ is continuous}\}$, and

$$\langle f, g \rangle = \int_0^1 f(x)\overline{g(x)}dx.$$

Properties (i)–(iii) are easily checked. For (iv) notice that $\langle f, f \rangle = \int_0^1 |f(x)|^2 dx \geq 0$, and as soon as $f(x)$ is not the zero function, by continuity $|f(x)|^2$ is positive on an interval (a, b), where $0 < a < b < 1$, so that $\langle f, f \rangle > 0$. This is the *standard inner product* on V.

Example 5.1.5 Let $V = \{f : [0,1] \to \mathbb{F} \ : \ f \text{ is continuous}\}$, and

$$\langle f, g \rangle = \int_0^1 (x^2 + 1)f(x)\overline{g(x)}dx.$$

Properties (i)–(iii) are easily checked. For (iv) notice that $\langle f, f \rangle = \int_0^1 (x^2 + 1)|f(x)|^2 dx \geq 0$, and as soon as $f(x)$ is not the zero function, by continuity $(x^2 + 1)|f(x)|^2$ is positive on an interval (a, b), where $0 < a < b < 1$, so that $\langle f, f \rangle > 0$.

Example 5.1.6 On $\mathbb{F}_n[X]$ define

$$\langle p(X), q(X) \rangle = p(x_1)\overline{q(x_1)} + \cdots + p(x_{n+1})\overline{q(x_{n+1})},$$

where $x_1, \ldots, x_{n+1} \in \mathbb{F}$ are different points chosen in advance. Properties (i)–(iii) are easily checked. For (iv) observe that $\langle p(X), p(X) \rangle = \sum_{j=1}^{n+1} |p(x_j)|^2 \geq 0$, and that $\langle p(X), p(X) \rangle = 0$ if and only if $p(x_1) = \cdots = p(x_{n+1}) = 0$. As a polynomial of degree $\leq n$ with $n + 1$ different roots must be the zero polynomial, we get that as soon as $p(X)$ is not the zero polynomial, then $\langle p(X), p(X) \rangle > 0$. Thus (iv) holds.

Example 5.1.7 On $\mathbb{F}_3[X]$ define

$$\langle p(X), q(X) \rangle = p(0)\overline{q(0)} + p(1)\overline{q(1)} + p(2)\overline{q(2)}.$$

Properties (i)–(iii) are easily checked. However, (iv) does not hold. If we let $p(X) = X(X - 1)(X - 2) \in \mathbb{F}_3[X]$, then $p(X)$ is not the zero polynomial, but $\langle p(X), p(X) \rangle = 0$.

Example 5.1.8 On $\mathbb{F}^{n \times m}$ define

$$\langle A, B \rangle = \mathrm{tr}(AB^*).$$

Properties (i)–(iii) are easily checked. For (iv) we observe that if $A = (a_{ij})_{i=1, j=1}^{n, m}$, then $\langle A, A \rangle = \sum_{i=1}^{n} \sum_{j=1}^{m} |a_{ij}|^2$, which is strictly positive as soon as $A \neq 0$. This is the *standard inner product* on $\mathbb{F}^{n \times m}$, where $\mathbb{F} = \mathbb{R}$ or $\mathbb{C}$.

Example 5.1.9 On $\mathbb{F}^{2 \times 2}$ define

$$\langle A, B \rangle = \mathrm{tr}(AWB^*),$$

where $W = \begin{pmatrix} 1 & 1 \\ 1 & 2 \end{pmatrix}$. Properties (i)–(iii) are easily checked. For (iv) we observe that if $A = (a_{ij})_{i,j=1}^{2}$, then

$$\langle A, A \rangle = (a_{11} + a_{12})\overline{a_{11}} + (a_{11} + 2a_{12})\overline{a_{12}} + (a_{21} + a_{22})\overline{a_{21}} + (a_{21} + 2a_{22})\overline{a_{22}} =$$

$$|a_{11} + a_{12}|^2 + |a_{12}|^2 + |a_{21} + a_{22}|^2 + |a_{22}|^2,$$

which is always nonnegative, and which can only equal zero when

$$a_{11} + a_{12} = 0, \ a_{12} = 0, \ a_{21} + a_{22} = 0, \ a_{22} = 0,$$

that is, when $A = 0$. Thus as soon as $A \neq 0$ we have $\langle A, A \rangle > 0$. Thus (iv) holds, and therefore $\langle \cdot, \cdot \rangle$ is an inner product.

Proposition 5.1.10 *(Cauchy–Schwarz inequality) For an inner product space $(V, \langle \cdot, \cdot \rangle)$, we have that*

$$|\langle \mathbf{x}, \mathbf{y} \rangle|^2 \leq \langle \mathbf{x}, \mathbf{x} \rangle \langle \mathbf{y}, \mathbf{y} \rangle \quad \text{for all } \mathbf{x}, \mathbf{y} \in V. \tag{5.1}$$

Moreover, equality in (5.1) holds if and only if $\{\mathbf{x}, \mathbf{y}\}$ is linearly dependent.

Proof. When $\mathbf{x} = \mathbf{0}$, inequality (5.1) clearly holds. Next, suppose that $\mathbf{x} \neq \mathbf{0}$. Put $\mathbf{z} = \mathbf{y} - \frac{\langle \mathbf{y}, \mathbf{x} \rangle}{\langle \mathbf{x}, \mathbf{x} \rangle} \mathbf{x}$. As $\langle \cdot, \cdot \rangle$ is an inner product, we have that $\langle \mathbf{z}, \mathbf{z} \rangle \geq 0$. This gives that

$$0 \leq \langle \mathbf{y}, \mathbf{y} \rangle - 2 \frac{|\langle \mathbf{y}, \mathbf{x} \rangle|^2}{\langle \mathbf{x}, \mathbf{x} \rangle} + \frac{|\langle \mathbf{y}, \mathbf{x} \rangle|^2}{\langle \mathbf{x}, \mathbf{x} \rangle^2} \langle \mathbf{x}, \mathbf{x} \rangle = \langle \mathbf{y}, \mathbf{y} \rangle - \frac{|\langle \mathbf{y}, \mathbf{x} \rangle|^2}{\langle \mathbf{x}, \mathbf{x} \rangle}.$$

But now (5.1) follows.

If $\{\mathbf{x}, \mathbf{y}\}$ is linearly dependent, it is easy to check that equality in (5.1) holds (as $\mathbf{x} = \mathbf{0}$ or $\mathbf{y}$ is a multiple of $\mathbf{x}$). Conversely, suppose that equality holds in (5.1). If $\mathbf{x} = \mathbf{0}$, then clearly $\{\mathbf{x}, \mathbf{y}\}$ is linearly dependent. Next, let us suppose that $\mathbf{x} \neq \mathbf{0}$. As before, put $\mathbf{z} = \mathbf{y} - \frac{\langle \mathbf{y}, \mathbf{x} \rangle}{\langle \mathbf{x}, \mathbf{x} \rangle} \mathbf{x}$. Using equality in (5.1) one computes directly that $\langle \mathbf{z}, \mathbf{z} \rangle = 0$. Thus $\mathbf{z} = \mathbf{0}$, showing that $\{\mathbf{x}, \mathbf{y}\}$ is linearly dependent. $\square$

Remark 5.1.11 Notice that in the first paragraph of the proof of Proposition 5.1.10 we did not use the full strength of property (iv) of an inner product. Indeed, we only needed to use

(v) $\langle \mathbf{x}, \mathbf{x} \rangle \geq 0$ for all $\mathbf{x} \in V$.

In Section 6.2 we will encounter so-called pre-inner products that only satisfy (i)–(iii) and (v). We will use in the proof of Proposition 6.2.12 that in such case the Cauchy–Schwarz inequality (5.1) still holds.

Next we define the notion of a norm. Let V be a vector space over $\mathbb{F} = \mathbb{R}$ or $\mathbb{F} = \mathbb{C}$. A function

$$\| \cdot \| : V \to \mathbb{R}$$

is called a *norm* if

(i) $\|\mathbf{x}\| \geq 0$ for all $\mathbf{x} \in V$, and $\|\mathbf{x}\| = 0$ if and only if $\mathbf{x} = \mathbf{0}$.

(ii) $\|c\mathbf{x}\| = |c| \|\mathbf{x}\|$ for all $\mathbf{x} \in V$ and $c \in \mathbb{F}$.

(iii) $\|\mathbf{x} + \mathbf{y}\| \leq \|\mathbf{x}\| + \|\mathbf{y}\|$ for all $\mathbf{x}, \mathbf{y} \in V$. *(Triangle inequality.)*

Every norm satisfies the following inequality.

Lemma 5.1.12 *Let V be a vector space with norm $\| \cdot \|$. Then for every $\mathbf{x}, \mathbf{y} \in V$ we have*

$$| \|\mathbf{x}\| - \|\mathbf{y}\| | \leq \|\mathbf{x} - \mathbf{y}\|. \tag{5.2}$$

Proof. Note that the triangle inequality implies

$$\|\mathbf{x}\| = \|\mathbf{x} - \mathbf{y} + \mathbf{y}\| \leq \|\mathbf{x} - \mathbf{y}\| + \|\mathbf{y}\|,$$

and thus

$$\|\mathbf{x}\| - \|\mathbf{y}\| \leq \|\mathbf{x} - \mathbf{y}\|. \tag{5.3}$$

Reversing the roles of $\mathbf{x}$ and $\mathbf{y}$, we also obtain that

$$\|\mathbf{y}\| - \|\mathbf{x}\| \leq \|\mathbf{y} - \mathbf{x}\| = \|\mathbf{x} - \mathbf{y}\|. \tag{5.4}$$

Combining (5.3) and (5.4) yields (5.2). $\qquad\square$

Example 5.1.13 On $\mathbb{F}^n$ define

$$\left\| \begin{pmatrix} x_1 \\ \vdots \\ x_n \end{pmatrix} \right\|_\infty = \max_{j=1,\ldots,n} |x_j|.$$

One easily checks that $\| \cdot \|_\infty$ is a norm.

Example 5.1.14 On $\mathbb{F}^n$ define

$$\left\|\begin{pmatrix} x_1 \\ \vdots \\ x_n \end{pmatrix}\right\|_1 = \sum_{j=1}^{n} |x_j|.$$

One easily checks that $\|\cdot\|_1$ is a norm.

Example 5.1.15 On $\mathbb{F}^2$ define

$$\left\|\begin{pmatrix} x_1 \\ x_2 \end{pmatrix}\right\| = 2|x_1| + 3|x_2|.$$

One easily checks that $\|\cdot\|$ is a norm.

Example 5.1.16 Let $V = \{f : [0,1] \to \mathbb{F} : f \text{ is continuous}\}$, and

$$\|f\|_\infty = \sup_{x \in [0,1]} |f(x)|.$$

One easily checks that $\|\cdot\|_\infty$ is a norm.

Example 5.1.17 On $\mathbb{F}_n[X]$ define

$$\left\|\sum_{j=0}^{n} p_j X^j\right\| = \sum_{j=0}^{n} |p_j|.$$

One easily checks that $\|\cdot\|$ is a norm.

Example 5.1.18 On $\mathbb{F}^{n \times m}$ define

$$\|(a_{ij})_{i=1,j=1}^{n,\ m}\| = \sum_{i=1}^{n} \sum_{j=1}^{m} |a_{ij}|.$$

One easily checks that $\|\cdot\|$ is a norm.

We are mostly interested in the norm associated with an inner product.

Theorem 5.1.19 *Let* $(V, \langle \cdot, \cdot \rangle)$ *be an inner product space. Define*

$$\|\mathbf{x}\| := \sqrt{\langle \mathbf{x}, \mathbf{x} \rangle}.$$

Then $\| \cdot \|$ *is a norm, which satisfies the parallelogram identity:*

$$\|\mathbf{x} + \mathbf{y}\|^2 + \|\mathbf{x} - \mathbf{y}\|^2 = 2\|\mathbf{x}\|^2 + \|\mathbf{y}\|^2 \quad \text{for all } \mathbf{x}, \mathbf{y} \in V. \tag{5.5}$$

Moreover,

$$\|\mathbf{x}_1 + \cdots + \mathbf{x}_n\| = \|\mathbf{x}_1\| + \cdots + \|\mathbf{x}_n\| \tag{5.6}$$

if and only if dim Span$\{\mathbf{x}_1, \ldots, \mathbf{x}_n\} \leq 1$ *and* $\langle \mathbf{x}_i, \mathbf{x}_j \rangle \geq 0$ *for all* $i, j = 1, \ldots, n$.

Proof. Conditions (i) and (ii) in the definition of a norm follow directly from the definition of an inner product. For (iii) we observe that

$$\|\mathbf{x} + \mathbf{y}\|^2 = \langle \mathbf{x} + \mathbf{y}, \mathbf{x} + \mathbf{y} \rangle = \langle \mathbf{x}, \mathbf{x} \rangle + 2\mathrm{Re}\,\langle \mathbf{x}, \mathbf{y} \rangle + \langle \mathbf{y}, \mathbf{y} \rangle \leq$$

$$\langle \mathbf{x}, \mathbf{x} \rangle + 2|\langle \mathbf{x}, \mathbf{y} \rangle| + \langle \mathbf{y}, \mathbf{y} \rangle \leq \langle \mathbf{x}, \mathbf{x} \rangle + 2\sqrt{\langle \mathbf{x}, \mathbf{x} \rangle}\sqrt{\langle \mathbf{y}, \mathbf{y} \rangle} + \langle \mathbf{y}, \mathbf{y} \rangle = (\|\mathbf{x}\| + \|\mathbf{y}\|)^2, \tag{5.7}$$

where we used the Cauchy–Schwarz inequality (5.1). Taking square roots on both sides proves (iii). Notice that if $\|\mathbf{x} + \mathbf{y}\| = \|\mathbf{x}\| + \|\mathbf{y}\|$, then we must have equality in (5.7). This then gives Re $\langle \mathbf{x}, \mathbf{y} \rangle = |\langle \mathbf{x}, \mathbf{y} \rangle| = \sqrt{\langle \mathbf{x}, \mathbf{x} \rangle}\sqrt{\langle \mathbf{y}, \mathbf{y} \rangle}$. In particular, we have equality in the Cauchy–Schwarz inequality, which by Proposition 5.1.10 implies that $\{\mathbf{x}, \mathbf{y}\}$ is linearly dependent. Moreover, Re $\langle \mathbf{x}, \mathbf{y} \rangle = |\langle \mathbf{x}, \mathbf{y} \rangle|$ yields that $\langle \mathbf{x}, \mathbf{y} \rangle \geq 0$. If (5.6) holds, we obtain

$$\|\mathbf{x}_1 + \cdots + \mathbf{x}_n\| = \|\mathbf{x}_1\| + \|\mathbf{x}_2 + \cdots + \mathbf{x}_n\| = \cdots =$$

$$\|\mathbf{x}_1\| + \cdots + \|\mathbf{x}_{n-2}\| + \|\mathbf{x}_{n-1} + \mathbf{x}_n\| = \|\mathbf{x}_1\| + \cdots + \|\mathbf{x}_n\|.$$

This gives that

$$\{\mathbf{x}_{n-1}, \mathbf{x}_n\}, \{\mathbf{x}_{n-2}, \mathbf{x}_{n-1} + \mathbf{x}_n\}, \ldots, \{\mathbf{x}_1, \mathbf{x}_2 + \cdots + \mathbf{x}_n\}$$

are all linearly dependent, which easily yields that dim Span$\{\mathbf{x}_1, \ldots, \mathbf{x}_n\} \leq 1$. In addition, we get that

$$\langle \mathbf{x}_{n-1}, \mathbf{x}_n \rangle \geq 0, \langle \mathbf{x}_{n-2}, \mathbf{x}_{n-1} + \mathbf{x}_n \rangle \geq 0, \ldots, \langle \mathbf{x}_1, \mathbf{x}_2 + \cdots + \mathbf{x}_n \rangle \geq 0.$$

Combining this with dim Span$\{\mathbf{x}_1, \ldots, \mathbf{x}_n\} \leq 1$ it is easy to deduce that $\langle \mathbf{x}_i, \mathbf{x}_j \rangle \geq 0$ for all $i, j = 1, \ldots, n$. The converse statement is straightforward.

To prove the parallelogram identity (5.5), one simply expands $\langle \mathbf{x} \pm \mathbf{y}, \mathbf{x} \pm \mathbf{y} \rangle$, and it follows immediately. $\qquad \square$

It is easy to see that the norm in Examples 5.1.13–5.1.18 do not satisfy the parallelogram identity (5.5), and thus these norms are not associated with an inner product.

Example 5.1.20 On $\mathbb{F}^n$ define

$$\left\| \begin{pmatrix} x_1 \\ \vdots \\ x_n \end{pmatrix} \right\|_2 = \sqrt{\sum_{j=1}^{n} |x_j|^2}.$$

This norm, sometimes referred to as the *Euclidean norm*, is the norm associated with the standard inner product on $\mathbb{F}^n$, where $\mathbb{F} = \mathbb{R}$ or $\mathbb{C}$.

Corollary 5.1.21 *Let $z_1, \ldots, z_n \in \mathbb{C}$. Then*

$$|z_1 + \cdots + z_n| = |z_1| + \cdots + |z_n|$$

if and only if there exists a $\theta \in \mathbb{R}$ so that $z_j = |z_j|e^{i\theta}$, $j = 1, \ldots, n$ (i.e., $z_1, \ldots, z_n$ all have the same argument).

Proof. If we view a complex number z as a vector $\begin{pmatrix} \mathrm{Re}\ z \\ \mathrm{Im}\ z \end{pmatrix} \in \mathbb{R}^2$ with the Euclidean norm, then $|z| = \left\| \begin{pmatrix} \mathrm{Re}\ z \\ \mathrm{Im}\ z \end{pmatrix} \right\|$. Apply now Theorem 5.1.19 to obtain the result. $\qquad\qquad\square$

Example 5.1.22 Let $V = \{f : [0,1] \to \mathbb{F} : f \text{ is continuous}\}$, and define

$$\|f\|_2 = \sqrt{\int_0^1 |f(x)|^2 dx}.$$

This "2-norm" on V is associated with the standard inner product on V.

Example 5.1.23 On $\mathbb{F}_n[X]$ define

$$\|p(X)\| = \sqrt{\sum_{j=1}^{n+1} |p(x_j)|^2},$$

where $x_1, \ldots, x_{n+1} \in \mathbb{F}$ are different points. This is the norm associated with the inner product defined in Example 5.1.6.

Example 5.1.24 On $\mathbb{F}^{n \times m}$ define

$$\|A\|_F = \sqrt{\mathrm{tr}(AA^*)}.$$

This norm is called the *Frobenius norm*, and is the norm associated with the inner product defined in Example 5.1.8.

Given a vector space V, we say that two norms $\|\cdot\|_a$ and $\|\cdot\|_b$ are *equivalent* if there exist constants $c, C > 0$ so that

$$c\|\mathbf{v}\|_a \leq \|\mathbf{v}\|_b \leq C\|\mathbf{v}\|_a \text{ for all } \mathbf{v} \in V. \tag{5.8}$$

Notice that if $\|\cdot\|_a$ and $\|\cdot\|_b$ are equivalent and $\|\cdot\|_b$ and $\|\cdot\|_c$ are equivalent, then $\|\cdot\|_a$ and $\|\cdot\|_c$ are equivalent.

Using the Heine–Borel theorem from analysis, along with the result that a continuous real-valued function defined on a compact set attains a maximum and a minimum, we can prove the following.

Theorem 5.1.25 *Let V be a finite-dimensional vector space over $\mathbb{F} = \mathbb{R}$ or $\mathbb{C}$, and let $\|\cdot\|_a$ and $\|\cdot\|_b$ be two norms. Then $\|\cdot\|_a$ and $\|\cdot\|_b$ are equivalent.*

Proof. Let $\mathcal{B} = \{\mathbf{b}_1, \ldots, \mathbf{b}_n\}$ be a basis for V, and define the norm

$$\|\mathbf{v}\|_c := \|[\mathbf{v}]_\mathcal{B}\|_\infty,$$

where $\|\cdot\|_\infty$ is as in Example 5.1.13. We will show that any other norm $\|\cdot\|$ on V is equivalent to $\|\cdot\|$. This will yield the result.

We first claim that $\|\cdot\| : V \to \mathbb{R}$ is a continuous function, where distance in V is measured using $\|\cdot\|_c$. In other words, we claim that for every $\epsilon > 0$ there exists a $\delta > 0$ so that

$$\|\mathbf{x} - \mathbf{y}\|_c \leq \delta \text{ implies } |\, \|\mathbf{x}\| - \|\mathbf{y}\| \,| \leq \epsilon.$$

Indeed, if $\epsilon > 0$ is given, we choose $\delta = \frac{\epsilon}{\sum_{i=1}^n \|\mathbf{b}_i\|}$. Let us write

$$[\mathbf{x}]_\mathcal{B} = \begin{pmatrix} x_1 \\ \vdots \\ x_n \end{pmatrix}, [\mathbf{y}]_\mathcal{B} = \begin{pmatrix} y_1 \\ \vdots \\ y_n \end{pmatrix}.$$

Then we get that

$$\|\mathbf{x} - \mathbf{y}\|_c = \max_{i=1,\ldots,n} |x_i - y_i| < \delta = \frac{\epsilon}{\sum_{i=1}^n \|\mathbf{b}_i\|}$$

yields that

$$|\, \|\mathbf{x}\| - \|\mathbf{y}\| \,| \leq \|\mathbf{x} - \mathbf{y}\| = \|\sum_{i=1}^n (x_i - y_i)\mathbf{b}_i\| \leq \sum_{i=1}^n |x_i - y_i|\|\mathbf{b}_i\| \leq$$

$$\left(\max_{i=1,\ldots,n} |x_i - y_i|\right) \sum_{i=1}^n \|\mathbf{b}_i\| < \frac{\epsilon}{\sum_{i=1}^n \|\mathbf{b}_i\|} \sum_{i=1}^n \|\mathbf{b}_i\| = \epsilon.$$

Consider now the set S of $\|\cdot\|_c$-unit vectors in V; thus $S = \{\mathbf{v} \in V : \|\mathbf{v}\|_c = 1\}$. By the Heine–Borel theorem (identifying V with $\mathbb{F}^n$) this set S is compact, as S is closed and bounded. As $\|\cdot\|$ is a real-valued continuous function on this set, we have that

$$c := \min_{\mathbf{v} \in S} \|\mathbf{v}\|, \, C = \max_{\mathbf{v} \in S} \|\mathbf{v}\|$$

exist, and as $\|\mathbf{v}\| > 0$ for all $\mathbf{v} \in S$, we get that $c, C > 0$. Take now an arbitrary nonzero $\mathbf{v} \in V$. Then $\frac{1}{\|\mathbf{v}\|_c} \mathbf{v} \in S$, and thus

$$c \le \|\frac{1}{\|\mathbf{v}\|_c}\mathbf{v}\| \le C,$$

which implies

$$c\|\mathbf{v}\|_c \le \|\mathbf{v}\| \le C\|\mathbf{v}\|_c.$$

Clearly, this inequality also holds for $\mathbf{v} = \mathbf{0}$, and thus the proof is complete.
□

Comparing, for instance, the norms $\|\cdot\|_\infty$ and $\|\cdot\|_2$ on $\mathbb{F}^n$, we have

$$\|\mathbf{x}\|_\infty \le \|\mathbf{x}\|_2 \le \sqrt{n}\|\mathbf{x}\|_\infty.$$

Notice that the upper bound (which is attained by the vector of all 1's) depends on the dimension n, and tends to ∞ as n goes to ∞. Therefore, one may expect Theorem 5.1.25 not to hold for infinite-dimensional vector spaces. This is confirmed by the following example.

Example 5.1.26 Let $V = \{f : [0,1] \to \mathbb{R} : f \text{ is continuous}\}$, and take the norms

$$\|f\|_2 = \sqrt{\int_0^1 |f(x)|^2 dx}, \, \|f\|_\infty = \max_{x \in [0,1]} |f(x)|.$$

Let $g_k \in V$, $k \in \mathbb{N}$, be defined by

$$g_k(x) = \begin{cases} 1 - kx, & \text{for } 0 \le x \le \frac{1}{k}, \\ 0, & \text{for } \frac{1}{k} < x \le 1. \end{cases}$$

Then

$$\|g_k\|_\infty = 1, \, \|g_k\|_2^2 = \int_0^{\frac{1}{k}} 1 - kx \, dx = \frac{1}{k} - \frac{k}{2}\frac{1}{k^2} = \frac{1}{2k}.$$

No constant $C > 0$ exists so that $1 \le C\sqrt{\frac{1}{2k}}$ for all $k \in \mathbb{N}$, and thus the norms $\|\cdot\|_2$ and $\|\cdot\|_\infty$ on V are not equivalent.

5.2 Orthogonal and orthonormal sets and bases

When a vector space has an inner product, it is natural to study objects that behave nicely with respect to the inner product. For bases this leads to the notions of *orthogonality* and *orthonormality*.

Given is an inner product space $(V, \langle \cdot, \cdot \rangle)$. When in an inner product space a norm $\| \cdot \|$ is used, then this norm is by default the associated norm $\| \cdot \| = \sqrt{\langle \cdot, \cdot \rangle}$ unless stated otherwise. We say that $\mathbf{v}$ and $\mathbf{w}$ are *orthogonal* if $\langle \mathbf{v}, \mathbf{w} \rangle = 0$, and we will denote this as $\mathbf{v} \perp \mathbf{w}$. Notice that $\mathbf{0}$ is orthogonal to any vector, and it is the only vector that is orthogonal to itself.

For $\emptyset \neq W \subseteq V$ we define

$$W^{\perp} = \{\mathbf{v} \in V : \langle \mathbf{v}, \mathbf{w} \rangle = 0 \text{ for all } \mathbf{w} \in W\} = \{\mathbf{v} \in V : \mathbf{v} \perp \mathbf{w} \text{ for all } \mathbf{w} \in W\}. \tag{5.9}$$

Notice that in this definition we do not require that W is a subspace, thus W can be any set of vectors of V.

Lemma 5.2.1 *For $\emptyset \neq W \subseteq V$ we have that $W^{\perp}$ is a subspace of V.*

Proof. Clearly $\mathbf{0} \in W^{\perp}$ as $\mathbf{0}$ is orthogonal to any vector, in particular to those in W. Next, let $\mathbf{x}, \mathbf{y} \in W^{\perp}$ and $c \in \mathbb{F}$. Then for every $\mathbf{w} \in W$ we have that $\langle \mathbf{x} + \mathbf{y}, \mathbf{w} \rangle = \langle \mathbf{x}, \mathbf{w} \rangle + \langle \mathbf{y}, \mathbf{w} \rangle = 0 + 0 = 0$, and $\langle c\mathbf{x}, \mathbf{w} \rangle = c\langle \mathbf{x}, \mathbf{w} \rangle = c0 = 0$. Thus $\mathbf{x} + \mathbf{y}, c\mathbf{x} \in W^{\perp}$, showing that $W^{\perp}$ is a subspace. $\square$

In Exercise 5.7.4 we will see that in case W is a subspace of a finite-dimensional space V, then

$$\dim W + \dim W^{\perp} = \dim V.$$

A set of vectors $\{\mathbf{v}_1, \ldots, \mathbf{v}_p\}$ is called *orthogonal* if $\mathbf{v}_i \perp \mathbf{v}_j$ for $i \neq j$. The set $\{\mathbf{v}_1, \ldots, \mathbf{v}_p\}$ is called *orthonormal* if $\mathbf{v}_i \perp \mathbf{v}_j$ for $i \neq j$ and $\|\mathbf{v}_i\| = 1$, $i = 1, \ldots, p$. For several reasons it will be convenient to work with orthogonal, or even better, orthonormal sets of vectors. We first show how any set of linearly independent vectors can be converted to an orthogonal or orthonormal set. Before we state the theorem, let us just see how it works with two vectors.

Example 5.2.2 Let $\{\mathbf{v}, \mathbf{w}\}$ be linearly independent, and let us make a new vector $\mathbf{z}$ of the form $\mathbf{z} = \mathbf{w} + c\mathbf{v}$ so that $\mathbf{z} \perp \mathbf{v}$. Thus we would like that

$$0 = \langle \mathbf{z}, \mathbf{v} \rangle = \langle \mathbf{w}, \mathbf{v} \rangle + c\langle \mathbf{v}, \mathbf{v} \rangle.$$

This is accomplished by taking

$$c = -\frac{\langle \mathbf{w}, \mathbf{v} \rangle}{\langle \mathbf{v}, \mathbf{v} \rangle}.$$

Note that we are not dividing by zero, as $\mathbf{v} \neq \mathbf{0}$. Thus, by putting

$$\mathbf{z} = \mathbf{w} - \frac{\langle \mathbf{w}, \mathbf{v} \rangle}{\langle \mathbf{v}, \mathbf{v} \rangle}\mathbf{v},$$

we obtain an orthogonal set $\{\mathbf{v}, \mathbf{z}\}$ so that $\text{Span}\{\mathbf{v}, \mathbf{z}\} = \text{Span}\{\mathbf{v}, \mathbf{w}\}$. If we want to convert it to an orthonormal set, we simply divide $\mathbf{v}$ and $\mathbf{z}$ by their respective lengths, obtaining the set $\{\frac{\mathbf{v}}{\|\mathbf{v}\|}, \frac{\mathbf{z}}{\|\mathbf{z}\|}\}$.

We can do the above process for a set of p linearly independent vectors as well. It is called the *Gram–Schmidt process*.

Theorem 5.2.3 *Let $(V, \langle \cdot, \cdot \rangle)$ be an inner product space, and let $\{\mathbf{v}_1, \ldots, \mathbf{v}_p\}$ be linearly independent. Construct $\{\mathbf{z}_1, \ldots, \mathbf{z}_p\}$ as follows:*

$$\mathbf{z}_1 = \mathbf{v}_1$$
$$\mathbf{z}_k = \mathbf{v}_k - \frac{\langle \mathbf{v}_k, \mathbf{z}_{k-1} \rangle}{\langle \mathbf{z}_{k-1}, \mathbf{z}_{k-1} \rangle}\mathbf{z}_{k-1} - \cdots - \frac{\langle \mathbf{v}_k, \mathbf{z}_1 \rangle}{\langle \mathbf{z}_1, \mathbf{z}_1 \rangle}\mathbf{z}_1, \quad k = 2, \ldots, p.$$

$$(5.10)$$

Then $\{\mathbf{z}_1, \ldots, \mathbf{z}_p\}$ is an orthogonal linearly independent set with the property that

$$\text{Span}\{\mathbf{v}_1, \ldots, \mathbf{v}_k\} = \text{Span}\{\mathbf{z}_1, \ldots, \mathbf{z}_k\} = \text{Span}\{\frac{\mathbf{z}_1}{\|\mathbf{z}_1\|}, \ldots, \frac{\mathbf{z}_k}{\|\mathbf{z}_k\|}\}, \quad k = 1, \ldots, p.$$

The set $\{\frac{\mathbf{z}_1}{\|\mathbf{z}_1\|}, \ldots, \frac{\mathbf{z}_p}{\|\mathbf{z}_p\|}\}$ is an orthonormal set.

The proof is straightforward, and left to the reader. It is important to note that none of the $\mathbf{z}_k$'s are zero, otherwise it would indicate that $\mathbf{v}_k \in \text{Span}\{\mathbf{v}_1, \ldots, \mathbf{v}_{k-1}\}$ (which is impossible due to the linear independence). If one applies the Gram–Schmidt process to a set that is not necessarily linearly independent, one may encounter a case where $\mathbf{z}_k = 0$. In that case, $\mathbf{v}_k \in \text{Span}\{\mathbf{v}_1, \ldots, \mathbf{v}_{k-1}\}$. In order to produce linearly independent $\mathbf{z}_j$'s, one would simple leave out $\mathbf{v}_k$ and $\mathbf{z}_k$, and continue with constructing $\mathbf{z}_{k+1}$ skipping over $\mathbf{v}_k$ and $\mathbf{z}_k$.

Example 5.2.4 Let $V = \mathbb{R}_2[X]$, with

$$\langle p, q \rangle = p(-1)q(-1) + p(0)q(0) + p(1)q(1).$$

Let $\{1, X, X^2\}$ be the linearly independent set. Applying Gram–Schmidt, we get

$$\mathbf{z}_1(X) = 1$$

$$\mathbf{z}_2(X) = X - \frac{\langle X, 1 \rangle}{\langle 1, 1 \rangle} 1 = X - 0 = X,$$

$$\mathbf{z}_3(X) = X^2 - \frac{\langle X^2, X \rangle}{\langle X, X \rangle} X - \frac{\langle X^2, 1 \rangle}{\langle 1, 1 \rangle} 1 = X^2 - \frac{2}{3}.$$

$$(5.11)$$

The orthonormal set would be $\{\frac{1}{\sqrt{3}}, \frac{X}{\sqrt{2}}, \frac{3}{\sqrt{6}} X^2 - \frac{2}{\sqrt{6}}\}$.

We call $\mathcal{B} = \{\mathbf{v}_1, \ldots, \mathbf{v}_n\}$ an *orthogonal/orthonormal basis* of V, if $\mathcal{B}$ is a basis and is orthogonal/orthonormal.

One of the reasons why it is easy to work with an orthonormal basis, is that it is easy to find the coordinates of a vector with respect to an orthonormal basis.

Lemma 5.2.5 *Let* $\mathcal{B} = \{\mathbf{v}_1, \ldots, \mathbf{v}_n\}$ *be an orthonormal basis of the inner product space* $(V, \langle \cdot, \cdot \rangle)$*. Let* $\mathbf{x} \in V$*. Then*

$$[\mathbf{x}]_{\mathcal{B}} = \begin{pmatrix} \langle \mathbf{x}, \mathbf{v}_1 \rangle \\ \vdots \\ \langle \mathbf{x}, \mathbf{v}_n \rangle \end{pmatrix}.$$

Proof. Let $\mathbf{x} = \sum_{i=1}^{n} c_i \mathbf{v}_i$. Then $\langle \mathbf{x}, \mathbf{v}_j \rangle = \sum_{i=1}^{n} c_i \langle \mathbf{v}_i, \mathbf{v}_j \rangle = c_j$, proving the lemma. $\qquad \square$

Proposition 5.2.6 *Let* $\mathcal{B} = \{\mathbf{v}_1, \ldots, \mathbf{v}_n\}$ *an orthonormal basis of the inner product space* $(V, \langle \cdot, \cdot \rangle_V)$*. Let* $\mathbf{x}, \mathbf{y} \in V$*, and write*

$$[\mathbf{x}]_{\mathcal{B}} = \begin{pmatrix} c_1 \\ \vdots \\ c_n \end{pmatrix}, [\mathbf{y}]_{\mathcal{B}} = \begin{pmatrix} d_1 \\ \vdots \\ d_n \end{pmatrix}.$$

Then

$$\langle \mathbf{x}, \mathbf{y} \rangle_V = c_1 \overline{d}_1 + \cdots + c_n \overline{d}_n = \langle [\mathbf{x}]_{\mathcal{B}}, [\mathbf{y}]_{\mathcal{B}} \rangle, \qquad (5.12)$$

where the last inner product is the standard (Euclidean) inner product for $\mathbb{F}^n$*.*

Proof. We have $\mathbf{x} = \sum_{i=1}^{n} c_i \mathbf{v}_i$, $\mathbf{y} = \sum_{j=1}^{n} d_j \mathbf{v}_j$, and thus

$$\langle \mathbf{x}, \mathbf{y} \rangle_V = \sum_{i=1}^{n} \sum_{j=1}^{n} c_i \bar{d}_j \langle \mathbf{v}_i, \mathbf{v}_j \rangle_V = \sum_{j=1}^{n} c_j \bar{d}_j,$$

where we used that $\langle \mathbf{v}_j, \mathbf{v}_j \rangle_V = 1$, and $\langle \mathbf{v}_i, \mathbf{v}_j \rangle_V = 0$ when $i \neq j$. $\square$

Let $(V, \langle \cdot, \cdot \rangle_V)$ and $(W, \langle \cdot, \cdot \rangle_W)$ be inner product spaces, and let $T : V \to W$ be linear. We call T an *isometry* if

$$\langle T(\mathbf{x}), T(\mathbf{y}) \rangle_W = \langle \mathbf{x}, \mathbf{y} \rangle_V \quad \text{for all } \mathbf{x}, \mathbf{y} \in V.$$

Two inner product spaces V and W are called *isometrically isomorphic* if there exists an isomorphism $T : V \to W$ that is also isometric.

Corollary 5.2.7 *Let $(V, \langle \cdot, \cdot \rangle_V)$ be an n-dimensional inner product space over $\mathbb{F}$, with $\mathbb{F}$ equal to $\mathbb{R}$ or $\mathbb{C}$. Then V is isometrically isomorphic to $\mathbb{F}^n$ with the standard inner product.*

Proof. Let $\mathcal{B} = \{\mathbf{v}_1, \ldots, \mathbf{v}_n\}$ be an orthonormal basis for V, and define the map $T : V \to \mathbb{F}^n$ via $T(\mathbf{v}) = [\mathbf{v}]_{\mathcal{B}}$. By Theorem 3.2.7, T is an isomorphism, and by Proposition 5.2.6, T is an isometry. Thus V and $\mathbb{F}^n$ are isometrically isomorphic. $\square$.

A consequence of Corollary 5.2.7 is that to understand finite-dimensional inner product spaces, it essentially suffices to study $\mathbb{F}^n$ with the standard inner product. We will gladly make use of this observation. It is important to remember, though, that to view an n-dimensional inner product space as $\mathbb{F}^n$ one needs to start by choosing an orthonormal basis and represent vectors with respect to this fixed chosen basis.

5.3   The adjoint of a linear map

Via the inner product, one can relate with a linear map another linear map (called the *adjoint*). On a vector space over the reals, the adjoint of multiplication with a matrix A corresponds to multiplication with the transpose A^T of the matrix A. Over the complex numbers, it also involves taking a complex conjugate. We now provide you with the definition.

Let $(V, \langle \cdot, \cdot \rangle_V)$ and $(W, \langle \cdot, \cdot \rangle_W)$ be inner product spaces, and let $T : V \to W$

be linear. We call a map $T^\star : W \to V$ the *adjoint* of T if

$$\langle T(\mathbf{v}), \mathbf{w}\rangle_W = \langle \mathbf{v}, T^\star(\mathbf{w})\rangle_V \quad \text{for all } \mathbf{v} \in V, \mathbf{w} \in W.$$

Notice that the adjoint is unique. Indeed if S is another adjoint for T, we get that $\langle \mathbf{v}, T^\star(\mathbf{w})\rangle_V = \langle \mathbf{v}, S(\mathbf{w})\rangle_V$ for all $\mathbf{v}, \mathbf{w}$. Choosing $\mathbf{v} = T^\star(\mathbf{w}) - S(\mathbf{w})$ yields

$$\langle T^\star(\mathbf{w}) - S(\mathbf{w}), T^\star(\mathbf{w}) - S(\mathbf{w})\rangle_V = 0,$$

and thus $T^\star(\mathbf{w}) - S(\mathbf{w}) = 0$. As this holds for all $\mathbf{w}$, we must have that $T^\star = S$.

Lemma 5.3.1 *If $T : V \to W$ is an isometry, then $T^\star T = id_V$.*

Proof. Since T is an isometry, we have that

$$\langle T(\mathbf{v}), T(\hat{\mathbf{v}})\rangle_W = \langle \mathbf{v}, \hat{\mathbf{v}}\rangle_V \quad \text{for all } \mathbf{v}, \hat{\mathbf{v}} \in V.$$

But then, we get that

$$\langle T^\star T(\mathbf{v}), \hat{\mathbf{v}}\rangle_W = \langle \mathbf{v}, \hat{\mathbf{v}}\rangle_V \quad \text{for all } \mathbf{v}, \hat{\mathbf{v}} \in V,$$

or equivalently,

$$\langle T^\star T(\mathbf{v}) - \mathbf{v}, \hat{\mathbf{v}}\rangle_W = \langle \mathbf{v}, \hat{\mathbf{v}}\rangle_V \quad \text{for all } \mathbf{v}, \hat{\mathbf{v}} \in V.$$

Letting $\hat{\mathbf{v}} = T^\star T(\mathbf{v}) - \mathbf{v}$, this yields $T^\star T(\mathbf{v}) - \mathbf{v} = 0$ for all $\mathbf{v} \in V$. Thus $T^\star T = id_V$. $\qquad\square$

We call T unitary, if T is an isometric isomorphism. In that case $T^\star$ is the inverse of T and we have $T^\star T = id_V$ and $TT^\star = id_W$.

A linear map $T : V \to V$ is called *self-adjoint* if

$$\langle T(\mathbf{v}), \hat{\mathbf{v}}\rangle_V = \langle \mathbf{v}, T(\hat{\mathbf{v}})\rangle_V \quad \text{for all } \mathbf{v}, \hat{\mathbf{v}} \in V.$$

In other words, T is *self-adjoint* if $T^\star = T$.

Example 5.3.2 Let $k \in \mathbb{N}$ and $V = \text{Span}\{\sin(x), \sin(2x), \ldots, \sin(kx)\}$, with the inner product

$$\langle f, g\rangle_V = \int_0^\pi f(x)g(x)dx.$$

Let $T = -\frac{d^2}{dx^2} : V \to V$. Notice that indeed $-\frac{d^2}{dx^2}\sin(mx) = m^2\sin(mx) \in V$, thus T is well-defined. We claim that T is self-adjoint. For this, we need to apply integration by parts twice, and it is important to note that for all f in V we have that $f(0) = f(\pi) = 0$. So let us compute

$$\langle T(f), g \rangle_V = \int_0^\pi -f''(x)g(x)dx = -f'(\pi)g(\pi) + f'(0)g(0) + \int_0^\pi f'(x)g'(x)dx =$$

$$f(\pi)g'(\pi) - f(0)g'(0) - \int_0^\pi f(x)g''(x)dx = \langle f, T(g) \rangle_V.$$

Theorem 5.3.3 *Let $(V, \langle \cdot, \cdot \rangle_V)$ and $(W, \langle \cdot, \cdot \rangle_W)$ be inner product spaces with orthonormal bases $\mathcal{B} = \{\mathbf{v}_1, \ldots, \mathbf{v}_n\}$ and $\mathcal{C} = \{\mathbf{w}_1, \ldots, \mathbf{w}_m\}$. Let $T : V \to W$ be linear. If*

$$[T]_{\mathcal{C} \leftarrow \mathcal{B}} = \begin{pmatrix} a_{11} & \cdots & a_{1n} \\ \vdots & & \vdots \\ a_{m1} & \cdots & a_{mn} \end{pmatrix},$$

then

$$[T^\star]_{\mathcal{B} \leftarrow \mathcal{C}} = \begin{pmatrix} \overline{a}_{11} & \cdots & \overline{a}_{m1} \\ \vdots & & \vdots \\ \overline{a}_{1n} & \cdots & \overline{a}_{mn} \end{pmatrix}.$$

In other words,

$$[T^\star]_{\mathcal{B} \leftarrow \mathcal{C}} = ([T]_{\mathcal{C} \leftarrow \mathcal{B}})^*, \tag{5.13}$$

where as before A^* is the conjugate transpose of the matrix A.

Proof. The matrix representation for T tells us, in conjunction with Lemma 5.2.5, that $a_{ij} = \langle T(\mathbf{v}_j), \mathbf{w}_i \rangle_W$. The (k, l)th entry of the matrix representation of $T^\star$ is, again by using the observation in Lemma 5.2.5, equal to

$$\langle T^*(\mathbf{w}_l), \mathbf{v}_k \rangle_V = \langle \mathbf{w}_l, T(\mathbf{v}_k) \rangle_W = \overline{\langle T(\mathbf{v}_k), \mathbf{w}_l \rangle_W} = \overline{a}_{lk},$$

proving the statement. $\square$

Thus, when we identify via a choice of an orthonormal basis a finite-dimensional inner product space V with $\mathbb{F}^{\dim V}$ endowed with the standard inner product, the corresponding matrix representation has the property that the adjoint corresponds to taking the conjugate transpose. One of the consequences of this correspondence is that any linear map between finite-dimensional vector spaces has an adjoint. Indeed, this follows from the observation that any matrix has a conjugate transpose. In what follows we will focus on using matrices and their conjugate transposes. Having the material of this and previous sections in mind, the results that follow may be interpreted on the level of general finite-dimensional inner product spaces. It is always good to remember that when adjoints appear, there are necessarily inner products in the background.

5.4   Unitary matrices, QR, and Schur triangularization

Unitary transformations are ones where a pair of vectors is mapped to a new pair of vectors without changing their lengths or the angle between them. Thus, one can think of a unitary transformation as viewing the vector space from a different viewpoint. Using unitary transformations (represented by unitary matrices) can be used to put general transformations in a simpler form. These simpler forms give rise to QR and Schur triangular decompositions.

Let $\mathbb{F} = \mathbb{R}$ or $\mathbb{C}$. We call a matrix $A \in \mathbb{F}^{n \times m}$ an *isometry* if $A^*A = I_m$. Notice that necessarily we need to have that $n \geq m$. The equation A^*A can also be interpreted as that the columns of A are orthonormal. When $A \in \mathbb{F}^{n \times n}$ is square, then automatically $A^*A = I_n$ implies that $AA^* = I_n$. Such a matrix is called *unitary*. Thus a square isometry is a unitary. From the Gram–Schmidt process we can deduce the following.

Theorem 5.4.1 *(QR factorization) Let $A \in \mathbb{F}^{n \times m}$ with $n \geq m$. Then there exists an isometry $Q \in \mathbb{F}^{n \times m}$ and an upper triangular matrix $R \in \mathbb{F}^{m \times m}$ with nonnegative entries on the diagonal, so that*

$$A = QR.$$

If A has rank equal to m, then the diagonal entries of R are positive, and R is invertible. If $n = m$, then Q is unitary.

Proof. First we consider the case when $\mathrm{rank}A = m$. Let $\mathbf{v}_1, \ldots, \mathbf{v}_m$ denote the columns of A, and let $\mathbf{z}_1, \ldots, \mathbf{z}_m$ denote the resulting vectors when we apply the Gram–Schmidt process to $\mathbf{v}_1, \ldots, \mathbf{v}_m$ as in Theorem 5.2.3. Let now Q be the matrix with columns $\frac{\mathbf{z}_1}{\|\mathbf{z}_1\|}, \ldots, \frac{\mathbf{z}_m}{\|\mathbf{z}_m\|}$. Then $Q^*Q = I_m$ as the columns of Q are orthonormal. Moreover, we have that

$$\mathbf{v}_k = \|\mathbf{z}_k\| \frac{\mathbf{z}_k}{\|\mathbf{z}_k\|} + \sum_{j=1}^{k-1} r_{kj} \mathbf{z}_j,$$

for some $r_{kj} \in \mathbb{F}$, $k > j$. Putting $r_{kk} = \|\mathbf{z}_k\|$, and $r_{kj} = 0$, $k < j$, and letting $R = (r_{kj})_{k,j=1}^m$, we get the desired upper triangular matrix R yielding $A = QR$.

When rank $< m$, apply the Gram–Schmidt process with those columns of A that do not lie in the span of the preceding columns. Place the vectors $\frac{\mathbf{z}}{\|\mathbf{z}\|}$

that are found in this way in the corresponding columns of Q. Next, one can fill up the remaining columns of Q with any vectors making the matrix an isometry. The upper triangular entries in R are obtained from writing the columns of A as linear combinations of the $\frac{\mathbf{z}}{\|\mathbf{z}\|}$'s found in the process above. $\square$

Let us illustrate the QR factorization on an example where the columns of A are linearly dependent.

Example 5.4.2 Let $A = \begin{pmatrix} 1 & 0 & 1 & 2 \\ 1 & -2 & 0 & -2 \\ 1 & 0 & 1 & 0 \\ 1 & -2 & 0 & 0 \end{pmatrix}$. Applying the Gram–Schmidt process we obtain,

$$\mathbf{z}_1 = \begin{pmatrix} 1 \\ 1 \\ 1 \\ 1 \end{pmatrix},$$

$$\mathbf{z}_2 = \begin{pmatrix} 0 \\ -2 \\ 0 \\ -2 \end{pmatrix} - \frac{-4}{4}\begin{pmatrix} 1 \\ 1 \\ 1 \\ 1 \end{pmatrix} = \begin{pmatrix} 1 \\ -1 \\ 1 \\ -1 \end{pmatrix},$$

$$\mathbf{z}_3 = \begin{pmatrix} 1 \\ 0 \\ 1 \\ 0 \end{pmatrix} - \frac{2}{4}\begin{pmatrix} 1 \\ 1 \\ 1 \\ 1 \end{pmatrix} - \frac{2}{4}\begin{pmatrix} 1 \\ -1 \\ 1 \\ -1 \end{pmatrix} = \begin{pmatrix} 0 \\ 0 \\ 0 \\ 0 \end{pmatrix}. \tag{5.14}$$

We thus notice that the third column of A is a linear combination of the first two columns of A, so we continue to compute $\mathbf{z}_4$ without using $\mathbf{z}_3$:

$$\mathbf{z}_4 = \begin{pmatrix} 2 \\ -2 \\ 0 \\ 0 \end{pmatrix} - 0\begin{pmatrix} 1 \\ 1 \\ 1 \\ 1 \end{pmatrix} - \frac{4}{4}\begin{pmatrix} 1 \\ -1 \\ 1 \\ -1 \end{pmatrix} = \begin{pmatrix} 1 \\ -1 \\ -1 \\ 1 \end{pmatrix}.$$

Dividing $\mathbf{z}_1, \mathbf{z}_2, \mathbf{z}_4$ by their respective lengths, and putting them in the matrix Q, we get the equality

$$A = \begin{pmatrix} 1 & 0 & 1 & 2 \\ 1 & -2 & 0 & -2 \\ 1 & 0 & 1 & 0 \\ 1 & -2 & 0 & 0 \end{pmatrix} = \begin{pmatrix} \frac{1}{2} & \frac{1}{2} & & \frac{1}{2} \\ \frac{1}{2} & -\frac{1}{2} & & -\frac{1}{2} \\ \frac{1}{2} & \frac{1}{2} & & -\frac{1}{2} \\ \frac{1}{2} & -\frac{1}{2} & & \frac{1}{2} \end{pmatrix}\begin{pmatrix} -2 & 2 & -1 & 0 \\ 0 & 2 & 1 & 2 \\ 0 & 0 & 0 & 0 \\ 0 & 0 & 0 & 2 \end{pmatrix},$$

where it remains to fill in the third column of Q. To make the columns of Q

orthonormal, we choose the third column to be $\left(\frac{1}{2} \quad \frac{1}{2} \quad -\frac{1}{2} \quad -\frac{1}{2}\right)^*$, so we get

$$Q = \begin{pmatrix} \frac{1}{2} & \frac{1}{2} & \frac{1}{2} & \frac{1}{2} \\ \frac{1}{2} & -\frac{1}{2} & \frac{1}{2} & -\frac{1}{2} \\ \frac{1}{2} & \frac{1}{2} & -\frac{1}{2} & -\frac{1}{2} \\ \frac{1}{2} & -\frac{1}{2} & -\frac{1}{2} & \frac{1}{2} \end{pmatrix}, R = \begin{pmatrix} -2 & 2 & -1 & 0 \\ 0 & 2 & 1 & 2 \\ 0 & 0 & 0 & 0 \\ 0 & 0 & 0 & 2 \end{pmatrix}.$$

Notice that finding the QR factorization only requires simple algebraic operations. Surprisingly, it can be used very effectively to find eigenvalues of a matrix. The QR algorithm is based on the following iteration scheme. Let $A \in \mathbb{F}^{n \times n}$ be given. Let $A_0 = A$, and perform the iteration:

find QR factorization $A_k = QR$, then put $A_{k+1} = RQ$.

Notice that $A_{k+1} = Q^{-1} A_k Q$, so that A_{k+1} and A_k have the same eigenvalues. As it turns out, A_k converges to an upper triangular matrix, manageable exceptions aside, and thus one can read the eigenvalues from the diagonal of this upper triangular limit. In a numerical linear algebra course, one studies the details of this algorithm. It is noteworthy, though, to remark that when one wants to find numerically the roots of a polynomial, it is often very effective to build the associated companion matrix, and subsequently use the QR algorithm to find the eigenvalues of this companion matrix, which coincide with the roots of the polynomial. Thus contrary to how we do things by hand, we rather find roots by computing eigenvalues than the other way around. We will discuss this further in Section 7.3.

By combining the Jordan canonical form theorem and the QR factorization theorem, we can prove the following result.

Theorem 5.4.3 *(Schur triangularization) Let $A \in \mathbb{F}^{n \times n}$ and suppose that all its eigenvalues are in $\mathbb{F}$. Then there exits a unitary $U \in \mathbb{F}^{n \times n}$ and an upper triangular $T \in \mathbb{F}^{n \times n}$ so that $A = UTU^*$.*

Proof. By Theorem 4.4.1 there exists an invertible nonsingular $P \in \mathbb{F}^{n \times n}$ such that $A = PJP^{-1}$, where the matrix J is a direct sum of Jordan blocks, and thus J is upper triangular. By Theorem 5.4.1 there exists a unitary Q and an invertible upper triangular R such that $P = QR$. Now,

$$A = PJP^{-1} = QRJ(QR)^{-1} = Q(RJR^{-1})Q^{-1} = Q(RJR^{-1})Q^*,$$

where $Q^{-1} = Q^*$ since Q is unitary. The inverse of an upper triangular matrix is upper triangular, and the product of upper triangular matrices is also upper triangular. It follows that $T := RJR^{-1}$ is upper triangular, and thus $A = QTQ^*$ with Q unitary and T upper triangular. $\square$

5.5   Normal and Hermitian matrices

In this section we study transformations that interact particularly nicely with respect to the inner product. A main feature of these normal and Hermitian transformations is that its eigenvectors can be used to form an orthonormal basis for the underlying space.

A matrix $A \in \mathbb{F}^{n \times n}$ is called *normal* if $A^*A = AA^*$.

Lemma 5.5.1 *(a) If U is unitary, then A is normal if and only if U^*AU is normal.*

(b) If T is upper triangular and normal, then T is diagonal.

Proof. (a). Let us compute $U^*AU(U^*AU)^* = U^*AUU^*A^*U = U^*AA^*U$, and $(U^*AU)^*U^*AU) = U^*A^*UU^*AU = U^*A^*AU$. The two are equal if and only if $AA^* = A^*A$ (where we used that U is invertible). This proves the first part.

(b). Suppose that $T = (t_{ij})_{i,j=1}^n$ is upper triangular. Thus $t_{ij} = 0$ for $i > j$. Since T is normal we have that $T^*T = TT^*$. Comparing the $(1,1)$ entry on both sides of this equation we get

$$|t_{11}|^2 = |t_{11}|^2 + |t_{12}|^2 + \cdots + |t_{1n}|^2.$$

This gives that $t_{12} = t_{13} = \cdots = t_{1n} = 0$. Next, comparing the $(2,2)$ entry on both sides of $T^*T = TT^*$ we get

$$|t_{22}|^2 = |t_{22}|^2 + |t_{23}|^2 + \cdots + |t_{2n}|^2.$$

This gives that $t_{23} = t_{24} = \cdots = t_{2n} = 0$. Continuing this way, we find that $t_{ij} = 0$ for all $i < j$. Thus T is diagonal. $\qquad\square$

Theorem 5.5.2 *(Spectral theorem for normal matrices) Let $A \in \mathbb{F}^{n \times n}$ be normal, and suppose that all eigenvalues of A lie in $\mathbb{F}$. Then there exists a unitary $U \in \mathbb{F}^{n \times n}$ and a diagonal $D \in \mathbb{F}^{n \times n}$ so that*

$$A = UDU^*.$$

Proof. By Theorem 5.4.3 we have that $A = UTU^*$, where U is unitary and T is upper triangular. By Lemma 5.5.1, since A is normal, so is T. Again, by

Lemma 5.5.1, as T is upper triangular and normal, we must have that T is diagonal. But then, with $D = T$ we have the desired factorization $A = UDU^*$. $\qquad\square$

Examples of normal matrices are the following:

- A is *Hermitian* if $A = A^*$.

- A is *skew-Hermitian* if $A = -A^*$.

- A is *unitary* if $AA^* = A^*A = I$.

Hermitian, skew-Hermitian, and unitary matrices are all normal. Hermitian and skew-Hermitian matrices have the following characterization.

Proposition 5.5.3 *Let* $A \in \mathbb{C}^{n \times n}$.

(i) *A is Hermitian if and only if* $\mathbf{x}^* A\mathbf{x} \in \mathbb{R}$ *for all* $\mathbf{x} \in \mathbb{C}^n$.

(ii) *A is skew-Hermitian if and only if* $\mathbf{x}^* A\mathbf{x} \in i\mathbb{R}$ *for all* $\mathbf{x} \in \mathbb{C}^n$.

Proof. (i) If $A = A^*$, then $(\mathbf{x} * A\mathbf{x})^* = \mathbf{x}^* A^* \mathbf{x} = \mathbf{x}^* A\mathbf{x}$, and thus $\mathbf{x}^* A\mathbf{x} \in \mathbb{R}$. Conversely, suppose that $\mathbf{x}^* A\mathbf{x} \in \mathbb{R}$ for all $\mathbf{x} \in \mathbb{C}^n$. Write $A = (a_{jk})_{j,k=1}^n$. First, let $\mathbf{x} = \mathbf{e}_j$. Then $\mathbf{e}_j^* A\mathbf{e}_j = a_{jj}$, so we get that $a_{jj} \in \mathbb{R}$, $j = 1, \ldots, n$. Next, let $\mathbf{x} = \mathbf{e}_j + \mathbf{e}_k$. Then $\mathbf{x}^* A\mathbf{x} = a_{jj} + a_{jk} + a_{kj} + a_{kk}$. Thus we get that $a_{jk} + a_{kj} \in \mathbb{R}$, and consequently Im $a_{jk} = -$Im a_{kj}. Finally, let $\mathbf{x} = \mathbf{e}_j + i\mathbf{e}_k$. Then $\mathbf{x}^* A\mathbf{x} = a_{jj} + ia_{jk} - ia_{kj} + a_{kk}$. Thus we get that $i(a_{jk} - a_{kj}) \in \mathbb{R}$, and thus Re $a_{jk} = $ Re a_{kj}. Thus $a_{jk} = \overline{a_{kj}}$. Thus $A = A^*$.

(ii) Replace A by iA and use (i). $\qquad\square$

We say that a matrix $A \in \mathbb{C}^{n \times n}$ is *positive semidefinite* if $\mathbf{x}^* A\mathbf{x} \geq 0$ for all $\mathbf{x} \in \mathbb{C}^n$. The matrix $A \in \mathbb{C}^{n \times n}$ is *positive definite* if $\mathbf{x}^* A\mathbf{x} > 0$ for all $\mathbf{x} \in \mathbb{C}^n \setminus \{\mathbf{0}\}$. Clearly, by Proposition 5.5.3(i), if A is positive (semi)definite, then A is Hermitian.

We have the following result.

Theorem 5.5.4 *Let* $A \in \mathbb{C}^{n \times n}$. *Then the following hold.*

(i) *A is Hermitian if and only if there exists a unitary U and a real diagonal D so that $A = UDU^*$. If $A \in \mathbb{R}^{n \times n}$, then U can be chosen to be real as well.*

(ii) *A is skew-Hermitian if and only if there exists a unitary U and a purely imaginary diagonal D so that $A = UDU^*$.*

(iii) *A is unitary if and only if there exists a unitary U and a diagonal $D = \operatorname{diag}(d_{ii})_{i=1}^n$ with $|d_{ii}| = 1$ so that $A = UDU^*$.*

(iv) *A is positive semidefinite if and only if there exists a unitary U and a nonnegative real diagonal D so that $A = UDU^*$. If $A \in \mathbb{R}^{n \times n}$, then U can be chosen to be real as well.*

(v) *A is positive semidefinite if and only if there exists a unitary U and a positive real diagonal D so that $A = UDU^*$. If $A \in \mathbb{R}^{n \times n}$, then U can be chosen to be real as well.*

Proof. It is easy to see that when $A = UDU^*$ with U unitary, then A is Hermitian/skew-Hermitian/unitary/positive (semi)definite if and only if D is Hermitian/skew-Hermitian/unitary/positive (semi)definite. Next, for a diagonal matrix one easily observes that D is Hermitian if and only if D is real, D is skew-Hermitian if and only if D is purely imaginary, D is unitary if and only if its diagonal entries have modulus 1, D is positive semidefinite if and only if D is nonnegative, and D is positive definite if and only if D is positive. Combining these observations with Theorem 5.5.2 yields the result. $\square$.

We end this section with Sylvester's Law of Inertia. Given a Hermitian matrix $A \in \mathbb{C}^{n \times n}$, its *inertia* $\operatorname{In}(A)$ is a triple $\operatorname{In}(A) = (i_+(A), i_-(A), i_0(A))$, where $i_+(A)$ is the number of positive eigenvalues of A (counting multiplicity), $i_-(A)$ is the number of negative eigenvalues of A (counting multiplicity), and $i_0(A)$ is the number of zero eigenvalues of A (counting multiplicity). For example,

$$\operatorname{In} \begin{pmatrix} 2 & 0 & 0 & 0 \\ 0 & 3 & 0 & 0 \\ 0 & 0 & 1 & 1 \\ 0 & 0 & 1 & 1 \end{pmatrix} = (3, 0, 1).$$

Note that $i_0(A) = \dim \operatorname{Ker}(A)$, and $i_+(A) + i_-(A) + i_0(A) = n$. We now have the following result.

Theorem 5.5.5 *(Sylvester's Law of Inertia) Let $A, B \in \mathbb{C}^{n \times n}$ be Hermitian. Then $\operatorname{In}(A) = \operatorname{In}(B)$ if and only if there exists an invertible S so that $A = SBS^*$.*

We will be using the following lemma.

Lemma 5.5.6 *Let $A \in \mathbb{C}^{n \times n}$ be Hermitian with $\operatorname{In}(A) = (\mu, \nu, \delta)$. Then there exists an invertible T so that*

$$A = T \begin{pmatrix} I_\mu & & \\ & -I_\nu & \\ & & 0 \end{pmatrix} T^*.$$

Proof. Let $A = U\Lambda U^*$ with U unitary, $\Lambda = \operatorname{diag}(\lambda_i)_{i=1}^n$, $\lambda_1, \ldots, \lambda_\mu > 0$, $\lambda_{\mu+1}, \ldots \lambda_{\mu+\nu} < 0$, and $\lambda_{\mu+\nu+1} = \cdots = \lambda_n = 0$. If we let

$$D = \begin{pmatrix} \sqrt{\lambda_1} & & & & & & \\ & \ddots & & & & & \\ & & \sqrt{\lambda_\mu} & & & & \\ & & & \sqrt{-\lambda_{\mu+1}} & & & \\ & & & & \ddots & & \\ & & & & & \sqrt{-\lambda_{\mu+\nu}} & \\ & & & & & & I_\delta \end{pmatrix}$$

and $T = UD$, then the lemma follows. $\qquad\square$

Proof of Theorem 5.5.5. First suppose that $\operatorname{In}(A) = (\mu, \nu, \delta) = \operatorname{In}(B)$. By Lemma 5.5.6 there exist invertible T and W so that

$$A = T \begin{pmatrix} I_\mu & & \\ & -I_\nu & \\ & & 0 \end{pmatrix} T^*, \quad B = W \begin{pmatrix} I_\mu & & \\ & -I_\nu & \\ & & 0 \end{pmatrix} W^*.$$

But then by letting $S = TW^{-1}$ we obtain that $A = SBS^*$.

Conversely, if $A = SBS^*$ for some invertible S. We first notice that $i_0(A) = \dim \operatorname{Ker}(A) = \dim \operatorname{Ker}(B) = i_0(B)$. By applying Lemma 5.5.6 to both A and B, and combining the results with $A = SBS^*$, we obtain that there exists an invertible W so that

$$\begin{pmatrix} I_{i_+(A)} & & \\ & -I_{i_-(A)} & \\ & & 0 \end{pmatrix} = W \begin{pmatrix} I_{i_+(B)} & & \\ & -I_{i_-(B)} & \\ & & 0 \end{pmatrix} W^*, \qquad (5.15)$$

where the diagonal zeros have equal size. Let us partition $W = (W_{ij})_{i,j=1}^3$ in an appropriately sized block matrix (so, for instance, W_{11} has size $i_+(A) \times i_+(B)$ and W_{22} has size $i_-(A) \times i_-(B)$). Then from the $(1,1)$ block entry of the equality (5.15) we get that

$$I_{i_+(A)} = W_{11} I_{i_+(B)} W_{11}^* - W_{12} I_{i_-(B)} W_{12}^*.$$

This gives that rank $W_{11} W_{11}^* \leq i_+(B)$ and $W_{11} W_{11} = I_{i_+(A)} + W_{12} W_{12}^*$ is positive definite of size $i_+(A) \times i_+(A)$, and thus rank $W_{11}W_{11}^* \geq i_+(A)$. Combining these observations, gives $i_+(B) \geq i_+(A)$. Reversing the roles of A and B, one can apply the same argument and arrive at the inequality $i_+(A) \geq i_+(B)$. But then $i_+(B) = i_+(A)$ follows. Finally,

$$i_-(A) = n - i_+(A) - i_0(A) = n - i_+(B) - i_0(B) = i_-(B),$$

and we are done.   □

5.6   Singular value decomposition

The singular values decomposition gives a way to write a general (typically, non-square) matrix A as the product $A = V\Sigma W^*$, where V and W^* are unitary and Σ is essentially diagonal with nonnegative entries. This means that by changing the viewpoint in both the domain and the co-domain, a linear transformation between finite-dimensional spaces can be viewed as multiplying with a relatively few (at most the dimension of the domain and/or co-domain) nonnegative numbers. One of the main applications of the singular value decomposition is that it gives an easy way to approximate a matrix with one that has a low rank. The advantage of a low rank matrix is that it requires less memory to store it. If you take a look at the solution of Exercise 5.7.31, you will see how a rank 524 matrix (the original image) is approximated by a rank 10, 30, and 50 one by using the singular value decomposition.

Here is the main result of this section.

Theorem 5.6.1 *Let $A \in \mathbb{F}^{n \times m}$ have rank k. Then there exist unitary matrices $V \in \mathbb{F}^{n \times n}$, $W \in \mathbb{F}^{m \times m}$, and a matrix $\Sigma \in \mathbb{F}^{n \times m}$ of the form*

$$\Sigma = \begin{pmatrix} \sigma_1 & 0 & \cdots & 0 & \cdots & 0 \\ 0 & \sigma_2 & \cdots & 0 & \cdots & 0 \\ \vdots & \vdots & \ddots & \vdots & & \vdots \\ 0 & 0 & \cdots & \sigma_k & \cdots & 0 \\ \vdots & \vdots & & \vdots & \ddots & \vdots \\ 0 & 0 & \cdots & 0 & \cdots & 0 \end{pmatrix}, \quad \sigma_1 \geq \sigma_2 \geq \ldots \geq \sigma_k > 0, \quad (5.16)$$

so that $A = V\Sigma W^*$.

Proof. As A^*A is positive semidefinite, there exists a unitary W and a diagonal matrix $\Lambda = (\lambda_i)_{i=1}^n$, with $\lambda_1 \geq \cdots \geq \lambda_k > 0 = \lambda_{k+1} = \cdots = \lambda_n$, so that $A^*A = W\Lambda W^*$. Notice that rank $A = \operatorname{rank} A^*A = k$. Put $\sigma_j = \sqrt{\lambda_j}$, $j = 1, \ldots, k$, and write $W = \begin{pmatrix} \mathbf{w}_1 & \cdots & \mathbf{w}_m \end{pmatrix}$. Next, put $\mathbf{v}_j = \frac{1}{\sigma_j} A\mathbf{w}_j$, $j = 1, \ldots, k$, and let $\{\mathbf{v}_{k+1}, \ldots, \mathbf{v}_n\}$ be an orthonormal basis for $\operatorname{Ker} A^*$. Put $V = \begin{pmatrix} \mathbf{v}_1 & \cdots & \mathbf{v}_n \end{pmatrix}$. First, let us show that V is unitary. When $i, j \in \{1, \ldots, k\}$, then

$$\mathbf{v}_j^* \mathbf{v}_i = \frac{1}{\sigma_i \sigma_j} \mathbf{w}_j A^* A \mathbf{w}_i = \frac{1}{\sigma_i \sigma_j} \mathbf{w}_j^* W \Lambda W^* \mathbf{w}_i =$$

$$\frac{1}{\sigma_i \sigma_j} \mathbf{e}_j^* \Lambda \mathbf{e}_i = \begin{cases} 0 & \text{when } i \neq j, \\ \frac{\lambda_j}{\sigma_j^2} = 1 & \text{when } i = j. \end{cases}$$

Next, when $j \in \{1, \ldots, k\}$ and $i \in \{k+1, \ldots, n\}$, we get that $\mathbf{v}_j^* \mathbf{v}_i = \frac{1}{\sigma_j} \mathbf{w}_j A^* \mathbf{v}_i = 0$ as $\mathbf{v}_i \in \operatorname{Ker} A^*$. Similarly, $\mathbf{v}_j^* \mathbf{v}_i = 0$ when $i \in \{1, \ldots, k\}$ and $j \in \{k+1, \ldots, n\}$. Finally, $\{\mathbf{v}_{k+1}, \ldots, \mathbf{v}_n\}$ is an orthonormal set, and thus we find that $V^* V = I_n$.

It remains to show that $A = V\Sigma W^*$, or equivalently, $AW = V\Sigma$. The equality in the first k columns follows from the definition of $\mathbf{v}_j$, $j = 1, \ldots, k$. In columns $k+1, \ldots, m$ we have $\mathbf{0}$ on both sides, and thus $AW = V\Sigma$ follows. $\qquad\square$

Alternative proof. First assume that $n = m$. Consider the Hermitian $2n \times 2n$ matrix

$$M = \begin{pmatrix} 0 & A \\ A^* & 0 \end{pmatrix}.$$

Observe that $M\mathbf{v} = \lambda\mathbf{v}$, where $\mathbf{v} = \begin{pmatrix} \mathbf{v}_1 \\ \mathbf{v}_2 \end{pmatrix}$, yields

$$\begin{pmatrix} 0 & A \\ A^* & 0 \end{pmatrix} \begin{pmatrix} \mathbf{v}_1 \\ \mathbf{v}_2 \end{pmatrix} = \lambda \begin{pmatrix} \mathbf{v}_1 \\ \mathbf{v}_2 \end{pmatrix}.$$

Then

$$\begin{pmatrix} 0 & A \\ A^* & 0 \end{pmatrix} \begin{pmatrix} \mathbf{v}_1 \\ -\mathbf{v}_2 \end{pmatrix} = -\lambda \begin{pmatrix} \mathbf{v}_1 \\ -\mathbf{v}_2 \end{pmatrix}.$$

Thus, if we denote the positive eigenvalues of M by $\sigma_1 \geq \sigma_2 \geq \ldots \geq \sigma_k > 0$, then $-\sigma_1, \ldots, -\sigma_k$ are also eigenvalues of M. Notice that when $\lambda = 0$, we can take a basis $\{\mathbf{v}_1^{(1)}, \ldots, \mathbf{v}_1^{(n-k)}\}$ of $\operatorname{Ker} A^*$, and a basis $\{\mathbf{v}_2^{(1)}, \ldots, \mathbf{v}_2^{(n-k)}\}$ of $\operatorname{Ker} A$, and then

$$\left\{ \begin{pmatrix} \mathbf{v}_1^{(1)} \\ \mathbf{v}_2^{(1)} \end{pmatrix}, \ldots, \begin{pmatrix} \mathbf{v}_1^{(n-k)} \\ \mathbf{v}_2^{(n-k)} \end{pmatrix}, \begin{pmatrix} \mathbf{v}_1^{(1)} \\ -\mathbf{v}_2^{(1)} \end{pmatrix}, \ldots, \begin{pmatrix} \mathbf{v}_1^{(n-k)} \\ -\mathbf{v}_2^{(n-k)} \end{pmatrix} \right\}$$

is a basis for $\operatorname{Ker} M$. By Theorem 5.5.4 there exists a unitary U and a diagonal D so that $M = UDU^*$, and by the previous observations we can organize it so that

$$U = \begin{pmatrix} X & X \\ Y & -Y \end{pmatrix}, D = \begin{pmatrix} \Sigma & 0 \\ 0 & -\Sigma \end{pmatrix}.$$

Now, we get

$$\begin{pmatrix} 0 & A \\ A^* & 0 \end{pmatrix} = \begin{pmatrix} X & X \\ Y & -Y \end{pmatrix} \begin{pmatrix} \Sigma & 0 \\ 0 & -\Sigma \end{pmatrix} \begin{pmatrix} X^* & Y^* \\ X^* & -Y^* \end{pmatrix},$$

and we also have that

$$\begin{pmatrix} X & X \\ Y & -Y \end{pmatrix} \begin{pmatrix} X^* & Y^* \\ X^* & -Y^* \end{pmatrix} = \begin{pmatrix} I_n & 0 \\ 0 & I_n \end{pmatrix}.$$

Writing out these equalities, we get that

$$A = (\sqrt{2}X)\Sigma(\sqrt{2}Y)^*,$$

with $\sqrt{2}X$ and $\sqrt{2}Y$ unitary.

When A is of size $n \times m$ with $n > m$, one can do a QR factorization $A = QR$. Next, obtain a singular value decomposition of the $m \times m$ matrix R: $R = \hat{V}\hat{\Sigma}\hat{W}^*$. Then $A = (Q\hat{V})\hat{\Sigma}\hat{W}^*$. The matrix $Q\hat{V}$ is isometric, and we can make it a square unitary by adding columns Q_2 so that the square matrix

$$V := \begin{pmatrix} Q\hat{V} & Q_2 \end{pmatrix}$$

has columns that form an orthonormal basis of $\mathbb{F}^n$; in other words, so that V is unitary. Next let

$$\Sigma = \begin{pmatrix} \hat{\Sigma} \\ 0 \end{pmatrix} \in \mathbb{F}^{n \times m}.$$

Then $A = V\Sigma W^*$ as desired.

Finally, when A is of size $n \times m$ with $n < m$, apply the previous paragraph to the $m \times n$ matrix A^*, to obtain $A^* = \hat{V}\hat{\Sigma}\hat{W}^*$. Then by letting $V = \hat{W}$, $\Sigma = \hat{\Sigma}^*$, and $W = \hat{V}$, we get the desired singular value decomposition $A = V\Sigma W^*$. $\qquad\square$

The values σ_j are called the *singular values* of A, and they are uniquely determined by A. We also denote them by $\sigma_j(A)$.

Proposition 5.6.2 *Let $A \in \mathbb{F}^{n \times m}$, and let $\|\cdot\|$ be the Euclidean norm. Then*

$$\sigma_1(A) = \max_{\|\mathbf{x}\|=1} \|A\mathbf{x}\|. \tag{5.17}$$

In particular, $\sigma_1(\cdot)$ is a norm of $\mathbb{F}^{n \times m}$. Finally, if $A \in \mathbb{F}^{n \times m}$ and $B \in \mathbb{F}^{m \times k}$ then

$$\sigma_1(AB) \le \sigma_1(A)\sigma_1(B). \tag{5.18}$$

Proof. Write $A = V\Sigma W^*$ in its singular value decomposition. For U unitary we have that $\|U\mathbf{v}\| = \|\mathbf{v}\|$ for all vectors $\mathbf{v}$. Thus $\|A\mathbf{x}\| = \|V\Sigma W^*\mathbf{x}\| = \|\Sigma W^*\mathbf{x}\|$. Let $\mathbf{u} = W^*\mathbf{x}$. Then $\|\mathbf{x}\| = \|W\mathbf{u}\| = \|\mathbf{u}\|$. Combining these observations, we have that

$$\max_{\|\mathbf{x}\|=1} \|A\mathbf{x}\| = \max_{\|\mathbf{u}\|=1} \|\Sigma\mathbf{u}\| = \max_{\|\mathbf{u}\|=1} \sqrt{\sigma_1^2|u_1|^2 + \cdots + \sigma_k^2|u_k|^2},$$

which is clearly bounded above by $\sqrt{\sigma_1^2|u_1|^2 + \cdots + \sigma_1^2|u_k|^2} = \sigma_1\|\mathbf{u}\| = \sigma_1$. When $\mathbf{u} = \mathbf{e}_1$, then we get that $\|\Sigma\mathbf{u}\| = \sigma_1$. Thus $\max_{\|\mathbf{u}\|=1} \|\Sigma\mathbf{u}\| = \sigma_1$ follows.

To check that $\sigma_1(\cdot)$ is a norm, the only condition that is not immediate is the triangle inequality. This now follows by observing that

$$\sigma_1(A + B) = \max_{\|\mathbf{x}\|=1} \|(A+B)\mathbf{x}\| \le \max_{\|\mathbf{x}\|=1} \|A\mathbf{x}\| + \max_{\|\mathbf{x}\|=1} \|B\mathbf{x}\| = \sigma_1(A) + \sigma_1(B).$$

To prove (5.18) we first observe that for every vector $\mathbf{v} \in \mathbb{F}^m$ we have that $\|A\mathbf{v}\| \le \sigma_1(A)\|\mathbf{v}\|$, as $\mathbf{w} := \frac{1}{\|\mathbf{v}\|}\mathbf{v}$ has norm 1, and thus $\|A\mathbf{w}\| \le \max_{\|\mathbf{x}\|=1} \|A\mathbf{x}\| = \sigma_1(A)$. Now, we obtain that

$$\sigma_1(AB) = \max_{\|\mathbf{x}\|=1} \|(AB)\mathbf{x}\| \le \max_{\|\mathbf{x}\|=1} \sigma_1(A)\|B\mathbf{x}\| =$$

$$\sigma_1(A) \max_{\|\mathbf{x}\|=1} \|B\mathbf{x}\| = \sigma_1(A)\sigma_1(B).$$

$\square$

An important application of the singular value decomposition is low rank approximation of matrices. The advantage of a low rank matrix is that it requires less memory to store a low rank matrix.

Proposition 5.6.3 *Let A have singular value decomposition $A = V\Sigma W^*$ with Σ as in (5.16). Let $l \le k$. Put $\hat{A} = V\hat{\Sigma}W^*$ with*

$$\hat{\Sigma} = \begin{pmatrix} \sigma_1 & 0 & \cdots & 0 & \cdots & 0 \\ 0 & \sigma_2 & \cdots & 0 & \cdots & 0 \\ \vdots & \vdots & \ddots & \vdots & & \vdots \\ 0 & 0 & \cdots & \sigma_l & \cdots & 0 \\ \vdots & \vdots & & \vdots & \ddots & \vdots \\ 0 & 0 & \cdots & 0 & \cdots & 0 \end{pmatrix}. \tag{5.19}$$

Then $\operatorname{rank} \hat{A} = l$, $\sigma_1(A - \hat{A}) = \sigma_{l+1}$, and for any matrix B with $\operatorname{rank}B \le l$ we have $\sigma_1(A - B) \ge \sigma_1(A - \hat{A})$.

Proof. Clearly rank $\hat{A} = l$, $\sigma_1(A - \hat{A}) = \sigma_{l+1}$. Next, let B with rank$B \le l$. Put $C = V^*BW$. Then rank$C = $ rank$B \le l$, and $\sigma_1(A - B) = \sigma_1(\Sigma - C)$. Notice that dim Ker $C \ge m - l$, and thus Ker $C \cap \text{Span}\{e_1, \ldots, e_{l+1}\}$ has dimension ≥ 1. Thus we can find a $v \in$ Ker $C \cap \text{Span}\{e_1, \ldots, e_{l+1}\}$ with $\|v\| = 1$. Then

$$\sigma_1(\Sigma - C) \ge \|(\Sigma - C)v\| = \|\Sigma v\| \ge \sigma_{l+1},$$

where in the last step we used that $v \in \text{Span}\{e_1, \ldots, e_{l+1}\}$. This proves the statement. □

Low rank approximations are used in several places, for instance in data compression and in search engines.

We end this section with an example where we compute the singular value decomposition of a matrix. For this it is useful to notice that if $A = V\Sigma W^*$, then $AA^* = V\Sigma\Sigma^*V^*$ and $A^*A = W\Sigma^*\Sigma W^*$. Thus the columns of V are eigenvectors of AA^*, and the diagonal elements σ_j^2 of the diagonal matrix $\Sigma\Sigma^*$ are the eigenvalues of AA^*. Thus the singular values can be found by computing the square roots of the nonzero eigenvalues of AA^*. Similarly, the columns of W are eigenvectors of A^*A, and the diagonal elements σ_j^2 of the diagonal matrix $\Sigma^*\Sigma$ are the nonzero eigenvalues of A^*A, as we have seen in the proof of Theorem 5.6.1.

Example 5.6.4 Let $A = \begin{pmatrix} 3 & 2 & 2 \\ 2 & 3 & -2 \end{pmatrix}$. Find the singular value decomposition of A.

Compute

$$AA^* = \begin{pmatrix} 17 & 8 \\ 8 & 17 \end{pmatrix},$$

which has eigenvalues 9 and 25. So the singular values of A are 3 and 5, and we get

$$\Sigma = \begin{pmatrix} 5 & 0 & 0 \\ 0 & 3 & 0 \end{pmatrix}.$$

To find V, we find unit eigenvectors of AA^*, giving

$$V = \begin{pmatrix} 1/\sqrt{2} & 1/\sqrt{2} \\ 1/\sqrt{2} & -1/\sqrt{2} \end{pmatrix}.$$

For W observe that $V^*A = \Sigma W^*$. Writing $W = \begin{pmatrix} w_1 & w_2 & w_3 \end{pmatrix}$, we get

$$\begin{pmatrix} 5/\sqrt{2} & 5/\sqrt{2} & 0 \\ 1/\sqrt{2} & -1/\sqrt{2} & 4/\sqrt{2} \end{pmatrix} = \begin{pmatrix} 5w_1^* \\ 3w_2^* \end{pmatrix}.$$

This yields $\mathbf{w}_1$ and $\mathbf{w}_2$. To find $\mathbf{w}_3$, we need to make sure that W is unitary, and thus $\mathbf{w}_3$ needs to be a unit vector orthogonal to $\mathbf{w}_1$ and $\mathbf{w}_2$. We find

$$W = \begin{pmatrix} 1/\sqrt{2} & 1/3\sqrt{2} & 2/3 \\ 1/\sqrt{2} & -1/3\sqrt{2} & -2/3 \\ 0 & 4/3\sqrt{2} & -1/3 \end{pmatrix}.$$

5.7 Exercises

Exercise 5.7.1 For the following, check whether $\langle \cdot, \cdot \rangle$ is an inner product.

(a) $V = \mathbb{R}^2$, $\mathbb{F} = \mathbb{R}$,

$$\left\langle \begin{pmatrix} x_1 \\ x_2 \end{pmatrix}, \begin{pmatrix} y_1 \\ y_2 \end{pmatrix} \right\rangle = 3x_1y_1 + x_1y_2 + x_2y_1 + 2x_2y_2.$$

(b) $V = \mathbb{C}^2$, $\mathbb{F} = \mathbb{C}$,

$$\left\langle \begin{pmatrix} x_1 \\ x_2 \end{pmatrix}, \begin{pmatrix} y_1 \\ y_2 \end{pmatrix} \right\rangle = 3x_1y_1 + x_1y_2 + x_2y_1 + 2x_2y_2.$$

(c) Let $V = \{f : [0,1] \to \mathbb{R} : f \text{ is continuous}\}$, $\mathbb{F} = \mathbb{R}$,

$$\langle f, g \rangle = f(0)g(0) + f(1)g(1) + f(2)g(2).$$

(d) Let $V = \mathbb{R}_2[X]$, $\mathbb{F} = \mathbb{R}$,

$$\langle f, g \rangle = f(0)g(0) + f(1)g(1) + f(2)g(2).$$

(e) Let $V = \{f : [0,1] \to \mathbb{C} : f \text{ is continuous}\}$, $\mathbb{F} = \mathbb{C}$,

$$\langle f, g \rangle = \int_0^1 f(x)\overline{g(x)}(x^2 + 1)dx.$$

Exercise 5.7.2 For the following, check whether $\| \cdot \|$ is a norm.

(a) $V = \mathbb{C}^2$, $\mathbb{F} = \mathbb{C}$,

$$\left\| \begin{pmatrix} x_1 \\ x_2 \end{pmatrix} \right\| = x_1^2 + x_2^2.$$

(b) $V = \mathbb{C}^2$, $\mathbb{F} = \mathbb{C}$,

$$\left\| \begin{pmatrix} x_1 \\ x_2 \end{pmatrix} \right\| = |x_1| + 2|x_2|.$$

(c) Let $V = \{ f : [0,2] \to \mathbb{R} \; : f \text{ is continuous} \}$, $\mathbb{F} = \mathbb{R}$,

$$\|f\| = \int_0^2 |f(x)|(1-x)dx.$$

(d) Let $V = \{ f : [0,1] \to \mathbb{R} \; : f \text{ is continuous} \}$, $\mathbb{F} = \mathbb{R}$,

$$\|f\| = \int_0^1 |f(x)|(1-x)dx.$$

Exercise 5.7.3 Let $\mathbf{v}_1, \ldots, \mathbf{v}_n$ be nonzero orthogonal vectors in an inner product space V. Show that $\{\mathbf{v}_1, \ldots, \mathbf{v}_n\}$ is linearly independent.

Exercise 5.7.4 Let V be an inner product space.

(a) Determine $\{\mathbf{0}\}^\perp$ and $V^\perp$.

(b) Let $V = \mathbb{C}^4$ and $W = \{ \begin{pmatrix} 1 \\ i \\ 1+i \\ 2 \end{pmatrix}, \begin{pmatrix} 0 \\ -i \\ 1+2i \\ 0 \end{pmatrix} \}$. Find a basis for $W^\perp$.

(c) In case V is finite dimensional and W is a subspace, show that $\dim W^\perp = \dim V - \dim W$. (Hint: start with an orthonormal basis for W and add vectors to it to obtain an orthonormal basis for V).

Exercise 5.7.5 Let $\langle \cdot, \cdot \rangle$ be the Euclidean inner product on $\mathbb{F}^n$, and $\| \cdot \|$ the associated norm.

(a) Let $\mathbb{F} = \mathbb{C}$. Show that $A \in \mathbb{C}^{n \times n}$ is the zero matrix if and only if $\langle A\mathbf{x}, \mathbf{x} \rangle = 0$ for all $\mathbf{x} \in \mathbb{C}^n$. (Hint: for $\mathbf{x}, \mathbf{y} \in \mathbb{C}$, use that $\langle A(\mathbf{x} + \mathbf{y}), \mathbf{x} + \mathbf{y} \rangle = 0 = \langle A(\mathbf{x} + i\mathbf{y}), \mathbf{x} + i\mathbf{y} \rangle$.)

(b) Show that when $\mathbb{F} = \mathbb{R}$, there exists nonzero matrices $A \in \mathbb{R}^{n \times n}$, $n > 1$, so that $\langle A\mathbf{x}, \mathbf{x} \rangle = 0$ for all $\mathbf{x} \in \mathbb{R}^n$.

(c) For $A \in \mathbb{C}^{n \times n}$ define

$$w(A) = \max_{\mathbf{x} \in \mathbb{C}^n, \|\mathbf{x}\|=1} |\langle A\mathbf{x}, \mathbf{x} \rangle|. \tag{5.20}$$

Show that $w(\cdot)$ is a norm on $\mathbb{C}^{n \times n}$. This norm is called the *numerical radius* of A.

(d) Explain why $\max_{\mathbf{x} \in \mathbb{R}^n, \|\mathbf{x}\|=1} |\langle A\mathbf{x}, \mathbf{x} \rangle|$ does not define a norm.

Exercise 5.7.6 Find an orthonormal basis for the subspace in $\mathbb{R}^4$ spanned by

$$\begin{pmatrix} 1 \\ 1 \\ 1 \\ 1 \end{pmatrix}, \begin{pmatrix} 1 \\ 2 \\ 1 \\ 2 \end{pmatrix}, \begin{pmatrix} 3 \\ 1 \\ 3 \\ 1 \end{pmatrix}.$$

Exercise 5.7.7

Let $V = \mathbb{R}[t]$ over the field $\mathbb{R}$. Define the inner product

$$\langle p, q \rangle := \int_{-1}^{1} p(t)q(t)dt.$$

For the following linear maps on V, determine whether they are self-adjoint.

(a) $Lp(t) := (t^2 + 1)p(t)$.

(b) $Lp(t) := \frac{dp}{dt}(t)$.

(c) $Lp(t) = -p(-t)$.

Exercise 5.7.8 Let $V = \mathbb{R}[t]$ over the field $\mathbb{R}$. Define the inner product

$$\langle p, q \rangle := \int_{0}^{2} p(t)q(t)dt.$$

For the following linear maps on V, determine whether they are unitary.

(a) $Lp(t) := tp(t)$.

(b) $Lp(t) = -p(2 - t)$.

Exercise 5.7.9 Let $U : V \to V$ be unitary, where the inner product on V is denoted by $\langle \cdot, \cdot \rangle$.

(a) Show that $|\langle \mathbf{x}, U\mathbf{x} \rangle| \leq \|\mathbf{x}\|^2$ for all $\mathbf{x}$ in V.

(b) Show that $|\langle \mathbf{x}, U\mathbf{x} \rangle| = \|\mathbf{x}\|^2$ for all $\mathbf{x}$ in V, implies that $U = \alpha I$ for some $|\alpha| = 1$.

Exercise 5.7.10 Let $V = \mathbb{C}^{n \times n}$, and define

$$\langle A, B \rangle = \operatorname{tr}(AB^*).$$

(a) Let $W = \operatorname{span}\left\{ \begin{pmatrix} 1 & 2 \\ 0 & 1 \end{pmatrix}, \begin{pmatrix} 1 & 0 \\ 2 & 1 \end{pmatrix} \right\}$. Find an orthonormal basis for W.

(b) Find a basis for $W^\perp := \{B \in V \; : \; B \perp C \text{ for all } C \in W\}$.

Exercise 5.7.11 Let $A \in \mathbb{C}^{n \times n}$. Show that if A is normal and $A^k = 0$ for some $k \in \mathbb{N}$, then $A = 0$.

Exercise 5.7.12 Let $A \in \mathbb{C}^{n \times n}$ and $a \in \mathbb{C}$. Show that A is normal if and only if $A - aI$ is normal.

Exercise 5.7.13 Show that the sum of two Hermitian matrices is Hermitian. How about the product?

Exercise 5.7.14 Show that the product of two unitary matrices is unitary. How about the sum?

Exercise 5.7.15 Is the product of two normal matrices is normal? How about the sum?

Exercise 5.7.16 Show that the following matrices are unitary.

(a) $\frac{1}{\sqrt{2}} \begin{pmatrix} 1 & 1 \\ 1 & -1 \end{pmatrix}$.

(b) $\frac{1}{\sqrt{3}} \begin{pmatrix} 1 & 1 & 1 \\ 1 & e^{\frac{2i\pi}{3}} & e^{\frac{4i\pi}{3}} \\ 1 & e^{\frac{4i\pi}{3}} & e^{\frac{8i\pi}{3}} \end{pmatrix}$.

(c) $\frac{1}{2} \begin{pmatrix} 1 & 1 & 1 & 1 \\ 1 & i & -1 & -i \\ 1 & -1 & 1 & -1 \\ 1 & -i & -1 & i \end{pmatrix}$.

(d) Can you guess the general rule? (Hint: the answer is in Proposition 7.4.3).

Exercise 5.7.17 For the following matrices A find the spectral decomposition UDU^* of A.

(a) $A = \begin{pmatrix} 2 & i \\ -i & 2 \end{pmatrix}$.

(b) $A = \begin{pmatrix} 2 & \sqrt{3} \\ \sqrt{3} & 4 \end{pmatrix}$.

(c) $A = \begin{pmatrix} 3 & 1 & 1 \\ 1 & 3 & 1 \\ 1 & 1 & 3 \end{pmatrix}$.

(d) $A = \begin{pmatrix} 0 & 1 & 0 \\ 0 & 0 & 1 \\ 1 & 0 & 0 \end{pmatrix}$.

Exercise 5.7.18 Let $A = \begin{pmatrix} 3 & 2i \\ -2i & 3 \end{pmatrix}$.

(a) Show that A is positive semidefinite.

(b) Find the positive *square root* of A; that is, find a positive semidefinite B so that $B^2 = A$. We denote B by $A^{\frac{1}{2}}$.

Exercise 5.7.19 Let $A \in \mathbb{C}^{n \times n}$ be positive semidefinite, and let $k \in \mathbb{N}$. Show that there exists a unique positive semidefinite B so that $B^k = A$. We call B the *kth root* of A and denote $B = A^{\frac{1}{k}}$.

Exercise 5.7.20 Let $A \in \mathbb{C}^{n \times n}$ be positive semidefinite. Show that

$$\lim_{k \to \infty} \operatorname{tr} A^{\frac{1}{k}} = \operatorname{rank} A.$$

(Hint: use that for $\lambda > 0$ we have that $\lim_{k \to \infty} \lambda^{\frac{1}{k}} = 1$.)

Exercise 5.7.21 Let $A = A^*$ be an $n \times n$ Hermitian matrix, with eigenvalues $\lambda_1 \geq \cdots \geq \lambda_n$.

(a) Show $tI - A$ is positive semidefinite if and only if $t \geq \lambda_1$.

(b) Show that $\lambda_{\max}(A) = \lambda_1 = \max_{\langle \mathbf{x}, \mathbf{x} \rangle = 1} \langle A\mathbf{x}, \mathbf{x} \rangle$, where $\langle \cdot, \cdot \rangle$ is the Euclidean inner product.

(c) Let $\hat{A}$ be the matrix obtained from A by removing row and column i. Then $\lambda_{\max}(\hat{A}) \le \lambda_{\max}(A)$.

Exercise 5.7.22 (a) Show that a square matrix A is Hermitian iff $A^2 = A^*A$.

(b) Let H be positive semidefinite, and write $H = A + iB$ where A and B are real matrices. Show that if A is singular, then H is singular as well.

Exercise 5.7.23 (a) Let A be positive definite. Show that $A + A^{-1} - 2I$ is positive semidefinite.

(b) Show that A is normal if and only if $A^* = AU$ for some unitary matrix U.

Exercise 5.7.24 Find a QR factorization of $\begin{pmatrix} 1 & 1 & 0 \\ 1 & 0 & 1 \\ 0 & 1 & 1 \end{pmatrix}$.

Exercise 5.7.25 Find the Schur factorization $A = UTU^*$, with U unitary and T triangular, for the matrix

$$A = \begin{pmatrix} -1 & -2 & 3 \\ 2 & 4 & -2 \\ 1 & -2 & 1 \end{pmatrix}.$$

Note: 2 is an eigenvalue of A.

Exercise 5.7.26 Let

$$T = \begin{pmatrix} A & B \\ C & D \end{pmatrix} \tag{5.21}$$

be a block matrix, and suppose that D is invertible. Define the *Schur complement* S of D in T by $S = A - BD^{-1}C$. Show that rank $T = \text{rank}(A - BD^{-1}C) + \text{rank } D$.

Exercise 5.7.27 Using Sylvester's law of inertia, show that if

$$M = \begin{pmatrix} A & B \\ B^* & C \end{pmatrix} = M^* \in \mathbb{C}^{(n+m)\times(n+m)}$$

with C invertible, then

$$\text{In } M = \text{In } C + \text{In}(A - BC^{-1}B^*). \tag{5.22}$$

(Hint: Let $S = \begin{pmatrix} I & 0 \\ -B^*A^{-1} & I \end{pmatrix}$ and compute SMS^*.)

Exercise 5.7.28 Determine the singular value decomposition of the following matrices.

(a) $A = \begin{pmatrix} 1 & 1 & 2\sqrt{2}i \\ -1 & -1 & 2\sqrt{2}i \\ \sqrt{2}i & -\sqrt{2}i & 0 \end{pmatrix}$.

(b) $A = \begin{pmatrix} -2 & 4 & 5 \\ 6 & 0 & -3 \\ 6 & 0 & -3 \\ -2 & 4 & 5 \end{pmatrix}$.

Exercise 5.7.29 Let A be a 4×4 matrix with spectrum $\sigma(A) = \{-2i, 2i, 3+i, 3+4i\}$ and singular values $\sigma_1 \geq \sigma_2 \geq \sigma_3 \geq \sigma_4$.

(a) Determine the product $\sigma_1 \sigma_2 \sigma_3 \sigma_4$.

(b) Show that $\sigma_1 \geq 5$.

(c) Assuming A is normal, determine $\text{tr}(A + AA^*)$.

Exercise 5.7.30 Let $A = \begin{pmatrix} P & Q \\ R & S \end{pmatrix} \in \mathbb{C}^{(k+l) \times (m+n)}$, where P is of size $k \times m$. Show that

$$\sigma_1(P) \leq \sigma_1(A).$$

Conclude that $\sigma_1(Q) \leq \sigma_1(A), \sigma_1(R) \leq \sigma_1(A), \sigma_1(S) \leq \sigma_1(A)$ as well.

Exercise 5.7.31 This is an exercise that uses MATLAB®[1], and its purpose is to show what happens with an image if you take a low rank approximation of it.

1. Take an image.

2. Load it into MATLAB® (using "imread"). This produces a matrix (three matrices (organized as a three-dimensional array for a color image). The elements are of type "uint8."

3. Convert the elements to type "double" (using the command "double"); otherwise you cannot do computations.

---

[1]MATLAB® is a trademark of TheMathWorks, Inc., and is used with permission. The-MathWorks does not warrant the accuracy of the text or exercises in this book. This book's use or discussion of MATLAB® software or related products does not constitute endorsement or sponsorship by TheMathWorks of a particular pedagogical approach or particular use of the MATLAB® software

4. Take a singular value decomposition (using "svd").

5. Keep only the first k largest singular values.

6. Compute the rank k approximation.

7. Look at the image (using "imshow").

Here are the commands I used on a color image (thus the array has three levels) with $k = 30$:

A=imread(Hugo2.png);

AA=double(cdata);

[U,S,V]=svd(AA(:,:,1));

[U2,S2,V2]=svd(AA(:,:,2));

[U3,S3,V3]=svd(AA(:,:,3));

H=zeros(size(S,1),size(S,2));

for i=1:30, H(i,i)=1; end;

Snew=S.*H;

Snew2=S2.*H;

Snew3=S3.*H;

Anew(:,:,1)=U*Snew*V';

Anew(:,:,2)=U2*Snew2*V2';

Anew(:,:,3)=U3*Snew3*V3';

Anew=uint8(Anew);

imshow(Anew)

Exercise 5.7.32 The *condition number* $\kappa(A)$ of an invertible $n \times n$ matrix A is given by $\kappa(A) = \frac{\sigma_1(A)}{\sigma_n(A)}$, where $\sigma_1(A) \geq \cdots \geq \sigma_n(A)$ are the singular values of A. Show that for all invertible matrices A and B, we have that $\kappa(AB) \leq \kappa(A)\kappa(B)$. (Hint: use that $\sigma_1(A^{-1}) = (\sigma_n(A))^{-1}$ and (5.18).)

Exercise 5.7.33 Prove that if X and Y are positive definite $n \times n$ matrices such that $Y - X$ is positive semidefinite, then $\det X \leq \det Y$. Moreover, $\det X = \det Y$ if and only if $X = Y$.

Exercise 5.7.34 *(Least squares solution)* When the equation $Ax = b$ does not have a solution, one may be interested in finding an x so that $\|Ax - b\|$ is minimal. Such an x is called a *least squares solution* to $Ax = b$. In this exercise we will show that if $A = QR$, with R invertible, then the least squares solution is given by $x = R^{-1}Q^*b$. Let $A \in \mathbb{F}^{n \times m}$ with rank $A = m$.

(a) Let $A = QR$ be a QR-factorization of A. Show that Ran $A =$ Ran Q.

(b) Observe that $QQ^*b \in$ Ran Q. Show that for all $v \in$ Ran Q we have $\|v - b\| \geq \|QQ^*b - b\|$ and that the inequality is strict if $v \neq QQ^*b$.

(c) Show that $x := R^{-1}Q^*b$ is the least squares solution to $Ax = b$.

(d) Let $A = \begin{pmatrix} 1 & 1 \\ 2 & 1 \\ 3 & 1 \end{pmatrix}$ and $b = \begin{pmatrix} 3 \\ 5 \\ 4 \end{pmatrix}$. Find the least squares solution to $Ax = b$.

(e) In trying to fit a line $y = cx + d$ through the points $(1,3)$, $(2,5)$, and $(3,4)$, one sets up the equations

$$3 = c + d, 5 = 2c + d, 4 = 3c + d.$$

Writing this in matrix form we get

$$A \begin{pmatrix} c \\ d \end{pmatrix} = b,$$

where A and b are as above. One way to get a "fitting line" $y = cx + d$, is to solve for c and d via least squares, as we did in the previous part. This is the most common way to find a so-called *regression line*. Plot the three points $(1,3)$, $(2,5)$, and $(3,4)$ and the line $y = cx + d$, where c and d are found via least squares as in the previous part.

Exercise 5.7.35 Let A, X be $m \times m$ matrices such that $A = A^*$ is invertible and

$$H := A - X^*AX \tag{5.23}$$

is positive definite.

(a) Show that X has no eigenvalues on the unit circle $\mathbb{T} = \{z \in \mathbb{C} : |z| = 1\}$.

(b) Show that A is positive definite if and only if X has all eigenvalues in $\mathbb{D} = \{z \in \mathbb{C} : |z| < 1\}$. (Hint: When X has all eigenvalues in $\mathbb{D}$, we have that $X^n \to 0$ as $n \to \infty$. Use this to show that $A = H + \sum_{k=1}^{\infty} X^{*k} H X^k$.)

Exercise 5.7.36 *(Honors)* On both $\mathbb{C}^{4\times4}$ and $\mathbb{C}^{6\times6}$, we have the inner product given via $\langle A, B \rangle = \operatorname{tr}(B^*A)$. Let $T : \mathbb{C}^{4\times4} \to \mathbb{C}^{6\times6}$ be given via

$$T\left((m_{ij})_{i,j=1}^{4}\right) := \begin{pmatrix} m_{11} & m_{12} & m_{13} & m_{14} & m_{13} & m_{14} \\ m_{21} & m_{22} & m_{23} & m_{24} & m_{23} & m_{24} \\ m_{31} & m_{32} & m_{33} & m_{34} & m_{33} & m_{34} \\ m_{41} & m_{42} & m_{43} & m_{44} & m_{43} & m_{44} \\ m_{31} & m_{32} & m_{33} & m_{34} & m_{33} & m_{34} \\ m_{41} & m_{42} & m_{43} & m_{44} & m_{43} & m_{44} \end{pmatrix}.$$

Determine the dual of T.

Exercise 5.7.37 *(Honors)* Let A have no eigenvalues on the unit circle, and let $C = -(A^* + I)(A^* - I)^{-1}$.

(a) Show that C is well-defined.

(b) Show that A satisfies the *Stein equation* $H - A^*HA = V$, with V positive definite, if and only if C satisfies a *Lyapunov equation* $CH + HC^* = G$ with G positive definite.

(c) With C as above, show that C has no purely imaginary eigenvalues.

(d) Show that H is positive definite if and only if C has all its eigenvalues in the right half-plane $\mathbb{H} = \{z \in \mathbb{C} : \operatorname{Re} z > 0\}$. (Hint: use Exercise 5.7.35.)

Example 5.7.38 *(Honors)* Let A have all its eigenvalues in the left half-plane $-\mathbb{H} = \{z \in \mathbb{C} : \operatorname{Re} z < 0\}$, and let C be a positive semidefinite matrix of the same size. Show that

$$X = \int_{0}^{\infty} e^{At} C e^{A^*t} dt$$

exists (where an integral of a matrix function is defined entrywise), is positive semidefinite, and satisfies the Lyapunov equation $XA + A^*X = -C$.

6

Constructing New Vector Spaces from Given Ones

CONTENTS

6.1	The Cartesian product	147
6.2	The quotient space	149
6.3	The dual space	157
6.4	Multilinear maps and functionals	166
6.5	The tensor product	168
6.6	Anti-symmetric and symmetric tensors	179
6.7	Exercises	189

In this chapter we study several useful constructions that yield a new vector space based on given ones. We also study how inner products and linear maps yield associated constructions.

6.1 The Cartesian product

Given vector spaces $V_1, \ldots, V_k$ over the same field $\mathbb{F}$, the *Cartesian product* vector space $V_1 \times \cdots \times V_k$ is defined via

$$V_1 \times \cdots \times V_k = \left\{ \begin{pmatrix} \mathbf{v}_1 \\ \vdots \\ \mathbf{v}_k \end{pmatrix} : \mathbf{v}_i \in V_i, i = 1, \ldots, k \right\},$$

$$\begin{pmatrix} \mathbf{v}_1 \\ \vdots \\ \mathbf{v}_k \end{pmatrix} + \begin{pmatrix} \mathbf{w}_1 \\ \vdots \\ \mathbf{w}_k \end{pmatrix} := \begin{pmatrix} \mathbf{v}_1 + \mathbf{w}_1 \\ \vdots \\ \mathbf{v}_k + \mathbf{w}_k \end{pmatrix},$$

and

$$c \begin{pmatrix} \mathbf{v}_1 \\ \vdots \\ \mathbf{v}_k \end{pmatrix} := \begin{pmatrix} c\mathbf{v}_1 \\ \vdots \\ c\mathbf{v}_k \end{pmatrix}.$$

Clearly, one may view $\mathbb{F}^k$ as the Cartesian product $\mathbb{F} \times \cdots \times \mathbb{F}$ (where $\mathbb{F}$ appears k times). Sometimes $V_1 \times \cdots \times V_k$ is viewed as a direct sum

$$(V_1 \times \{\mathbf{0}\} \times \cdots \times \{\mathbf{0}\}) \dot{+} (\{\mathbf{0}\} \times V_2 \times \{\mathbf{0}\} \times \cdots \times \{\mathbf{0}\}) \dot{+} \cdots \cdots \dot{+} (\{\mathbf{0}\} \times \cdots \times \{\mathbf{0}\} \times V_k).$$

It is not hard to determine the dimension of a Cartesian product.

Proposition 6.1.1 *Let $V_1, \ldots, V_k$ be finite-dimensional vector sapces. Then*

$$\dim(V_1 \times \cdots \times V_k) = \dim V_1 + \cdots + \dim V_k.$$

Proof. Let $\mathcal{B}_i$ be a basis for V_i, $i = 1, \ldots, k$. Put

$$\mathcal{B} = \{ \begin{pmatrix} \mathbf{b}_1 \\ 0 \\ \vdots \\ 0 \end{pmatrix} : \mathbf{b}_1 \in \mathcal{B}_1 \} \cup \{ \begin{pmatrix} 0 \\ \mathbf{b}_2 \\ \vdots \\ 0 \end{pmatrix} : \mathbf{b}_2 \in \mathcal{B}_2 \} \cup \ldots \cup \{ \begin{pmatrix} 0 \\ \vdots \\ 0 \\ \mathbf{b}_k \end{pmatrix} : \mathbf{b}_k \in \mathcal{B}_k \}.$$

It is easy to check that $\mathcal{B}$ is a basis for $V_1 \times \cdots \times V_k$.   □

When V_i has an inner product $\langle \cdot, \cdot \rangle_i$, $i = 1, \ldots, k$, then it is straightforward to check that

$$\langle \begin{pmatrix} \mathbf{v}_1 \\ \vdots \\ \mathbf{v}_k \end{pmatrix}, \begin{pmatrix} \mathbf{w}_1 \\ \vdots \\ \mathbf{w}_k \end{pmatrix} \rangle := \sum_{i=1}^{k} \langle \mathbf{v}_i, \mathbf{w}_i \rangle_i$$

defines an inner product on $V_1 \times \cdots \times V_k$. While this is the default way to make an inner product on the Cartesian product, one can also take for instance $\sum_{i=1}^{k} \beta_i \langle \mathbf{v}_i, \mathbf{w}_i \rangle_i$, where $\beta_i > 0$, $i = 1, \ldots, k$.

When V_i has a norm $\| \cdot \|_i$, $i = 1, \ldots, k$, then there are infinitely many ways to make a norm on $V_1 \times \cdots \times V_k$. For instance, one can take any $p \geq 1$, and put

$$\| \begin{pmatrix} \mathbf{v}_1 \\ \vdots \\ \mathbf{v}_k \end{pmatrix} \|_p := \sqrt[p]{\sum_{i=1}^{k} \| \mathbf{v}_i \|_i^p}.$$

It takes some effort to prove that this is a norm, and we will outline the proof in Exercise 6.7.1. Also,

$$\| \begin{pmatrix} \mathbf{v}_1 \\ \vdots \\ \mathbf{v}_k \end{pmatrix} \|_\infty := \max_{i=1,\ldots,k} \| \mathbf{v}_i \|$$

defines a norm on the Cartesian product.

Finally, when $A_{ij} : V_j \to V_i$, $1 \leq i, j \leq k$, are linear maps, then

$$A := \begin{pmatrix} A_{11} & \cdots & A_{1k} \\ \vdots & & \vdots \\ A_{k1} & \cdots & A_{kk} \end{pmatrix} : V_1 \times \cdots \times V_k \to V_1 \times \cdots \times V_k$$

defines a linear map via usual block matrix multiplication

$$A \begin{pmatrix} \mathbf{v}_1 \\ \vdots \\ \mathbf{v}_k \end{pmatrix} = \begin{pmatrix} \sum_{j=1}^{k} A_{1j}\mathbf{v}_j \\ \vdots \\ \sum_{j=1}^{k} A_{kj}\mathbf{v}_j \end{pmatrix}.$$

A similar construction also works when $A_{ij} : V_j \to W_i$, $1 \leq i \leq l$, $1 \leq j \leq k$. Then $A = (A_{ij})_{i=1, j=1}^{l \quad k}$ acts $V_1 \times \cdots \times V_k \to W_1 \times \cdots \times W_l$.

6.2 The quotient space

Let V be a vector space over $\mathbb{F}$ and $W \subseteq V$ a subspace. We define the relation $\sim$ via

$$\mathbf{v}_1 \sim \mathbf{v}_2 \iff \mathbf{v}_1 - \mathbf{v}_2 \in W.$$

Then $\sim$ is an equivalence relation:

  (i) *Reflexivity:* $\mathbf{v} \sim \mathbf{v}$ for all $\mathbf{v} \in V$, since $\mathbf{v} - \mathbf{v} = \mathbf{0} \in W$.

 (ii) *Symmetry:* Suppose $\mathbf{v}_1 \sim \mathbf{v}_2$. Then $\mathbf{v}_1 - \mathbf{v}_2 \in W$. Thus $-(\mathbf{v}_1 - \mathbf{v}_2) = \mathbf{v}_2 - \mathbf{v}_1 \in W$, which yields $\mathbf{v}_2 \sim \mathbf{v}_1$.

(iii) *Transitivity:* Suppose $\mathbf{v}_1 \sim \mathbf{v}_2$ and $\mathbf{v}_2 \sim \mathbf{v}_3$. Then $\mathbf{v}_1 - \mathbf{v}_2 \in W$ and $\mathbf{v}_2 - \mathbf{v}_3 \in W$. Thus $\mathbf{v}_1 - \mathbf{v}_3 = (\mathbf{v}_1 - \mathbf{v}_2) + (\mathbf{v}_2 - \mathbf{v}_3) \in W$. This yields $\mathbf{v}_1 \sim \mathbf{v}_3$.

As $\sim$ is an equivalence relation, it has equivalence classes, which we will denote as $\mathbf{v} + W$:

$$\mathbf{v} + W := \{\hat{\mathbf{v}} : \mathbf{v} \sim \hat{\mathbf{v}}\} = \{\hat{\mathbf{v}} : \mathbf{v} - \hat{\mathbf{v}} \in W\} =$$

$$\{\hat{\mathbf{v}} : \text{there exists } \mathbf{w} \in W \text{ with } \hat{\mathbf{v}} = \mathbf{v} + \mathbf{w}\}.$$

Any member of an equivalence class is called a *representative* of the equivalence class.

Example 6.2.1 Let $V = \mathbb{R}^2$ and $W = \text{Span}\{\mathbf{e}_1\}$. Then the equivalence class of $\mathbf{v} = \begin{pmatrix} v_1 \\ v_2 \end{pmatrix}$ is the horizontal line through $\mathbf{v}$. In this example it is simple to see how one would add two equivalence classes. Indeed, to add the horizontal line through $\begin{pmatrix} 0 \\ c \end{pmatrix}$ to the horizontal line through $\begin{pmatrix} 0 \\ d \end{pmatrix}$, would result in the horizontal line through $\begin{pmatrix} 0 \\ c+d \end{pmatrix}$. Or, what is equivalent, to add the horizontal line through $\begin{pmatrix} 5 \\ c \end{pmatrix}$ to the horizontal line through $\begin{pmatrix} 10 \\ d \end{pmatrix}$, would result in the horizontal line through $\begin{pmatrix} 15 \\ c+d \end{pmatrix}$. Similarly, one can define scalar multiplication for these equivalence classes. We give the general definition below.

The set of equivalence classes is denoted by V/W:

$$V/W := \{\mathbf{v} + W : \mathbf{v} \in V\}.$$

We define addition and scalar multiplication on V/W via

$$(\mathbf{v}_1 + W) + (\mathbf{v}_2 + W) := (\mathbf{v}_1 + \mathbf{v}_2) + W, \qquad (6.1)$$

$$c(\mathbf{v} + W) := (c\mathbf{v}) + W. \qquad (6.2)$$

These two operations are defined via representatives (namely, $\mathbf{v}_1, \mathbf{v}_2$, and $\mathbf{v}$) of the equivalence classes, so we need to make sure that if we had chosen different representatives for the same equivalence classes, the outcome would be the same. We do this in the following lemma.

Lemma 6.2.2 *The addition and scalar multiplication on V/W as defined in (6.1) and (6.2) are well-defined.*

Proof. Suppose $\mathbf{v}_1 + W = \mathbf{x}_1 + W$ and $\mathbf{v}_2 + W = \mathbf{x}_2 + W$. Then $\mathbf{v}_1 - \mathbf{x}_1 \in W$ and $\mathbf{v}_2 - \mathbf{x}_2 \in W$. As W is a subspace, it follows that

$$\mathbf{v}_1 + \mathbf{v}_2 - (\mathbf{x}_1 + \mathbf{x}_2) = \mathbf{v}_1 - \mathbf{x}_1 + \mathbf{v}_2 - \mathbf{x}_2 \in W.$$

Thus $(\mathbf{v}_1 + \mathbf{v}_2) + W = (\mathbf{x}_1 + \mathbf{x}_2) + W$ follows.

Next, suppose $\mathbf{v} + W = \mathbf{x} + W$. Then $\mathbf{v} - \mathbf{x} \in W$ and as W is a subspace, it follows that

$$c\mathbf{v} - c\mathbf{x} = c(\mathbf{v} - \mathbf{x}) \in W.$$

Thus $(c\mathbf{v}) + W = (c\mathbf{x}) + W$. $\qquad \square$

It is now a straightforward (and tedious) exercise to show that V/W with addition and scalar multiplication defined via (6.1) and (6.2), yields a vector space, called the *quotient space*. Let us next determine its dimension.

Proposition 6.2.3 *Let V be a finite-dimensional vector space and W a subspace. Then*

$$\dim V/W = \dim V - \dim W.$$

Proof. Choose a basis $\{\mathbf{w}_1, \ldots, \mathbf{w}_l\}$ for W, and complement this linearly independent set with vectors $\{\mathbf{v}_1, \ldots, \mathbf{v}_k\}$ in V, so that the resulting set

$$\{\mathbf{w}_1, \ldots, \mathbf{w}_l, \mathbf{v}_1, \ldots, \mathbf{v}_k\}$$

is a basis for V (see in Exercise 2.6.8 why this is always possible). We now claim that

$$\mathcal{B} = \{\mathbf{v}_1 + W, \ldots, \mathbf{v}_k + W\}$$

is a basis for V/W, which then proves the proposition.

First, let us prove that $\mathcal{B}$ is a linearly independent set. Suppose that

$$c_1(\mathbf{v}_1 + W) + \cdots + c_k(\mathbf{v}_k + W) = \mathbf{0} + W$$

(where we use the observation that $\mathbf{0} + W$ is the neutral element for addition in V/W). Then

$$c_1\mathbf{v}_1 + \cdots + c_k\mathbf{v}_k - \mathbf{0} \in W.$$

Thus there exist $d_1, \ldots, d_l$ so that

$$c_1\mathbf{v}_1 + \cdots + c_k\mathbf{v}_k = d_1\mathbf{w}_1 + \cdots + d_l\mathbf{w}_l.$$

This gives that

$$c_1\mathbf{v}_1 + \cdots + c_k\mathbf{v}_k + (-d_1)\mathbf{w}_1 + \cdots + (-d_l)\mathbf{w}_l = \mathbf{0}.$$

As $\{\mathbf{w}_1, \ldots, \mathbf{w}_l, \mathbf{v}_1, \ldots, \mathbf{v}_k\}$ is a linearly independent set, we get that $c_1 = \cdots = c_k = d_1 = \ldots = d_l = 0$. Thus, in particular $c_1 = \cdots = c_k = 0$, yielding that $\mathcal{B}$ is a linearly independent set.

Next, we need to show that $\mathcal{B}$ spans V/W. Let $\mathbf{v} + W \in V/W$. As $\mathbf{v} \in V$, there exist $c_1, \ldots, c_k, d_1, \ldots, d_l \in \mathbb{F}$ so that

$$\mathbf{v} = c_1\mathbf{v}_1 + \cdots + c_k\mathbf{v}_k + d_1\mathbf{w}_1 + \cdots + d_l\mathbf{w}_l.$$

But then

$$\mathbf{v} - (c_1\mathbf{v}_1 + \cdots + c_k\mathbf{v}_k) = d_1\mathbf{w}_1 + \cdots + d_l\mathbf{w}_l \in W,$$

and thus

$$\mathbf{v} + W = (c_1\mathbf{v}_1 + \cdots + c_k\mathbf{v}_k) + W = c_1(\mathbf{v}_1 + W) + \cdots + c_k(\mathbf{v}_k + W).$$

$\square$

In case a finite-dimensional vector space V has an inner product $\langle \cdot, \cdot \rangle$, then the spaces V/W and

$$W^{\perp} = \{\mathbf{v} \in V : \langle \mathbf{v}, \mathbf{w} \rangle = 0 \text{ for every } \mathbf{w} \in W\}$$

are isomorphic. This follows immediately from a dimension count (see Exercise 5.7.4), but let us elaborate and provide the explicit isomorphism in the proof below.

Proposition 6.2.4 *Let V be a finite-dimensional vector space with an inner product $\langle \cdot, \cdot \rangle$, and let $W \subseteq V$ be a subspace. Then V/W and $W^{\perp}$ are isomorphic.*

Proof. Let $T : W^{\perp} \to V/W$ be defined by $T(\mathbf{v}) = \mathbf{v} + W$. We claim that T is an isomorphism. Clearly, T is linear. Next, let $\mathbf{v} \in W^{\perp}$ be so that $\mathbf{v} + W = \mathbf{0} + W$. Then $\mathbf{v} = \mathbf{v} - \mathbf{0} \in W$. We also have that $\mathbf{v} \in W^{\perp}$, and thus $\mathbf{v}$ is orthogonal to itself: $\langle \mathbf{v}, \mathbf{v} \rangle = 0$. But then $\mathbf{v} = \mathbf{0}$ follows, and thus $\mathrm{Ker} T = \{\mathbf{0}\}$. As the dimensions of V/W and $W^{\perp}$ are the same, we also obtain that T is onto. $\square$

When V has a norm $\|\cdot\|$, we say that a subset $W \subseteq V$ is *closed* with respect to the norm $\|\cdot\|$, if $\mathbf{w}_n \in W$, $n \in \mathbb{N}$, and $\lim_{n\to\infty} \|\mathbf{w}_n - \mathbf{v}\|$ implies that $\mathbf{v} \in W$. In finite-dimensional vector spaces all subspaces are closed, as the following proposition shows.

Proposition 6.2.5 *If V is a finite-dimensional vector space with a norm $\|\cdot\|$, and W is a subspace, then W is closed.*

Proof. Let $\{\mathbf{v}_1, \ldots, \mathbf{v}_k\}$ be a basis for W, and extend it to a basis $\mathcal{B} = \{\mathbf{v}_1, \ldots, \mathbf{v}_n\}$ for V. Define the norm $\|\cdot\|_V$ via

$$\|\mathbf{v}\|_V = \|[\mathbf{v}]_{\mathcal{B}}\|_E,$$

where $\|\cdot\|_E$ is the Euclidean norm on $\mathbb{F}^n$. In other words, if $\mathbf{v} = \sum_{i=1}^n c_i \mathbf{v}_i$, then $\|\mathbf{v}\|_V = \sqrt{\sum_{i=1}^n |c_i|^2}$.

Let $\mathbf{w}^{(m)}$, $m = 1, 2, \ldots$, be vectors in W, and suppose that $\lim_{m\to\infty} \|\mathbf{w}^{(m)} - \mathbf{v}\| = 0$ for some $\mathbf{v} \in V$. By Theorem 5.1.25 any two norms on a finite-dimensional space are equivalent, thus we also have $\lim_{m\to\infty} \|\mathbf{w}^{(m)} - \mathbf{v}\|_V = 0$. We need to prove that $\mathbf{v} \in W$. As $\{\mathbf{v}_1, \ldots, \mathbf{v}_k\}$ is a basis for W, we obtain $\mathbf{w}^{(m)} = c_1^{(m)} \mathbf{v}_1 + \cdots + c_k^{(m)} \mathbf{v}_k$ for some scalars $c_1^{(m)}, \ldots, c_k^{(m)}$. In addition, we have $\mathbf{v} = \sum_{i=1}^n c_i \mathbf{v}_i$ for some scalars c_i. Then for $j = k+1, \ldots, n$ we observe

$$|c_j|^2 \leq \sum_{i=1}^k |c_i^{(m)} - c_i|^2 + \sum_{i=k+1}^n |c_i|^2 = \|\mathbf{w}^{(m)} - \mathbf{v}\|_V^2,$$

and thus $|c_j| \leq \lim_{m \to \infty} \|\mathbf{w}^{(m)} - \mathbf{v}\|_V = 0$. Consequently, $c_{k+1} = \cdots = c_n = 0$, yielding that $\mathbf{v} \in W$. $\qquad \square$

The following example shows that in infinite dimensions, not all subspaces are closed.

Example 6.2.6 Let

$$V = \{\mathbf{x} = (x_j)_{j=1}^{\infty} : \|\mathbf{x}\|_V := \sum_{j=1}^{\infty} |x_j| < \infty\}.$$

Thus V consists of vectors with infinitely many entries whose absolute values have a finite sum. As an example, since $\sum_{j=1}^{\infty} \frac{1}{j^2} (= \frac{\pi^2}{6}) < \infty$,

$$\mathbf{v} = (1, \frac{1}{4}, \frac{1}{9}, \frac{1}{16}, \dots) = (\frac{1}{j^2})_{j=1}^{\infty} \in V.$$

The addition and scalar multiplication are defined entrywise. Thus

$$(x_j)_{j=1}^{\infty} + (y_j)_{j=1}^{\infty} = (x_j + y_j)_{j=1}^{\infty}, \quad c(x_j)_{j=1}^{\infty} = (cx_j)_{j=1}^{\infty}.$$

With these definitions, V is a vector space and $\|\cdot\|_V$ is a norm on V. Let now

$$W = \{\mathbf{x} = (x_j)_{j=1}^{\infty} \in V : \text{only finitely many } x_j \text{ are nonzero}\}.$$

It is clear that W is closed under addition and scalar multiplication, and that $\mathbf{0} = (0, 0, \dots) \in W$. Thus W is a subspace. Moreover, if we let

$$\mathbf{v}_k = (1, \frac{1}{4}, \dots, \frac{1}{k^2}, 0, 0, \dots),$$

then $\mathbf{v}_k \in W$, $k \in \mathbb{N}$. Also, $\lim_{k \to \infty} \|\mathbf{v}_k - \mathbf{v}\|_V = 0$, where $\mathbf{v}$ is as above. However, $\mathbf{v} \notin W$, and thus W is not closed with respect to the norm $\|\cdot\|_V$.

When V has a norm $\|\cdot\|_V$ and the subspace $W \subseteq V$ is closed with respect to $\|\cdot\|_V$, one defines a norm on V/W as follows:

$$\|\mathbf{v} + W\| = \inf_{\mathbf{w} \in W} \|\mathbf{v} + \mathbf{w}\|_V. \tag{6.3}$$

Let us show that this is indeed a norm.

Proposition 6.2.7 *Let V have a norm $\|\cdot\|_V$ and let the subspace $W \subseteq V$ be closed with respect to $\|\cdot\|_V$. Then $\|\cdot\|$ defined via (6.3) defines a norm on V/W.*

Proof. Clearly $\|\mathbf{v} + W\| \geq 0$ for all $\mathbf{v} + W \in V/W$. Next, suppose that $\|\mathbf{v} + W\| = 0$. Then $\inf_{\mathbf{w} \in W} \|\mathbf{v} + \mathbf{w}\|_V = 0$, and thus for every $n \in \mathbb{N}$, there exists a $\mathbf{w}_n \in W$ so that $\|\mathbf{v} + \mathbf{w}_n\|_V < \frac{1}{n}$. Thus $\lim_{n \to \infty} \|\mathbf{v} - (-\mathbf{w}_n)\|_V = 0$, and since $-\mathbf{w}_n \in W$, we use that W is closed to conclude that $\mathbf{v} \in W$. But then $\mathbf{v} + W = \mathbf{0} + W$, taking care of the first property of a norm.

Next,

$$\|c(\mathbf{v} + W)\| = \inf_{\mathbf{w} \in W} \|c\mathbf{v} + \mathbf{w}\|_V = \inf_{\hat{\mathbf{w}} \in W} \|c(\mathbf{v} + \hat{\mathbf{w}})\|_V =$$

$$\inf_{\hat{\mathbf{w}} \in W} |c| \|\mathbf{v} + \hat{\mathbf{w}}\|_V = |c| \|\mathbf{v} + W\|.$$

Finally, for the triangle inequality let $\mathbf{v} + W$ and $\hat{\mathbf{v}} + W$ be in V/W. We show that for every $\epsilon > 0$ we can find $\mathbf{w} \in W$ so that

$$\|\mathbf{v} + \hat{\mathbf{v}} + \mathbf{w}\|_V \leq \|\mathbf{v} + W\| + \|\hat{\mathbf{v}} + W\| + \epsilon. \tag{6.4}$$

Indeed, let $\mathbf{w}_1$ be so that $\|\mathbf{v} + \mathbf{w}_1\|_V \leq \|\mathbf{v} + W\| + \frac{\epsilon}{2}$ and let $\mathbf{w}_2$ be so that $\|\hat{\mathbf{v}} + \mathbf{w}_2\|_V \leq \|\hat{\mathbf{v}} + W\| + \frac{\epsilon}{2}$. Put $\mathbf{w} = \mathbf{w}_1 + \mathbf{w}_2$, and then (6.4) holds. As ϵ was arbitrary, we obtain that

$$\|(\mathbf{v}+W)+(\hat{\mathbf{v}}+W)\| = \|(\mathbf{v}+\hat{\mathbf{v}})+W\| = \inf_{\mathbf{w} \in W} \|\mathbf{v}+\hat{\mathbf{v}}+\mathbf{w}\|_V \leq \|\mathbf{v}+W\|+\|\hat{\mathbf{v}}+W\|.$$

$\square$

Next, we see how a linear map $A : V \to \hat{V}$ induces a map acting $V/W \to \hat{V}/\hat{W}$, provided $A[W] := \{A\mathbf{w} : \mathbf{w} \in W\}$ is a subset of $\hat{W}$.

Proposition 6.2.8 *Let* $A : V \to \hat{V}$ *be linear, and suppose that* W *is a subspace of* V, $\hat{W}$ *a subspace* $\hat{V}$, *so that* $A[W] \subseteq \hat{W}$. *Then* $A_\sim(\mathbf{v} + W) := A\mathbf{v} + \hat{W}$ *defines a linear map* $A_\sim : V/W \to \hat{V}/\hat{W}$.

Proof. We need to check that if $\mathbf{v} + W = \mathbf{x} + W$, then $A\mathbf{v} + \hat{W} = A\mathbf{x} + \hat{W}$. As $\mathbf{v} - \mathbf{x} \in W$, we have that $A(\mathbf{v} - \mathbf{x}) \in \hat{W}$, and thus $A\mathbf{v} + \hat{W} = A\mathbf{x} + \hat{W}$ follows. This makes $A_\sim$ well-defined. The linearity of $A_\sim$ is straightforward to check. $\square$

Typically, the induced map $A_\sim$ is simply denoted by A again. While this is a slight abuse of notation, it usually does not lead to any confusion. We will adopt this convention as well.

The techniques introduced in this section provide a useful way to look at the Jordan canonical form. Let us return to Theorem 4.2.1 and have a nilpotent $A \in \mathbb{F}^n$. The crucial subspaces of $\mathbb{F}^n$ here are

$$W_j := \text{Ker} A^j, j = 0, \ldots, n,$$

as we observed before. We have

$$\{0\} = W_0 \subseteq W_1 \subseteq \cdots \subseteq W_n = \mathbb{F}^n.$$

In addition, the following holds.

Proposition 6.2.9 *We have that* $A[W_{j+1}] \subseteq W_j$. *Moreover, the induced map*

$$A : W_{j+1}/W_j \to W_j/W_{j-1}$$

is one-to-one.

Proof. Let $\mathbf{x} \in W_{l+1}$. Then $A^{l+1}\mathbf{x} = \mathbf{0}$. Thus $A^l(A\mathbf{x}) = \mathbf{0}$, yielding that $A\mathbf{x} \in W_l$. Thus with $V = W_{j+1}$, $\hat{V} = W_j = W$, $\hat{W} = W_{j-1}$, we satisfy the conditions of Proposition 6.2.8, and thus the induced map $A : W_{j+1}/W_j \to W_j/W_{j-1}$ is well-defined.

Next, suppose that $\mathbf{x} + W_j \in W_{j+1}/W_j$ is so that $A(\mathbf{x} + W_j) = \mathbf{0} + W_{j-1}$. Then $A\mathbf{x} \in W_{j-1}$, and thus $\mathbf{0} = A^{j-1}(A\mathbf{x}) = A^j\mathbf{x}$. This gives that $\mathbf{x} \in W_j$, and thus $\mathbf{x} + W_j = \mathbf{0} + W_j$. This proves that $A : W_{j+1}/W_j \to W_j/W_{j-1}$ is one-to-one. $\square$

We let $w_j = \dim W_j/W_{j-1}$. As a consequence of Proposition 6.2.9 we have that a when $\mathcal{B}_{j+1} = \{\mathbf{b}_1^{(j+1)} + W_j, \ldots, \mathbf{b}_{w_{j+1}}^{(j+1)} + W_j\}$ is a basis for W_{j+1}/W_j, then

$$\{A\mathbf{b}_1^{(j+1)} + W_{j-1}, \ldots, A\mathbf{b}_{w_{j+1}}^{(j+1)} + W_{j-1}\}$$

is a linearly independent set in W_j/W_{j-1}. This set can be complemented by vectors $\{\mathbf{x}_{j,1} + W_{j-1}, \ldots, \mathbf{x}_{j,s_j} + W_{j-1}\}$, where $s_j = w_j - w_{j+1}$, so that

$$\mathcal{B}_j := \{\mathbf{x}_{j,1} + W_{j-1}, \ldots, \mathbf{x}_{j,s_j} + W_{j-1}, A\mathbf{b}_1^{(j+1)} + W_{j-1}, \ldots, A\mathbf{b}_{w_{j+1}}^{(j+1)} + W_{j-1}\}$$

is a basis for W_j/W_{j-1}. Starting with a basis for W_n/W_{n-1} and repeating the iteration outlined in this paragraph, one ultimately arrives at bases $\mathcal{B}_j$ for W_j/W_{j-1}, $j = 1, \ldots, n$. Picking the specific representatives of these basis elements (thus by taking the vector $\mathbf{x}$ when $\mathbf{x} + W_{j-1}$ appears in $\mathcal{B}_j$), one arrives at the desired basis for $\mathbb{F}^n$ giving the Jordan canonical form of A. These observations form the essence of the construction in the proof of Theorem 4.2.1.

A scenario where the quotient space shows up, is in the case we have a vector space V with a Hermitian form $[\cdot, \cdot]$ that satisfies $[\mathbf{v}, \mathbf{v}] \geq 0$ for all $\mathbf{v} \in V$. Such a Hermitian form is sometimes called a *pre-inner product*. It is not an inner product as $[\mathbf{x}, \mathbf{x}] = 0$ does not necessarily imply $\mathbf{x} = \mathbf{0}$, but all the other rules of an inner product are satisfied. The following example is the type of setting where this may occur.

Example 6.2.10 Let

$$V = \{f : [0,1] \to \mathbb{R} : f \text{ is continuous except at a finite number of points}\}.$$

Define

$$[f, g] := \int_0^1 f(t)g(t)dt.$$

Then $[\cdot, \cdot]$ is a Hermitian form and $[f, f] = \int_0^1 f(t)^2 dt \geq 0$. However, there are nonzero functions f in V so that $[f, f] = 0$; for instance,

$$f(x) = \begin{cases} 0 & \text{if } x \neq \frac{1}{2}, \\ 1 & \text{if } x = \frac{1}{2}, \end{cases}$$

satisfies $[f, f] = 0$. Thus $[\cdot, \cdot]$ is a pre-inner product, but not an inner product.

So, what prevents a pre-inner product $[\cdot, \cdot]$ from being an inner product, is that $W := \{\mathbf{v} \in V : [\mathbf{v}, \mathbf{v}] = 0\}$ contains nonzero elements. It turns out that this set W is a subspace.

Lemma 6.2.11 *Let the vector space V over $\mathbb{F} = \mathbb{R}$ or $\mathbb{C}$, have an pre-inner product $[\cdot, \cdot]$. Then $W = \{\mathbf{v} \in V : [\mathbf{v}, \mathbf{v}] = 0\}$ is a subspace.*

Proof. Let $\mathbf{x}, \mathbf{y} \in W$. As $[\cdot, \cdot]$ is a pre-inner product, we have that for all $c \in \mathbb{F}$ the inequality $[\mathbf{x} + c\mathbf{y}, \mathbf{x} + c\mathbf{y}] \geq 0$ holds. Thus

$$0 \leq [\mathbf{x} + c\mathbf{y}, \mathbf{x} + c\mathbf{y}] = [\mathbf{x}, \mathbf{x}] + c[\mathbf{y}, \mathbf{x}] + \bar{c}[\mathbf{x}, \mathbf{y}] + |c|^2[\mathbf{y}, \mathbf{y}] = 2\mathrm{Re}(c[\mathbf{y}, \mathbf{x}]).$$

By choosing $c = -\overline{[\mathbf{y}, \mathbf{x}]}$, we get that $-|[\mathbf{y}, \mathbf{x}]|^2 \geq 0$, and thus $[\mathbf{y}, \mathbf{x}] = 0$. But then it follows that $\mathbf{x} + \mathbf{y} \in W$, proving that W is closed under addition.

Since $\mathbf{0} \in W$ and W is clearly closed under scalar multiplication, we obtain that W is a subspace. □

By considering the vector space V/W we can turn a pre-inner product into an inner product, as we see next.

Proposition 6.2.12 *Let the vector space V over $\mathbb{F} = \mathbb{R}$ or $\mathbb{C}$, have an pre-inner product $[\cdot, \cdot]$. Let W be the subspace $W = \{\mathbf{v} \in V : [\mathbf{v}, \mathbf{v}] = 0\}$, and define $\langle \cdot, \cdot \rangle$ on V/W via*

$$\langle \mathbf{x} + W, \mathbf{y} + W \rangle := [\mathbf{x}, \mathbf{y}].$$

Then $\langle \cdot, \cdot \rangle$ defines an inner product on V/W.

Proof. First we need to show that $\langle \cdot, \cdot \rangle$ is well-defined. Assume that $\mathbf{x} + W = \hat{\mathbf{x}} + W$ and let us show that

$$\langle \mathbf{x} + W, \mathbf{y} + W \rangle = \langle \hat{\mathbf{x}} + W, \mathbf{y} + W \rangle. \tag{6.5}$$

We have $\mathbf{x} - \hat{\mathbf{x}} \in W$. As $[\cdot, \cdot]$ satisfies the Cauchy–Schwarz inequality (see Remark 5.1.11) we have that

$$|[\mathbf{x} - \hat{\mathbf{x}}, \mathbf{y}]|^2 \leq [\mathbf{x} - \hat{\mathbf{x}}, \mathbf{x} - \hat{\mathbf{x}}][\mathbf{y}, \mathbf{y}] = 0,$$

since $\mathbf{x} - \hat{\mathbf{x}} \in W$. Thus (6.5) follows. Similarly, when $\mathbf{y} + W = \hat{\mathbf{y}} + W$, we have $\langle \hat{\mathbf{x}} + W, \mathbf{y} + W \rangle = \langle \hat{\mathbf{x}} + W, \hat{\mathbf{y}} + W \rangle$. But then, when $\mathbf{x} + W = \hat{\mathbf{x}} + W$ and $\mathbf{y} + W = \hat{\mathbf{y}} + W$, we find that $\langle \mathbf{x} + W, \mathbf{y} + W \rangle = \langle \hat{\mathbf{x}} + W, \mathbf{y} + W \rangle = \langle \hat{\mathbf{x}} + W, \hat{\mathbf{y}} + W \rangle$, showing that $\langle \cdot, \cdot \rangle$ is well-defined.

That $\langle \cdot, \cdot \rangle$ defines a pre-inner product on V/W is easily checked, so let us just address the definiteness property. Assume that $\langle \mathbf{x} + W, \mathbf{x} + W \rangle = 0$. Then $[\mathbf{x}, \mathbf{x}] = 0$, and thus $\mathbf{x} \in W$. This gives that $\mathbf{x} + W = \mathbf{0} + W$, which is exactly what we were after. $\qquad \square$

Getting back to Example 6.2.10, studying V/W instead of V, means that we are identifying functions whose values only differ in a finite number of points. In a setting of a vector space consisting of function, and where the interest lies in taking integrals, this is a common feature. In a Functional Analysis course this idea will be pursued further.

6.3 The dual space

Let V be a vector space over the field $\mathbb{F}$. We call a linear map $f : V \to \mathbb{F}$ that takes values in the underlying field, a *linear functional*. Linear functionals, as all function with values in a field, allow for addition among them, as well as scalar multiplication:

$$(f + g)(\mathbf{v}) := f(\mathbf{v}) + g(\mathbf{v}), (cf)(\mathbf{x}) := cf(\mathbf{x}).$$

With these operations the linear functions form a vector space V', the *dual space* of V. Thus

$$V' = \{f : V \to \mathbb{F} : f \text{ is linear}\}.$$

The first observation is that the dual space of a finite-dimensional space V has the same dimension as V.

Proposition 6.3.1 *Let V be a finite-dimensional space, and V' be its dual space. Then*

$$\dim V = \dim V'.$$

When $\{\mathbf{v}_1, \ldots, \mathbf{v}_n\}$ is a basis for V, then a basis for V' is given by $\{f_1, \ldots, f_n\}$, where $f_j \in V'$, $j = 1, \ldots, n$, is so that

$$f_j(\mathbf{v}_k) = \begin{cases} 0 & \text{if } k \neq j, \\ 1 & \text{if } k = j. \end{cases}$$

The basis $\{f_1, \ldots, f_n\}$ above is called the *dual basis* of $\{\mathbf{v}_1, \ldots, \mathbf{v}_n\}$.

Proof. When $\mathbf{v} = \sum_{k=1}^n c_k \mathbf{v}_k$, then $f_j(\mathbf{v}) = c_j$, yielding a well-defined linear functional on V. Let us show that $\{f_1, \ldots, f_n\}$ is linearly independent. For this, suppose that $d_1 f_1 + \cdots + d_n f_n = \mathbf{0}$. Then

$$0 = \mathbf{0}(\mathbf{v}_k) = \left(\sum_{j=1}^n d_j f_j \right)(\mathbf{v}_k) = \sum_{j=1}^n d_j f_j(\mathbf{v}_k) = d_k, \ k = 1, \ldots, n,$$

showing linear independence.

Next, we need to show that $\text{Span}\{f_1, \ldots, f_n\} = V'$, so let $f \in V'$ be arbitrary. We claim that

$$f = f(\mathbf{v}_1) f_1 + \cdots + f(\mathbf{v}_n) f_n. \tag{6.6}$$

Indeed, for $k = 1, \ldots, n$, we have that

$$f(\mathbf{v}_k) = f(\mathbf{v}_k) f_k(\mathbf{v}_k) = \sum_{j=1}^n f(\mathbf{v}_j) f_j(\mathbf{v}_k).$$

Thus the functionals in the left- and right-hand sides of (6.6) coincide on the basis elements $\mathbf{v}_k$, $k = 1, \ldots, n$. But then, by linearity, the functionals in the left- and right-hand sides of (6.6) coincide for all $\mathbf{v} \in V$. $\square$

When $\langle \cdot, \cdot \rangle$ is an inner product, then for a fixed $\mathbf{v} \in V$, the function $f_{\mathbf{v}} = \langle \cdot, \mathbf{v} \rangle$ defined via

$$f_{\mathbf{v}}(\mathbf{x}) = \langle \mathbf{x}, \mathbf{v} \rangle$$

is a linear functional; that is, $f_{\mathbf{v}} \in V'$. In the case of finite-dimensional inner product vector spaces, these functionals $f_{\mathbf{v}}$ comprise all of V'.

Theorem 6.3.2 *(Riesz representation theorem) Let V be a finite-dimensional vector space with inner product $\langle \cdot, \cdot \rangle$. Then for every*

$f \in V'$ there exists a $\mathbf{v} \in V$ so that $f = f_{\mathbf{v}}$; that is, $f(\mathbf{x}) = \langle \mathbf{x}, \mathbf{v} \rangle$, for all $\mathbf{x} \in V$. Moreover, we have that

$$\|f_{\mathbf{v}}\|_{V'} := \sup_{\|\mathbf{x}\|_V \leq 1} |f_{\mathbf{v}}(\mathbf{x})| = \|\mathbf{v}\|_V,$$

where $\|\mathbf{x}\|_V = \sqrt{\langle \mathbf{x}, \mathbf{x} \rangle}$.

Proof. Let $\mathcal{B} = \{\mathbf{e}_1, \ldots, \mathbf{e}_n\}$ be an orthonormal basis for V. Given $f \in V'$, let $\mathbf{v} = \overline{f(\mathbf{e}_1)}\mathbf{e}_1 + \cdots + \overline{f(\mathbf{e}_n)}\mathbf{e}_n$. Then $f = f_{\mathbf{v}}$. Indeed, if $\mathbf{x} = \sum_{j=1}^{n} c_j \mathbf{e}_j$, then

$$f_{\mathbf{v}}(\mathbf{x}) = \langle \sum_{j=1}^{n} c_j \mathbf{e}_j, \sum_{k=1}^{n} \overline{f(\mathbf{e}_k)}\mathbf{e}_k \rangle = \sum_{k=1}^{n} c_k f(\mathbf{e}_k) = f(\sum_{k=1}^{n} c_k \mathbf{e}_k) = f(\mathbf{x}).$$

Next, suppose that $\|\mathbf{x}\|_V \leq 1$. Then, by the Cauchy–Schwarz inequality (5.1),

$$|f_{\mathbf{v}}(\mathbf{x})| = |\langle \mathbf{x}, \mathbf{v} \rangle| \leq \sqrt{\langle \mathbf{x}, \mathbf{x} \rangle}\sqrt{\langle \mathbf{v}, \mathbf{v} \rangle} = \|\mathbf{x}\|_V \|\mathbf{v}\|_V \leq \|\mathbf{v}\|_V.$$

As for $\mathbf{v} \neq \mathbf{0}$,

$$|f_{\mathbf{v}}(\frac{1}{\|\mathbf{v}\|_V}\mathbf{v})| = \frac{\langle \mathbf{v}, \mathbf{v} \rangle}{\|\mathbf{v}\|_V} = \|\mathbf{v}\|_V,$$

we obtain that $\|f_{\mathbf{v}}\|_{V'} = \|\mathbf{v}\|_V$ (an equality that trivially holds for $\mathbf{v} = \mathbf{0}$ as well). $\qquad\square$

One may define a map $\Phi : V \to V'$ via

$$\Phi(\mathbf{v}) = f_{\mathbf{v}} = \langle \cdot, \mathbf{v} \rangle. \tag{6.7}$$

Notice that

$$\Phi(\mathbf{v} + \hat{\mathbf{v}}) = f_{\mathbf{v}+\hat{\mathbf{v}}} = f_{\mathbf{v}} + f_{\hat{\mathbf{v}}} = \Phi(\mathbf{v}) + \Phi(\hat{\mathbf{v}}), \tag{6.8}$$

and

$$\Phi(c\mathbf{v}) = f_{c\mathbf{v}} = \overline{c}f_{\mathbf{v}} = \overline{c}\Phi(\mathbf{v}). \tag{6.9}$$

Thus, when the underlying field is $\mathbb{C}$, the map Φ is not linear, due to the complex conjugate showing up in (6.9). A map Φ satisfying

$$\Phi(\mathbf{v} + \hat{\mathbf{v}}) = \Phi(\mathbf{v}) + \Phi(\hat{\mathbf{v}}), \quad \Phi(c\mathbf{v}) = \overline{c}\Phi(\mathbf{v})$$

is called a *conjugate linear map*. Thus, for a finite-dimensional vector space, the map Φ defined in (6.7) is a bijective conjugate linear map. Moreover, $\|\Phi(\mathbf{v})\|_{V'} = \|\mathbf{v}\|_V$, so Φ also has an isometry property. For infinite-dimensional, so-called, Hilbert spaces, the same result is true (provided we only bounded linear functionals), but this requires more analysis results than we are ready to address here. The following example shows that in the infinite-dimensional case, one indeed needs to proceed with caution.

Example 6.3.3 Let $V = \{f : [0,1] \to \mathbb{R} : f$ is continuous$\}$, and

$$\langle f, g \rangle := \int_0^1 f(t)g(t)dt,$$

which defines an inner product on V. Let $L : V \to \mathbb{R}$ be defined by $L(f) = f(0)$. Then $L \in V'$. However, there is no function $g \in V$ so that

$$f(0) = \int_0^1 f(t)g(t)dt \text{ for all } f \in V. \tag{6.10}$$

Indeed, if (6.10) holds then by Cauchy–Schwarz,

$$|L(f)| = |\langle f, g \rangle| \le \sqrt{\langle f, f \rangle}\sqrt{\langle g, g \rangle} \text{ for all } f \in V. \tag{6.11}$$

For $n \in \mathbb{N}$ we define the function $f_n \in V$ via

$$f_n(t) = \begin{cases} \sqrt{n - n^2 t} & \text{if } 0 \le t \le \frac{1}{n}, \\ 0 & \text{if } \frac{1}{n} \le t \le 1. \end{cases}$$

Then $L(f_n) = \sqrt{n}$ and

$$\langle f_n, f_n \rangle = \int_0^1 f_n(t)^2 dt = \int_0^{\frac{1}{n}} n - n^2 t\, dt = 1 - n^2 \frac{t^2}{2}\Big|_{t=0}^{t=\frac{1}{n}} = 1 - \frac{1}{2} = \frac{1}{2}.$$

If (6.10) holds we would need by (6.11) that $\sqrt{\langle g, g \rangle} \ge \sqrt{2n}$ for all $n \in \mathbb{N}$, which is clearly impossible as $\langle g, g \rangle$ is a real number that does not depend on n.

When V has a norm $\| \cdot \|_V$, we define $\| \cdot \|_{V'}$ on V' via

$$\|f\|_{V'} := \sup_{\|\mathbf{x}\|_V \le 1} |f(\mathbf{x})|, \ f \in V'.$$

If the supremum is finite we say that f is a *bounded functional*. As we will see, in finite dimensions every linear functional is bounded. However, as the previous example shows, this is not true in infinite dimensions. We therefore introduce

$$V'_{\text{bdd}} = \{f \in V' : \|f\|_{V'} < \infty\} = \{f \in V' : f \text{ is bounded}\}.$$

Proposition 6.3.4 *Let V have a norm $\| \cdot \|_V$. Then $\| \cdot \|_{V'}$ defined above is a norm on the vector space V'_{bdd}. When $\dim V < \infty$, then $V' = V'_{\text{bdd}}$.*

Proof. First suppose that $f, g \in V'_{\text{bdd}}$, thus $\|f\|_{V'}, \|g\|_{V'} < \infty$. Then

$$\|f + g\|_{V'} = \sup_{\|\mathbf{x}\|_V \le 1} |(f + g)(\mathbf{x})| \le \sup_{\|\mathbf{x}\|_V \le 1} |f(\mathbf{x})| + |g(\mathbf{x})| \le$$

$$\sup_{\|\mathbf{x}\|_V \le 1} |f(\mathbf{x})| + \sup_{\|\mathbf{x}\|_V \le 1} |g(\mathbf{x})| = \|f\|_{V'} + \|g\|_{V'},$$

and thus $f + g \in V'_{\text{bdd}}$. Next, $\|cf\|_{V'} = |c|\|f\|_{V'}$ follows immediately by using the corresponding property of $\| \cdot \|_V$. Thus V'_{bdd} is closed under scalar multiplication. As the zero functional also belongs to V'_{bdd}, we obtain that V'_{bdd} is a vector space.

To show that $\|f\|_{V'}$ is a norm, it remains to show that item (i) in the definition of a norm is satisfied. Clearly, $\|f\|_{V'} \ge 0$. Next, if $\|f\|_{V'} = 0$, then $|f(\mathbf{x})| = 0$ for all $\|\mathbf{x}\| \le 1$. Thus $f(\mathbf{x}) = 0$ for all $\|\mathbf{x}\| \le 1$, and thus by scaling $f(\mathbf{x}) = 0$ for all $\mathbf{x} \in V$.

In the case that $\dim V = n < \infty$, we may choose a basis in V and identify V with $\mathbb{F}^n$. Defining the standard inner product on $\mathbb{F}^n$, we obtain also an inner product $\langle \cdot, \cdot \rangle$ on V. Using Theorem 6.3.2 we obtain that for every $f \in V'$ we have that

$$\sup_{\langle \mathbf{x}, \mathbf{x} \rangle \le 1} |f(\mathbf{x})| < \infty,$$

as $f = f_\mathbf{v}$ for some $\mathbf{v} \in V$ and $\sup_{\langle \mathbf{x}, \mathbf{x} \rangle \le 1} |f_\mathbf{v}(\mathbf{x})| \le \sqrt{\langle \mathbf{v}, \mathbf{v} \rangle}$ (by the Cauchy–Schwarz inequality). Using Theorem 5.1.25, we have that $\sqrt{\langle \cdot, \cdot \rangle}$ and $\| \cdot \|_V$ are equivalent norms. From this $\|f\|_{V'} < \infty$ now easily follows. $\qquad\square$

If $A : V \to W$ is a linear map, then the induced map $A' : W' \to V'$ is given by

$$A'g = f, \text{ where } f(\mathbf{v}) = g(A\mathbf{v}).$$

Note that indeed g acts on elements of W while f acts on elements of V. We show next that if the matrix representation of A with respect to some bases is B, then the matrix representation of A' with respect to the corresponding dual bases is B^T, the transpose of B.

Proposition 6.3.5 *Let $A : V \to W$ be linear and let $\mathcal{B}$ and $\mathcal{C}$ be bases for V and W, respectively. Let $\mathcal{B}'$ and $\mathcal{C}'$ be the dual bases of $\mathcal{B}$ and $\mathcal{C}$, respectively. Then*

$$([A]_{\mathcal{C}\leftarrow\mathcal{B}})^T = [A']_{\mathcal{B}'\leftarrow\mathcal{C}'}.$$

Proof. Let us denote $\mathcal{B} = \{\mathbf{b}_1, \ldots, \mathbf{b}_n\}$, $\mathcal{C} = \{\mathbf{c}_1, \ldots, \mathbf{c}_m\}$, $\mathcal{B}' = \{f_1, \ldots, f_n\}$, $\mathcal{C} = \{g_1, \ldots, g_n\}$. Also let

$$B = (b_{ij})_{i=1, j=1}^{m,\ n} = [A]_{\mathcal{C}\leftarrow\mathcal{B}}.$$

Let us compute $A'g_k$. For $\mathbf{v} = \sum_{l=1}^n d_l \mathbf{b}_l$ we have

$$A'g_k(\mathbf{v}) = A'g_k(\sum_{l=1}^n d_l \mathbf{b}_l) = g_k(A(\sum_{l=1}^n d_l \mathbf{b}_l)) = g_k(\sum_{l=1}^n d_l A\mathbf{b}_l) =$$

$$g_k(\sum_{l=1}^{n} d_l(\sum_{i=1}^{n} b_{il}\mathbf{c}_i)) = \sum_{l=1}^{n} d_l(\sum_{i=1}^{n} b_{il}g_k(\mathbf{c}_i)) = \sum_{l=1}^{n} d_l b_{kl}.$$

Observing that $d_l = f_l(\sum_{i=1}^{n} d_j\mathbf{b}_j) = f_l(\mathbf{v})$, we thus obtain that

$$A'g_k(\mathbf{v}) = \sum_{l=1}^{n} b_{kl} f_l(\mathbf{v}) \text{ for all } \mathbf{v} \in V.$$

Consequently,

$$A'g_k = \sum_{l=1}^{n} b_{kl} f_l,$$

and thus the kth column of $[A']_{\mathcal{B}' \leftarrow \mathcal{C}'}$ equals

$$\begin{pmatrix} b_{k1} \\ \vdots \\ b_{kn} \end{pmatrix},$$

which is the transpose of the kth row of B.   □

As V' is a vector space, we can study its dual space

$$V'' = \{E : V' \to \mathbb{F} : E \text{ linear}\},$$

also referred to as the *double dual* of V. One way to generate an element of V'' is to introduce the *evaluation map* $E_\mathbf{v}$ at $\mathbf{v} \in V$ as follows:

$$E_\mathbf{v}(f) = f(\mathbf{v}).$$

Clearly, $E_\mathbf{v}(f + g) = E_\mathbf{v}(f) + E_\mathbf{v}(g)$ and $E_\mathbf{v}(cf) = cE_\mathbf{v}(f)$, and thus $E_\mathbf{v}$ is indeed linear. In case V is finite dimensional, we have that every element of V'' corresponds to an evaluation map.

Proposition 6.3.6 *Let V be finite dimensional, and consider the map $\Phi : V \to V''$ defined by*

$$\Phi(\mathbf{v}) = E_\mathbf{v}.$$

Then Φ is an isomorphism.

Proof. First we observe that
$E_{\mathbf{v}+\mathbf{w}}(f) = f(\mathbf{v} + \mathbf{w}) = f(\mathbf{v}) + f(\mathbf{w}) = E_\mathbf{v}(f) + E_\mathbf{w}(f)$ and
$E_{c\mathbf{v}}(f) = f(c\mathbf{v}) = cf(\mathbf{v}) = cE_\mathbf{v}(f)$. Thus Φ is linear.

As $\dim V = \dim V' = \dim V''$, it suffices to show that Φ is one-to-one. Suppose that $\mathbf{v} \neq \mathbf{0}$. Then we can choose a basis $\mathcal{B} = \{\mathbf{v}, \mathbf{v}_2, \ldots, \mathbf{v}_n\}$ of V (where $\dim V = n$). Let now $f \in V'$ be so that $f(\mathbf{v}) = 1$ and $f(\mathbf{v}_j) = 0$,

$j = 2, \ldots, n$. Then $E_{\mathbf{v}}(f) = f(\mathbf{v}) = 1$, and thus $E_{\mathbf{v}} \neq 0$. This shows that $\mathbf{v} \neq \mathbf{0}$ yields that $\Phi(\mathbf{v}) \neq 0$. Thus Φ is one-to-one. $\qquad\square$

The notion of a dual space is useful in the context of optimization. For instance, let

$$f : \mathbb{R} \to \mathbb{R}^n, \ f(t) = \begin{pmatrix} f_1(t) \\ \vdots \\ f_n(t) \end{pmatrix}$$

be a differentiable function. With the Euclidean norm on $\mathbb{R}^n$ we have that

$$\frac{d}{dt}\|f(t)\|^2 = \frac{d}{dt}(f_1(t)^2 + \cdots + f_n(t)^2) = 2(f_1'(t)f_1(t) + \cdots + f_n'(t)f_n(t)) =$$

$$2 \begin{pmatrix} f_1'(t) & \cdots & f_n'(t) \end{pmatrix} \begin{pmatrix} f_1(t) \\ \vdots \\ f_n(t) \end{pmatrix}.$$

The row vector

$$\nabla f(t) = \begin{pmatrix} f_1'(t) & \cdots & f_n'(t) \end{pmatrix}$$

is called the *gradient* of f at t. In a more general setting, where $f : \mathbb{F} \to V$, it turns out that viewing $\nabla f(t)$ as an element of the dual space (or, equivalently, viewing ∇f as a function acting $\mathbb{F} \to V'$) is a natural way to develop a solid theory.

While we focused in the section on the vector space of linear functionals, one can, in more generality, study the vector space

$$\mathcal{L}(V, W) = \{T : V \to W : T \text{ is linear}\},$$

with the usual definition of adding linear maps and multiplying them with a scalar. In finite dimensions, we have seen that after choosing bases $\mathcal{B}$ and $\mathcal{C}$ in V and W, respectively, every linear map $T : V \to W$ is uniquely identified by its matrix representation $[T]_{\mathcal{C} \leftarrow \mathcal{B}}$. Using this, one immediately sees that

$$\dim \mathcal{L}(V, W) = (\dim V)(\dim W).$$

The main item we would like to address here is when V and W have norms $\|\cdot\|_V$ and $\|\cdot\|_W$, respectively. In this case there is a natural norm on $\mathcal{L}(V, W)$, as follows:

$$\|T\|_{\mathcal{L}(V,W)} := \sup_{\|\mathbf{v}\|_V = 1} \|T(\mathbf{v})\|_W. \tag{6.12}$$

When V and W are finite dimensional, this supremum is always finite and thus $\|T\|_{\mathcal{L}(V,W)}$ is a nonnegative real number. We say that $\|\cdot\|_{\mathcal{L}(V,W)}$ is the *induced operator norm*, as its definition relies on the norms on V and W and on the property of T as a linear operator.

Proposition 6.3.7 . *Let V and W be finite-dimensional vector spaces with norms $\| \cdot \|_V$ and $\| \cdot \|_W$, respectively. Then $\| \cdot \|_{\mathcal{L}(V,W)}$ defines a norm on $\mathcal{L}(V,W)$. In addition, for every $\mathbf{v} \in V$, we have that*

$$\|T(\mathbf{v})\|_W \leq \|T\|_{\mathcal{L}(V,W)} \|\mathbf{v}\|_V. \tag{6.13}$$

Proof. Since V and W are finite dimensional, the set $\{T\mathbf{v} : \|\mathbf{v}\|_V = 1\}$ is a compact set, and thus $\| \cdot \|_W$ attains a maximum on this set. This gives that the supremum in (6.12) is in fact a maximum, and is finite. Next, clearly $\|T\|_{\mathcal{L}(V,W)} \geq 0$. Next, suppose that $\|T\|_{\mathcal{L}(V,W)} = 0$. This implies that for every $\mathbf{v} \in V$ with $\mathbf{v}\|_V = 1$, we have that $\|T(\mathbf{v})\|_W = 0$, and thus $T(\mathbf{v}) = \mathbf{0}$. But then $T = 0$.

When $c \in \mathbb{F}$, we have that

$$\|cT\|_{\mathcal{L}(V,W)} = \sup_{\|\mathbf{v}\|_V = 1} \|cT(\mathbf{v})\|_W = \sup_{\|\mathbf{v}\|_V = 1} |c| \|T(\mathbf{v})\|_W = |c| \|T\|_{\mathcal{L}(V,W)}.$$

Next, note that for $T_1, T_2 \in \mathcal{L}(V,W)$, we have that

$$\|(T_1 + T_2)(\mathbf{v})\|_W = \|T_1(\mathbf{v}) + T_2(\mathbf{v})\|_W \leq \|T_1(\mathbf{v})\|_W + \|T_2(\mathbf{v})\|_W.$$

Using this it is straightforward to see that

$$\|T_1 + T_2\|_{\mathcal{L}(V,W)} \leq \|T_1\|_{\mathcal{L}(V,W)} + \|T_2\|_{\mathcal{L}(V,W)}.$$

Finally, if $\mathbf{v} \neq \mathbf{0}$, then $\frac{\mathbf{v}}{\|\mathbf{v}\|_V}$ has norm 1, and thus

$$\|T(\frac{\mathbf{v}}{\|\mathbf{v}\|_V})\|_W \leq \|T\|_{\mathcal{L}(V,W)}.$$

Multiplying both sides with $\|\mathbf{v}\|_V$, and using the norm properties, yields (6.13). When $\mathbf{v} = \mathbf{0}$, then (6.13) obviously holds as well. □

Example 6.3.8 Let $T : \mathbb{C}^n \to \mathbb{C}^m$ be the linear map given by multiplication with the matrix $A = (a_{ij})_{i=1,j=1}^{m,\ n}$. Let the norm on both V and W be given by $\| \cdot \|_1$, as in Example 5.1.14. Then

$$\|T\|_{\mathcal{L}(V,W)} = \max_j |a_{1j}| + \cdots + |a_{mj}|. \tag{6.14}$$

Indeed, if we take $\mathbf{e}_j \in \mathbb{C}^n$, which is a unit vector in the $\| \cdot \|_1$ norm, then $T(\mathbf{e}_j) = (a_{ij})_{i=1}^n$, and thus we find

$$\|T(\mathbf{e}_j)\|_W = \|T(\mathbf{e}_j)\|_1 = |a_{1j}| + \cdots |a_{mj}|.$$

Thus the inequality $\geq$ holds in (6.14). To prove the other inequality, we observe that for $\mathbf{x} = \sum_{j=1}^{n} x_j \mathbf{e}_j$ with $\sum_{j=1}^{n} |x_j| = 1$, we have that $\|T(\mathbf{x})\|_W$ equals

$$\|\sum_{j=1}^{n} x_j T(\mathbf{e}_j)\|_W \leq \sum_{j=1}^{n} |x_j| \|T(\mathbf{e}_j)\|_W \leq$$

$$(\sum_{j=1}^{n} |x_j|)(\max_{j=1,\ldots,n} \|T(\mathbf{e}_j)\|_W) = \max_j |a_{1j}| + \cdots + |a_{mj}|.$$

Example 6.3.9 Let $T : \mathbb{C}^n \to \mathbb{C}^m$ be the linear map given by multiplication with the matrix $A = (a_{ij})_{i=1,j=1}^{m,\;n}$. Let the norm on both V and W be given by the Euclidean norm $\|\cdot\|_2$. Then

$$\|T\|_{\mathcal{L}(V,W)} = \sigma_1(A). \tag{6.15}$$

This was already observed in Proposition 5.6.2.

When the vector spaces are not finite dimensional, it could happen that a linear map does not have a finite norm. When this happens, we say that the linear map is *unbounded*. A typical example of an unbounded linear map is taking the derivative. We provide the details next.

Example 6.3.10 Let

$$V = \{f : (0,1) \to \mathbb{R} : f \text{ is bounded and differentiable with } f' \text{ bounded}\}$$

and

$$W = \{f : (0,1) \to \mathbb{R} : f \text{ is bounded}\}.$$

On both spaces let

$$\|f\|_\infty = \sup_{t \in (0,1)} |f(t)|$$

be the norm. Note that f being bounded means exactly that $\|f\|_\infty < \infty$. Let $T = \frac{d}{dt} : V \to W$ be the differentiation map. Then T is linear. Let now $f_n(t) = t^n$, $n \in \mathbb{N}$. Then $\|f_n\|_\infty = 1$ for all $n \in \mathbb{N}$. However, $(Tf_n)(t) = f_n'(t) = nt^{n-1}$ has the norm equal to $\|Tf_n\|_\infty = n$, $n \in \mathbb{N}$. Thus, it follows that

$$\sup_{\|f\|_\infty=1} \|Tf\|_\infty \geq \|Tf_n\|_\infty = n$$

for all $n \in \mathbb{N}$, and thus T is unbounded.

We end this section with the following norm of a product inequality.

Proposition 6.3.11 *Let V, W and X be finite-dimensional vector spaces with norms $\|\cdot\|_V$, $\|\cdot\|_W$, and $\|\cdot\|_X$, respectively. Let $T : V \to W$ and $S : W \to X$ be linear maps. Then*

$$\|ST\|_{\mathcal{L}(V,X)} \le \|S\|_{\mathcal{L}(W,X)} \|T\|_{\mathcal{L}(V,W)}. \qquad (6.16)$$

Proof. Let $\mathbf{v} \in V$ with $\|\mathbf{v}\|_V = 1$. By (6.13) applied to the vector $T(\mathbf{v})$ and the map S we have that

$$\|S(T(\mathbf{v}))\|_X \le \|S\|_{\mathcal{L}(W,X)} \|T(\mathbf{v})\|_W.$$

Next we use (6.13) again, and obtain that

$$\|S(T(\mathbf{v}))\|_X \le \|S\|_{\mathcal{L}(W,X)} \|T(\mathbf{v})\|_W l \le$$

$$\|S\|_{\mathcal{L}(W,X)} \|T\|_{\mathcal{L}(V,W)} \|\mathbf{v}\|_V = \|S\|_{\mathcal{L}(W,X)} \|T\|_{\mathcal{L}(V,W)}.$$

Thus $\|S\|_{\mathcal{L}(W,X)} \|T\|_{\mathcal{L}(V,W)}$ is an upper bound for $\|S(T(\mathbf{v}))\|_X$ for all unit vectors $\mathbf{v}$ in V, and therefore the least upper bound is at most $\|S\|_{\mathcal{L}(W,X)} \|T\|_{\mathcal{L}(V,W)}$. $\qquad\square$

6.4   Multilinear maps and functionals

Let $V_1, \dots, V_k, W$ be vector spaces over a field $\mathbb{F}$. We say that a function

$$\phi : V_1 \times \cdots \times V_k \to W$$

is *multilinear* if the function is linear in each coordinate. Thus, for each $i \in \{1, \dots, k\}$, if we fix $\mathbf{v}_j \in V_j$, $j \ne i$, we require that the map

$$\mathbf{u} \mapsto \phi(\mathbf{v}_1, \dots, \mathbf{v}_{i-1}, \mathbf{u}, \mathbf{v}_{i+1}, \dots, \mathbf{v}_n)$$

is linear. Thus

$$\phi(\mathbf{v}_1, \dots, \mathbf{v}_{i-1}, \mathbf{u} + \hat{\mathbf{u}}, \mathbf{v}_{i+1}, \dots, \mathbf{v}_n) = \phi(\mathbf{v}_1, \dots, \mathbf{v}_{i-1}, \mathbf{u}, \mathbf{v}_{i+1}, \dots, \mathbf{v}_n) +$$

$$\phi(\mathbf{v}_1, \dots, \mathbf{v}_{i-1}, \hat{\mathbf{u}}, \mathbf{v}_{i+1}, \dots, \mathbf{v}_n)$$

and

$$\phi(\mathbf{v}_1, \dots, \mathbf{v}_{i-1}, c\mathbf{u}, \mathbf{v}_{i+1}, \dots, \mathbf{v}_n) = c\phi(\mathbf{v}_1, \dots, \mathbf{v}_{i-1}, \mathbf{u}, \mathbf{v}_{i+1}, \dots, \mathbf{v}_n).$$

When $W = \mathbb{F}$ we call ϕ a *multilinear functional*. When $k = 2$, we say that ϕ is *bilinear*.

Example 6.4.1 Let $\Phi : \mathbb{F}^k \to \mathbb{F}$ be defined by $\phi \begin{pmatrix} x_1 \\ \vdots \\ x_k \end{pmatrix} = x_1 x_2 \cdots x_k$. Then ϕ is a multilinear functional.

Example 6.4.2 Let $\phi : \mathbb{F}^k \times \cdots \times \mathbb{F}^k \to \mathbb{F}$ be defined by

$$\Phi(\mathbf{v}_1, \ldots, \mathbf{v}_k) = \det \begin{pmatrix} \mathbf{v}_1 & \cdots & \mathbf{v}_k \end{pmatrix}.$$

Then Φ is a multilinear functional.

Example 6.4.3 Let $\Phi : \mathbb{R}^3 \times \mathbb{R}^3 \to \mathbb{R}^3$ be defined by

$$\Phi\left(\begin{pmatrix} x_1 \\ x_2 \\ x_3 \end{pmatrix}, \begin{pmatrix} y_1 \\ y_2 \\ y_3 \end{pmatrix}\right) = \begin{pmatrix} x_2 y_3 - x_3 y_2 \\ x_3 y_1 - x_1 y_3 \\ x_1 y_3 - x_3 y_1 \end{pmatrix} =: \begin{pmatrix} x_1 \\ x_2 \\ x_3 \end{pmatrix} \times \begin{pmatrix} y_1 \\ y_2 \\ y_3 \end{pmatrix}. \tag{6.17}$$

Then Φ is a bilinear map, which corresponds to the so-called *cross product* in $\mathbb{R}^3$. Typically, the cross product of $\mathbf{x}$ and $\mathbf{y}$ is denoted as $\mathbf{x} \times \mathbf{y}$.

Example 6.4.4 Given matrices $A_j \in \mathbb{F}^{n_j \times m_j}$, $j = 0, \ldots, k$. Define

$$\Phi : \mathbb{F}^{m_0 \times n_1} \times \cdots \times \mathbb{F}^{m_{k-1} \times n_k} \to \mathbb{F}^{n_0 \times m_k},$$

$$\Phi(X_1, \ldots, X_k) = A_0 X_1 A_1 X_2 A_2 \cdots A_{k-1} X_k A_k.$$

Then Φ is a multilinear map.

If we let

$$M = \{\phi : V_1 \times \cdots \times V_k \to W : \phi \text{ is multilinear}\},$$

then by usual addition and scalar multiplication of functions, we have that M is a vector space over $\mathbb{F}$. When the vector spaces $V_1, \ldots, V_k$ have inner products, $\langle \cdot, \cdot \rangle_1, \ldots, \langle \cdot, \cdot \rangle_k$, respectively, then for fixed $\mathbf{u}_1 \in V_1, \ldots, \mathbf{u}_k \in V_k$ the map

$$\phi_{\mathbf{u}_1, \ldots, \mathbf{u}_k}(\mathbf{v}_1, \ldots, \mathbf{v}_k) := \langle \mathbf{v}_1, \mathbf{u}_1 \rangle_1 \cdots \langle \mathbf{v}_k, \mathbf{u}_k \rangle_k = \prod_{j=1}^{k} \langle \mathbf{v}_j, \mathbf{u}_j \rangle_j$$

is a multilinear functional acting $V_1 \times \cdots \times V_k \to \mathbb{F}$. Notice, that due to the Cauchy–Schwarz inequality, we have

$$|\phi_{\mathbf{u}_1, \ldots, \mathbf{u}_k}(\mathbf{v}_1, \ldots, \mathbf{v}_k)| = \prod_{j=1}^{k} |\langle \mathbf{v}_j, \mathbf{u}_j \rangle_j| \leq \prod_{j=1}^{k} \|\mathbf{u}_j\|_j \prod_{j=1}^{k} \|\mathbf{v}_j\|_j. \tag{6.18}$$

In finite dimensions, any multilinear functional is a linear combination of $\phi_{\mathbf{u}_1, \ldots, \mathbf{u}_k}$, $\mathbf{u}_1 \in V_1, \ldots, \mathbf{u}_k \in V_k$, as we will now see.

Proposition 6.4.5 *Let $V_1, \ldots, V_k$ be finite-dimensional vector spaces with inner products $\langle \cdot, \cdot \rangle_1, \ldots, \langle \cdot, \cdot \rangle_k$, respectively. Then every multilinear functional on $V_1 \times \cdots \times V_k$ is a linear combination of multilinear functionals $\phi_{\mathbf{u}_1, \ldots, \mathbf{u}_k}$, where $\mathbf{u}_1 \in V_1, \ldots, \mathbf{u}_k \in V_k$.*

Proof. Let ϕ be a multilinear functional on $V_1 \times \cdots \times V_k$, and let $\{\mathbf{e}_1^{(j)}, \ldots, \mathbf{e}_{n_j}^{(j)}\}$ be an orthonormal basis for V_j, $j = 1, \ldots, k$. Writing $\mathbf{v}_j = \sum_{r=1}^{n_j} \langle \mathbf{v}_j, \mathbf{e}_r^{(j)} \rangle_j \mathbf{e}_r^{(j)}$, we obtain that

$$\phi(\mathbf{v}_1, \ldots, \mathbf{v}_k) = \sum_{r_1=1}^{n_1} \cdots \sum_{r_k=1}^{n_k} \langle \mathbf{v}_1, \mathbf{e}_{r_1}^{(1)} \rangle_1 \cdots \langle \mathbf{v}_k, \mathbf{e}_{r_k}^{(k)} \rangle_k \phi(\mathbf{e}_{r_1}^{(1)}, \ldots, \mathbf{e}_{r_k}^{(k)}).$$

Thus ϕ is a linear combination of $\phi_{\mathbf{e}_{r_1}^{(1)}, \ldots, \mathbf{e}_{r_k}^{(k)}}$, $r_j = 1, \ldots, n_j$, $j = 1, \ldots, k$. $\square$

When $\| \cdot \|_j$ is a norm on V_j, $j = 1, \ldots, k$, and $\| \cdot \|_W$ a norm on W, then we say that ϕ is bounded if

$$\sup_{\|\mathbf{v}_1\|_1 \leq 1, \ldots, \|\mathbf{v}_k\|_k \leq 1} \|\phi(\mathbf{v}_1, \ldots, \mathbf{v}_k)\|_W < \infty.$$

Similar to the proof of Proposition 6.3.4, one can show that if $V_1, \ldots, V_k$ are finite dimensional and $W = \mathbb{F}$, then ϕ is automatically bounded. Indeed, if the norms come from inner products, one can use Proposition 6.4.5 and (6.18) to see that ϕ is bounded. Next, using that on finite-dimensional spaces any two norms are equivalent, one obtains the boundedness with respect to any norms on $V_1, \ldots, V_k$.

For a detailed study of multilinear functionals, it is actually useful to introduce tensor products. We will do this is the next section.

6.5   The tensor product

Given two vector spaces V_1 and V_2 over a field $\mathbb{F}$, we introduce a tensor product $\otimes : V_1 \times V_2 \to V_1 \otimes V_2$ with the properties

$$(\mathbf{x} + \mathbf{y}) \otimes \mathbf{v} = \mathbf{x} \otimes \mathbf{v} + \mathbf{y} \otimes \mathbf{v} \quad \text{for all } \mathbf{x}, \mathbf{y} \in V_1, \mathbf{v} \in V_2, \tag{6.19}$$

$$\mathbf{x} \otimes (\mathbf{v} + \mathbf{w}) = \mathbf{x} \otimes \mathbf{v} + \mathbf{x} \otimes \mathbf{w} \quad \text{for all } \mathbf{x} \in V_1, \mathbf{v}, \mathbf{w} \in V_2, \tag{6.20}$$

and

$$(c\mathbf{x}) \otimes \mathbf{v} = c(\mathbf{x} \otimes \mathbf{v}) = \mathbf{x} \otimes (c\mathbf{v}) \quad \text{for all } \mathbf{x} \in V_1, \mathbf{v} \in V_2, c \in \mathbb{F}. \tag{6.21}$$

The set $V_1 \otimes V_2$ is defined by

$$V_1 \otimes V_2 = \{\mathbf{0}\} \cup \{\sum_{j=1}^{m} c_j(\mathbf{x}_j \otimes \mathbf{v}_j) \ : \ m \in \mathbb{N}_0, c_j \in \mathbb{F}, \mathbf{x}_j \in V_1, \mathbf{v}_j \in V_2\},$$

where we say that two elements in $V_1 \otimes V_2$ are equal, if by applying rules (6.19)–(6.21) one element can be converted into the other.

Example 6.5.1 We have that the elements

$$(\mathbf{x}_1 + \mathbf{x}_2) \otimes (\mathbf{v}_1 + \mathbf{v}_2) - 2(\mathbf{x}_1 \otimes \mathbf{v}_2)$$

and

$$(\mathbf{x}_1 - \mathbf{x}_2) \otimes (\mathbf{v}_1 - \mathbf{v}_2) + 2(\mathbf{x}_2 \otimes \mathbf{v}_1)$$

are equal. Indeed, applying rules (6.19)–(6.21), we get

$$(\mathbf{x}_1 + \mathbf{x}_2) \otimes (\mathbf{v}_1 + \mathbf{v}_2) - 2(\mathbf{x}_1 \otimes \mathbf{v}_2) = \mathbf{x}_1 \otimes \mathbf{v}_1 + \mathbf{x}_2 \otimes \mathbf{v}_1 - \mathbf{x}_1 \otimes \mathbf{v}_2 + \mathbf{x}_2 \otimes \mathbf{v}_2$$

and

$$(\mathbf{x}_1 - \mathbf{x}_2) \otimes (\mathbf{v}_1 - \mathbf{v}_2) + 2(\mathbf{x}_2 \otimes \mathbf{v}_1) = \mathbf{x}_1 \otimes \mathbf{v}_1 + \mathbf{x}_2 \otimes \mathbf{v}_1 - \mathbf{x}_1 \otimes \mathbf{v}_2 + \mathbf{x}_2 \otimes \mathbf{v}_2.$$

It is convenient to allow $m = 0$ in the expression $\sum_{j=1}^{m} c_j(\mathbf{x}_j \otimes \mathbf{v}_j)$, in which case the sum should just be interpreted as $\mathbf{0}$. We define addition and scalar multiplication on $V_1 \otimes V_2$ by

$$\sum_{j=1}^{m} c_j(\mathbf{x}_j \otimes \mathbf{v}_j) + \sum_{j=m+1}^{l} c_j(\mathbf{x}_j \otimes \mathbf{v}_j) = \sum_{j=1}^{l} c_j(\mathbf{x}_j \otimes \mathbf{v}_j),$$

and

$$d \sum_{j=1}^{m} c_j(\mathbf{x}_j \otimes \mathbf{v}_j) = \sum_{j=1}^{m} (dc_j)(\mathbf{x}_j \otimes \mathbf{v}_j).$$

With these operations, one can easily check that $V_1 \otimes V_2$ is a vector space. An element of the form $\mathbf{x} \otimes \mathbf{v}$ is called a *simple tensor*. In general, the elements of $V_1 \otimes V_2$ are linear combinations of simple tensors. This definition of the tensor product of two vector spaces is perhaps the most abstract notion in this book. The elements of this space are just sums of a set of symbols, and then we have equality when we can convert one sum to the other by using the rules (6.19)–(6.21). We intend to make things more concrete in the remainder of this section.

First, let us figure out a way to determine whether an equality like

$$\sum_{j=1}^{m} c_j(\mathbf{x}_j \otimes \mathbf{v}_j) = \sum_{j=1}^{l} d_j(\mathbf{y}_j \otimes \mathbf{w}_j)$$

holds. For this, the following proposition is helpful.

Proposition 6.5.2 *Consider the vector space* $V_1 \otimes V_2$ *over* $\mathbb{F}$, *and let* $\sum_{j=1}^{m} c_j(\mathbf{x}_j \otimes \mathbf{v}_j) \in V_1 \otimes V_2$. *Let* $W_1 = \text{Span}\{\mathbf{x}_1, \ldots, \mathbf{x}_m\}$, *and* $W_2 = \text{Span}\{\mathbf{v}_1, \ldots, \mathbf{v}_m\}$ *The following are equivalent:*

(i) $\sum_{j=1}^{m} c_j(\mathbf{x}_j \otimes \mathbf{v}_j) = \mathbf{0}$,

(ii) *for all bilinear maps* $F : W_1 \times W_2 \to W$ *we have that*
$\sum_{j=1}^{m} c_j F(\mathbf{x}_j, \mathbf{v}_j) = \mathbf{0}_W$,

(iii) *for all bilinear functionals* $f : W_1 \times W_2 \to \mathbb{F}$ *we have that*
$\sum_{j=1}^{m} c_j f(\mathbf{x}_j, \mathbf{v}_j) = 0$.

Proof. (i) → (ii): Let $F : W_1 \times W_2 \to W$ be bilinear. It is clear that if we apply F to the left-hand side of (6.19) and to the right-hand side of (6.19), we get the same outcome; that is

$$F(\mathbf{x} + \mathbf{y}, \mathbf{v}) = F(\mathbf{x}, \mathbf{v}) + F(\mathbf{y}, \mathbf{v}).$$

The same holds for (6.20) and (6.21). Thus if the expression $\sum_{j=1}^{m} c_j(\mathbf{x}_j \otimes \mathbf{v}_j)$ can be converted to $\mathbf{0}$ by applying (6.19)–(6.21), then we must have that $\sum_{j=1}^{m} c_j F(\mathbf{x}_j, \mathbf{v}_j) = F(\mathbf{0}, \mathbf{0}) = \mathbf{0}_W$. (It could be that in the conversion of $\sum_{j=1}^{m} c_j(\mathbf{x}_j \otimes \mathbf{v}_j)$ to $\mathbf{0}$, one encounters vectors in V_1 that do not lie in W_1 and/or vectors in V_2 that do not lie in W_2. In this case, one needs to extend the definition of F to a larger space $\hat{W}_1 \times \hat{W}_2$. In the end, one can restrict again to $W_1 \times W_2$, as in the equality $\sum_{j=1}^{m} c_j F(\mathbf{x}_j, \mathbf{v}_j) = F(\mathbf{0}, \mathbf{0})$, the bilinear map F acts only on $W_1 \times W_2$.)

(ii) → (iii): Note that (iii) is just a special case of (ii), by taking $W = \mathbb{F}$, and thus (iii) holds when (ii) holds.

(iii) → (i): We prove the contrapositive, so we assume that (i) does not hold. Suppose that $\sum_{j=1}^{m} c_j(\mathbf{x}_j \otimes \mathbf{v}_j) \neq \mathbf{0}$. Thus the expression $\sum_{j=1}^{m} c_j(\mathbf{x}_j \otimes \mathbf{v}_j)$ cannot be converted to $\mathbf{0}$ by rules (6.19)–(6.21). Let $\mathcal{B} = \{\mathbf{y}_1, \ldots, \mathbf{y}_s\}$ be a basis for $W_1 = \text{Span}\{\mathbf{x}_1, \ldots, \mathbf{x}_m\}$, and $\mathcal{C} = \{\mathbf{w}_1, \ldots, \mathbf{w}_t\}$ be a basis for $W_2 = \text{Span}\{\mathbf{v}_1, \ldots, \mathbf{v}_m\}$. We introduce the $s \times m$ matrix $S = (s_{ij})$ and the $t \times m$ matrix $T = (t_{ij})$ as follows:

$$S = \left([c_1\mathbf{x}_1]_{\mathcal{B}} \quad \cdots \quad [c_m\mathbf{x}_m]_{\mathcal{B}}\right), T = \left([\mathbf{v}_1]_{\mathcal{C}} \quad \cdots \quad [\mathbf{v}_m]_{\mathcal{C}}\right).$$

We now claim that $ST^T \neq 0$. Indeed, note that by applying (6.19)–(6.21) we may write

$$\sum_{j=1}^{m}(c_j\mathbf{x}_j) \otimes \mathbf{v}_j = \sum_{j=1}^{m}[(\sum_{l=1}^{s} s_{lj}\mathbf{y}_l) \otimes (\sum_{n=1}^{t} t_{nj}\mathbf{w}_n)] = \sum_{l=1}^{s}\sum_{n=1}^{t}(\sum_{j=1}^{m} s_{lj}t_{nj})\mathbf{y}_l \otimes \mathbf{w}_n.$$

The number $\sum_{j=1}^{m} s_{lj}t_{nj}$ is exactly the (l, n)th entry of ST^T, so if $ST^T = 0$, it would mean that $\sum_{j=1}^{m} c_j(\mathbf{x}_j \otimes \mathbf{v}_j) = \mathbf{0}$.

As $ST^T \neq 0$, we have that some entry of it is nonzero. Say, entry (p, q) of ST^T is nonzero. Let now $g : W_1 \to \mathbb{F}$ be linear so that $g(\mathbf{y}_p) = 1$ and $g(\mathbf{y}_j) = 0$, $j \neq p$. Thus $g \in W_1'$. Similarly, let $h \in W_2'$ be so that $h(\mathbf{y}_q) = 1$ and $h(\mathbf{y}_j) = 0$, $j \neq q$. Let now $f : W_1 \times W_2 \to \mathbb{F}$ be defined by $f(\mathbf{x}, \mathbf{v}) = g(\mathbf{x})h(\mathbf{v})$. Then f is bilinear. Furthermore,

$$\sum_{j=1}^{m} f(c_j \mathbf{x}_j, \mathbf{v}_j) = \sum_{j=1}^{m} f(\sum_{l=1}^{t} s_{lj} \mathbf{y}_j, \sum_{n=1}^{t} t_{nj} \mathbf{w}_n) =$$

$$\sum_{j=1}^{m} g(\sum_{l=1}^{t} s_{lj} \mathbf{y}_j) h(\sum_{n=1}^{t} t_{nj} \mathbf{w}_n) = \sum_{j=1}^{m} s_{pj} t_{qj} \neq 0,$$

as this number is exactly equal to the (p, q) entry of ST^T. This finishes the proof. $\qquad \square$

The proof of Proposition 6.5.2 provides a way for checking whether an element $\sum_{j=1}^{m} c_j(\mathbf{x}_j \otimes \mathbf{v}_j) \in V_1 \otimes V_2$ equals $\mathbf{0}$ or not. Indeed, we would produce that matrices S and T as in the proof, and check whether $ST^T = 0$ or not. Let us do an example.

Example 6.5.3 In $\mathbb{Z}_5^3 \otimes \mathbb{Z}_5^2$ consider the element

$$\begin{pmatrix} 1 \\ 2 \\ 3 \end{pmatrix} \otimes \begin{pmatrix} 1 \\ 2 \end{pmatrix} + \begin{pmatrix} 1 \\ 1 \\ 1 \end{pmatrix} \otimes \begin{pmatrix} 0 \\ 1 \end{pmatrix} + \begin{pmatrix} 3 \\ 4 \\ 0 \end{pmatrix} \otimes \begin{pmatrix} 1 \\ 1 \end{pmatrix} + \begin{pmatrix} 3 \\ 3 \\ 3 \end{pmatrix} \otimes \begin{pmatrix} 2 \\ 1 \end{pmatrix}. \qquad (6.22)$$

We choose

$$\mathcal{B} = \{ \begin{pmatrix} 1 \\ 2 \\ 3 \end{pmatrix}, \begin{pmatrix} 1 \\ 1 \\ 1 \end{pmatrix} \}, \mathcal{C} = \{ \begin{pmatrix} 0 \\ 1 \end{pmatrix}, \begin{pmatrix} 1 \\ 1 \end{pmatrix} \},$$

and find that

$$S = \begin{pmatrix} 1 & 0 & 1 & 0 \\ 0 & 1 & 2 & 3 \end{pmatrix}, T = \begin{pmatrix} 1 & 1 & 0 & 4 \\ 1 & 0 & 1 & 2 \end{pmatrix}.$$

Compute now

$$ST^T = \begin{pmatrix} 1 & 2 \\ 3 & 3 \end{pmatrix}.$$

Thus (6.22) is not $\mathbf{0}$. Using any factorization of ST^T, for instance

$$ST^T = \begin{pmatrix} 1 & 2 \\ 3 & 3 \end{pmatrix} \begin{pmatrix} 1 & 0 \\ 0 & 1 \end{pmatrix} = \begin{pmatrix} 1 \\ 3 \end{pmatrix} \begin{pmatrix} 1 & 0 \end{pmatrix} + \begin{pmatrix} 2 \\ 3 \end{pmatrix} \begin{pmatrix} 0 & 1 \end{pmatrix},$$

we can write (6.22) differently. Indeed, choose $\mathbf{x}_1$, $\mathbf{x}_2$, $\mathbf{v}_1$ and $\mathbf{v}_2$ so that

$$[\mathbf{x}_1]_{\mathcal{B}} = \begin{pmatrix} 1 \\ 3 \end{pmatrix}, [\mathbf{x}_2]_{\mathcal{B}} = \begin{pmatrix} 2 \\ 3 \end{pmatrix}, [\mathbf{v}_1]_{\mathcal{C}} = \begin{pmatrix} 1 \\ 0 \end{pmatrix}, [\mathbf{v}_2]_{\mathcal{C}} = \begin{pmatrix} 0 \\ 1 \end{pmatrix}.$$

Thus

$$\mathbf{x}_1 = \begin{pmatrix} 4 \\ 0 \\ 1 \end{pmatrix}, \mathbf{x}_2 = \begin{pmatrix} 0 \\ 2 \\ 4 \end{pmatrix}, \mathbf{v}_1 = \begin{pmatrix} 0 \\ 1 \end{pmatrix}, \mathbf{v}_2 = \begin{pmatrix} 1 \\ 1 \end{pmatrix}.$$

Then (6.22) equals

$$\mathbf{x}_1 \otimes \mathbf{v}_1 + \mathbf{x}_2 \otimes \mathbf{v}_2 = \begin{pmatrix} 4 \\ 0 \\ 1 \end{pmatrix} \otimes \begin{pmatrix} 0 \\ 1 \end{pmatrix} + \begin{pmatrix} 0 \\ 2 \\ 4 \end{pmatrix} \otimes \begin{pmatrix} 1 \\ 1 \end{pmatrix}. \qquad (6.23)$$

We can now also determine the dimension of $V_1 \otimes V_2$.

Proposition 6.5.4 *Let V_1 and V_2 be finite-dimensional spaces. Then*

$$\dim V_1 \otimes V_2 = (\dim V_1)(\dim V_2).$$

More specifically, if $\mathcal{B} = \{\mathbf{x}_1, \ldots, \mathbf{x}_n\}$ is a basis for V_1 and $\mathcal{C} = \{\mathbf{v}_1, \ldots, \mathbf{v}_m\}$ be a basis for V_2, then $\{\mathbf{x}_i \otimes \mathbf{v}_j : 1 \le i \le n, 1 \le j \le m\}$ is a basis for $V_1 \otimes V_2$.

Proof. For any $\mathbf{y} \otimes \mathbf{w} \in V_1 \otimes V_2$, we can write $\mathbf{y} = \sum_{i=1}^n c_i \mathbf{x}_i$ and $\mathbf{w} = \sum_{j=1}^m \mathbf{v}_j$. Then

$$\mathbf{y} \otimes \mathbf{w} = \sum_{i=1}^n \sum_{j=1}^m c_i d_j \mathbf{x}_i \otimes \mathbf{v}_j.$$

For a linear combination $\sum_{r=1}^k a_r \mathbf{y}_r \otimes \mathbf{w}_r$ we can write each term as a linear combination of $\{\mathbf{x}_i \otimes \mathbf{v}_j : 1 \le i \le n, 1 \le j \le m\}$, and thus this linear combination also lies in $\text{Span}\{\mathbf{x}_i \otimes \mathbf{v}_j : 1 \le i \le n, 1 \le j \le m\}$. This shows that $\{\mathbf{x}_i \otimes \mathbf{v}_j : 1 \le i \le n, 1 \le j \le m\}$ spans $V_1 \otimes V_2$.

To show that $\{\mathbf{x}_i \otimes \mathbf{v}_j : 1 \le i \le n, 1 \le j \le m\}$ is linearly independent, suppose that $\sum_{i=1}^n \sum_{j=1}^m a_{ij} \mathbf{x}_i \otimes \mathbf{v}_j = \mathbf{0}$. Performing the procedure in the proof of Proposition 6.5.2 with $\mathcal{B}$ and $\mathcal{C}$ as above we obtain that

$$ST^T = \begin{pmatrix} a_{11} & \cdots & a_{1m} \\ \vdots & & \vdots \\ a_{n1} & \cdots & a_{nm} \end{pmatrix}.$$

Thus $\sum_{i=1}^n \sum_{j=1}^m a_{ij} \mathbf{x}_i \otimes \mathbf{v}_j = \mathbf{0}$ holds if and only if $a_{ij} = 0$ for all i and j. This proves linear independence.   □

By Proposition 6.5.4 we have that $\mathbb{F}^n \otimes \mathbb{F}^m$ has dimension nm. Thus $\mathbb{F}^n \otimes \mathbb{F}^m$ is isomorphic to $\mathbb{F}^{nm}$. This isomorphism can be obtained via a bijection between the basis

$$\{\mathbf{e}_i \otimes \mathbf{e}_j : 1 \le i \le n, 1 \le j \le m\}$$

of $\mathbb{F}^n \otimes \mathbb{F}^m$ and the basis

$$\{\mathbf{e}_i : 1 \leq i \leq nm\}$$

of $\mathbb{F}^{nm}$. The canonical way to do this is to order
$\{\mathbf{e}_i \otimes \mathbf{e}_j : 1 \leq i \leq n, 1 \leq j \leq m\}$ lexicographically. The ordering on pairs
(i, j) is *lexicographical* if

$$(i, j) \leq (k, l) \Leftrightarrow i < k \text{ or } (i = j \text{ and } j \leq k).$$

For example, ordering $\{1, 2, 3\} \times \{1, 2\}$ lexicographically results in

$$(1, 1) \leq (1, 2) \leq (1, 3) \leq (2, 1) \leq (2, 2) \leq (2, 3).$$

In this example we would match the bases by

$$\mathbf{e}_1 \otimes \mathbf{e}_1 \leftrightarrow \mathbf{e}_1, \mathbf{e}_1 \otimes \mathbf{e}_2 \leftrightarrow \mathbf{e}_2, \mathbf{e}_1 \otimes \mathbf{e}_3 \leftrightarrow \mathbf{e}_3, \mathbf{e}_2 \otimes \mathbf{e}_1 \leftrightarrow \mathbf{e}_4, \mathbf{e}_2 \otimes \mathbf{e}_2 \leftrightarrow \mathbf{e}_5, \mathbf{e}_2 \otimes \mathbf{e}_3 \leftrightarrow \mathbf{e}_6.$$

In general, we match

$$\mathbf{e}_i \otimes \mathbf{e}_j \in \mathbb{F}^n \otimes \mathbb{F}^m \leftrightarrow \mathbf{e}_{(j-1)n+i} \in \mathbb{F}^{nm}, \ 1 \leq i \leq n, 1 \leq j \leq m.$$

For a general $\mathbf{x} = (x_j)_{j=1}^n \in \mathbb{F}^n$ and $\mathbf{v} \in \mathbb{F}^m$ we now get the correspondence

$$\mathbf{x} \otimes \mathbf{v} \in \mathbb{F}^n \otimes \mathbb{F}^m \leftrightarrow \begin{pmatrix} x_1 \mathbf{v} \\ \vdots \\ x_n \mathbf{v} \end{pmatrix} \in \mathbb{F}^{nm}.$$

For example

$$\begin{pmatrix} 1 \\ 2 \\ 3 \end{pmatrix} \otimes \begin{pmatrix} 4 \\ 5 \end{pmatrix} \mathbb{R}^3 \otimes \mathbb{R}^2 \leftrightarrow \begin{pmatrix} 1 \begin{pmatrix} 4 \\ 5 \end{pmatrix} \\ 2 \begin{pmatrix} 4 \\ 5 \end{pmatrix} \\ 3 \begin{pmatrix} 4 \\ 5 \end{pmatrix} \end{pmatrix} = \begin{pmatrix} 4 \\ 5 \\ 8 \\ 10 \\ 12 \\ 15 \end{pmatrix} \in \mathbb{R}^6.$$

In other words, if we define

$$\Phi : \mathbb{F}^n \otimes \mathbb{F}^m \to \mathbb{F}^{nm} \text{ by } \Phi(\mathbf{e}_i \otimes \mathbf{e}_j) = \mathbf{e}_{(j-1)n+i},$$

or equivalently, by

$$\Phi(\mathbf{x} \otimes \mathbf{v}) = \begin{pmatrix} x_1 \mathbf{v} \\ \vdots \\ x_n \mathbf{v} \end{pmatrix}$$

and extend it to the full space by linear extension

$$\Phi\left(\sum_{j=1}^m c_j(\mathbf{x}_j \otimes \mathbf{v}_j)\right) = \sum_{j=1}^m c_j \Phi(\mathbf{x}_j \otimes \mathbf{v}_j),$$

then Φ is an isomorphism. We call this the *canonical isomorphism* between
$\mathbb{F}^n \otimes \mathbb{F}^m$ and $\mathbb{F}^{nm}$.

Example 6.5.5 For the vector in $\mathbb{Z}_5^3 \otimes \mathbb{Z}_5^2$

$$\mathbf{f} = \begin{pmatrix} 1 \\ 2 \\ 3 \end{pmatrix} \otimes \begin{pmatrix} 1 \\ 2 \end{pmatrix} + \begin{pmatrix} 1 \\ 1 \\ 1 \end{pmatrix} \otimes \begin{pmatrix} 0 \\ 1 \end{pmatrix} + \begin{pmatrix} 3 \\ 4 \\ 0 \end{pmatrix} \otimes \begin{pmatrix} 1 \\ 1 \end{pmatrix} + \begin{pmatrix} 3 \\ 3 \\ 3 \end{pmatrix} \otimes \begin{pmatrix} 2 \\ 1 \end{pmatrix} \qquad (6.24)$$

from (6.22), we have that

$$\Phi(\mathbf{f}) = \begin{pmatrix} 1 \\ 2 \\ 2 \\ 4 \\ 3 \\ 1 \end{pmatrix} + \begin{pmatrix} 0 \\ 1 \\ 0 \\ 1 \\ 0 \\ 1 \end{pmatrix} + \begin{pmatrix} 3 \\ 3 \\ 4 \\ 4 \\ 0 \\ 0 \end{pmatrix} + \begin{pmatrix} 1 \\ 3 \\ 1 \\ 3 \\ 1 \\ 3 \end{pmatrix} = \begin{pmatrix} 0 \\ 4 \\ 2 \\ 2 \\ 4 \\ 0 \end{pmatrix}.$$

If we apply Φ to the vector in (6.23) we obtain

$$\begin{pmatrix} 0 \\ 4 \\ 0 \\ 0 \\ 0 \\ 1 \end{pmatrix} + \begin{pmatrix} 0 \\ 0 \\ 2 \\ 2 \\ 4 \\ 4 \end{pmatrix} = \begin{pmatrix} 0 \\ 4 \\ 2 \\ 2 \\ 4 \\ 0 \end{pmatrix},$$

which is the same vector in $\mathbb{Z}_5^6$ as expected.

When V_1 and V_2 have inner products, the tensor product space $V_1 \otimes V_2$ has a natural associated inner product, as follows.

Proposition 6.5.6 *Let V_1 and V_2 have inner products $\langle \cdot, \cdot \rangle_1$ and $\langle \cdot, \cdot \rangle_2$, respectively. Define $\langle \cdot, \cdot \rangle$ on $V_1 \otimes V_2$ via*

$$\langle \mathbf{x} \otimes \mathbf{v}, \mathbf{y} \otimes \mathbf{w} \rangle = \langle \mathbf{x}, \mathbf{y} \rangle_1 \langle \mathbf{v} \otimes \mathbf{w} \rangle_2,$$

and extend $\langle \cdot, \cdot \rangle$ via the rules of a Hermitian form to all of $V_1 \otimes V_2$. Then $\langle \cdot, \cdot \rangle$ is an inner product.

By the extension via the rules of a Hermitian form, we mean that we set

$$\langle \sum_{i=1}^{n} c_i \mathbf{x}_i \otimes \mathbf{v}_i, \sum_{j=1}^{m} d_j \mathbf{y}_j \otimes \mathbf{w}_j \rangle = \sum_{i=1}^{n} \sum_{j=1}^{m} \bar{c}_i d_j \langle \mathbf{x}_i \otimes \mathbf{v}_i, \mathbf{y}_j \otimes \mathbf{w}_j \rangle =$$

$$\sum_{i=1}^{n} \sum_{j=1}^{m} \bar{c}_i d_j \langle \mathbf{x}_i, \mathbf{y}_j \rangle_1 \langle \mathbf{v}_i \otimes \mathbf{w}_j \rangle_2.$$

Proof. The only tricky part is to check that when $\mathbf{f} = \sum_{i=1}^{n} c_i \mathbf{x}_i \otimes \mathbf{v}_i$ has the property that from $\langle \mathbf{f}, \mathbf{f} \rangle = 0$ we obtain $\mathbf{f} = \mathbf{0}$. For this, we choose an orthonormal basis $\{\mathbf{z}_1, \ldots, \mathbf{z}_k\}$ for $\mathrm{Span}\{\mathbf{v}_1, \ldots, \mathbf{v}_n\}$, and rewrite $\mathbf{f}$ as

$$\mathbf{f} = \sum_{j=1}^{k} d_j \mathbf{y}_j \otimes \mathbf{z}_j.$$

This can always be done by writing $\mathbf{v}_i$ as linear combinations of $\{\mathbf{z}_1, \ldots, \mathbf{z}_k\}$, and reworking the expression for $\mathbf{f}$ using the rules (6.19)–(6.21). From $\langle \mathbf{f}, \mathbf{f} \rangle = 0$, we now obtain that

$$0 = \sum_{i=1}^{k} \sum_{j=1}^{k} \bar{d}_i d_j \langle \mathbf{y}_i, \mathbf{y}_j \rangle_1 \langle \mathbf{z}_i \otimes \mathbf{z}_j \rangle_2 = \sum_{i=1}^{k} \langle d_i \mathbf{y}_i, d_i \mathbf{y}_i \rangle_1,$$

yielding for each i that $\langle d_i \mathbf{y}_i, d_i \mathbf{y}_i \rangle_1 = 0$, and thus $d_i \mathbf{y}_i = \mathbf{0}$. This gives that $\mathbf{f} = \mathbf{0}$.

It is straightforward to check that $\langle \cdot, \cdot \rangle$ satisfies all the other rules of an inner product, and we will leave this to the reader. □

When V_1 and V_2 have norms $\| \cdot \|_1$ and $\| \cdot \|_2$, it is possible to provide $V_1 \otimes V_2$ with an associated norm as well. However, there are many ways of doing this. One way is to define

$$\|\mathbf{f}\| := \inf \sum_{j=1}^{k} |c_j| \|\mathbf{x}_j\|_1 \|\mathbf{v}_j\|_2,$$

where the infimum is taken over all possible ways of writing $\mathbf{f}$ as $\mathbf{f} = \sum_{j=1}^{k} c_j \mathbf{x}_j \otimes \mathbf{v}_j$. We will not further pursue this here.

When we have linear maps $A : V_1 \to W_1$ and $B : V_2 \to W_2$, one can define a linear map $A \otimes B : V_1 \otimes V_2 \to W_1 \otimes W_2$ via

$$(A \otimes B)(\mathbf{x} \otimes \mathbf{v}) := (A\mathbf{x}) \otimes (B\mathbf{v}),$$

and extend by linearity. Thus

$$(A \otimes B)(\sum_{j=1}^{n} \mathbf{x}_j \otimes \mathbf{v}_j) := \sum_{j=1}^{n} (A\mathbf{x}_j) \otimes (B\mathbf{v}_j).$$

Since

$$(A \otimes B)[(\mathbf{x} + \mathbf{y}) \otimes \mathbf{v}] = (A \otimes B)(\mathbf{x} \otimes \mathbf{v} + \mathbf{y} \otimes \mathbf{v}) \qquad (6.25)$$

$$(A \otimes B)[\mathbf{x} \otimes (\mathbf{v} + \mathbf{w})] = (A \otimes B)(\mathbf{x} \otimes \mathbf{v} + \mathbf{x} \otimes \mathbf{w}), \qquad (6.26)$$

and

$$(A \otimes B)[(c\mathbf{x}) \otimes \mathbf{v}] = (A \otimes B)[c(\mathbf{x} \otimes \mathbf{v})] = (A \otimes B)[\mathbf{x} \otimes (c\mathbf{v})], \qquad (6.27)$$

$A \otimes B$ is well-defined. Let us see how this "tensor" map works on a small example.

Example 6.5.7 Consider the linear maps given by matrix multiplication with the matrices

$$A = \begin{pmatrix} a_{11} & a_{12} \\ a_{21} & a_{22} \end{pmatrix} : \mathbb{F}^2 \to \mathbb{F}^2, B = \begin{pmatrix} b_{11} & b_{12} \\ b_{21} & b_{22} \end{pmatrix} : \mathbb{F}^2 \to \mathbb{F}^2.$$

Then

$$(A \otimes B)(\mathbf{e}_1 \otimes \mathbf{e}_1) = (a_{11}\mathbf{e}_1 + a_{21}\mathbf{e}_2) \otimes (b_{11}\mathbf{e}_1 + b_{21}\mathbf{e}_2) =$$

$$a_{11}b_{11}\mathbf{e}_1 \otimes \mathbf{e}_1 + a_{11}b_{21}\mathbf{e}_1 \otimes \mathbf{e}_2 + a_{21}b_{11}\mathbf{e}_2 \otimes \mathbf{e}_1 + a_{21}b_{21}\mathbf{e}_2 \otimes \mathbf{e}_2.$$

Similarly,

$$(A \otimes B)(\mathbf{e}_1 \otimes \mathbf{e}_2) = a_{11}b_{12}\mathbf{e}_1 \otimes \mathbf{e}_1 + a_{11}b_{22}\mathbf{e}_1 \otimes \mathbf{e}_2 + a_{21}b_{12}\mathbf{e}_2 \otimes \mathbf{e}_1 + a_{21}b_{22}\mathbf{e}_2 \otimes \mathbf{e}_2,$$

$$(A \otimes B)(\mathbf{e}_2 \otimes \mathbf{e}_1) = a_{12}b_{11}\mathbf{e}_1 \otimes \mathbf{e}_1 + a_{12}b_{21}\mathbf{e}_1 \otimes \mathbf{e}_2 + a_{22}b_{11}\mathbf{e}_2 \otimes \mathbf{e}_1 + a_{22}b_{21}\mathbf{e}_2 \otimes \mathbf{e}_2,$$

$$(A \otimes B)(\mathbf{e}_2 \otimes \mathbf{e}_2) = a_{12}b_{12}\mathbf{e}_1 \otimes \mathbf{e}_1 + a_{12}b_{22}\mathbf{e}_1 \otimes \mathbf{e}_2 + a_{22}b_{12}\mathbf{e}_2 \otimes \mathbf{e}_1 + a_{22}b_{22}\mathbf{e}_2 \otimes \mathbf{e}_2.$$

Thus, if we take the canonical basis $\mathcal{E} = \{\mathbf{e}_1 \otimes \mathbf{e}_1, \mathbf{e}_1 \otimes \mathbf{e}_2, \mathbf{e}_2 \otimes \mathbf{e}_1, \mathbf{e}_2 \otimes \mathbf{e}_2\}$, we obtain that

$$[A \otimes B]_{\mathcal{E} \leftarrow \mathcal{E}} = \begin{pmatrix} a_{11}b_{11} & a_{11}b_{12} & a_{12}b_{11} & a_{12}b_{12} \\ a_{11}b_{21} & a_{11}b_{22} & a_{12}b_{21} & a_{12}b_{22} \\ a_{21}b_{11} & a_{21}b_{12} & a_{22}b_{11} & a_{22}b_{12} \\ a_{21}b_{21} & a_{21}b_{22} & a_{22}b_{21} & a_{22}b_{22} \end{pmatrix}.$$

Note that we may write this as

$$[A \otimes B]_{\mathcal{E} \leftarrow \mathcal{E}} = \begin{pmatrix} a_{11}B & a_{12}B \\ a_{21}B & a_{22}B \end{pmatrix}.$$

The above example indicates how find a matrix representation for $T \otimes S$ in general.

Proposition 6.5.8 *Let* V_1, V_2, W_1, W_2 *be vector spaces over* $\mathbb{F}$ *with bases* $\mathcal{B}_1 = \{\mathbf{x}_j : j = 1, \ldots, n_1\}, \mathcal{B}_1 = \{\mathbf{v}_j : j = 1, \ldots, n_2\},$ $\mathcal{C}_1 = \{\mathbf{y}_j : j = 1, \ldots, m_1\}, \mathcal{C}_2 = \{\mathbf{w}_j : j = 1, \ldots, m_2\}, $ *respectively. For* $V_1 \otimes V_2$ *and* $W_1 \otimes W_2$, *we choose the bases*

$$\mathcal{E} = \{\mathbf{x}_j \otimes \mathbf{v}_l : j = 1, \ldots, n_1, l = 1, \ldots, n_2\},$$

$$\mathcal{F} = \{\mathbf{y}_j \otimes \mathbf{w}_l : j = 1, \ldots, m_1, l = 1, \ldots, m_2\},$$

respectively, where we order the elements lexicographically. If $T : V_1 \to W_1$ and $S : V_2 \to W_2$ are linear maps, with matrix representations

$$A = (a_{jl})_{j=1,l=1}^{m_1,\ n_1} = [T]_{\mathcal{C}_1 \leftarrow \mathcal{B}_1}, B = [S]_{\mathcal{C}_2 \leftarrow \mathcal{B}_2},$$

then the matrix representation for $T \otimes S$ is given by the $(m_1 m_2) \times (n_1 n_2)$ matrix

$$[T \otimes S]_{\mathcal{F} \leftarrow \mathcal{E}} = \begin{pmatrix} a_{11} B & \cdots & a_{1,n_1} B \\ \vdots & & \vdots \\ a_{m_1,1} B & \cdots & a_{m_1,n_1} B \end{pmatrix}. \tag{6.28}$$

Remark 6.5.9 Sometimes the matrix in (6.28) is taken as the definition of $A \otimes B$. It is important to realize that this particular form of the matrix is due to the chosen (lexicographically ordered) bases of the underlying spaces, and that changing the convention for choosing these bases will also change the matrix.

Proof of Proposition 6.5.8. Writing $B = (b_{ij})_{i=1,j=1}^{m_2,\ n_2}$, we have that

$$(T \otimes S)(\mathbf{x}_j \otimes \mathbf{v}_l) = \sum_{r=1}^{m_1} \sum_{s=1}^{m_2} a_{rj} b_{sl} \mathbf{y}_r \otimes \mathbf{w}_s, j = 1, \ldots, n_1, l = 1, \ldots, n_2.$$

Organizing this information appropriately in the representation matrix, we find that (6.28) holds. □

Several important properties of linear maps carry over to their tensor products. We first note the following.

Lemma 6.5.10 *If $T : V_1 \to W_1$, $\hat{T} : W_1 \to Z_1$, $S : V_2 \to W_2$, $\hat{S} : W_2 \to Z_2$ are linear maps. Then*

$$(\hat{T} \otimes \hat{S})(T \otimes S) = (\hat{T}T) \otimes (\hat{S}S).$$

Proof. For a simple tensor $\mathbf{x} \otimes \mathbf{v}$ we clearly have that

$$(\hat{T} \otimes \hat{S})(T \otimes S)(\mathbf{x} \otimes \mathbf{v}) = (\hat{T} \otimes \hat{S})(T\mathbf{x} \otimes S\mathbf{v}) = (\hat{T}T\mathbf{x}) \otimes (\hat{S}S\mathbf{v}) = (\hat{T}T) \otimes (\hat{S}S)(\mathbf{x} \otimes \mathbf{v}).$$

But then $(\hat{T} \otimes \hat{S})(T \otimes S)$ and $(\hat{T}T) \otimes (\hat{S}S)$ also act the same on linear combinations of simple tensors. Thus the lemma follows. □

Proposition 6.5.11 *Let $T : V_1 \to W_1$ and $S : V_2 \to W_2$ be linear, where the vector spaces are over $\mathbb{F}$. Then the following hold:*

(i) *If T and S are invertible, then so is $T \otimes S$ and $(T \otimes S)^{-1} = T^{-1} \otimes S^{-1}$.*

(ii) *If $V_1 = W_1$ and $V_2 = W_2$, and $\mathbf{x}$ and $\mathbf{v}$ are eigenvectors for T and S with eigenvalues λ and μ, respectively, then $\mathbf{x} \otimes \mathbf{v}$ is an eigenvector for $T \otimes S$, with eigenvalue $\lambda\mu$; thus $(T \otimes S)(\mathbf{x} \otimes \mathbf{v}) = \lambda\mu(\mathbf{x} \otimes \mathbf{v})$.*

For the remaining parts, the vector spaces are assumed to be inner product spaces (and thus necessarily, $\mathbb{F} = \mathbb{R}$ or $\mathbb{C}$), and the inner product on the tensor product is given via the construction in Proposition 6.5.6.

(iii) $(T \otimes S)^\star = T^\star \otimes S^\star$.

(iv) *If T and S are isometries, then so is $T \otimes S$.*

(v) *If T and S are unitary, then so is $T \otimes S$.*

(vi) *If T and S are normal, then so is $T \otimes S$.*

(vii) *If T and S are Hermitian, then so is $T \otimes S$.*

(viii) *If T and S are positive (semi-)definite, then so is $T \otimes S$.*

Proof. The proof is straightforward. For instance, using Lemma 6.5.10,

$$(T \otimes S)(T^{-1} \otimes S^{-1}) = (TT^{-1}) \otimes (SS^{-1}) = id_{V_1} \otimes id_{V_2} = id_{V_1 \otimes V_2},$$

and

$$(T^{-1} \otimes S^{-1})(T \otimes S) = (T^{-1}T) \otimes (S^{-1}S) = id_{W_1} \otimes id_{W_2} = id_{W_1 \otimes W_2},$$

proving (i).

For parts (iii)–(viii) it is important to observe that

$$\langle (T \otimes S)(\mathbf{x} \otimes \mathbf{v}), \mathbf{y} \otimes \mathbf{w} \rangle = \langle T\mathbf{x} \otimes S\mathbf{v}, \mathbf{y} \otimes \mathbf{w} \rangle = \langle T\mathbf{x}, \mathbf{y} \rangle \langle S\mathbf{v}, \mathbf{w} \rangle =$$

$$\langle \mathbf{x}, T^\star\mathbf{y} \rangle \langle \mathbf{v}, S^\star\mathbf{w} \rangle = \langle \mathbf{x} \otimes \mathbf{v}, T^\star\mathbf{y} \otimes S^\star\mathbf{w} \rangle = \langle \mathbf{x} \otimes \mathbf{v}, (T^\star \otimes S^\star)(\mathbf{y} \otimes \mathbf{w}) \rangle.$$

This equality extends to linear combinations of simple tensors, showing (iii).

The remaining details of the proof are left to the reader. For part (viii) use that T is positive semidefinite if and only if $T = CC^*$ for some C, which can be chosen to be invertible when T is positive definite.   □

The theory we developed in this section for two vector spaces, can also be extended to a tensor product $V_1 \otimes \cdots \otimes V_k$ of k vector spaces. In that case $V_1 \otimes \cdots \otimes V_k$ is generated by elements

$$\mathbf{v}_1 \otimes \cdots \otimes \mathbf{v}_k,$$

where $\mathbf{v}_1 \in V_1, \ldots, \mathbf{v}_k \in V_k$. The tensor product needs to satisfy the rules

$$(\mathbf{v}_1 \otimes \cdots \otimes \mathbf{v}_r \otimes \cdots \otimes \mathbf{v}_k) + (\mathbf{v}_1 \otimes \cdots \otimes \hat{\mathbf{v}}_r \otimes \cdots \otimes \mathbf{v}_k) = \mathbf{v}_1 \otimes \cdots \otimes (\mathbf{v}_r + \hat{\mathbf{v}}_r) \otimes \cdots \otimes \mathbf{v}_k,$$
(6.29)

and

$$\mathbf{v}_1 \otimes \cdots \otimes (c\mathbf{v}_r) \otimes \cdots \otimes \mathbf{v}_k = c(\mathbf{v}_1 \otimes \cdots \otimes \mathbf{v}_r \otimes \cdots \otimes \mathbf{v}_k).$$
(6.30)

Alternatively, one can first construct $V_1 \otimes V_2$, and then $(V_1 \otimes V_2) \otimes V_3$ and so forth, arriving at a vector space generated by elements

$$(\cdot((\mathbf{v}_1 \otimes \mathbf{v}_2) \otimes \mathbf{v}_3) \otimes \cdots) \otimes \mathbf{v}_k.$$

These vector spaces $V_1 \otimes \cdots \otimes V_k$ and $(\cdot((V_1 \otimes V_2) \otimes V_3) \otimes \cdots) \otimes V_k$ are isomorphic, by introducing the isomorphism Φ via

$$\Phi(\mathbf{v}_1 \otimes \cdots \otimes \mathbf{v}_k) = (\cdot((\mathbf{v}_1 \otimes \mathbf{v}_2) \otimes \mathbf{v}_3) \otimes \cdots) \otimes \mathbf{v}_k.$$

As these vector spaces are isomorphic, we will not draw a distinction between them and treat the tensor product as an associative operation, so that for instance

$$(\mathbf{v} \otimes \mathbf{w}) \otimes \mathbf{x} = \mathbf{v} \otimes \mathbf{w} \otimes \mathbf{x} = \mathbf{v} \otimes (\mathbf{w} \otimes \mathbf{x}).$$

In the following section, we will use the tensor product of k vector spaces, where each vector space is the same vector space. In other words, $V_1 = \cdots = V_k = V$. In this case we write

$$V_1 \otimes \cdots \otimes V_k = V \otimes \cdots \otimes V =: \otimes^k V.$$

6.6   Anti-symmetric and symmetric tensors

In this section we define two important subspaces of $V \otimes \cdots \otimes V =: \otimes^k V$, the vector space obtained by taking a vector space V and taking the kth tensor product of itself. Elements in $\otimes^k V$ are linear combinations of vectors $\mathbf{v}_1 \otimes \cdots \otimes \mathbf{v}_k$, where $\mathbf{v}_1, \ldots, \mathbf{v}_k \in V$.

The *anti-symmetric tensor product* of vectors $\mathbf{v}_1, \ldots, \mathbf{v}_k \in V$ is defined to be the vector

$$\mathbf{v}_1 \wedge \cdots \wedge \mathbf{v}_k = \sum_{\sigma \in S_k} \text{sign}\sigma \; \mathbf{v}_{\sigma(1)} \otimes \cdots \otimes \mathbf{v}_{\sigma(k)},$$

where S_k denotes the set of all permutations on $\{1, \ldots, k\}$ and $\text{sign}\sigma = 1$ when σ is an even permutation and $\text{sign}\sigma = -1$ when σ is an odd permutation.

Example 6.6.1 In $\mathbb{F}^2$, we have

$$\begin{pmatrix} 1 \\ 0 \end{pmatrix} \wedge \begin{pmatrix} 0 \\ 1 \end{pmatrix} = \begin{pmatrix} 1 \\ 0 \end{pmatrix} \otimes \begin{pmatrix} 0 \\ 1 \end{pmatrix} - \begin{pmatrix} 0 \\ 1 \end{pmatrix} \otimes \begin{pmatrix} 1 \\ 0 \end{pmatrix} \leftrightarrow \begin{pmatrix} 0 \\ 1 \\ -1 \\ 0 \end{pmatrix} \in \mathbb{F}^4.$$

In $\mathbb{F}^3$, we have

$$\begin{pmatrix} 1 \\ 0 \\ 0 \end{pmatrix} \wedge \begin{pmatrix} 0 \\ 1 \\ 0 \end{pmatrix} \leftrightarrow \begin{pmatrix} 0 \\ 1 \\ 0 \\ -1 \\ 0 \\ 0 \\ 0 \\ 0 \\ 0 \end{pmatrix}, \quad \begin{pmatrix} 1 \\ 0 \\ 0 \end{pmatrix} \wedge \begin{pmatrix} 0 \\ 0 \\ 1 \end{pmatrix} \leftrightarrow \begin{pmatrix} 0 \\ 0 \\ 1 \\ 0 \\ 0 \\ 0 \\ -1 \\ 0 \\ 0 \end{pmatrix}, \quad \begin{pmatrix} 0 \\ 1 \\ 0 \end{pmatrix} \wedge \begin{pmatrix} 0 \\ 0 \\ 1 \end{pmatrix} \leftrightarrow \begin{pmatrix} 0 \\ 0 \\ 0 \\ 0 \\ 0 \\ 1 \\ 0 \\ -1 \\ 0 \end{pmatrix}.$$

Lemma 6.6.2 *The anti-symmetric tensor is linear in each of its parts; that is*

$$\mathbf{v}_1 \wedge \cdots \wedge (c\mathbf{v}_i + d\hat{\mathbf{v}}_i) \wedge \cdots \wedge \mathbf{v}_k = c(\mathbf{v}_1 \wedge \cdots \wedge \mathbf{v}_i \wedge \cdots \wedge \mathbf{v}_k) + d(\mathbf{v}_1 \wedge \cdots \wedge \hat{\mathbf{v}}_i \wedge \cdots \wedge \mathbf{v}_k).$$

Proof. Follows immediately from the corresponding property of the tensor product. ☐

Proposition 6.6.3 *If two vectors in an anti-symmetric tensor are switched, it will change sign; that is,*

$$\mathbf{v}_1 \wedge \cdots \wedge \mathbf{v}_i \wedge \cdots \wedge \mathbf{v}_j \wedge \cdots \wedge \mathbf{v}_k = -\mathbf{v}_1 \wedge \cdots \wedge \mathbf{v}_j \wedge \cdots \wedge \mathbf{v}_i \wedge \cdots \wedge \mathbf{v}_k.$$

Proof. Let $\tau = (i \ j)$ be the permutation that switches i and j. Then

$$\mathbf{v}_1 \wedge \cdots \wedge \mathbf{v}_j \wedge \cdots \wedge \mathbf{v}_i \wedge \cdots \wedge \mathbf{v}_k = \sum_{\sigma \in S_k} \text{sign}\sigma \ \mathbf{v}_{\sigma(\tau(1))} \otimes \cdots \otimes \mathbf{v}_{\sigma(\tau(k))} =$$

$$-\sum_{\sigma \in S_k} \text{sign}(\sigma\tau) \ \mathbf{v}_{\sigma(\tau(1))} \otimes \cdots \otimes \mathbf{v}_{\sigma(\tau(k))} = -\sum_{\hat{\sigma} \in S_k} \text{sign}\hat{\sigma} \ \mathbf{v}_{\hat{\sigma}(1)} \otimes \cdots \otimes \mathbf{v}_{\hat{\sigma}(k)} =$$

$$-\mathbf{v}_1 \wedge \cdots \wedge \mathbf{v}_i \wedge \cdots \wedge \mathbf{v}_j \wedge \cdots \wedge \mathbf{v}_k,$$

where we used that if τ runs through all of S_k, then so does $\hat{\sigma} = \sigma\tau$. ☐

An immediate consequence is the following.

Corollary 6.6.4 *If a vector appears twice in an anti-symmetric tensor, then the anti-symmetric tensor is zero; that is,*

$$\mathbf{v}_1 \wedge \cdots \wedge \mathbf{v}_i \wedge \cdots \wedge \mathbf{v}_i \wedge \cdots \wedge \mathbf{v}_k = \mathbf{0}.$$

Proof. When $\mathbb{F}$ is so that $2 \neq 0$, it is a consequence of Proposition 6.6.3 as follows. Let $\mathbf{f} = \mathbf{v}_1 \wedge \cdots \wedge \mathbf{v}_i \wedge \cdots \wedge \mathbf{v}_i \wedge \cdots \wedge \mathbf{v}_k$. By Proposition 6.6.3 we have that $\mathbf{f} = -\mathbf{f}$. Thus $2\mathbf{f} = \mathbf{0}$. As $2 \neq 0$, we obtain that $\mathbf{f} = \mathbf{0}$.

When $\mathbb{F}$ is so that $1 + 1 = 0$ (which is also referred to as a field of characteristic 2), we have the following argument which actually works for any field. Let $E_k = \{\sigma \in S_k : \sigma$ is even$\}$. Let i and j be the two locations where $\mathbf{v}_i$ appears, and let $\tau = (i \; j)$. Then all odd permutations in S_k are of the form $\sigma\tau$ with $\sigma \in E_k$. Thus

$$\mathbf{v}_1 \wedge \cdots \wedge \mathbf{v}_i \wedge \cdots \wedge \mathbf{v}_i \wedge \cdots \wedge \mathbf{v}_k =$$

$$\sum_{\sigma \in E_k} \left(\text{sign}\sigma \; \mathbf{v}_{\sigma(1)} \otimes \cdots \otimes \mathbf{v}_{\sigma(k)} + \text{sign}(\sigma\tau) \; \mathbf{v}_{\sigma(\tau(1))} \otimes \cdots \otimes \mathbf{v}_{\sigma(\tau(k))} \right).$$

As $\text{sign}(\sigma\tau) = -\text{sign}\sigma$ and

$$\mathbf{v}_{\sigma(1)} \otimes \cdots \otimes \mathbf{v}_{\sigma(k)} = \mathbf{v}_{\sigma(\tau(1))} \otimes \cdots \otimes \mathbf{v}_{\sigma(\tau(k))},$$

we get that all terms cancel. $\qquad\square$

We define

$$\wedge^k V := \text{Span}\{\mathbf{v}_1 \wedge \cdots \wedge \mathbf{v}_k : \mathbf{v}_j \in V_j, j = 1, \ldots, k\}.$$

Then $\wedge^k V$ is a subspace of $\otimes^k V$.

Proposition 6.6.5 *When* $\dim V = n$, *then* $\dim \wedge^k V = \binom{n}{k}$. *In fact, if* $\{\mathbf{v}_1, \ldots, \mathbf{v}_n\}$ *is a basis of* V, *then*
$\mathcal{E} = \{\mathbf{v}_{i_1} \wedge \cdots \wedge \mathbf{v}_{i_k} : 1 \leq i_1 < \cdots < i_k \leq n\}$ *is a basis of* $\wedge^k V$.

Proof. Let $\mathbf{f} \in \wedge^k V$. Then $\mathbf{f}$ is a linear combination of elements of the form $\mathbf{x}_1 \wedge \cdots \wedge \mathbf{x}_k$ where $\mathbf{x}_j \in V$, $j = 1, \ldots, k$. Each $\mathbf{x}_j$ is a linear combination of $\mathbf{v}_1, \ldots, \mathbf{v}_n$, thus $\mathbf{x}_j = \sum_{l=1}^{n} c_{lj}\mathbf{v}_l$ for some scalars c_{lj}. Plugging this into $\mathbf{x}_1 \wedge \cdots \wedge \mathbf{x}_k$ and using Lemma 6.6.2 we get that

$$\mathbf{x}_1 \wedge \cdots \wedge \mathbf{x}_k = \sum_{l_1, \ldots, l_k = 1}^{n} c_{l_1,1} \cdots c_{l_k,k} \mathbf{v}_{l_1} \wedge \cdots \wedge \mathbf{v}_{l_k}.$$

When $l_r = l_s$ for some $r \neq s$, we have that $\mathbf{v}_{l_1} \wedge \cdots \wedge \mathbf{v}_{l_k} = 0$, so we only have nonzero terms when all $l_1, \ldots, l_k$ are different. Moreover, by applying

Proposition 6.6.3, we can always do several switches so that $\mathbf{v}_{l_1} \wedge \cdots \wedge \mathbf{v}_{l_k}$ turns into $\mathbf{v}_{i_1} \wedge \cdots \wedge \mathbf{v}_{i_k}$ or $-\mathbf{v}_{i_1} \wedge \cdots \wedge \mathbf{v}_{i_k}$, where now $i_1 < \cdots < i_k$ (and $\{i_1, \ldots, i_k\} = \{l_1, \ldots, l_k\}$). Putting these observations together, we obtain that $\mathbf{f} \in \mathrm{Span}\,\mathcal{E}$.

For linear independence, suppose that
$\sum_{1 \le i_1 < \cdots < i_k \le n} c_{i_1, \ldots, i_k} \mathbf{v}_{i_1} \wedge \cdots \wedge \mathbf{v}_{i_k} = \mathbf{0}$ for some scalars $c_{i_1, \ldots, i_k} \in \mathbb{F}$,
$1 \le i_1 < \cdots < i_k \le n$. Putting in the definition of $\mathbf{v}_{i_1} \wedge \cdots \wedge \mathbf{v}_{i_k}$, we arrive at the equality

$$\sum_{l_1, \ldots, l_k = 1}^{n} a_{l_1, \ldots, l_k} \mathbf{v}_{l_1} \otimes \cdots \otimes \mathbf{v}_{i_k} = \mathbf{0}, \tag{6.31}$$

where either $a_{l_1, \ldots, l_k} = 0$ (when $l_r = l_s$ for some $r \ne s$) or where $a_{l_1, \ldots, l_k}$ equals one of the numbers $\pm c_{i_1, \ldots, i_k}$. As the tensors
$\{\mathbf{v}_{l_1} \otimes \cdots \otimes \mathbf{v}_{i_k} : l_j = 1, \ldots, n, j = 1, \ldots, k\}$ form a basis of $\otimes^k V$, we have that (6.31) implies that $a_{l_1, \ldots, l_k}$ are all equal to 0. This implies that all $c_{i_1, \ldots, i_k}$ are equal to 0. This shows that $\mathcal{E}$ is linearly independent.

It remains to observe that the number of elements of $\mathcal{E}$ corresponds to the number of ways one can choose k numbers from $\{1, \ldots, n\}$, which equals $\binom{n}{k}$.
$\square$

Remark 6.6.6 Note that when V has an inner product, and when $\{\mathbf{v}_1, \ldots, \mathbf{v}_n\}$ is chosen to be an orthonormal basis of V, the basis $\mathcal{E}$ is not orthonormal. It is however an orthogonal basis, and thus all that needs to be done is to make the elements of $\mathcal{E}$ of unit length. As all have length $\sqrt{k!}$, one obtains an orthonormal basis for $\wedge^k V$ by taking

$$\mathcal{E}_{\mathrm{on}} = \{\frac{1}{\sqrt{k!}}(\mathbf{v}_{i_1} \wedge \cdots \wedge \mathbf{v}_{i_k}) : 1 \le i_1 < \cdots < i_k \le n\}.$$

We next analyze how linear operators and anti-symmetric tensors interact.

Proposition 6.6.7 *Let $T : V \to W$ be a linear map. Denote $\otimes^k T = T \otimes \cdots \otimes T : \otimes^k V \to \otimes^k W$. Then $\otimes^k T[\wedge^k V] \subseteq \wedge^k W$.*

Proof. Let $\mathbf{v}_1 \wedge \cdots \wedge \mathbf{v}_k \in \wedge^k V$. Then

$$\otimes^k T(\mathbf{x}_1 \wedge \cdots \wedge \mathbf{x}_k) = \otimes^k T\big(\sum_{\sigma \in S_k} \mathrm{sign}\sigma\, \mathbf{v}_{\sigma(1)} \otimes \cdots \otimes \mathbf{v}_{\sigma(k)}\big) =$$

$$\sum_{\sigma \in S_k} \mathrm{sign}\sigma\, \otimes^k T(\mathbf{v}_{\sigma(1)} \otimes \cdots \otimes \mathbf{v}_{\sigma(k)}) = \sum_{\sigma \in S_k} \mathrm{sign}\sigma\, (T\mathbf{v}_{\sigma(1)}) \otimes \cdots \otimes (T\mathbf{v}_{\sigma(k)}) =$$

$$(T\mathbf{v}_1) \wedge \cdots \wedge (T\mathbf{v}_k) \in \wedge^k W.$$

As every element of $\wedge^k V$ is a linear combination of elements of the type $\mathbf{v}_1 \wedge \cdots \wedge \mathbf{v}_k$, we obtain the result. $\qquad\square$

The restriction of $\otimes^k T$ to the subspace $\otimes^k V$ is denoted as $\wedge^k T$. Equivalently, $\wedge^k T : \wedge^k V \to \wedge^k W$ is defined via

$$\wedge^k T(\mathbf{v}_1 \wedge \cdots \wedge \mathbf{v}_k) = (T\mathbf{v}_1) \wedge \cdots \wedge (T\mathbf{v}_k).$$

Let us find $\wedge^k T$ in the following example.

Example 6.6.8 Let $T : \mathbb{F}^3 \to \mathbb{F}^3$ be multiplication with the matrix $A = (a_{ij})_{i,j=1}^3$. Thus $T(\mathbf{x}) = A\mathbf{x}$. The standard basis on $\wedge^2 \mathbb{F}^3$ is $\mathcal{E} = \{\mathbf{e}_1 \wedge \mathbf{e}_2, \mathbf{e}_1 \wedge \mathbf{e}_3, \mathbf{e}_2 \wedge \mathbf{e}_3\}$. Let us compute $[\wedge^2 T]_{\mathcal{E}\leftarrow\mathcal{E}}$. We apply $\wedge^2 T$ to the first basis element:

$$\wedge^2 T(\mathbf{e}_1 \wedge \mathbf{e}_2) = (T\mathbf{e}_1) \wedge (T\mathbf{e}_2) = (a_{11}\mathbf{e}_1 + a_{21}\mathbf{e}_2 + a_{31}\mathbf{e}_3) \wedge (a_{12}\mathbf{e}_1 + a_{22}\mathbf{e}_2 + a_{32}\mathbf{e}_3) =$$

$$(a_{11}a_{22} - a_{21}a_{12})\mathbf{e}_1 \wedge \mathbf{e}_2 + (a_{11}a_{32} - a_{31}a_{32})\mathbf{e}_1 \wedge \mathbf{e}_3 + (a_{21}a_{32} - a_{31}a_{22})\mathbf{e}_2 \wedge \mathbf{e}_3.$$

Continuing, we find

$$[\wedge^2 T]_{\mathcal{E}\leftarrow\mathcal{E}} = \begin{pmatrix} a_{11}a_{22} - a_{21}a_{12} & a_{11}a_{23} - a_{21}a_{13} & a_{12}a_{23} - a_{22}a_{13} \\ a_{11}a_{32} - a_{31}a_{12} & a_{11}a_{33} - a_{31}a_{13} & a_{12}a_{33} - a_{32}a_{13} \\ a_{21}a_{32} - a_{31}a_{22} & a_{21}a_{33} - a_{31}a_{23} & a_{22}a_{33} - a_{33}a_{23} \end{pmatrix}.$$

If we denote $A[I.J] = (a_{ij})_{i \in I, j \in J}$, then we obtain

$$[\wedge^2 T]_{\mathcal{E}\leftarrow\mathcal{E}} = \begin{pmatrix} \det A[\{1,2\},\{1,2\}] & \det A[\{1,2\},\{1,3\}] & \det A[\{1,2\},\{2,3\}] \\ \det A[\{1,3\},\{1,2\}] & \det A[\{1,3\},\{1,3\}] & \det A[\{1,3\},\{2,3\}] \\ \det A[\{2,3\},\{1,2\}] & \det A[\{2,3\},\{1,3\}] & \det A[\{2,3\},\{2,3\}] \end{pmatrix}.$$

$$(6.32)$$

The matrix on the right-hand side of (6.32) is called the second compound matrix of A. In general, for a $n \times n$ matrix the *kth compound matrix* of A is an $\binom{n}{k} \times \binom{n}{k}$ matrix whose entries are $A[I, J]$, where I and J, run through all subsets of $\{1, \ldots, n\}$ with k elements. This matrix corresponds to $[\wedge^k T]_{\mathcal{E}\leftarrow\mathcal{E}}$ when $T(\mathbf{x}) = A\mathbf{x}$ and $\mathcal{E} = \{\mathbf{e}_{i_1} \wedge \cdots \wedge \mathbf{e}_{i_k} : 1 \le i_1 < \cdots < i_k \le n\}$.

Lemma 6.6.9 *Let $T : V \to W$ and $S : X \to Y$ be linear, where the vector spaces are over $\mathbb{F}$. Then*

$$(\wedge^k T)(\wedge^k S) = \wedge^k (TS).$$

Proof. Since $(\otimes^k T)(\otimes^k S) = \otimes^k (TS)$, and as in general $\wedge^k U$ is defined as $\otimes^k U$ on the subspace of anti-symmetric tensors, the lemma follows. $\qquad\square$

Proposition 6.6.10 *Let $T : V \to W$ be linear, where the vector spaces are over $\mathbb{F}$. Then the following hold:*

(i) *If T is invertible, then so is $\wedge^k T$ and $(\wedge^k T)^{-1} = \wedge^k(T^{-1})$.*

(ii) *If $V = W$ and $\mathbf{x}_1, \ldots, \mathbf{x}_k$ are linearly independent eigenvectors for T with eigenvalues $\lambda_1, \ldots, \lambda_k$, respectively, then $\mathbf{x}_1 \wedge \cdots \wedge \mathbf{x}_k$ is an eigenvector for $\wedge^k T$, with eigenvalue $\prod_{i=1}^k \lambda_i = \lambda_1 \cdots \lambda_k$; thus $\wedge^k T(\mathbf{x}_1 \wedge \cdots \wedge \mathbf{x}_k) = \lambda_1 \cdots \lambda_k(\mathbf{x}_1 \wedge \cdots \wedge \mathbf{x}_k)$.*

For the remaining parts, V is assumed to be an inner product space (and thus necessarily, $\mathbb{F} = \mathbb{R}$ or $\mathbb{C}$), and the inner product $\wedge^k V$ is inherited from the inner product on the tensor product given via the construction in Proposition 6.5.6.

(iii) *$(\wedge^k T)^\star = \wedge^k T^\star$.*

(iv) *If T is an isometry, then so is $\wedge^k T$.*

(v) *If T is unitary, then so is $\wedge^k T$.*

(vi) *If T is normal, then so is $\wedge^k T$.*

(vii) *If T is Hermitian, then so is $\wedge^k T$.*

(viii) *If T is positive (semi-)definite, then so is $\wedge^k T$.*

Proof. Use Lemma 6.6.9 and Proposition 6.5.11. $\square$

We next switch to the symmetric tensor product. The development is very similar to the anti-symmetric tensor product, where now determinants are replaced by permanents (see (6.34) for the definition). The *symmetric tensor product* of vectors $\mathbf{v}_1, \ldots, \mathbf{v}_k \in V$ is defined to be the vector

$$\mathbf{v}_1 \vee \cdots \vee \mathbf{v}_k = \sum_{\sigma \in S_k} \mathbf{v}_{\sigma(1)} \otimes \cdots \otimes \mathbf{v}_{\sigma(k)}.$$

Thus, the difference with the anti-symmetric tensor product is the absence of the factor signσ.

Example 6.6.11 In $\mathbb{F}^2$ with $k = 2$, we have

$$\begin{pmatrix} 1 \\ 0 \end{pmatrix} \vee \begin{pmatrix} 1 \\ 0 \end{pmatrix} = \begin{pmatrix} 1 \\ 0 \end{pmatrix} \otimes \begin{pmatrix} 1 \\ 0 \end{pmatrix} + \begin{pmatrix} 1 \\ 0 \end{pmatrix} \otimes \begin{pmatrix} 1 \\ 0 \end{pmatrix} \leftrightarrow \begin{pmatrix} 2 \\ 0 \\ 0 \\ 0 \end{pmatrix} \in \mathbb{F}^4$$

and

$$\begin{pmatrix} 1 \\ 0 \end{pmatrix} \vee \begin{pmatrix} 0 \\ 1 \end{pmatrix} = \begin{pmatrix} 1 \\ 0 \end{pmatrix} \otimes \begin{pmatrix} 0 \\ 1 \end{pmatrix} + \begin{pmatrix} 0 \\ 1 \end{pmatrix} \otimes \begin{pmatrix} 1 \\ 0 \end{pmatrix} \leftrightarrow \begin{pmatrix} 0 \\ 1 \\ 1 \\ 0 \end{pmatrix} \in \mathbb{F}^4.$$

In $\mathbb{F}^2$ with $k = 3$, we have

$$\begin{pmatrix} 1 \\ 0 \end{pmatrix} \vee \begin{pmatrix} 1 \\ 0 \end{pmatrix} \vee \begin{pmatrix} 1 \\ 0 \end{pmatrix} \leftrightarrow \begin{pmatrix} 6 \\ 0 \\ 0 \\ 0 \\ 0 \\ 0 \\ 0 \\ 0 \end{pmatrix} \in \mathbb{F}^8, \begin{pmatrix} 1 \\ 0 \end{pmatrix} \vee \begin{pmatrix} 1 \\ 0 \end{pmatrix} \vee \begin{pmatrix} 0 \\ 1 \end{pmatrix} \leftrightarrow \begin{pmatrix} 0 \\ 2 \\ 2 \\ 0 \\ 2 \\ 0 \\ 0 \\ 0 \end{pmatrix} \in \mathbb{F}^8.$$

Lemma 6.6.12 *The symmetric tensor is linear in each of its parts; that is*

$$\mathbf{v}_1 \vee \cdots \vee (c\mathbf{v}_i + d\hat{\mathbf{v}}_i) \vee \cdots \vee \mathbf{v}_k = c(\mathbf{v}_1 \vee \cdots \vee \mathbf{v}_i \vee \cdots \vee \mathbf{v}_k) + d(\mathbf{v}_1 \vee \cdots \vee \hat{\mathbf{v}}_i \vee \cdots \vee \mathbf{v}_k).$$

Proof. Follows immediately from the corresponding property of the tensor product. $\qquad\square$

Proposition 6.6.13 *If two vectors in a symmetric tensor are switched, it will not change the symmetric tensor; that is,*

$$\mathbf{v}_1 \vee \cdots \vee \mathbf{v}_i \vee \cdots \vee \mathbf{v}_j \vee \cdots \vee \mathbf{v}_k = \mathbf{v}_1 \vee \cdots \vee \mathbf{v}_j \vee \cdots \vee \mathbf{v}_i \vee \cdots \vee \mathbf{v}_k.$$

Proof. Let $\tau = (i\ j)$ be the permutation that switches i and j. Then

$$\mathbf{v}_1 \vee \cdots \vee \mathbf{v}_j \vee \cdots \vee \mathbf{v}_i \vee \cdots \vee \mathbf{v}_k = \sum_{\sigma \in S_k} \mathbf{v}_{\sigma(\tau(1))} \otimes \cdots \otimes \mathbf{v}_{\sigma(\tau(k))} =$$

$$\sum_{\sigma \in S_k} \mathbf{v}_{\sigma(\tau(1))} \otimes \cdots \otimes \mathbf{v}_{\sigma(\tau(k))} = \sum_{\hat{\sigma} \in S_k} \mathbf{v}_{\hat{\sigma}(1)} \otimes \cdots \otimes \mathbf{v}_{\hat{\sigma}(k)} =$$

$$\mathbf{v}_1 \vee \cdots \vee \mathbf{v}_i \vee \cdots \vee \mathbf{v}_j \vee \cdots \vee \mathbf{v}_k,$$

where we used that if τ runs through all of S_k, then so does $\hat{\sigma} = \sigma\tau$. $\qquad\square$

We define

$$\vee^k V := \mathrm{Span}\{\mathbf{v}_1 \vee \cdots \vee \mathbf{v}_k : \mathbf{v}_j \in V_j, j = 1, \ldots, k\}.$$

Then $\vee^k V$ is a subspace of $\otimes^k V$.

Proposition 6.6.14 *When* $\dim V = n$, *then* $\dim \vee^k V = \binom{n+k-1}{k}$. *In fact, if* $\{\mathbf{v}_1, \ldots, \mathbf{v}_n\}$ *is a basis of* V, *then* $\mathcal{F} = \{\mathbf{v}_{i_1} \vee \cdots \vee \mathbf{v}_{i_k} : 1 \leq i_1 \leq \cdots \leq i_k \leq n\}$ *is a basis of* $\vee^k V$.

Proof. Let $\mathbf{f} \in \vee^k V$. Then $\mathbf{f}$ is a linear combination of elements of the form $\mathbf{x}_1 \vee \cdots \vee \mathbf{x}_k$ where $\mathbf{x}_j \in V$, $j = 1, \ldots, k$. Each $\mathbf{x}_j$ is a linear combination of $\mathbf{v}_1, \ldots, \mathbf{v}_n$, thus $x_j = \sum_{l=1}^n c_{lj} \mathbf{v}_l$ for some scalars c_{lj}. Plugging this into $\mathbf{x}_1 \vee \cdots \vee \mathbf{x}_k$ and using Lemma 6.6.12 we get that

$$\mathbf{x}_1 \vee \cdots \vee \mathbf{x}_k = \sum_{l_1, \ldots, l_k = 1}^n c_{l_1,1} \cdots c_{l_k,k} \mathbf{v}_{l_1} \vee \cdots \vee \mathbf{v}_{l_k}.$$

By applying Proposition 6.6.3, we can always do several switches so that $\mathbf{v}_{l_1} \vee \cdots \vee \mathbf{v}_{l_k}$ turns into $\mathbf{v}_{i_1} \vee \cdots \vee \mathbf{v}_{i_k}$, where now $i_1 \leq \cdots \leq i_k$. Putting these observations together, we obtain that $\mathbf{f} \in \text{Span } \mathcal{F}$.

For linear independence, suppose that $\sum_{1 \leq i_1 \leq \cdots \leq i_k \leq n} c_{i_1, \ldots, i_k} \mathbf{v}_{i_1} \vee \cdots \vee \mathbf{v}_{i_k} = \mathbf{0}$ for some scalars $c_{i_1, \ldots, i_k} \in \mathbb{F}$, $1 \leq i_1 \leq \cdots \leq i_k \leq n$. Putting in the definition of $\mathbf{v}_{i_1} \vee \cdots \vee \mathbf{v}_{i_k}$, we arrive at the equality

$$\sum_{l_1, \ldots, l_k = 1}^n a_{l_1, \ldots, l_k} \mathbf{v}_{l_1} \otimes \cdots \otimes \mathbf{v}_{l_k} = \mathbf{0}, \tag{6.33}$$

where $a_{l_1, \ldots, l_k}$ equals one of the numbers $c_{i_1, \ldots, i_k}$ or a positive integer multiple of it. As the tensors $\{\mathbf{v}_{l_1} \otimes \cdots \otimes \mathbf{v}_{l_k} : l_j = 1, \ldots, n, j = 1, \ldots, k\}$ form a basis of $\otimes^k V$, we have that (6.33) implies that $a_{l_1, \ldots, l_k}$ are all equal to 0. This implies that all $c_{i_1, \ldots, i_k}$ are equal to 0. This shows that $\mathcal{F}$ is linearly independent.

It remains to observe that the number of elements of $\mathcal{F}$ equals $\binom{n+k-1}{k}$. To see this, one chooses among the numbers $2, 3, \ldots, n + k$ numbers $m_1 < \cdots < m_k$. This choice corresponds in a one-to-one way to a choice $1 \leq i_1 \leq \cdots \leq i_k \leq n$, by letting $i_j = m_j - j$, $j = 1, \ldots, k$. $\quad\square$

Remark 6.6.15 Note that when V has an inner product, and when $\{\mathbf{v}_1, \ldots, \mathbf{v}_n\}$ is chosen to be an orthonormal basis of V, the basis $\mathcal{F}$ is not orthonormal. It is however an orthogonal basis, and thus all that needs to be done is to make the elements of $\mathcal{E}$ of unit length. The elements of $\mathcal{F}$ have different lengths, so some care needs to be taken in doing this.

We next analyze how linear operators and symmetric tensors interact.

Proposition 6.6.16 *Let $T : V \to W$ be a linear map. Denote $\otimes^k T = T \otimes \cdots \otimes T : \otimes^k V \to \otimes^k W$. Then $\otimes^k T[\vee^k V] \subseteq \vee^k W$.*

Proof. Let $\mathbf{v}_1 \vee \cdots \vee \mathbf{v}_k \in \vee^k V$. Then

$$\otimes^k T(\mathbf{x}_1 \vee \cdots \vee \mathbf{x}_k) = \otimes^k T\left(\sum_{\sigma \in S_k} \mathbf{v}_{\sigma(1)} \otimes \cdots \otimes \mathbf{v}_{\sigma(k)}\right) = \sum_{\sigma \in S_k} \otimes^k T(\mathbf{v}_{\sigma(1)} \otimes \cdots \otimes \mathbf{v}_{\sigma(k)}) =$$

$$\sum_{\sigma \in S_k} (T\mathbf{v}_{\sigma(1)}) \otimes \cdots \otimes (T\mathbf{v}_{\sigma(k)}) = (T\mathbf{v}_1) \vee \cdots \vee (T\mathbf{v}_k) \in \vee^k W.$$

As every element of $\vee^k V$ is a linear combination of elements of the type $\mathbf{v}_1 \vee \cdots \vee \mathbf{v}_k$, we obtain the result. $\qquad \square$

The restriction of $\otimes^k T$ to the subspace $\otimes^k V$ is denoted as $\vee^k T$. Thus, $\vee^k T : \vee^k V \to \vee^k W$ is defined via

$$\vee^k T(\mathbf{v}_1 \vee \cdots \vee \mathbf{v}_k) = (T\mathbf{v}_1) \vee \cdots \vee (T\mathbf{v}_k).$$

For a matrix $B = (b_{ij})_{i,j=1}^n$ we define its *permanent* by

$$\text{per } B = \sum_{\sigma \in S_n} b_{1,\sigma(1)} \cdots b_{n,\sigma(n)}. \qquad (6.34)$$

This is almost the same expression as for the determinant except that $\text{sign}\sigma$ does not appear. Thus all terms have a $+$. For instance,

$$\text{per} \begin{pmatrix} b_{11} & b_{12} \\ b_{21} & b_{22} \end{pmatrix} = b_{11}b_{22} + b_{21}b_{12}.$$

Example 6.6.17 Let $T : \mathbb{F}^2 \to \mathbb{F}^2$ be matrix multiplication with the matrix $A = (a_{ij})_{i,j=1}^2$. The standard basis on $\vee^2 \mathbb{F}^2$ is $\mathcal{F} = \{\mathbf{e}_1 \vee \mathbf{e}_1, \mathbf{e}_1 \vee \mathbf{e}_2, \mathbf{e}_2 \vee \mathbf{e}_2\}$. Let us compute $[\vee^2 T]_{\mathcal{F} \leftarrow \mathcal{F}}$. We apply $\vee^2 T$ to the first basis element:

$$\vee^2 T(\mathbf{e}_1 \vee \mathbf{e}_1) = (T\mathbf{e}_1) \vee (T\mathbf{e}_1) = (a_{11}\mathbf{e}_1 + a_{21}\mathbf{e}_2) \vee (a_{11}\mathbf{e}_1 + a_{21}\mathbf{e}_2) =$$

$$a_{11}^2 \mathbf{e}_1 \vee \mathbf{e}_1 + 2a_{11}a_{21}\mathbf{e}_1 \vee \mathbf{e}_2 + a_{21}^2 \mathbf{e}_2 \vee \mathbf{e}_2.$$

Similarly,

$$\vee^2 T(\mathbf{e}_1 \vee \mathbf{e}_2) = (T\mathbf{e}_1) \vee (T\mathbf{e}_2) = (a_{11}\mathbf{e}_1 + a_{21}\mathbf{e}_2) \vee (a_{12}\mathbf{e}_1 + a_{22}\mathbf{e}_2) =$$

$$a_{11}a_{12}\mathbf{e}_1 \vee \mathbf{e}_1 + (a_{11}a_{22} + a_{21}a_{12})\mathbf{e}_1 \vee \mathbf{e}_2 + a_{21}a_{22}\mathbf{e}_2 \vee \mathbf{e}_2.$$

Continuing, we find

$$[\vee^2 T]_{\mathcal{F} \leftarrow \mathcal{F}} = \begin{pmatrix} a_{11}^2 & a_{11}a_{12} & a_{12}^2 \\ 2a_{11}a_{21} & a_{11}a_{22} + a_{21}a_{12} & 2a_{12}a_{22} \\ a_{21}^2 & a_{21}a_{22} & a_{22}^2 \end{pmatrix}. \qquad (6.35)$$

Notice that the $(2,2)$ element is equal to $\text{per}A[\{1,2\},\{1,2\}]$.

Lemma 6.6.18 *Let $T : V \to W$ and $S : X \to Y$ be linear, where the vector spaces are over $\mathbb{F}$. Then*

$$(\vee^k T)(\vee^k S) = \vee^k (TS).$$

Proof. Since $(\otimes^k T)(\otimes^k S) = \otimes^k(TS)$, and as in general $\vee^k U$ is defined as $\otimes^k U$ on the subspace of symmetric tensors, the lemma follows. $\qquad\square$

Proposition 6.6.19 *Let* $T : V \to W$ *be linear, where the vector spaces are over* $\mathbb{F}$. *Then the following hold:*

(i) *If* T *is invertible, then so is* $\vee^k T$ *and* $(\vee^k T)^{-1} = \vee^k(T^{-1})$.

(ii) *If* $V = W$ *and* $\mathbf{x}_1, \ldots, \mathbf{x}_k$ *are eigenvectors for* T *with eigenvalues* $\lambda_1, \ldots, \lambda_k$, *respectively, then* $\mathbf{x}_1 \vee \cdots \vee \mathbf{x}_k$ *is an eigenvector for* $\vee^k T$, *with eigenvalue* $\lambda_1 \cdots \lambda_k$; *thus*
$$\vee^k T(\mathbf{x}_1 \vee \cdots \vee \mathbf{x}_k) = \lambda_1 \cdots \lambda_k(\mathbf{x}_1 \vee \cdots \vee \mathbf{x}_k).$$

For the remaining parts, V *is assumed to be an inner product space (and thus necessarily,* $\mathbb{F} = \mathbb{R}$ *or* $\mathbb{C}$), *and the inner product* $\vee^k V$ *is inherited from the inner product on the tensor product given via the construction in Proposition 6.5.6.*

(iii) $(\vee^k T)^\star = \vee^k T^\star$.

(iv) *If* T *is an isometry, then so is* $\vee^k T$.

(v) *If* T *is unitary, then so is* $\vee^k T$.

(vi) *If* T *is normal, then so is* $\vee^k T$.

(vii) *If* T *is Hermitian, then so is* $\vee^k T$.

(viii) *If* T *is positive (semi-)definite, then so is* $\vee^k T$.

Proof. Use Lemma 6.6.18 and Proposition 6.5.11. $\qquad\square$

Remark 6.6.20 It should be noted that when $A = (a_{ij})_{i,j=1}^2$ is a Hermitian matrix, the matrix in (6.35) is not (necessarily) Hermitian. Why is this not in contradiction with Proposition 6.6.19 (vii)? The reason is that that the basis $\mathcal{F} = \{\mathbf{e}_1 \vee \mathbf{e}_1, \mathbf{e}_1 \vee \mathbf{e}_2, \mathbf{e}_2 \vee \mathbf{e}_2\}$ is **not** orthonormal. For the Hermitian property to be necessarily reflected in the matrix representation, we need the basis to be orthonormal. If we introduce
$$\mathcal{F}_{\text{on}} = \{\tfrac{1}{2}(\mathbf{e}_1 \vee \mathbf{e}_1), \frac{1}{\sqrt{2}}(\mathbf{e}_1 \vee \mathbf{e}_2), \tfrac{1}{2}(\mathbf{e}_2 \vee \mathbf{e}_2)\},$$
then
$$[\vee^2 T]_{\mathcal{F}_{\text{on}} \leftarrow \mathcal{F}_{\text{on}}} = \begin{pmatrix} a_{11}^2 & \sqrt{2}a_{11}a_{12} & a_{12}^2 \\ \sqrt{2}a_{11}a_{21} & a_{11}a_{22} + a_{21}a_{12} & \sqrt{2}a_{12}a_{22} \\ a_{21}^2 & \sqrt{2}a_{21}a_{22} & a_{22}^2 \end{pmatrix}, \qquad (6.36)$$

which now is Hermitian when A is. The same remark holds for the other properties in Proposition 6.6.19 (iii)–(viii).

6.7 Exercises

Exercise 6.7.1 The purpose of this exercise is to show (the vector form of) *Minkowski's inequality*, which says that for complex numbers x_i, y_i, $i = 1, \ldots, n$, and $p \geq 1$, we have

$$\left(\sum_{i=1}^{n} |x_i + y_i|\right)^{\frac{1}{p}} \leq \left(\sum_{i=1}^{n} |y_i|\right)^{\frac{1}{p}} + \left(\sum_{i=1}^{n} |y_i|\right)^{\frac{1}{p}}. \tag{6.37}$$

Recall that a real-valued function f defined on an interval in $\mathbb{R}$ is called *convex* if for all c, d in the domain of f, we have that

$$f(tc + (1-t)d) \leq tf(c) + (1-t)f(d), 0 \leq t \leq 1.$$

(a) Show that $f(x) = -\log x$ is a convex function on $(0, \infty)$. (One can do this by showing that $f''(x) \geq 0$.)

(b) Use (a) to show that for $a, b > 0$ and $p, q \geq 1$, with $\frac{1}{p} + \frac{1}{q} = 1$, we have $ab \leq \frac{a^p}{p} + \frac{b^q}{q}$. This inequality is called *Young's inequality*.

(c) Show *Hölder's inequality*: when $a_i, b_i \geq 0$, $i = 1, \ldots, n$, then

$$\sum_{i=1}^{n} a_i b_i \leq \left(\sum_{i=1}^{n} a_i^p\right)^{\frac{1}{p}} \left(\sum_{i=1}^{n} b_i^q\right)^{\frac{1}{q}}.$$

(Hint: Let $\lambda = (\sum_{i=1}^{n} a_i^p)^{\frac{1}{p}}$ and $\mu = (\sum_{i=1}^{n} b_i^q)^{\frac{1}{q}}$, and divide on both sides a_i by λ and b_i by μ. Use this to argue that it is enough to prove the inequality when $\lambda = \mu = 1$. Next use (b)).

(d) Use (c) to prove (6.37) in the case when $x_i, y_i \geq 0$. (Hint: Write $(x_i + y_i)^p = x_i(x_i + y_i)^{p-1} + y_i((x_i + y_i)^{p-1}$, take the sum on both sides, and now apply Hölder's inequality to each of the terms on the right-hand side. Rework the resulting inequality, and use that $p + q = pq$.)

(e) Prove Minkowski's inequality (6.37).

(f) Show that when V_i has a norm $\| \cdot \|_i$, $i = 1, \ldots, k$, then for $p \geq 1$ we have that

$$\left\| \begin{pmatrix} \mathbf{v}_1 \\ \vdots \\ \mathbf{v}_k \end{pmatrix} \right\|_p := \left(\sum_{i=1}^{k} \|\mathbf{v}_i\|_i^p \right)^{\frac{1}{p}}$$

defines a norm on $V_1 \times \cdots \times V_k$.

Exercise 6.7.2 Let V and Z be vector spaces over $\mathbb{F}$ and $T : V \to Z$ be linear. Suppose $W \subseteq \text{Ker } T$. Show there exists a linear transformation $S : V/W \to \text{Ran } T$ such that $S(\mathbf{v} + W) = T\mathbf{v}$ for $\mathbf{v} \in V$. Show that S is surjective and that Ker S is isomorphic to $(\text{Ker } T)/W$.

Exercise 6.7.3 Consider the vector space $\mathbb{F}^{n \times m}$, where $\mathbb{F} = \mathbb{R}$ or $\mathbb{F} = \mathbb{C}$, and let $\| \cdot \|$ be norm on $\mathbb{F}^{n \times m}$.

(a) Let $A = (a_{ij})_{i=1,j=1}^{n}{}_{,j=1}^{m}$, $A_k = (a_{ij}^{(k)})_{i=1,j=1}^{n}{}_{,j=1}^{m}$, $k = 1, 2, \ldots$, be matrices in $\mathbb{F}^{n \times m}$. Show that $\lim_{k \to \infty} \|A_k - A\| = 0$ if and only if $\lim_{k \to \infty} |a_{ij}^{(k)} - a_{ij}| = 0$ for every $i = 1, \ldots, n$ and $j = 1, \ldots, m$.

(b) Let $n = m$. Show that $\lim_{k \to \infty} \|A_k - A\| = 0$ and $\lim_{k \to \infty} \|B_k - B\| = 0$ imply that $\lim_{k \to \infty} \|A_k B_k - AB\| = 0$.

Exercise 6.7.4 Given $A \in \mathbb{C}^{n \times n}$, we define its *similarity orbit* to be the set of matrices

$$\mathcal{O}(A) = \{SAS^{-1} : S \in \mathbb{C}^{n \times n} \text{ is invertible}\}.$$

Thus the similarity orbit of a matrix A consists of all matrices that are similar to A.

(a) Show that if A is diagonalizable, then its similarity orbit $\mathcal{O}(A)$ is closed. (Hint: notice that due to A being diagonalizable, we have that $B \in \mathcal{O}(A)$ if and only if $m_A(B) = 0$.)

(b) Show that if A is not diagonalizable, then its similarity orbit is not closed.

Exercise 6.7.5 Suppose that V is an infinite-dimensional vector space with basis $\{\mathbf{v}_j\}_{j \in J}$. Let $f_j \in V'$, $j \in J$, be so that $f_j(\mathbf{v}_j) = 1$ and $f_j(\mathbf{v}_k) = 0$ for $k \neq j$. Show that $\{f_j\}_{j \in J}$ is a linearly independent set in V' but is not a basis of V'.

Exercise 6.7.6 Describe the linear functionals on $\mathbb{C}_n[X]$ that form the dual basis of $\{1, X, \ldots, X^n\}$.

Exercise 6.7.7 Let $a_0, \ldots, a_n$ be different complex numbers, and define $E_j \in (\mathbb{C}_n[X])'$, $j = 0, \ldots, n$, via $E_j(p(X)) = p(a_j)$. Find a basis of $\mathbb{C}_n[X]$ for which $\{E_0, \ldots, E_n\}$ is the dual basis.

Exercise 6.7.8 Let $V = W \dotplus X$.

(a) Show how given $f \in W'$ and $g \in X'$, one can define $h \in V'$ so that $h(\mathbf{w}) = f(\mathbf{w})$ for $\mathbf{w} \in W$ and $h(\mathbf{x}) = g(\mathbf{x})$ for $\mathbf{x} \in X$.

(b) Using the construction in part (a), show that $V' = W' \dotplus X'$. Here it is understood that we view W' as a subspace of V', by letting $f \in W'$ be defined on all of V by putting $f(\mathbf{w} + \mathbf{x}) = f(\mathbf{w})$, when $\mathbf{w} \in W$ and $\mathbf{x} \in X$. Similarly, we view X' as a subspace of V', by letting $g \in W'$ be defined on all of V by putting $g(\mathbf{w} + \mathbf{x}) = g(\mathbf{x})$, when $\mathbf{w} \in W$ and $\mathbf{x} \in X$.

Exercise 6.7.9 Let W be a subspace of V. Define

$$W_{\text{ann}} = \{f \in V' : f(\mathbf{w}) = 0 \text{ for all } \mathbf{w} \in W\},$$

the *annihilator* of W.

(a) Show that W_{ann} is a subspace of V'.

(b) Determine the annihilator of $\text{Span}\{\begin{pmatrix} 1 \\ -1 \\ 2 \\ -2 \end{pmatrix}, \begin{pmatrix} 1 \\ 0 \\ 1 \\ 0 \end{pmatrix}\} \subseteq \mathbb{C}^4$.

(c) Determine the annihilator of $\text{Span}\{1 + 2X, X + X^2\} \subseteq \mathbb{R}_3[X]$.

Exercise 6.7.10 Let V be a finite-dimensional vector space over $\mathbb{R}$, and let $\{\mathbf{v}_1, \ldots, \mathbf{v}_k\}$ be linearly independent. We define

$$\mathcal{C} = \{\mathbf{v} \in V : \text{there exist } c_1, \ldots, c_k \geq 0 \text{ so that } \mathbf{v} = \sum_{i=1}^{k} c_i \mathbf{v}_i\}.$$

Show that $\mathbf{v} \in \mathcal{C}$ if and only if for all $f \in V'$ with $f(\mathbf{v}_j) \geq 0$, $j = 1, \ldots, k$, we have that $f(\mathbf{v}) \geq 0$.

Remark. The statement is also true when $\{\mathbf{v}_1, \ldots, \mathbf{v}_k\}$ are not linearly independent, but in that case the proof is more involved. The corresponding result is the Farkas–Minkowski Theorem, which plays an important role in linear programming.

Exercise 6.7.11 Let V and W be finite-dimensional vector spaces and $A : V \to W$ a linear map. Show that $A\mathbf{v} = \mathbf{w}$ has a solution if and only if for all $f \in (\mathrm{Ran}A)_{\mathrm{ann}}$ we have that $f(\mathbf{w}) = 0$. Here the definition of the annihilator is used as defined in Exercise 6.7.9.

Exercise 6.7.12 For $\mathbf{x}, \mathbf{y} \in \mathbb{R}^3$, let the cross product $\mathbf{x} \times \mathbf{y}$ be defined as in (6.17).

(a) Show that $\langle \mathbf{x}, \mathbf{x} \times \mathbf{y} \rangle = \langle \mathbf{y}, \mathbf{x} \times \mathbf{y} \rangle = 0$.

(b) Show that $\mathbf{x} \times \mathbf{y} = -\mathbf{y} \times \mathbf{x}$.

(c) Show that $\mathbf{x} \times \mathbf{y} = \mathbf{0}$ if and only if $\{\mathbf{x}, \mathbf{y}\}$ is linearly dependent.

Exercise 6.7.13 Let

$$
A = \begin{pmatrix} i & 1-i & 2-i \\ 1+i & -2 & -3+i \end{pmatrix}, B = \begin{pmatrix} -1 & 0 \\ -2 & 5 \\ 1 & 3 \end{pmatrix}.
$$

Compute $A \otimes B$ and $B \otimes A$, and show that they are similar via a permutation matrix.

Exercise 6.7.14 Let $A \in \mathbb{F}^{n \times n}$ and $B \in \mathbb{F}^{m \times m}$.

(a) Show that $\mathrm{tr}(A \otimes B) = (\mathrm{tr}\, A)(\mathrm{tr}\, B)$.

(b) Show that $\mathrm{rank}(A \otimes B) = (\mathrm{rank}\, A)(\mathrm{rank}\, B)$.

Exercise 6.7.15 Given Schur triangularization decompositions for A and B, find a Schur triangularization decomposition for $A \otimes B$. Conclude that if $\lambda_1, \ldots, \lambda_n$ are the eigenvalues for A and $\mu_1, \ldots, \mu_m$ are the eigenvalues for B, then $\lambda_i \mu_j$, $i = 1, \ldots, n$, $j = 1, \ldots, m$, are the nm eigenvalues of $A \otimes B$.

Exercise 6.7.16 Given singular value decompositions for A and B, find a singular value decomposition for $A \otimes B$. Conclude that if $\sigma_1, \ldots, \sigma_k$ are the nonzero singular values for A and $\hat{\sigma}_1, \ldots, \hat{\sigma}_l$ are the nonzero singular values for B, then $\sigma_i \hat{\sigma}_j$, $i = 1, \ldots, k$, $j = 1, \ldots, l$, are the kl nonzero singular values of $A \otimes B$.

Exercise 6.7.17 Show that $\det(I \otimes A + A \otimes I) = (-1)^n \det p_A(-A)$, where $A \in \mathbb{C}^{n \times n}$.

Exercise 6.7.18 Show that if A is a matrix and f a function, so that $f(A)$ is well-defined, then $f(I_m \otimes A)$ is well-defined as well, and $f(I_m \otimes A) = I_m \otimes f(A)$.

Exercise 6.7.19 For a diagonal matrix $A = \operatorname{diag}(\lambda_i)_{i=1}^n$, find matrix representations for $A \wedge A$ and $A \vee A$ using the canonical (lexicographically ordered) bases for $\mathbb{F}^n \wedge \mathbb{F}^n$ and $\mathbb{F}^n \vee \mathbb{F}^n$, respectively.

Exercise 6.7.20 Show that
$$\langle \mathbf{v}_1 \wedge \cdots \wedge \mathbf{v}_k, \mathbf{w}_1 \wedge \cdots \wedge \mathbf{w}_k \rangle = k! \det(\langle \mathbf{v}_i, \mathbf{w}_j \rangle)_{i,j=1}^k.$$

Exercise 6.7.21 Find an orthonormal basis for $\vee^2 \mathbb{C}^3$.

Exercise 6.7.22 (a) Let $A = (a_{ij})_{i=1,j=1}^{2,m} \in \mathbb{F}^{2 \times m}$ and $B = (b_{ij})_{i=1,j=1}^{m,2} \in \mathbb{F}^{m \times 2}$. Find the matrix representations for $A \wedge A$, $B \wedge B$ and $AB \wedge AB$ using the canonical (lexicographically ordered) bases for $\wedge^k \mathbb{F}^n$, $k = 2$, $n = 2, m, 1$, respectively.

(b) Show that the equality $AB \wedge AB = (A \wedge A)(B \wedge B)$ implies that

$$\left(\sum_{j=1}^m a_{1j} b_{j1} \right) \left(\sum_{j=1}^m a_{2j} b_{j2} \right) - \left(\sum_{j=1}^m a_{1j} b_{j2} \right) \left(\sum_{j=1}^m a_{2j} b_{j1} \right) =$$

$$\sum_{1 \leq j < k \leq m} (a_{1j} a_{2k} - a_{1k} a_{2j})(b_{1j} b_{2k} - b_{1k} b_{2j}). \tag{6.38}$$

(c) Let $M = \{1, \ldots, m\}$ and $P = \{1, \ldots, p\}$. For $A \in \mathbb{F}^{p \times m}$ and $B \in \mathbb{F}^{m \times p}$, show that

$$\det AB = \sum_{S \subseteq M, |S| = p} \det(A[P, S]) \det(B[S, P]). \tag{6.39}$$

(Hint: Use that $(\wedge^p A)(\wedge^p B) = \wedge^p (AB) = \det AB$.)

Remark: Equation (6.39) is called the *Cauchy–Binet identity*. When $p = 2$ it reduces to (6.38), which when $B = A^T$ (or $B = A^*$ when $\mathbb{F} = \mathbb{C}$) is called the *Lagrange identity*.

Exercise 6.7.23 For $\mathbf{x}, \mathbf{y} \in \mathbb{R}^3$, let the cross product $\mathbf{x} \times \mathbf{y}$ be defined as in (6.17). Show, using (6.38) (with $B = A^T$), that

$$\|\mathbf{x} \times \mathbf{y}\|^2 = \|\mathbf{x}\|^2 \|\mathbf{y}\|^2 - (\langle \mathbf{x}, \mathbf{y} \rangle)^2. \tag{6.40}$$

Notice that this equality implies the Cauchy–Schwarz inequality.

Exercise 6.7.24 *(Honors)* Let V be a vector space and let $W \subseteq Y \subseteq V$ be subspaces.

(a) Show that $(V/W)/(Y/W)$ is isomorphic to V/Y.

(b) Show that $\dim(V/W) = \dim(V/Y) + \dim(Y/W)$, assuming that $\dim(V/W)$ is finite.

Exercise 6.7.25 *(Honors)*

(a) Show that when $a \neq 0$, the Jordan canonical form of $J_s(a) \otimes J_t(0)$ is given by $\oplus_{i=1}^{s} J_t(0)$.

(b) Show that when $b \neq 0$, the Jordan canonical form of $J_s(0) \otimes J_t(b)$ is given by $\oplus_{i=1}^{t} J_s(0)$.

(c) Show that when $t \geq s$, the Jordan canonical form of $J_s(0) \otimes J_t(0)$ is given by
$$[\oplus_{i=1}^{t-s+1} J_s(0)] \oplus [\oplus_{i=1}^{s-1}(J_{s-i}(0) \oplus J_{s-i}(0))].$$

(d) Show that when $a, b \neq 0$ and $t \geq s$, the Jordan canonical form of $J_s(a) \otimes J_t(b)$ is given by

$$J_{t+s-1}(ab) \oplus J_{t+s-3}(ab) \oplus \cdots \oplus J_{t+s-(2s-3)}(ab) \oplus J_{t+s-(2s-1)}(ab).$$

This is also the Jordan canonical form of $J_t(a) \otimes J_s(b)$.

Using the above information one can now find the Jordan canonical form of $A \otimes B$, when one is given the Jordan canonical forms of A and B.

7

How to Use Linear Algebra

CONTENTS

7.1   Matrices you can't write down, but would still like to use 196
7.2   Algorithms based on matrix vector products 198
7.3   Why use matrices when computing roots of polynomials? 203
7.4   How to find functions with linear algebra? 209
7.5   How to deal with incomplete matrices 217
7.6   Solving millennium prize problems with linear algebra 222
      7.6.1   The Riemann hypothesis 223
      7.6.2   P vs. NP 225
7.7   How secure is RSA encryption? 229
7.8   Quantum computation and positive maps 232
7.9   Exercises 238
      Bibliography for Chapter 7 245

In this chapter we would like to give you an idea how creative thinking led to some very useful ideas to exploit the power of linear algebra. The hope is that it will inspire you to think of new ways to use linear algebra in areas of your interest. It would be great if one day we would be remiss by not including your ideas in this chapter. So, go for it!

This chapter has a somewhat different flavor than the other chapters. As applications use mathematics from different fields, we will be mentioning and use some results from other areas of mathematics without proofs. In addition, not everything will have a complete theory. Some of the algorithms described may be based on heuristic arguments and do not necessarily have a full theoretical justification. It is natural that these things happen: mathematics is a discipline with several different, often useful, aspects and continues to develop as a discipline. There will always be mathematical research continuing to improve on existing results.

7.1 Matrices you can't write down, but would still like to use

In previous chapters we have done computations with matrices to learn the concepts, and they were all small matrices (at most 8×8). Bigger matrices (say, number of rows and columns in the thousands) you may not want to deal with by hand, but working with them in a spreadsheet or other software seems doable. But what do we do when matrices are simply too big to store anywhere (say, if the number of rows or columns run in the billions), or if it is simply impossible to gather all the data? Can we still work with the matrix?

Here are two examples to begin with, both used in search engines:

- A matrix P where there is a row and column for every existing web page, and the (i, j)th entry p_{ij} represents the probability that you go from web page i to web page j. Currently (October 2015), there are about 4.76 billion indexed web pages, so this matrix is huge. However, if you have a way of looking at a page i and determining all the probabilities p_{ij}, then determining a row of this matrix is not a big deal.

- A matrix M where there is a row for every web page, and a column for every search word. The (i, j)th entry m_{ij} of this matrix is set to be 1 if search word j appears on page i, and 0 otherwise. Again, this matrix is huge, but determining row i is easily done by looking at this particular page.

One big difference between these two matrices is obvious: P is square and M is not. Thus P has eigenvectors, and M does not. In fact, it is the eigenvector of P^T at the eigenvalue 1 that is of interest. Notice that for these matrices it may not be convenient to use numbers $1, 2, \ldots$ as indices for the rows and columns, as we usually do. Rather one may just use the name of the web page or the actual search word as the index. So, for instance, we would write

$$p_{\text{www.linear\_algebra.edu,www.woerdeman.edu}} = \frac{1}{10},$$

$$m_{\text{www.linear\_algebra.edu,Hugo}} = 1.$$

Notice that this means that the rows and columns are not ordered in a natural way (although we can order them if we have to), and thus anything meaningful that we should be looking for should not depend on any particular order. In the case of P, though, the rows and columns are indexed by the same index set, so if any ordering is chosen for the rows we should use the same for the columns. Let us also observe that any vector **x** for which we would to consider the product $P\mathbf{x}$, needs to be indexed by the same index set as is used for the columns of P. Thus **x** would have entries like

$$x_{\text{www.linear\_algebra.edu}} \quad \text{and} \quad x_{\text{www.woerdeman.edu}}.$$

Here are some more matrices that may be of interest:

- A matrix K where the columns represent the products you sell and the rows represent your customers. The entry K_{ij} is the rating customer i gives to product j. So for instance

  $$K_{\text{Hugo Woerdeman , Advanced Linear Algebra by Woerdeman}} = \star\ \star\ \star\ \star\ \star.$$

  Notice that the ratings, one through five stars, do not form a field (we can make it a matrix with entries in $\mathbb{Z}_5$, but it would not be meaningful in this context). Still, as it turns out, it is useful to consider this as a matrix over $\mathbb{R}$. Why the real numbers? Because the real numbers have a natural ordering, and the ratings are ordered as well. In fact, the ordering of the ratings is the only thing we care about! The main problem with this matrix is that you will never know all of its entries (unless you are running a really small business), and the ones you think you know may not be accurate.

- A matrix C where both the rows and columns represent all the people (in the worlds, in a country, in a community), and the entries c_{ij} are 1 or 0 depending whether person i knows person j. If we believe the "six degrees of separation" theory, the matrix $C + C^2 + \cdots + C^6$ will only have positive entries. (The matrix C is an adjacency matrix; see Exercise 7.9.16 for the definition.)

- A matrix H where each row represents the genetic data of each known (DNA) virus. Does this even make sense? Can anything be done with this? The entries would be letters (A, C, G, T), without any (obvious) addition and multiplication to make it into a meaningful field.

- (Make up your own.)

There are at least two types of techniques that can help in dealing with these types of matrices:

1. Techniques where one just needs to multiply a matrix with a vector. In the case of matrices P and M we can figure out the rows fairly easily, so if our techniques just involve multiplying with row vectors on the left, preferably with ones that only have few nonzero entries, then we can apply this technique.

2. If we can assume that the matrix is low rank, then knowing and/or storing just part of the matrix gives us enough to work with the whole matrix.

We will explore these ideas further in the next sections.

7.2 Algorithms based on matrix vector products

If we are in a situation where it is hard to deal with the whole square matrix A, but we are able to compute a vector product $A\mathbf{v}$, are we still able to compute eigenvalues of A, or solve an equation $A\mathbf{x} = \mathbf{b}$. Examples of such a situation include

- A *sparse* matrix A; that is, a matrix with relatively few nonzero entries. While the matrix may be huge, computing a product $A\mathbf{v}$ may be doable.

- A situation where the matrix A represents the action of some system in which we can give inputs and measure outputs. If the input is $\mathbf{u}$ and the output is $\mathbf{y}$, then by giving the system the input $\mathbf{u}$ and by measuring the output $\mathbf{y}$ we would in effect be computing the product $\mathbf{y} = A\mathbf{u}$. In this situation we would not know the (complete) inner workings of this system, but assume (or just guess as a first try) that the system can be modeled/approximated by a simple matrix multiplication.

- The matrices M and P from Section 7.1.

Here is a first algorithm that computes the eigenvalue of the largest modulus in case it has geometric multiplicity 1.

Theorem 7.2.1 *(Power method) Let $A \in \mathbb{C}^{n \times n}$ have eigenvalues $\{\lambda_1, \ldots, \lambda_n\}$ with $\lambda_1 > \max_{j=2,\ldots,n} |\lambda_j|$. Let $\mathbf{v}$ be so that $\mathbf{v} \notin \operatorname{Ker} \prod_{j=2}^{n}(A - \lambda_j)$. Then the iteration*

$$\mathbf{v}_0 := \mathbf{v}, \ , \mathbf{v}_{k+1} = \frac{1}{\|A\mathbf{v}_k\|} A\mathbf{v}_k, \mu_k := \frac{\mathbf{v}_k^* A\mathbf{v}_k}{\mathbf{v}_k^* \mathbf{v}_k}, k = 1, 2, \ldots,$$

has the property that $\lambda_1 = \lim_{k \to \infty} \mu_k$ and $\mathbf{w} := \lim_{k \to \infty} \mathbf{v}_k$ is a unit eigenvector for at λ_1, thus $A\mathbf{w} = \lambda_1 \mathbf{w}$ and $\|\mathbf{w}\| = 1$.

Example 7.2.2 For illustration, let us see how the algorithm works on the matrix

$$A = \begin{pmatrix} 3 & 0 & 0 \\ 0 & 2 & 0 \\ 0 & 0 & 1 \end{pmatrix}$$

with initial vector

$$\mathbf{v}_0 = \begin{pmatrix} 1 \\ 1 \\ 1 \end{pmatrix}.$$

Then

$$\mathbf{v}_1 = \frac{1}{\sqrt{3^2 + 2^2 + 1^2}} \begin{pmatrix} 3 \\ 2 \\ 1 \end{pmatrix}, \mathbf{v}_2 = \frac{1}{\sqrt{3^4 + 2^4 + 1^4}} \begin{pmatrix} 9 \\ 4 \\ 1 \end{pmatrix},$$

$$\mathbf{v}_k = \frac{1}{\sqrt{3^{2k} + 2^{2k} + 1^{2k}}} \begin{pmatrix} 3^k \\ 2^k \\ 1 \end{pmatrix}, k \in \mathbb{N}.$$

Notice that

$$\frac{3^k}{\sqrt{3^{2k} + 2^{2k} + 1^{2k}}} = \frac{1}{\sqrt{1 + (\frac{2}{3})^{2k} + (\frac{1}{3})^{2k}}} \to 1,$$

$$\frac{2^k}{\sqrt{3^{2k} + 2^{2k} + 1^{2k}}} = \frac{(\frac{2}{3})^k}{\sqrt{1 + (\frac{2}{3})^{2k} + (\frac{1}{3})^{2k}}} \to 0,$$

$$\frac{1^k}{\sqrt{3^{2k} + 2^{2k} + 1^{2k}}} = \frac{(\frac{1}{3})^k}{\sqrt{1 + (\frac{2}{3})^{2k} + (\frac{1}{3})^{2k}}} \to 0,$$

so that $\mathbf{v}_k \to \mathbf{e}_1$ as $k \to \infty$. In addition,

$$\lim_{k \to \infty} \mu_k = \lim_{k \to \infty} \frac{3^{2k+1} + 2^{2k+1} + 1^{2k+1}}{3^{2k} + 2^{2k} + 1^{2k}} = 3.$$

As long as the initial vector $\mathbf{v}_0$ does not have a 0 as the first entry, we will always have that $\lim_{k \to \infty} \mathbf{v}_k = \mathbf{e}_1$.

In order to prove Theorem 7.2.1 we need to show that powers of a Jordan block with an eigenvalue of modulus less than 1, converge to the zero matrix.

Lemma 7.2.3 *Let $|\mu| < 1$ and $k \in \mathbb{N}$. Then $\lim_{m \to \infty} J_k(\mu)^m = 0$.*

Proof. Let $1 \le i \le j \le k$. Then the (i, j)th entry of $J_k(\mu)^m$ equals $\binom{m}{j-i} \mu^m$. Notice that

$$\binom{n}{j-i} = \frac{m(m-1) \cdots (m - (j-i) + 1)}{(j-i)!}$$

is a polynomial $p(m)$ of degree $j - i \le k - 1$ in m. But then

$$\lim_{m \to \infty} \binom{m}{j - i} \mu^m = \lim_{m \to \infty} p(m) \mu^m = 0,$$

where we used that $|\mu| < 1$. For details on the last step, please see Exercise 7.9.1. □

For $A \in \mathbb{C}^{n \times n}$ we define its *spectral radius* $\rho(A)$ via

$$\rho(A) = \max\{|\lambda| : \lambda \text{ is an eigenvalue of } A\} = \max_{\lambda \in \sigma(A)} |\lambda|.$$

Corollary 7.2.4 *Let $A \in \mathbb{C}^{n \times n}$. Then $\lim_{m \to \infty} A^m = 0$ if and only if $\rho(A) < 1$.*

Proof. Let $\rho(A) < 1$. Then $A = SJS^{-1}$ with $J = \oplus_{j=1}^{s} J_{n_j}(\lambda_j)$, where $|\lambda_j| < 1$, $j = 1, \ldots, s$. Then, by Lemma 7.2.3, $J^m = \oplus_{j=1}^{s}(J_{n_j}(\lambda_j))^m \to 0$ as $m \to \infty$. But then $A^m = SJ^m S^{-1} \to 0$ as $m \to \infty$.

Next, suppose that $\rho(A) \ge 1$. Then A has an eigenvalue λ with $|\lambda| \ge 0$. Let $\mathbf{x}$ be a corresponding eigenvector. Then $A^m \mathbf{x} = \lambda^m \mathbf{x} \nrightarrow \mathbf{0}$ as $m \to \infty$. But then it follows that $A^m \nrightarrow 0$ as $m \to \infty$. □

Proof of Theorem 7.2.1. Use Theorem 4.4.1 to write $A = SJS^{-1}$, with

$$J = \begin{pmatrix} \lambda_1 & 0 & \cdots & 0 \\ 0 & J(\lambda_2) & \cdots & 0 \\ \vdots & \vdots & \ddots & \vdots \\ 0 & 0 & \cdots & J(\lambda_m) \end{pmatrix},$$

where we use that λ_1 has only one Jordan block of size 1. Denote

$$S^{-1}\mathbf{v} = \begin{pmatrix} c_1 \\ \vdots \\ c_n \end{pmatrix}.$$

Since $\mathbf{v} \notin \operatorname{Ker} \prod_{j=2}^{n}(A - \lambda_j)$, we have that $c_1 \ne 0$. Put

$$\mathbf{w}_k = \frac{1}{\lambda_1^k} A^k \mathbf{v} = S \begin{pmatrix} 1 & 0 & \cdots & 0 \\ 0 & \frac{1}{\lambda_1^k} J(\lambda_2)^k & \cdots & 0 \\ \vdots & \vdots & \ddots & \vdots \\ 0 & 0 & \cdots & \frac{1}{\lambda_1^k} J(\lambda_m)^k \end{pmatrix} \begin{pmatrix} c_1 \\ \vdots \\ c_n \end{pmatrix}, \; k = 0, 1, \ldots.$$

Then $\mathbf{v}_k = \frac{\mathbf{w}_k}{\|\mathbf{w}_k\|}$. Also, using Lemma 7.2.3, we see that for $j > 1$, we have that

$$\frac{1}{\lambda_1^k} J(\lambda_j)^k = \mathrm{diag}(\lambda_1^{r-1})_{r=1}^n J(\frac{\lambda_j}{\lambda_1})^k \mathrm{diag}(\lambda_1^{-r+1})_{r=1}^n \to 0 \text{ when } k \to \infty.$$

Thus

$$\mathbf{w}_k \to S \begin{pmatrix} 1 & 0 & \cdots & 0 \\ 0 & 0 & \cdots & 0 \\ \vdots & \vdots & \ddots & \vdots \\ 0 & 0 & \cdots & 0 \end{pmatrix} \begin{pmatrix} c_1 \\ \vdots \\ \vdots c_n \end{pmatrix} = S \begin{pmatrix} c_1 \\ 0 \\ \vdots \\ 0 \end{pmatrix} =: \mathbf{x} \text{ when } k \to \infty.$$

Notice that $\mathbf{x}$, a multiple of the first column of S, is an eigenvector of A at λ_1. We now get that $\mathbf{v}_k = \frac{\mathbf{w}_k}{\|\mathbf{w}_k\|} \to \frac{\mathbf{x}}{\|\mathbf{x}\|} =: \mathbf{w}$ is a unit eigenvector of A at λ_1, and

$$\mu_k = \frac{\mathbf{v}_k^* A \mathbf{v}_k}{\mathbf{v}_k^* \mathbf{v}_k} = \frac{\mathbf{w}_k^* A \mathbf{w}_k}{\mathbf{w}_k^* \mathbf{w}_k} \to \frac{\mathbf{x}^* A \mathbf{x}}{\mathbf{x}^* \mathbf{x}} = \mathbf{w}^* A \mathbf{w} = \lambda_1 \mathbf{w}^* \mathbf{w} = \lambda_1 \text{ when } k \to \infty.$$

$\square$

If one is interested in more than just one eigenvalue of the matrix, one can introduce so-called *Krylov spaces*:

$$\mathrm{Span}\{\mathbf{v}, A\mathbf{v}, A^2\mathbf{v}, \ldots, A^k\mathbf{v}\}.$$

Typically one finds an orthonormal basis for this space, and then studies how powers of the matrix A act on this space. In this way one can approximate more than one eigenvalue of A.

Another problem of interest is to find a solution $\mathbf{x}$ to the equation $A\mathbf{x} = \mathbf{b}$, where we expect the equation to have a solution $\mathbf{x}$ with only few nonzero entries. In this case, A typically has far more columns than rows, so that solutions to the equation are never unique. We are however interested in the solution that only has a few nonzero entries, say at most s nonzero entries. The system typically is of the form

$$\begin{pmatrix} & & \\ & A & \\ & & \end{pmatrix} \begin{pmatrix} 0 \\ \vdots \\ 0 \\ * \\ 0 \\ \vdots \\ 0 \\ * \\ 0 \\ \vdots \end{pmatrix} = \mathbf{b},$$

where the $*$'s indicate the few nonzero entries in the desired solution $\mathbf{x}$. It is important to realize that the location of the nonzero entries in $\mathbf{x}$ are not known; otherwise one can simply remove all the columns in A that correspond to a 0 in $\mathbf{x}$ and solve the much smaller system.

To solve the above problem one needs to use some non-linear operations. One possibility is to use the *hard thresholding* operator $H_s : \mathbb{C}^n \to \mathbb{C}^n$, which keeps the s largest (in magnitude) entries of a vector $\mathbf{x}$ and sets the other entries equal to zero. For instance

$$H_2 \begin{pmatrix} 3+i \\ 2-8i \\ 2-i \\ 10 \end{pmatrix} = \begin{pmatrix} 0 \\ 2-8i \\ 0 \\ 10 \end{pmatrix}, H_3 \begin{pmatrix} 1 \\ 5 \\ -20 \\ 2 \\ 11 \\ -7 \end{pmatrix} = \begin{pmatrix} 0 \\ 0 \\ -20 \\ 0 \\ 11 \\ -7 \end{pmatrix}.$$

Notice that these hard thresholding operators are not linear; for instance

$$H_1 \begin{pmatrix} 3 \\ 1 \end{pmatrix} + H_1 \begin{pmatrix} -2 \\ 1 \end{pmatrix} = \begin{pmatrix} 3 \\ 0 \end{pmatrix} + \begin{pmatrix} -2 \\ 0 \end{pmatrix} = \begin{pmatrix} 1 \\ 0 \end{pmatrix} \neq \begin{pmatrix} 0 \\ 2 \end{pmatrix} = H_1 (\begin{pmatrix} 3 \\ 1 \end{pmatrix} + \begin{pmatrix} -2 \\ 1 \end{pmatrix}).$$

Notice that H_s is actually not well-defined on vectors where the sth largest element and the $(s+1)$th largest element have the same magnitude. For instance, is

$$H_2 \begin{pmatrix} 3+i \\ 3-i \\ 2-i \\ 10 \end{pmatrix} = \begin{pmatrix} 3+i \\ 0 \\ 0 \\ 10 \end{pmatrix} \text{ or } H_2 \begin{pmatrix} 3+i \\ 3-i \\ 2-i \\ 10 \end{pmatrix} = \begin{pmatrix} 0 \\ 3-i \\ 0 \\ 10 \end{pmatrix} ?$$

When the algorithm below is used, this scenario either does not show up, or the choice one makes does not affect the outcome, so this detail is usually ignored. Of course, it may cause a serious problem in some future application, at which point one needs to rethink the algorithm. There are other thresholding functions where some of the values are diminished, but not quite set to 0. The fact that one completely annihilates some elements (by setting them to 0, thus completely ignoring their value) gives it the term "hard" in hard thresholding.

The hard thresholding algorithm is now as follows:

Let $A \in \mathbb{C}^{m \times n}$ so that $\sigma_1(A) < 1$.

1. Let $\mathbf{x}_0 = \mathbf{0}$.

2. Put $\mathbf{x}_{n+1} = H_s(\mathbf{x}_n + A^*(\mathbf{b} - A\mathbf{x}_n))$.

3. Stop when $\|\mathbf{x}_{n+1} - \mathbf{x}_n\| < \epsilon$.

The above algorithm (without stopping it) converges to a local minimum of the problem

$$\min_{\mathbf{x}} \|\mathbf{b} - A\mathbf{x}\| \text{ subject to } H_s(\mathbf{x}) = \mathbf{x}.$$

Finding a solution $\mathbf{x}$ to $A\mathbf{x} = \mathbf{b}$ that is sparse (only few entries nonzero), is referred as a *compressed sensing* problem. It has been successfully applied in several settings. For instance, in [T. Zhang, J. M. Pauly, S. S. Vasanawala and M. Lustig, 2013] one can see how compressed sensing was used in reducing MRI acquisition time substantially.

7.3 Why use matrices when computing roots of polynomials?

We saw in Section 5.4 that in order to compute the QR factorization of a matrix only simple arithmetic computations are required. Indeed, one only needs addition, subtraction, multiplication, division, and taking square roots to find the QR factorization of a matrix. Amazingly, doing it repeatedly in a clever way provides an excellent way to compute eigenvalues of a matrix. This is surprising since finding roots of a polynomial is not as easy as performing simple algebraic operations (other than for degree 1, 2, 3, 4 polynomials, using the quadratic formula (for degree 2) and its generalizations; for polynomials of degree 5 and higher it was shown by Niels Hendrik Abel in 1823 that no algebraic formula exists for its roots). In fact, it works so well that for finding roots of a polynomial one can just build its corresponding companion matrix, and subsequently apply the QR algorithm to compute its roots. Let us give an example.

Example 7.3.1 Let $p(t) = t^3 - 6t^2 + 11t - 6 \ (= (t-1)(t-2)(t-3))$. Its companion matrix is

$$A = \begin{pmatrix} 0 & 0 & 6 \\ 1 & 0 & -11 \\ 0 & 1 & 6 \end{pmatrix}.$$

Computing its QR factorization, we find

$$A = QR = \begin{pmatrix} 0 & 0 & 1 \\ 1 & 0 & 0 \\ 0 & 1 & 0 \end{pmatrix} \begin{pmatrix} 1 & 0 & -11 \\ 0 & 1 & 6 \\ 0 & 0 & 6 \end{pmatrix}.$$

If we now let $A_1 = RQ = Q^{-1}QRQ = Q^{-1}AQ$, then A_1 has the same eigenvalues as A. We find

$$
A_1 = \begin{pmatrix} 1 & 0 & -11 \\ 0 & 1 & 6 \\ 0 & 0 & 6 \end{pmatrix} \begin{pmatrix} 0 & 0 & 1 \\ 1 & 0 & 0 \\ 0 & 1 & 0 \end{pmatrix} = \begin{pmatrix} 0 & -11 & 1 \\ 1 & 6 & 0 \\ 0 & 6 & 0 \end{pmatrix}.
$$

Again, we do a QR factorization of $A_1 = Q_1 R_1$, and let $A_2 = R_1 Q_1 (= Q_1^{-1} A_1 Q_1)$. We find

$$
A_2 = \begin{pmatrix} 6.0000 & -0.8779 & 0.4789 \\ 12.5300 & -0.4204 & -0.7707 \\ 0 & 0.2293 & 0.4204 \end{pmatrix}.
$$

After 8 more iterations $(A_i = Q_i R_i, A_{i+1} := R_i Q_i)$ we find that

$$
A_{10} = \begin{pmatrix} 3.0493 & -10.9830 & 7.5430 \\ 0.0047 & 1.9551 & -1.8346 \\ 0 & 0.0023 & 0.9956 \end{pmatrix}.
$$

Notice that the entries below the diagonal are relatively small. In addition, the diagonal entries are not too far off from the eigenvalues of the matrix: 1,2,3. Let us do another 20 iterations. We find

$$
A_{30} = \begin{pmatrix} 3.0000 & -10.9697 & 7.5609 \\ 0.0000 & 2.0000 & -1.8708 \\ 0 & 0.0000 & 1.0000 \end{pmatrix}.
$$

As A_{30} is upper triangular we obtain that its diagonal entries 3,2,1 are the eigenvalues of A_{30}, and therefore they are also the eigenvalues of A.

The QR algorithm converges to an upper triangular matrix for large classes of matrices. We provide the proof for the following class of Hermitian matrices.

Theorem 7.3.2 *If $A = A^* \in \mathbb{C}^{n \times n}$ has eigenvalues*

$$
|\lambda_1| > |\lambda_2| > \cdots > |\lambda_n| > 0,
$$

and $A = V \Lambda V^*$ where $\Lambda = \operatorname{diag}(\lambda_i)_{i=1}^n$ and V is unitary with $V^* = LU$ where L is lower triangular and U is upper triangular, then the iteration

$$
A_1 = A, \quad A_i = Q_i R_i, A_{i+1} = R_i Q_i, i = 1, 2, \ldots,
$$

with $Q_i Q_i^* = I_n$ and R_i upper triangular with positive diagonal entries, gives that

$$
\lim_{k \to \infty} A_k = \Lambda.
$$

We first need a lemma.

Lemma 7.3.3 *Let $V_k \in \mathbb{C}^{n \times n}$, $k \in \mathbb{N}$, be unitary matrices and $U_k \in \mathbb{C}^{n \times n}$, $k \in \mathbb{N}$, be upper triangular matrices with positive diagonal entries. Suppose that $\lim_{k \to \infty} V_k U_k = I_n$. Then $\lim_{k \to \infty} V_k = I_n$ and $\lim_{k \to \infty} U_k = I_n$.*

Proof. Let us write

$$V_k = \left(\mathbf{v}_1^{(k)} \quad \cdots \quad \mathbf{v}_n^{(k)} \right), U_k = (u_{ij}^{(k)})_{i,j=1}^n.$$

Then, looking at the first column of the equality $\lim_{k \to \infty} V_k U_k = I_n$, we have that

$$u_{11}^{(k)} \mathbf{v}_1^{(k)} \to \mathbf{e}_1, \tag{7.1}$$

and thus

$$u_{11}^{(k)} = u_{11}^{(k)} \| \mathbf{v}_1^{(k)} \| = \| u_{11}^{(k)} \mathbf{v}_1^{(k)} \| \to \| \mathbf{e}_1 \| = 1,$$

giving that $\lim_{k \to \infty} u_{11}^{(k)} = 1$. Combining this with (7.1) gives that $\mathbf{v}_1^{(k)} \to \mathbf{e}_1$. Next, from the second column of the equality $\lim_{k \to \infty} V_k U_k = I_n$, we have that

$$u_{12}^{(k)} \mathbf{v}_1^{(k)} + u_{22}^{(k)} \mathbf{v}_2^{(k)} \to \mathbf{e}_2. \tag{7.2}$$

Taking the inner product with $\mathbf{v}_1^{(k)}$ gives

$$u_{12}^{(k)} = \langle u_{12}^{(k)} \mathbf{v}_1^{(k)} + u_{22}^{(k)} \mathbf{v}_2^{(k)}, \mathbf{v}_1^{(k)} \rangle \to \langle \mathbf{e}_2, \mathbf{e}_1 \rangle = 0. \tag{7.3}$$

Then by (7.49) we find that $u_{22}^{(k)} \mathbf{v}_2^{(k)} \to \mathbf{e}_2$, which in a similar manner as before implies that $u_{22}^{(k)} \to 1$ and $\mathbf{v}_2^{(k)} \to \mathbf{e}_2$. Continuing this way, we find that $u_{ij}^{(k)} \to 0$, $i < j$, and $u_{ii}^{(k)} \to 1$, $i = 1, \ldots, n$, and $\mathbf{v}_j^{(k)} \to \mathbf{e}_j$, $j = 1, \ldots, n$. This proves the result. □

Proof of Theorem 7.3.2. Notice that $A^2 = Q_1 R_1 Q_1 R_1 = Q_1 Q_2 R_2 R_1$, and that in general we have that

$$A^k = Q_1 Q_2 \cdots Q_k R_k \cdots R_2 R_1.$$

In addition,

$$A^k = V \Lambda^k V^* = V \Lambda^k L U.$$

Notice that we may choose for L to have diagonal elements equal to 1. Combining we obtain

$$\Lambda^k L = (V^* Q_1 Q_2 \cdots Q_k)(R_k \cdots R_2 R_1 U^{-1}),$$

and thus

$$\Lambda^k L \Lambda^{-k} = (V^* Q_1 Q_2 \cdots Q_k)(R_k \cdots R_2 R_1 U^{-1} \Lambda^{-k}).$$

Write $L = (l_{ij})_{i,j=1}^n$ with $l_{ii} = 1$, $i = 1, \ldots, n$, and $l_{ij} = 0$ for $i < j$. We now have that $\Lambda^k L \Lambda^{-k}$ is lower triangular with a unit diagonal, and with (i,j)th entry $l_{ij}(\frac{\lambda_i}{\lambda_j})^k$, $i < j$, in the lower triangular part. As $|\frac{\lambda_i}{\lambda_j}| < 1$, $i > j$, we have that $\lim_{k \to \infty} l_{ij}(\frac{\lambda_i}{\lambda_j})^k = 0$, and thus $\lim_{k \to \infty} \Lambda^k L \Lambda^{-k} = I_n$. Let

$$\Delta = \mathrm{diag}(\frac{\lambda_i}{|\lambda_i|})_{i=1}^n, E = \mathrm{diag}(\frac{u_{ii}}{|u_{ii}|})_{i=1}^n,$$

where $U = (u_{ij})_{i,j=1}^n$. Let

$$W_k = V^* Q_1 Q_2 \cdots Q_k E^{-1} \Delta^{-k}, U_k = \Delta^k E R_k \cdots R_2 R_1 U^{-1} \Lambda^{-k}. \qquad (7.4)$$

Then W_k is unitary, U_k is upper triangular with positive diagonal entries, and $W_k U_k \to I_n$. By Lemma 7.3.3 it now follows that $W_k \to I_n$ and $U_k \to I_n$. Now

$$A_k = Q_k R_k = E^* \Delta^{-(k-1)} W_{k-1}^* W_k \Delta^k E \Delta^{-k} E^* U_k \Lambda U_{k-1}^{-1} E \Delta^{k-1} \to \Lambda.$$

Indeed, if we write $W_k = I + G_k$ and $U_k = I + H_k$, then $G_k \to 0$ and $H_k \to 0$. Reworking the expression

$$E^* \Delta^{-(k-1)} W_{k-1}^* W_k \Delta^k E \Delta^{-k} E^* U_k \Lambda U_{k-1}^* E \Delta^{k-1} - \Lambda \qquad (7.5)$$

gives that each term has at least one of $G_k, G_{k-1}^*, H_k, H_{k-1}^*$ in it, while multiplying with diagonal unitaries E and Δ does not affect the norms of the expression. This show that (7.5) converges to 0 as $k \to \infty$. $\qquad \square$

While Theorem 7.3.2 only addresses the case of Hermitian matrices, the convergence result goes well beyond this case. In particular, it works for large classes of companion matrices. Due to the structure of companion matrices, one can set up the algorithm quite efficiently, so that one can actually compute roots of polynomials of very high degree accurately. In Figure 7.1, we give an example of degree 10,000.

Concerns with large matrices (say, $10^4 \times 10^4 = 10^8$ entries) are (i) how do you update them quickly? (ii) how do you store them? As it happens, companion matrices have a lot of structure that can be maintained throughout the QR algorithm. First observe that a companion matrix has zeros in the lower triangular part under the subdiagonal. The terminology is as follows. We say that $A = (a_{ij})_{i,j=1}^n$ is *upper Hessenberg* if $a_{ij} = 0$ when $i > j + 1$. The upper Hessenberg structure is maintained throughout the QR algorithm, as we will see now.

Proposition 7.3.4 *If A is upper Hessenberg, Q is unitary and R is upper triangular, and $A = QR$, then RQ is upper Hessenberg as well.*

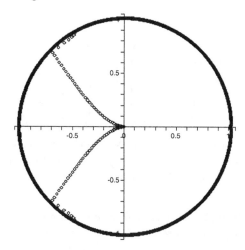

Figure 7.1: These are the roots of the polynomial $\sum_{k=1}^{10,000} p_k(10,000)x^k$, where $p_k(n)$ is the number of partitions of n in k parts, which is the number of ways n can be written as the sum of k positive integers.

Proof. As the jth column of Q is a linear combination of columns $1, \ldots, j$ of A, the jth column of Q has zeroes in positions $j + 2, \ldots, n$. This gives that Q is upper Hessenberg. As R is upper triangular, the ith row of RQ is a linear combination of rows $i, \ldots, n$ of Q, and thus the ith row of RQ has zeros in positions $1, \ldots, i - 2$. This gives that RQ is upper Hessenberg. $\qquad\square$

Corollary 7.3.5 *If A is upper Hessenberg, then its iterates in the QR algorithm are also upper Hessenberg.*

Proof. Follows directly from Proposition 7.3.4. $\qquad\square$

Aside from the upper Hessenberg property, a companion matrix has more structure: it is the sum of a unitary matrix and a rank 1 matrix. Indeed, the companion matrix

$$
C = \begin{pmatrix}
0 & 0 & \cdots & 0 & -a_0 \\
1 & 0 & \cdots & 0 & -a_1 \\
\vdots & & \ddots & & \vdots \\
0 & \cdots & 1 & 0 & -a_{n-2} \\
0 & \cdots & 0 & 1 & -a_{n-1}
\end{pmatrix},
$$

can be written as

$$
C = Z + \mathbf{x}\mathbf{y}^*,
$$

where

$$Z = \begin{pmatrix} 0 & 0 & \cdots & 0 & e^{i\theta} \\ 1 & 0 & \cdots & 0 & 0 \\ \vdots & & \ddots & & \vdots \\ 0 & \cdots & 1 & 0 & 0 \\ 0 & \cdots & 0 & 1 & 0 \end{pmatrix}, \mathbf{x} = \begin{pmatrix} -a_0 - e^{i\theta} \\ -a_1 \\ \vdots \\ -a_{n-2} \\ -a_{n-1} \end{pmatrix}, \mathbf{y} = \mathbf{e}_n.$$

Here Z is unitary and $\mathbf{x}\mathbf{y}^*$ has rank 1. Notice that θ can be chosen to be any real number. The property of being the sum of a unitary and a rank 1 is maintained throughout the QR algorithm, as we prove next.

Proposition 7.3.6 *If $A = Z + K$ with Z unitary and* rank $K = 1$, *then its iterates in the QR algorithm are also the sum of a unitary matrix and a rank 1 matrix.*

Proof. Let $A = Z + K$ and $A = QR$. Then $R = Q^*Z + Q^*K$, and thus $RQ = Q^*ZQ + Q^*KQ$. As Q^*ZQ is unitary, and rank $Q^*KQ =$ rank $K = 1$, we find that the first iterate has the required form. But then repeating the argument we get that the same follows for every iterate. $\square$

Combining the observations in Corollary 7.3.5 and Proposition 7.3.6 it is clear that when starting with a companion matrix, all its iterates continue to have a lot of structure that can be used to perform computations and store them efficiently. Taking advantage of this can lower the number of arithmetic operations required in each iteration, as well as the amount of storage required to store the information. As a result, one can deal with high-degree polynomials in this way.

Let us observe that in finding roots of non-linear systems one often still relies on linear algebra. Indeed, Newton's method is based on the idea that if we would like to find a root of a function f, we start at a first guess, and if this is not a root, we pretend that the graph at this point is a line (the tangent line) and find the root of that line. This is our next guess for our root of f. If the guess is right, we stop. If not, we continue as before by computing the root of the tangent line there, and repeat this process iteratively.

There are many iterative linear schemes that solve a nonlinear problem. One such example is an image enhancement scheme that was used in law enforcement. Such methods need to be defendable in court, convincing a jury that the information extracted was there to begin with rather than that the program "invented" information. In the riots in Los Angeles in 1992 one of the convictions was based on the enhancement of video images taken from a helicopter. Indeed, after enhancement of these images a tattoo became recognizable leading to the identity of one of the rioters.

7.4 How to find functions with linear algebra?

Many scenarios require finding a function based on partial data:

- In medical imaging, one is looking for a function $f(x, y, z)$ which describes the material density of one's body at a point (x, y, z). To do this, one sends radiation through the body and measures on the other side the intensities at different locations. These intensities will be different based on the different intensities the rays of radiation encountered in the body. Mathematically, one measures integrals $\int f g_L$ along lines L (here g_L is the function that takes on the value 1 on the line L and is zero elsewhere), this being the data one collects from which one would like to reconstruct the function.

- In prediction theory, one tries to predict what will happen in the future based on measurements in the past and present. In this situation, one has data $f(w_1), \ldots, f(w_{n+1})$, and one would like to find values $f(w_{n+2}), f(w_{n+3}), \ldots$.

These are the problems we will focus on in this section: reconstruct a function f based on either interpolating data $f(w_1), \ldots, f(w_n)$, or integral data $\int f g_1, \ldots, \int f g_n$. As the maps

$$f \mapsto f(w) \quad \text{and} \quad f \mapsto \int fg$$

are linear, linear algebra plays a very useful role here. In both cases we will restrict the discussion to collecting a finite number of data points. For more general data collection, one would need some tools from functional analysis to set up a robust theory.

We are thus considering the problem: Given a linear map

$$E : \mathbb{F}^X \to \mathbb{F}^n$$

and a vector $\mathbf{v} \in \mathbb{F}^n$, find a function $f \in \mathbb{F}^X$ so that $E(f) = \mathbf{v}$. In the case of interpolation data the field $\mathbb{F}$ can be any field, while in the case of integral data the underlying field is $\mathbb{R}$ or $\mathbb{C}$. Certainly in the last two cases the vector space $\mathbb{F}^X$ is infinite dimensional, and one typically would like to restrict the question to a finite-dimensional subspace W of $\mathbb{F}^X$. Thus, rather than trying to find just any type of function, one restricts the attention to a (finite-dimensional) subspace. This leads to an important question: What subspace W makes the most sense in your application? When one deals with

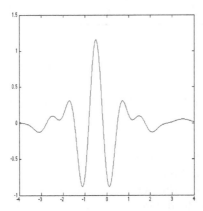

Figure 7.2: A Meyer wavelet.

sound signals, cosines and sines are great functions to work with. In this case one could take

$$W = \text{Span}\{1, \cos x, \sin x, \cos 2x, \sin 2x, \ldots, \cos Nx, \sin Nx\}.$$

The number k in $\cos kx, \sin kx$ is referred to as the *frequency*, and our ear hears a higher tone when the frequency is higher. In addition, the range that the human ear can hear is between 20 Hz and 20,000 Hz (with Hz corresponding to 1 cycle per second). Thus, when it is about sounds the human ear can hear, it makes perfectly sense to use a finite-dimensional subspace. As

$$e^{ix} = \cos x + i \sin x, e^{-ix} = \cos x - i \sin x$$

one can also deal with the subspace

$$W = \text{Span}\{e^{-iNx}, e^{i(N-1)x}, \ldots, e^{i(N-1)x}, e^{iNx}\},$$

often simplifying the calculations (which may sound counterintuitive when you are still getting used to complex numbers, but for instance, simple rules like $e^a e^b = e^{a+b}$ are easier to work with than the formulas for $\cos(a+b)$ and $\sin(a+b)$). In some cases it is better to work with functions that are nonzero only on a finite interval (which is not true for cos and sin), and so-called wavelet functions were invented to have this property while still keeping some advantages that cos and sin have. In Figure 7.2 is an example of a wavelet. Once we have settled on a finite dimensional subspace of functions, we can start to use linear algebra. We begin the exposition using polynomials.

Example 7.4.1 Consider $\mathbb{F}_{n-1}(X)$ with basis $\mathcal{B} = \{1, X, X^2, \ldots, X^{n-1}\}$. Let $\{x_1, \ldots, x_n\} \subseteq \mathbb{F}$ and $E : W \to \mathbb{F}^n$ be given by

$$E(p(X)) = \begin{pmatrix} p(x_1) \\ \vdots \\ p(x_n) \end{pmatrix}.$$

Then

$$[E]_{\mathcal{E} \leftarrow \mathcal{B}} = \begin{pmatrix} 1 & x_1 & \cdots & x_1^{n-1} \\ 1 & x_2 & \cdots & x_2^{n-1} \\ \vdots & \vdots & & \vdots \\ 1 & x_n & \cdots & x_n^{n-1} \end{pmatrix} =: V(x_1, \ldots, x_n), \tag{7.6}$$

where $\mathcal{E}$ is the standard basis of $\mathbb{F}^n$. The matrix $V(x_1, \ldots, x_n)$ is called the *Vandermonde matrix*. Thus interpolation with polynomials leads to a system of equations with a Vandermonde matrix.

Proposition 7.4.2 *The Vandermonde matrix $V(x_1, \ldots, x_n)$ satisfies*

$$\det V(x_1, \ldots, x_n) = \prod_{1 \leq j < i \leq n} (x_i - x_j). \tag{7.7}$$

In particular, $V(x_1, \ldots, x_n)$ is invertible as soon as $x_i \neq x_j$ when $i \neq j$.

Proof. We prove this by induction. When $n = 2$ we have

$$\det V(x_1, x_2) = \det \begin{pmatrix} 1 & x_1 \\ 1 & x_2 \end{pmatrix} = x_2 - x_1.$$

Next, suppose the satement holds for $V(w_1, \ldots, w_{n-1})$. We now take $V(x_1, \ldots, x_n)$ and subtract row 1 from all the other rows, leaving the determinant unchanged and arriving at the matrix

$$\begin{pmatrix} 1 & x_1 & \cdots & x_1^{n-1} \\ 0 & x_2 - x_1 & \cdots & x_2^{n-1} - x_1^{n-1} \\ \vdots & \vdots & & \vdots \\ 0 & x_n - x_1 & \cdots & x_n^{n-1} - x_1^{n-1} \end{pmatrix}.$$

Next, we subtract, in order, x_1 times column $n-1$ from column n, x_1 times column $n-2$ from column $n-1$, and so on, until we subtract x_1 times column 1 from column 2. This again leaves the determinant unchanged, and leads to the matrix

$$\begin{pmatrix} 1 & 0 & 0 & \cdots & 0 & 0 \\ 0 & x_2 - x_1 & (x_2 - x_1)x_2 & \cdots & (x_2 - x_1)x_2^{n-3} & (x_2 - x_1)x_2^{n-2} \\ \vdots & \vdots & \vdots & & \vdots & \vdots \\ 0 & x_n - x_1 & (x_n - x_1)x_n & \cdots & (x_n - x_1)x_n^{n-3} & (x_n - x_1)x_n^{n-2} \end{pmatrix}.$$

This matrix equals

$$\begin{pmatrix} 1 & 0 & \cdots & 0 \\ 0 & x_2 - x_1 & \cdots & 0 \\ \vdots & \vdots & \ddots & \vdots \\ 0 & 0 & \cdots & x_n - x_1 \end{pmatrix} \begin{pmatrix} 1 & 0 & 0 & \cdots & 0 \\ 0 & 1 & x_2 & \cdots & x_2^{n-2} \\ \vdots & \vdots & \vdots & & \vdots \\ 0 & 1 & x_n & \cdots & x_n^{n-2} \end{pmatrix}.$$

So we find that

$$\det V(x_1, \ldots, x_n) = [\prod_{j=2}^{n}(x_j - x_1)] \det V(x_2, \ldots, x_n),$$

and (7.7) follows by using the induction assumption. $\qquad\square$

A particular useful Vandermonde matrix is the *Fourier matrix*

$$F_n = V(1, \alpha, \alpha^2, \ldots, \alpha^{n-1}), \alpha = e^{\frac{2\pi i}{n}}.$$

Proposition 7.4.3 *The matrix $\frac{1}{\sqrt{n}}F_n$ is unitary. In particular, $F_n^{-1} = \frac{1}{n}F_n^*$.*

Proof. Notice that for $k \in \{1, \ldots, n-1\}$, we have that

$$0 = 1 - (\alpha^k)^n = (1 - \alpha^k)(1 + \alpha^k + (\alpha^k)^2 + \cdots + (\alpha^k)^{n-1}).$$

As $\alpha^k \neq 1$, we get that

$$1 + \alpha^k + (\alpha^k)^2 + \cdots + (\alpha^k)^{n-1} = 0.$$

Now one can easily check that $F_n F_n^* = nI_n$. $\qquad\square$

Aside from having an easily computable inverse, the Fourier matrix (when n is a power of 2) also has the advantage that it factors in simpler matrices. This makes multiplication with the Fourier matrix easy (and fast!) to compute. We just illustrate the idea for $n = 4$ and $n = 8$:

$$F_4 = \begin{pmatrix} 1 & 1 & 1 & 1 \\ 1 & i & -1 & -i \\ 1 & -1 & 1 & -1 \\ 1 & -i & -1 & i \end{pmatrix} = \begin{pmatrix} 1 & 1 & 0 & 0 \\ 0 & 0 & 1 & i \\ 1 & -1 & 0 & 0 \\ 0 & 0 & 1 & -i \end{pmatrix} \begin{pmatrix} 1 & 0 & 1 & 0 \\ 0 & 1 & 0 & 1 \\ 1 & 0 & -1 & 0 \\ 0 & 1 & 0 & -1 \end{pmatrix},$$

$$F_8 = \begin{pmatrix} 1 & 1 & 0 & 0 & 0 & 0 & 0 & 0 \\ 0 & 0 & 1 & \alpha & 0 & 0 & 0 & 0 \\ 0 & 0 & 0 & 0 & 1 & \alpha^2 & 0 & 0 \\ 0 & 0 & 0 & 0 & 0 & 0 & 1 & \alpha^3 \\ 1 & \alpha^4 & 0 & 0 & 0 & 0 & 0 & 0 \\ 0 & 0 & 1 & \alpha^5 & 0 & 0 & 0 & 0 \\ 0 & 0 & 0 & 0 & 1 & \alpha^6 & 0 & 0 \\ 0 & 0 & 0 & 0 & 0 & 0 & 1 & \alpha^7 \end{pmatrix} \begin{pmatrix} 1 & 0 & 1 & 0 & 0 & 0 & 0 & 0 \\ 0 & 1 & 0 & 1 & 0 & 0 & 0 & 0 \\ 0 & 0 & 0 & 0 & 1 & 0 & \alpha^2 & 0 \\ 0 & 0 & 0 & 0 & 0 & 1 & 0 & \alpha^2 \\ 1 & 0 & \alpha^4 & 0 & 0 & 0 & 0 & 0 \\ 0 & 1 & 0 & \alpha^4 & 0 & 0 & 0 & 0 \\ 0 & 0 & 0 & 0 & 1 & 0 & \alpha^6 & 0 \\ 0 & 0 & 0 & 0 & 0 & 1 & 0 & \alpha^6 \end{pmatrix} \times$$

$$\begin{pmatrix} 1 & 0 & 0 & 0 & 1 & 0 & 0 & 0 \\ 0 & 1 & 0 & 0 & 0 & 1 & 0 & 0 \\ 0 & 0 & 1 & 0 & 0 & 0 & 1 & 0 \\ 0 & 0 & 0 & 1 & 0 & 0 & 0 & 1 \\ 1 & 0 & 0 & 0 & -1 & 0 & 0 & 0 \\ 0 & 1 & 0 & 0 & 0 & -1 & 0 & 0 \\ 0 & 0 & 1 & 0 & 0 & 0 & -1 & 0 \\ 0 & 0 & 0 & 1 & 0 & 0 & 0 & -1 \end{pmatrix}, \text{where } \alpha = e^{\frac{2\pi i}{8}}.$$

Notice that when multiplying with one of these simpler matrices, for each entry one only needs to do one multiplication and one addition. Thus multiplying with F_8 requires 24 multiplications and 24 additions. In general, we have that multiplying with F_n requires $n \log_2 n$ multiplications and $n \log_2 n$ additions. This is a lot better than n^2 multiplications and $(n-1)n$ additions, which one has with a regular $n \times n$ matrix vector multiplication (this number can be reduced somewhat, but still for a general matrix it is of the order n^2).

Interpolation techniques are also useful over finite fields. An example where this is used is secret sharing.

Example 7.4.4 *(Shamir's secret sharing)* Suppose that we have a secret number and we would like, among N people, that every k of them can piece together the secret, making sure that if only $k-1$ people get together they cannot figure out the secret. An example would be of a bank, where with any 3 people of the upper management, one would like to be able to open the vault, but not with just 2 of them. We explain the idea in an example. Suppose that $m = 1432$ is our secret number. Let us choose a prime number $p > m$, for instance $p = 2309$ (which happens to be a primorial prime). Suppose that $N = 10$ and $k = 3$. We choose a degree $3 - 1 = 2$ polynomial

$$a(x) = a_0 + a_1 x + a_2 x^2,$$

where $a_0 = m = 1432$ and a_1 and a_2 are some other numbers in $\mathbb{Z}_p \setminus \{0\}$. For instance, $a_1 = 132$ and $a_2 = 547$. Now we generate interpolation data, for instance:

x	$f(x)$
1	2111
2	1575
3	2133
4	1476
5	1913
6	1135
7	1451
8	552
9	747
10	2036

If now three people get together, one will be able to reconstruct the polynomial $a(x)$, and thus the secret $a(0)$. With only two people (thus, with only two interpolation points), one will not be able to reconstruct the secret code. For instance, with the data $(2, 1575), (5, 1913), (9, 747)$, one finds the secret by computing

$$a_0 = \begin{pmatrix} 1 & 0 & 0 \end{pmatrix} \begin{pmatrix} 1 & 2 & 4 \\ 1 & 5 & 25 \\ 1 & 9 & 81 \end{pmatrix}^{-1} \begin{pmatrix} 1575 \\ 1913 \\ 747 \end{pmatrix}, \tag{7.8}$$

where one is working over the field $\mathbb{Z}_{2309}$. The calculation (7.8) can be programmed so that those holding the interpolation points do not need to know the prime number p. When the three data points are known, but the prime p is unknown, one still will not be able to reconstruct the secret, providing some protection when someone listening in is able to get 3 interpolation points. This secret sharing scheme was introduced by Adi Shamir.

We will next explain how one arrives at problems where a function is to be found satisfying certain integral conditions. We start by explaining the ideas behind the *Galerkin method*. Let X and Y be vector spaces of functions, and $\Phi : X \to Y$ a linear operator. Consider the problem of solving the equation $\Phi(f) = g$. Typically, X and Y are infinite-dimensional spaces. If, however, Y has a Hermitian form $\langle \cdot, \cdot \rangle$ (which on a function space is often given via an integral), and $w_1, \ldots, w_n \in Y$, we can instead solve:

$$\langle \Phi(f), w_i \rangle = \langle g, w_i \rangle, i = 1, \ldots, n.$$

In addition, we can take a finite-dimensional subspace $U = \mathrm{Span}\{u_1, \ldots, u_n\}$, and seek a solution f in this subspace, thus $f = \sum_{j=1}^{n} a_j u_j$ for some scalars $a_1, \ldots, a_n$. Now we obtain the system of equations

$$\langle \Phi(f), w_i \rangle = \langle \Phi(\sum_{j=1}^{n} a_j u_j), w_i \rangle = \sum_{j=1}^{n} a_j \langle \Phi(u_j), w_i \rangle = \langle g, w_i \rangle, i = 1, \ldots, n.$$

If we let B be the matrix $B = (\langle \Phi(u_j), w_i \rangle)_{i,j=1}^{n}$, then we obtain the equation

$$B \begin{pmatrix} a_1 \\ \vdots \\ a_n \end{pmatrix} = \begin{pmatrix} \langle \Phi(u_1), w_1 \rangle & \cdots & \langle \Phi(u_n), w_1 \rangle \\ \vdots & & \vdots \\ \langle \Phi(u_1), w_n \rangle & \cdots & \langle \Phi(u_n), w_n \rangle \end{pmatrix} \begin{pmatrix} a_1 \\ \vdots \\ a_n \end{pmatrix} = \begin{pmatrix} \langle g, w_1 \rangle \\ \vdots \\ \langle g, w_n \rangle \end{pmatrix}. \tag{7.9}$$

Now we are in a position to solve for $a_1, \ldots, a_n$, and build $f = \sum_{j=1}^{n} a_j u_j$. Clearly, whether this is a meaningful solution to our original problem all depends on whether we made good choices for $u_1, \ldots, u_n \in X$ and $w_1, \ldots, w_n \in Y$ (and, potentially, also on our choice for the Hermitian form $\langle \cdot, \cdot \rangle$ on Y). One particular construction involves dividing the domain up in small subdomains (elements) and having functions that are patched together by taking, on each of these subdomains, a very simple function (linear, quadratic, etc.). This is the main idea behind the finite element method.

Next, let us compute the matrix B in an important example that involves the Laplace operator.

Example 7.4.5 Let $\Omega \subseteq \mathbb{R}^2$ be a bounded region with boundary $\partial\Omega$. One can think of Ω being the inside of a circle, an ellipse, a rectangle, or some other shape. We consider real-valued functions defined on the set $\Omega \cup \partial\Omega$. We let Φ be the Laplace operator $\frac{\partial^2}{\partial x^2} + \frac{\partial^2}{\partial y^2}$, arriving at the Poisson equation

$$\frac{\partial^2 f}{\partial x^2} + \frac{\partial^2 f}{\partial y^2} = g,$$

and let us add the zero boundary condition

$$f = 0 \quad \text{on} \quad \partial\Omega.$$

Thus our vector space X consists of functions that are differentiable twice with respect to each of the variables x and y, and that are zero on the boundary $\partial\Omega$. We introduce the Hermitian form

$$\langle k, h \rangle := \int_\Omega \int k(x,y) h(x,y) \, dx \, dy,$$

which is actually an inner product as we are dealing with continuous functions. Let us choose functions $u_1(x,y), \ldots, u_n(x,y) \in X$, and let $w_i(x,y) = u_i(x,y)$, $i = 1, \ldots, n$, be the same set of functions. Now the matrix $B = (b_{ij})_{i,j=1}^n$ in (7.9) is given by

$$b_{ij} = \langle \frac{\partial^2 u_j}{\partial x^2} + \frac{\partial^2 u_j}{\partial y^2}, u_i \rangle = \int_\Omega \int \left(\frac{\partial^2 u_j}{\partial x^2} + \frac{\partial^2 u_j}{\partial y^2} \right) u_i \, dx \, dy.$$

Performing partial integration, and using the zero boundary condition, we arrive at

$$b_{ij} = \int_\Omega \int \frac{\partial u_j}{\partial x} \frac{\partial u_i}{\partial x} + \frac{\partial u_j}{\partial y} \frac{\partial u_i}{\partial y} \, dx \, dy = \langle \frac{\partial u_j}{\partial x}, \frac{\partial u_i}{\partial x} \rangle + \langle \frac{\partial u_j}{\partial y}, \frac{\partial u_i}{\partial y} \rangle.$$

Note that B is symmetric, and when u_i, $i = 1, \ldots, n$ are chosen so that $\{ \frac{\partial u_1}{\partial x}, \ldots, \frac{\partial u_n}{\partial x} \}$ or $\{ \frac{\partial u_1}{\partial y}, \ldots, \frac{\partial u_n}{\partial y} \}$ is linearly independent, we have that B is positive definite. This guarantees that one can solve for $a_1, \ldots, a_n$ in (7.9) and thus construct a solution f.

Another widely used construction involves the Fourier transform.

Example 7.4.6 Given a function $f : \mathbb{R} \to \mathbb{C}$, we define its Fourier transform $\hat{f}$ via

$$\hat{f}(\omega) = \int_{-\infty}^{\infty} f(x) e^{-2\pi i \omega x} dx,$$

where $\omega \in \mathbb{R}$. Of course, one needs to worry whether $\hat{f}$ is well-defined (which it is if, for instance, f is continuous and $\int_{-\infty}^{\infty} |f(x)| dx < \infty$), but we will not go into a detailed discussion about this. The quantity $\hat{f}(\omega)$ measures intuitively how well $f(x)$ matches the function $e^{-2\pi i \omega x}$. The variable ω is referred to as the frequency, and as mentioned in the beginning of this section, this is a meaningful notion in sound. For instance, if f represents a noisy recording of a conversation, one could take its Fourier transform and analyze which frequencies correspond to the noise (typically the high frequencies) and which frequencies correspond to the actual conversation. By keeping only the frequencies corresponding to the conversation, and performing an inverse Fourier transform, one obtains a noise-free conversation. This process is referred to as *filtering* and can be done in real time (as opposed to first having to record the full conversation). In many of our communication devices filters are being used. Filters have their flaws, of course, and can for instance create an echo. A signal processing course would explain all this in detail.

Example 7.4.7 Blurring of an image is represented by an integral. If $f(x, y)$, $(x, y) \in \Omega$ represents the image (at each location (x, y) there is an intensity), then the blurred image will look like

$$Bf(x, y) = \int_\Omega \int f(x - s, y - t) g(s, t) \, ds \, dt,$$

which is a so-called convolution integral. The function g will have the following shape. The effect of the convolution integral is that the value

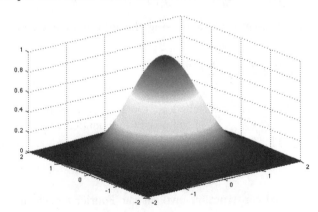

Figure 7.3: Blurring function.

$Bf(x, y)$ is a weighted average of the values of f in a region around the point (x, y). To deblur a picture, one would start with $Bf(x, y)$ and try to solve for $f(x, y)$. As blurring is like taking averages, the deblurring is not going to be perfect. The following shows some typical effects.

Figure 7.4: The original image (of size $3000 \times 4000 \times 3$).

(a) Blurred image.

(b) Deblurred image.

7.5 How to deal with incomplete matrices

In 2006 Netflix put out a one-million-dollar challenge: improve on their existing movie recommendation scheme. They provided anonymous rating data, and the assignment was to predict ratings by customers 10% better than Netflix's program Cinematch® did based on the same training set. In September 2009 the \$1M grand prize was awarded to team Bellkor's Pragmatic Chaos, after a nail-biting finish as The Ensemble submitted their solution only 19 minutes and 54 seconds after the winners did. These teams were groups joining together after they submitted solutions that were in the 9% range, not yet quite achieving the desired 10%. An important ingredient in these solutions is the idea of minimal rank completions, which we will explain in this section.

A *partial matrix* over $\mathbb{F}$ is a matrix with some entries in $\mathbb{F}$ given and others unknown. For instance

$$\mathcal{A} = \begin{pmatrix} 1 & 0 & ? \\ ? & 1 & ? \end{pmatrix} \tag{7.10}$$

is a 2×3 partial matrix with entries $(1,3), (2,1)$ and $(2,3)$ unknown. When convenient, we indicate the unknowns by variables:

$$\mathcal{A} = \begin{pmatrix} 1 & 0 & x_{13} \\ x_{21} & 1 & x_{23} \end{pmatrix}.$$

We view the unknown as variables x_{ij} that take value in the field $\mathbb{F}$. The set of locations $J \subseteq \{1, \ldots, n\} \times \{1, \ldots, m\}$ of known entries is called the *pattern* of the partial matrix. For instance, for the partial matrix (7.10) the pattern is $\{(1,1), (1,2), (2,2)\}$. A *completion* of a partial matrix is obtained by choosing values in $\mathbb{F}$ for the unknowns. For instance, if $\mathbb{F} = \mathbb{R}$, then

$$A = \begin{pmatrix} 1 & 0 & 1 \\ 10 & 1 & -7 \end{pmatrix}, A = \begin{pmatrix} 1 & 0 & \pi \\ e^2 & 1 & \sqrt{\frac{5}{17}} \end{pmatrix}$$

are completions of the partial matrix (7.10). We will denote partial matrices by $\mathcal{A}, \mathcal{B}$, etc., and their completions by A, B, etc.

Going back to the Netflix challenge, a partial matrix corresponding to ratings data may look like

$$\mathcal{A} = \begin{pmatrix} 1 & ? & 4 \\ ? & ? & 3 \\ 5 & 2 & ? \\ 3 & ? & ? \end{pmatrix},$$

where each customer is represented by a row and each movie is represented by a column. So, for instance, customer 1 rated movie 3 with 4 stars, while customer 3 did not rate movie 3.

Given a partial matrix $\mathcal{A}$, we call a completion A a *minimal rank completion* of $\mathcal{A}$ if among all completions B of $\mathcal{A}$ the rank of A is minimal. Thus

$$\operatorname{rank} A = \min_{B \text{ a completion of } \mathcal{A}} \operatorname{rank} B.$$

The *minimal rank* of a partial matrix $\mathcal{A}$ is defined to be the rank of a minimal rank completion of $\mathcal{A}$. In other words

$$\min \operatorname{rank} \mathcal{A} = \min_{B \text{ a completion of } \mathcal{A}} \operatorname{rank} B.$$

For instance,

$$\min \operatorname{rank} \begin{pmatrix} 1 & 0 & ? \\ ? & 1 & ? \end{pmatrix} = 2, \quad \min \operatorname{rank} \begin{pmatrix} 1 & 1 & ? \\ ? & 1 & ? \end{pmatrix} = 1.$$

Indeed, independent of the choice for x_{13}, x_{21} and x_{23}, we have that

$$\text{rank} \begin{pmatrix} 1 & 0 & x_{13} \\ x_{21} & 1 & x_{23} \end{pmatrix} = 2,$$

while any completion of $\mathcal{B} = \begin{pmatrix} 1 & 1 & ? \\ ? & 1 & ? \end{pmatrix}$ has rank at least 1, and

$$\begin{pmatrix} 1 & 1 & 1 \\ 1 & 1 & 1 \end{pmatrix}$$

is a completion of $\mathcal{B}$ with rank 1.

With the partial ranking data, one obtains a large matrix (say, of size $1,000,000,000 \times 100,000$) where only a small percentage of the values are known. It turned out that looking for (an approximation of) a minimal rank completion was a good move. Apparently, a model where our individual movie rankings are a linear combination of the ranking of a relatively few number of people provides a reasonable way to predict a person's movie rankings. Of course, a minimal rank completion of a partial matrix that has entries in the set $\{1, 2, 3, 4, 5\}$ will not necessarily have its entries in this set, so additional steps need to be taken to get ranking predictions.

So, how does one find a minimal rank completion? Here we discuss one algorithm, which assumes that $\mathbb{F} = \mathbb{R}$ or $\mathbb{C}$, based on an initial guess of an upper bound of the minimal rank. For

$$\Sigma = \begin{pmatrix} \sigma_1 & 0 & \cdots & 0 & \cdots & 0 \\ 0 & \sigma_2 & \cdots & 0 & \cdots & 0 \\ \vdots & \vdots & \ddots & \vdots & & \vdots \\ 0 & 0 & \cdots & \sigma_m & \cdots & 0 \\ \vdots & \vdots & & \vdots & \ddots & \vdots \\ 0 & 0 & \cdots & 0 & \cdots & 0 \end{pmatrix}, \quad \sigma_1 \geq \cdots \geq \sigma_m, \tag{7.11}$$

and $k \leq m$, let us define

$$H_k(\Sigma) := \begin{pmatrix} \sigma_1 & 0 & \cdots & 0 & \cdots & 0 \\ 0 & \sigma_2 & \cdots & 0 & \cdots & 0 \\ \vdots & \vdots & \ddots & \vdots & & \vdots \\ 0 & 0 & \cdots & \sigma_k & \cdots & 0 \\ \vdots & \vdots & & \vdots & \ddots & \vdots \\ 0 & 0 & \cdots & 0 & \cdots & 0 \end{pmatrix},$$

thus just keeping the k largest singular values. Notice that the operation H_k is like the hard thresholding operator introduced in Section 7.2.

The algorithm to find a minimal rank completion is now as follows.

Given are a real or complex partial matrix $\mathcal{A}$ with pattern J, an integer k, and a tolerance $\epsilon > 0$.

1. Choose a completion A_0 of $\mathcal{A}$.

2. While $s_{k+1}(A_i) \geq \epsilon$, do the following:

   (i) Find a singular value decomposition $A_i = U_i \Sigma_i V_i^*$ of A_i. Compute
   $B_i = U_i H_k(\Sigma_i) V_i^*$.

   (ii) Let A_{i+1} be defined by

$$(A_{i+1})_{rs} = \begin{cases} (A_i)_{rs} & \text{if } (r,s) \in J, \\ (B_i)_{rs} & \text{if } (r,s) \notin J. \end{cases}$$

3. If the algorithm fails to stop in a reasonable time, raise the integer k.

For this algorithm to work, one needs to be able to find a (good approximation of a) singular value decomposition of a large matrix. Such algorithms have been developed, and are used for instance in search engines.

Another area where incomplete matrices appear involve distance matrices. A matrix $D = (d_{ij})_{i,j=1}^n$ is called a (*Euclidean*) *distance matrix* if there exist an $n \in \mathbb{N}$ and vectors $v_1, \ldots, v_k \in \mathbb{R}^n$ such that $d_{ij} = \|v_i - v_j\|^2$, where $\|\cdot\|$ denotes the Euclidean distance.

Example 7.5.1 Let $v_1 = \begin{pmatrix} 0 & 1 & 1 \end{pmatrix}^T$, $v_2 = \begin{pmatrix} 1 & -1 & 1 \end{pmatrix}^T$, and $v_3 = \begin{pmatrix} 0 & 0 & 2 \end{pmatrix}^T$. Then the corresponding distance matrix is given by

$$\begin{pmatrix} 0 & 5 & 2 \\ 5 & 0 & 3 \\ 2 & 3 & 0 \end{pmatrix}.$$

The following result gives a characterization of distance matrices.

Theorem 7.5.2 *A real symmetric matrix* $D = (d_{ij})_{i,j=1}^k$, *with* $d_{ii} = 0$, $i = 1, \ldots, k$, *is a distance matrix if and only if the* $(k+1) \times (k+1)$ *bordered matrix*

$$B = (b_{ij})_{i,j=1}^{k+1} := \begin{pmatrix} 0 & e^T \\ e & D \end{pmatrix} \tag{7.12}$$

has only one positive eigenvalue. Here, e *is the vector with all of its entries* *equal to 1. In that case, the minimal dimension n for which there exists vectors $\mathbf{v}_1, \ldots, \mathbf{v}_k \in \mathbb{R}^n$ such that $d_{ij} = \|\mathbf{v}_i - \mathbf{v}_j\|^2$, $i, j = 1, \ldots, k$, is given by the rank of the matrix*

$$S = B_{22} - B_{21}B_{11}^{-1}B_{12}, \tag{7.13}$$

where

$$B_{11} = (b_{ij})_{i,j=1}^2 = \begin{pmatrix} 0 & 1 \\ 1 & 0 \end{pmatrix}, B_{12} = B_{21}^T = (b_{ij})_{i=1, j=3}^{2, k+1}, B_{22} = (b_{ij})_{i,j=3}^{k+1}.$$

Proof. We first note that

$$\begin{pmatrix} I_2 & 0 \\ -B_{21}B_{11}^{-1} & I_{k-1} \end{pmatrix} \begin{pmatrix} B_{11} & B_{12} \\ B_{21} & B_{22} \end{pmatrix} \begin{pmatrix} I_2 & 0 \\ -B_{21}B_{11}^{-1} & I_{k-1} \end{pmatrix}^T = \begin{pmatrix} B_{11} & 0 \\ 0 & S \end{pmatrix}.$$

Thus, by Theorem 5.5.5, we obtain

$$\text{In } B = \text{In } \begin{pmatrix} B_{11} & 0 \\ 0 & S \end{pmatrix} = \text{In } B_{11} + \text{In } S = (1, 1, 0) + \text{In } S. \tag{7.14}$$

Assume without loss of generality that $\mathbf{v}_1 = 0$ (by replacing $\mathbf{v}_j$ by $\mathbf{v}_j - \mathbf{v}_1$, $j = 1, \ldots, j$, which does not affect the matrix) and consider the distance matrix

$$D = \begin{pmatrix} 0 & \|\mathbf{v}_2\|^2 & \cdots & \|\mathbf{v}_k\|^2 \\ \|\mathbf{v}_2\|^2 & 0 & \cdots & \|\mathbf{v}_2 - \mathbf{v}_k\|^2 \\ \|\mathbf{v}_3\|^2 & \|\mathbf{v}_3 - \mathbf{v}_2\|^2 & \cdots & \|\mathbf{v}_3 - \mathbf{v}_k\|^2 \\ \vdots & \vdots & & \vdots \\ \|\mathbf{v}_k\|^2 & \|\mathbf{v}_k - \mathbf{v}_2\|^2 & \cdots & 0 \end{pmatrix}.$$

Computing the matrix S in (7.13), one obtains

$$\begin{pmatrix} 0 & \|\mathbf{v}_2 - \mathbf{v}_3\|^2 & \cdots & \|\mathbf{v}_2 - \mathbf{v}_k\|^2 \\ \|\mathbf{v}_3 - \mathbf{v}_2\|^2 & 0 & \cdots & \|\mathbf{v}_3 - \mathbf{v}_k\|^2 \\ \vdots & \vdots & \ddots & \vdots \\ \|\mathbf{v}_k - \mathbf{v}_2\|^2 & \|\mathbf{v}_k - \mathbf{v}_3\|^2 & \cdots & 0 \end{pmatrix}$$

$$- \begin{pmatrix} 1 & \|\mathbf{v}_2\|^2 \\ 1 & \|\mathbf{v}_3\|^2 \\ \vdots & \vdots \\ 1 & \|\mathbf{v}_k\|^2 \end{pmatrix} \begin{pmatrix} 0 & 1 \\ 1 & 0 \end{pmatrix} \begin{pmatrix} 1 & 1 & \cdots & 1 \\ \|\mathbf{v}_2\|^2 & \|\mathbf{v}_3\|^2 & \cdots & \|\mathbf{v}_k\|^2 \end{pmatrix},$$

which equals the matrix

$$\left(\|\mathbf{v}_i - \mathbf{v}_j\|^2 - \|\mathbf{v}_i\|^2 - \|\mathbf{v}_j\|^2\right)_{i,j=2}^k = -2\left(\mathbf{v}_i^T\mathbf{v}_j\right)_{i,j=2}^k$$

$$= -2\begin{pmatrix}\mathbf{v}_2^T\\\mathbf{v}_3^T\\\vdots\\\mathbf{v}_k^T\end{pmatrix}\begin{pmatrix}\mathbf{v}_2 & \mathbf{v}_3 & \cdots & \mathbf{v}_k\end{pmatrix},$$

which is negative semidefinite of rank n, where n is the dimension of Span$\{\mathbf{v}_i : i = 2, \ldots, k\}$. Thus In $S = (0, n, k - 1 - n)$, and by using (7.14) we find that In $B = (1, 1, 0) + (0, n, k - 1 - n) = (1, n + 1, k - 1 - n)$ and thus B has only one positive eigenvalue.

Conversely, if B has only one positive eigenvalue, then In $S = $ In $B - (1, 1, 0)$, gives that S has no positive eigenvalues. Thus $-S$ is positive semidefinite. Let us write $-\frac{1}{2}S = Q^T Q$, with Q of size $n \times k - 1$, where $n = $ rank S. Write

$$Q = \begin{pmatrix}\mathbf{q}_2 & \cdots & \mathbf{q}_k\end{pmatrix},$$

where $\mathbf{q}_2, \ldots, \mathbf{q}_k \in \mathbb{R}^n$. Put $\mathbf{q}_1 = \mathbf{0}$. We claim that $d_{ij} = \|\mathbf{q}_i - \mathbf{q}_j\|^2$. From $-\frac{1}{2}S = Q^T Q$ we obtain that

$$d_{ij} - d_{i1} - d_{1j} = -2\mathbf{q}_i^T\mathbf{q}_j = \|\mathbf{q}_i - \mathbf{q}_j\|^2 - \|\mathbf{q}_i\|^2 - \|\mathbf{q}_j\|^2, i, j = 2, \ldots, k. \quad (7.15)$$

Letting $i = j \in \{2, \ldots, k\}$ gives that $d_{i1} = \|\mathbf{q}_i\|^2$, $i = 2, \ldots, k$. Using this (7.15) now gives that $d_{ij} = \|\mathbf{q}_i - \mathbf{q}_j\|^2$, $i, j = 2, \ldots, k$, finishing the proof. $\square$

To know the interatomic distances in a molecule is important in understanding the molecule and its chemical behavior. By using nuclear magnetic resonance (NMR) data, one would like to determine these interatomic distances. Clearly, this is a challenge as the distances are so small, so unavoidably there are errors in the measurements, and moreover one may not be able to determine all the distances from the data. Now, we do know that the data comes from a three-dimensional space, so when one writes down the corresponding distance matrix, it should have the property that the matrix S in (7.13) has rank 3. This gives the opportunity to fill in some missing data, as well as correct some inaccurate data.

7.6 Solving millennium prize problems with linear algebra

The Clay Mathematics Institute (CMI) of Cambridge, Massachusetts, established the Millennium Prize Problems, seven problems for which the solution carries a \$1 million prize payable by CMI, not to mention with a place in the (mathematics) history books. The prizes were announced at a meeting in Paris, held on May 24, 2000 at the Collège de France. In this section we will discuss two of these problems from a linear algebra perspective.

7.6.1   The Riemann hypothesis

The Riemann zeta function is defined by

$$\zeta(s) := \sum_{n=1}^{\infty} \frac{1}{n^s}.$$

This infinite sum (a series) is defined by letting $s_k = \sum_{n=1}^{k} \frac{1}{n^s}$ and when $\lim_{k \to \infty} s_k$ exists, we say that the series converges and call its limit the sum of the series $\sum_{n=1}^{\infty} \frac{1}{n^s}$. As it turns out, $\zeta(s)$ is well-defined when s is a complex number with Re $s > 1$. The convergence when $s = 2, 3, \ldots$, thus for

$$\zeta(2) = \frac{1}{1^2} + \frac{1}{2^2} + \frac{1}{3^2} + \cdots, \quad \zeta(3) = \frac{1}{1^3} + \frac{1}{2^3} + \frac{1}{3^3} + \cdots, \quad \zeta(4) = \frac{1}{1^4} + \frac{1}{2^4} + \frac{1}{3^4} + \cdots,$$

is typically addressed in a first treatment on series. Riemann showed that a (necessarily unique) analytic function exists (also denoted by ζ) defined on $\mathbb{C} \setminus \{1\}$ that coincides with $\zeta(s)$ on the domain $\{s \in \mathbb{C} : \text{Re } s > 1\}$. If you are not familiar with the notion of a function being analytic, one can think of this property as being complex differentiable k times for every $k \in \mathbb{N}$ (also, referred to as *infinitely complex differentiable*). The Riemann hypothesis can now be formulated as follows.

Riemann hypothesis *If s is a zero of $\zeta(s)$, then either s is a negative even integer $-2, -4, \ldots$ or s has a real part equal to $\frac{1}{2}$.*

The negative even integers are considered to be the *trivial zeros* of $\zeta(s)$, so the Riemann hypothesis can also be stated as *the non-trivial zeros of the Riemann zeta function have a real part $\frac{1}{2}$*. There is a lot to say about the Riemann hypothesis as the vast literature on the subject shows. A good place to start to read up on it would be the website of the Clay Mathematics Institute. In this subsection we would just like to introduce a linear algebra problem, the solution of which would imply the Riemann hypothesis.

Define $n \times n$ matrices $D_n = (d_{ij})_{i,j=1}^{n}$ and $C_n = (c_{ij})_{i,j=1}^{n}$ by

$$d_{ij} = \begin{cases} i & \text{if } i \text{ divides j} \\ 0 & \text{otherwise,} \end{cases}$$

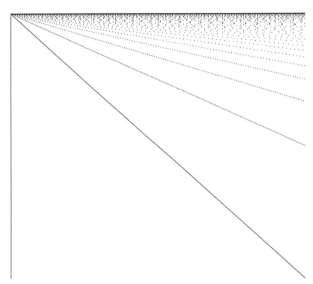

Figure 7.5: The Redheffer matrix of size 500×500.

and

$$C_n = (\mathbf{e}_2 + \cdots + \mathbf{e}_n)^T \mathbf{e}_1.$$

Let $A_n = D_n + C_n$, which is called the Redheffer matrix, after its inventor. So, for instance

$$A_6 = \begin{pmatrix} 1 & 1 & 1 & 1 & 1 & 1 \\ 1 & 1 & 0 & 1 & 0 & 1 \\ 1 & 0 & 1 & 0 & 0 & 1 \\ 1 & 0 & 0 & 1 & 0 & 0 \\ 1 & 0 & 0 & 0 & 1 & 0 \\ 1 & 0 & 0 & 0 & 0 & 1 \end{pmatrix}.$$

In Figure 7.5 one can see what A_{500} looks like.

We now have the following result:

The Riemann hypothesis holds if and only if for every $\epsilon > 0$ there exist $M, N > 0$ so that $|\det A_n| \leq M n^{\frac{1}{2}+\epsilon}$ for all $n \geq N$.

If you are familiar with *big O notation*, then you will recognize that the last statement can be written as $|\det A_n| = O(n^{\frac{1}{2}+\epsilon})$ as $n \to \infty$. The proof of this result requires material beyond the scope of this book; please see [Redheffer, 1977] for more information. While this formulation may be an interesting way to familiarize oneself with the Riemann hypothesis, the machinery to solve this problem will most likely tap into many fields of mathematics. Certainly the solution has been elusive to many mathematicians since the problem was introduced in 1859, and continues to capture the interest of many.

7.6.2 P vs. NP

A major unresolved problem in computational complexity theory is the P versus NP problem. The way to solve this problem is to find a polynomial time algorithm for one of the problems that are identified as NP hard. In this section we will discuss the NP hard problem MaxCut. By a polynomial time algorithm we mean an algorithm for which the running time can be bounded above by a polynomial expression in the size of the input for the algorithm. The P versus NP problem was formally introduced in 1971 by Stephen Cook in his paper "The complexity of theorem proving procedures," but earlier versions go back at least to a 1956 letter written by Kurt Gődel to John von Neumann.

An *(undirected) graph* is an ordered pair $G = (V, E)$ comprising a set V of vertices (or nodes) together with a set $E \subseteq V \times V$, the elements of which are called *edges*. The set E is required to be symmetric, that is $(i, j) \in E$ if and only if $(j, i) \in E$. For this reason we write $\{i, j\}$ instead of both (i, j) and (j, i). In addition, when we count edges we count $\{i, j\}$ only once. The edges are depicted as lines between the corresponding vertices, so $\{1, 2\} \in E$ means that a line (edge) is drawn between vertex 1 and vertex 2. An example with $V = \{1, 2, 3, 4, 5, 6\}$, $E = \Big\{ \{1, 2\}, \{1, 5\}, \{2, 5\}, \{2, 3\}, \{3, 4\}, \{4, 5\}, \{4, 6\} \Big\}$ is:

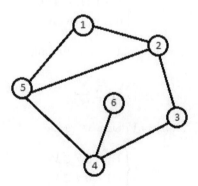

Figure 7.6: A sample graph.

A *cut* in a graph is a disjoint union $V = V_1 \cup V_2$ of the vertices; that is $V_1 \cap V_2 = \emptyset$. The size $s(V_1, V_2)$ of the cut (V_1, V_2) is the number of edges where one endpoint lies in V_1 and the other in V_2. So, for instance, for the graph above and the choice $V_1 = \{1, 3, 6\}$, $V_2 = \{2, 4, 5\}$, the size equals $s(V_1, V_2) = 5$. A *maximum cut* of a graph G is a cut whose size is at least the size of any other cut of G.

MaxCut problem: Given a graph G, find a maximum cut for G.

A graph G with n vertices has 2^{n-1} cuts, and thus to find a maximum cut one may simply check the size of each cut and pick one for which the size is maximal. There is one major problem with this approach: there are simply too many cuts to check. For instance, if $n = 100$ and one can check 100,000 cuts per second, it will take more than 10^{17} years to finish the search. The main problem is that the time it takes is proportional to 2^{n-1}, which is an exponential function of n. We would rather have an algorithm taking time that is proportional to a polynomial $p(n)$ in n. We call such an algorithm a *polynomial time algorithm*. As an example of a problem that can be solved in polynomial time, putting n numbers in order from smallest to largest can be done in a time proportional to n^2, for instance by using the Quicksort algorithm. The MaxCut problem is one of many for which no polynomial time has been established. We will describe now a polynomial time algorithm that finds a cut with a size of at least 0.878 times the maximal cut size. The development of a polynomial time algorithm that would bring this number up to $\frac{16}{17} \approx 0.941$ would show that **P=NP** and thus solve one of the millennium prize problems.

Let $G = (V, E)$ with $V = \{1, \ldots, n\}$. Introduce the symmetric matrix $W = (w_{ij})_{i,j=1}^n$ given by

$$w_{ij} = \begin{cases} 1 & \text{if } \{i, j\} \in E \\ 0 & \text{otherwise.} \end{cases} \tag{7.16}$$

The MaxCut problem may be rephrased as finding

$$\text{mc}(G) = \max_{y_i \in \{-1,1\}} \frac{1}{2} \sum_{i<j} w_{ij}(1 - y_i y_j). \tag{7.17}$$

Indeed, with a cut $\{1, \ldots, n\} = V_1 \cup V_2$ we set

$$y_i = \begin{cases} 1 & \text{if } i \in V_1, \\ -1 & \text{if } i \in V_2. \end{cases}$$

Notice that

$$1 - y_i y_j = \begin{cases} 0 & \text{if } i, j \in V_1 \text{ or } i, j \in V_2, \\ 2 & \text{if } i \in V_1, j \in V_2 \text{ or } i \in V_2, j \in V_1. \end{cases}$$

Thus $\frac{1}{2} \sum_{i<j} w_{ij}(1 - y_i y_j)$ indeed corresponds to the size of the cut

$\{1, \ldots, n\} = V_1 \cup V_2$, and thus (7.17) corresponds to the MaxCut problem. Notice that if $\mathbf{y} = (y_i)_{i=1}^n$ is a vector, then the matrix $Y = (y_{ij})_{i,j=1}^n := \mathbf{y}\mathbf{y}^T$ is positive semidefinite and has diagonal entries equal to 1. In addition, Y has rank 1. If we drop the rank 1 condition, we arrive at a larger set of matrices Y. This now leads to the following problem.

Problem: Find $Y = (y_{ij})_{i,j=1}^n$ positive semidefinite with $y_{ii} = 1$, $i = 1, \ldots, n$, which maximizes

$$\text{mcr}(G) = \max_Y \frac{1}{2} \sum_{i<j} w_{ij}(1 - y_{ij}). \tag{7.18}$$

As we showed, when we have a solution $\mathbf{y} = (y_i)_{i=1}^n$ to (7.17), we have that $Y = (y_{ij})_{i,j=1}^n := \mathbf{y}\mathbf{y}^T$ is one of the matrices we are maximizing over in (7.18), and thus we find that $\text{mc}(G) \leq \text{mcr}(G)$. If the optimum for (7.18) is achieved at a Y of rank 1, then $Y = \mathbf{y}\mathbf{y}^T$ with $\mathbf{y} = (y_i)_{i=1}^n$ and $y_i \in \{-1, 1\}$, so in that case we find that $\text{mc}(G) = \text{mcr}(G)$. The usefulness of considering problem (7.18) is that given an accuracy $\epsilon > 0$, it can be solved in polynomial time within the accuracy ϵ. The algorithm to solve (7.18) is based on semidefinite programming, which is a generalization of linear programming where instead of maximizing over the set of vectors with nonnegative entries, one now optimizes over the set of positive semidefinite matrices. In the remainder of this section we will explain why

$$0.878576 \, \text{mcr}(G) \leq \text{mc}(G).$$

Let $S^n = \{\mathbf{v} \in \mathbb{R}^n : \|\mathbf{v}\| = 1\}$ be the unit sphere in $\mathbb{R}^n$. We define a function $\text{sign} : \mathbb{R} \to \{-1, 1\}$ by

$$\text{sign}(x) = \begin{cases} 1 & \text{if } x \geq 0, \\ -1 & \text{if } x < 0. \end{cases}$$

Lemma 7.6.1 *Let $\mathbf{v}_1, \mathbf{v}_2 \in S^n$. Choose $\mathbf{r} \in S^n$ randomly (uniform distribution). Then the probability that $\text{sign} \, \mathbf{r}^T \mathbf{v}_1 \neq \text{sign} \, \mathbf{r}^T \mathbf{v}_2$ equals $\frac{\arccos(\mathbf{v}_1^T \mathbf{v}_2)}{\pi}$.*

Proof. The subspace $\{\mathbf{x} \in \mathbb{R}^n : \mathbf{r}^T \mathbf{x} = 0\}$ cuts the unit sphere S^n in two equal half-spheres. The chance that $\mathbf{v}_1$ and $\mathbf{v}_2$ end up in different half-spheres is proportional to the angle $\arccos(\mathbf{v}_1^T \mathbf{v}_2)$ between the vectors $\mathbf{v}_1$ and $\mathbf{v}_2$. When this angle is π (thus $\mathbf{v}_1 = -\mathbf{v}_2$) the chance is 1 that they end up in different half-spheres, while if the angle is 0 (thus $\mathbf{v}_1 = \mathbf{v}_2$) the chance is 0 that they end up in different half-spheres. The proportionality now leads to the chance of ending up in different half-spheres being in general equal to $\frac{\arccos(\mathbf{v}_1^T \mathbf{v}_2)}{\pi}$. $\square$

We now propose the following algorithm.

Given is a graph $G = (V, E)$, with $V = \{1, \ldots, n\}$.

1. Define the matrix W via (7.16).

2. Solve (7.18), resulting in a maximizing matrix Y.

3. Factor $Y = Q^T Q$, where $Q = \begin{pmatrix} \mathbf{v}_1 & \cdots & \mathbf{v}_n \end{pmatrix}$.

4. Choose $\mathbf{r} \in S^n$ randomly (uniform distribution).

5. Set $V_1 = \{i \in V : \text{sign } \mathbf{r}^T \mathbf{v}_i = 1\}$ and $V_2 = \{i \in V : \text{sign } \mathbf{r}^T \mathbf{v}_i = -1\}$.

This leads to a cut $V = V_1 \cup V_2$.

The following proposition gives an estimate for the expected size of a cut obtained by the above algorithm.

Proposition 7.6.2 *Using the above algorithm, the expected size of a cut is* $\geq 0.87856 \, \text{mcr}(G)$.

Proof. We let $\mathbf{E}$ denote the expectation (of a probability distribution). Then

$$\mathbf{E}[\text{size of a cut}] = \frac{1}{2} \sum_{i,j} w_{ij} \text{Probability}(\text{sign } \mathbf{r}^T \mathbf{v}_i \neq \text{sign } \mathbf{r}^T \mathbf{v}_j) =$$

$$\frac{1}{2} \sum_{i,j} w_{ij} \frac{\arccos(\mathbf{v}_i^T \mathbf{v}_j)}{\pi} = \frac{1}{2} \sum_{i,j} w_{ij} \frac{\arccos(y_{ij})}{\pi} =$$

$$\frac{1}{4} \sum_{i,j} w_{ij}(1 - y_{ij}) \frac{2}{\pi} \frac{\arccos(y_{ij})}{1 - y_{ij}} \geq \frac{1}{4} \sum_{i,j} w_{ij}(1 - y_{ij}) \min_{-1 \leq t \leq 1} \frac{2}{\pi} \frac{\arccos(t)}{1 - t} =$$

$$\text{mcr}(G) \min_{0 \leq \theta \leq \pi} \frac{2}{\pi} \frac{\theta}{1 - \cos \theta} \geq 0.87856 \, \text{mcr}(G),$$

where we used the substitution $t = \cos \theta$, and where the last step is the result of a calculus exercise (of determining the minimum of $\frac{\theta}{1 - \cos \theta}$ on $[0, \pi]$). $\qquad \square$

Corollary 7.6.3 *For a graph G, we have*

$$0.87856 \, \text{mcr}(G) \leq \text{mc}(G) \leq \text{mcr}(G).$$

Proof. Indeed, using Proposition 7.6.2, we obtain

$$\text{mc}(G) \geq \mathbf{E}[\text{size of a cut}] \geq 0.87856 \, \text{mcr}(G).$$

The inequality $\text{mcr}(G) \geq \text{mc}(G)$ holds, as we observed before. $\qquad \square$

Thus the outcome of the polynomial time algorithm provides an answer that bounds the value of the NP hard problem MaxCut within 12.2% accuracy. It has been proven that if the approximation ratio can be made better than $\frac{16}{17} \approx 0.941$, then a polynomial time algorithm for MaxCut can be obtained. Thus if a polynomial time algorithm can be found achieving this approximation ratio of 0.941 (instead of 0.87856), one obtains that **P=NP** (and a million US dollars).

7.7   How secure is RSA encryption?

Algorithms in public-key cryptography are only secure if the mathematical steps to crack the code are impossible to perform within a time frame to take advantage of the solution. The RSA (Rivest-Shamir-Adleman) encryption scheme is based the challenge to find a factorization $n = pq$ of a given number n that is known to be the product of two primes. For instance, the 250-digit number RSA-250

214032465024074496126442307283933356300861471514475501779775492088
141802344714013664334551909580467961099285187247091458768739626192
155736304745477052080511905649310668769159001975940569345745223058
932597669747168173806936489469987157849497593749793793097937

is known to be the product of two primes p and q, but finding this factorization is a hard task. Trying out all possibilities will require time that is beyond our lifetime. So how to do it much much faster? We will explain some of the ideas behind one method, which ultimately leads to finding nonzero vectors in the kernel of a large sparse matrix over $\mathbb{Z}_2$. One may think in terms of a matrix with 100,000 columns, with on average 15 nonzero entries in each column.

The approach we describe is based on a so-called quadratic sieve. The first observation is that with p and q odd we have that

$$n = pq = (\frac{p+q}{2})^2 - (\frac{p-q}{2})^2$$

is a difference of squares. So, if we want to factor n as a product pq, as a first step, one can try to take integers $s > \sqrt{n}$ and see if $s^2 - n$ happens to be a square. In that case we would choose p and q so that $s = (p+q)/2$ and $\sqrt{s^2 - n} = (p-q)/2$ and a factorization of n would follow. Most likely this will not work. However, adjusting this idea (initiated by Carl Pomerance) leads to the following. Let Φ be the function $\Phi(x) = x^2 - n$. Suppose that

$x_1, \ldots, x_k$ are integers so that

$$\Phi(x_1) \cdots \Phi(x_k)$$

is a square. Say, we have that $v^2 = \Phi(x_1) \cdots \Phi(x_k)$. Let now $u = x_1 \cdots x_k$, and compute

$$u^2 = x_1^2 \cdots x_k^2 = \Phi(x_1) \cdots \Phi(x_k) + rn = v^2 + rn,$$

for some integer $r \in \mathbb{Z}$. Thus n divides $u^2 - v^2 = (u + v)(u - v)$. Let now $s = \gcd(n, u - v)$. Then there is a chance that x is a nontrivial factor of n. The question now becomes: Given $\{\Phi(x_1), \ldots, \Phi(x_k)\}$, how do we find $J \subseteq \{1, \ldots, l\}$ so that $\prod_{j \in J} \Phi(x_j)$ is a square? For this we factorize $\Phi(x_j)$ in primes

$$\Phi(x_j) = p_1^{n_1(j)} \cdots p_l^{n_l(j)}, j = 1 \ldots, k.$$

To obtain a square we need to see to it that we choose $J \subseteq \{1, \ldots, k\}$ so that $\sum_{j \in J} n_i(j)$ is even for all $i = 1, \ldots, k$. If we set up the matrix $b_{ij} = n_i(j) \pmod 2$ we get a $l \times k$ matrix A with entries in $\mathbb{Z}_2$ for which we would like to find a vector $\mathbf{x} \in \mathbb{Z}_2^k$ so that $A\mathbf{x} = \mathbf{0}$. We now choose $J = \{j : x_j = 1\}$.

Let us consider an example.

Example 7.7.1 Let the numbers $3675, 7865, 165, 231, 7007$ be given. Can we find a square among the different products of these numbers? For this we first do a prime factorization of each of them:

$$3675 = 3 \cdot 5^2 \cdot 7^2, \ 7865 = 5 \cdot 11^2 \cdot 13, \ 165 = 3 \cdot 5 \cdot 11, \ 231 = 3 \cdot 7 \cdot 11, \ 7007 = 7^2 \cdot 11 \cdot 13.$$

Notice that these are all products of the primes 3, 5, 7, 11, and 13. For each of these numbers we make a column consisting of the power of these primes modulo 2. For instance, the column corresponding to 3675 is $\begin{pmatrix} 1 & 0 & 0 & 0 & 0 \end{pmatrix}^T$; as only the power of 3 in the prime factorization of 3675 is odd, we only get a 1 in the first position (the row corresponding to the prime 3). Doing it for all 5 numbers we get the matrix

$$A = \begin{pmatrix} 1 & 0 & 1 & 1 & 0 \\ 0 & 1 & 1 & 0 & 0 \\ 0 & 0 & 0 & 1 & 0 \\ 0 & 0 & 1 & 1 & 1 \\ 0 & 1 & 0 & 0 & 1 \end{pmatrix}.$$

Taking the vector $\mathbf{x} = \begin{pmatrix} 1 & 1 & 1 & 0 & 1 \end{pmatrix}^T$, we get that $A\mathbf{x} = \mathbf{0}$. This now gives that the product

$$3675 \cdot 7865 \cdot 165 \cdot 7007 = 3^2 \cdot 5^4 \cdot 7^4 \cdot 11^4 \cdot 13^2$$

is a square.

To find a solution of $A\mathbf{x} = \mathbf{0}$ one may use Gaussian elimination, but in doing so we will lose the sparse structure of the matrix, and too much storage may be required to do it effectively. The following algorithm, due to Douglas H. Wiedemann, has smaller storage requirements.

We present an algorithm to find a nonzero vector $\mathbf{x} \in \mathbb{Z}_2^m$ as the solution of the homogeneous system $A\mathbf{x} = \mathbf{0}$, where $A \in \mathbb{Z}_2^{m \times m}$.

1. Randomly choose $\mathbf{0} \neq \mathbf{y} \in \mathbb{Z}_2^m$. If $\mathbf{u} = A\mathbf{y} = \mathbf{0}$ we are done.

2. Randomly choose $\mathbf{0} \neq \mathbf{v} \in \mathbb{Z}_2^m$.

3. Compute $s_i = \mathbf{v}^T A^i \mathbf{u}, i = 0, 1, \ldots..$

4. Compute a nonzero polynomial $m(X) = m_0 + m_1 X + \cdots + m_d X^d$ so that $\sum_{j=0}^{d} s_{k+j} m_j = 0$ for all k. Equivalently, find a nontrivial solution of the equation

$$\begin{pmatrix} s_0 & s_1 & s_2 & \cdots & s_{d-1} & s_d \\ s_1 & s_2 & s_3 & \cdots & s_d & s_{d+1} \\ s_2 & s_3 & s_4 & \cdots & s_{d+1} & s_{d+2} \\ \vdots & \vdots & \vdots & & \vdots & \vdots \end{pmatrix} \begin{pmatrix} m_0 \\ m_1 \\ m_2 \\ \vdots \\ m_{d-1} \\ m_d \end{pmatrix} = \mathbf{0}. \tag{7.19}$$

5. Let j be so that $m_0 = \cdots = m_{j-1} = 0$ and $m_j \neq 0$. Put $p(X) = \frac{m(X)}{m_j X^j}$.

6. Let now $\mathbf{x} = p(A)\mathbf{y}$, and check whether $A^{l-1}\mathbf{x} \neq \mathbf{0}$ and $A^l \mathbf{x} = \mathbf{0}$, for some $l = 1, \ldots, j$. If so, $A^{l-1}\mathbf{x} \neq \mathbf{0}$ is the desired vector. If not, start over with another random vector $\mathbf{y}$.

Note that most steps only require a vector matrix multiplication.

It should be noted that Step 4 in the above algorithm can actually be used to find the minimal polynomial of a matrix. Let us illustrate this on a small example over $\mathbb{R}$.

Example 7.7.2 Let $A = \begin{pmatrix} 3 & 0 & 0 & 0 \\ 0 & 2 & 1 & 0 \\ 0 & 0 & 2 & 0 \\ 0 & 0 & 0 & 2 \end{pmatrix}$. If we take

$\mathbf{v} = \mathbf{u} = \begin{pmatrix} 1 & 1 & 1 & 1 \end{pmatrix}^T$, one easily calulates that

$$(s_j)_{j \in \mathbb{N}_0} = (\mathbf{v}^T A^j \mathbf{u})_{j \in \mathbb{N}_0} = (3^j + 3 \cdot 2^j + j \cdot 2^{j-1})_{j \in \mathbb{N}_0} =$$

$$4, 10, 25, 63, 161, 419, 1113, 3019, \ldots.$$

Setting up Equation (7.19) (with $d = 3$), we get

$$\begin{pmatrix} 4 & 10 & 25 & 63 \\ 10 & 25 & 63 & 161 \\ 25 & 63 & 161 & 419 \\ 63 & 161 & 419 & 1113 \\ 161 & 419 & 1113 & 3019 \end{pmatrix} \begin{pmatrix} m_0 \\ m_1 \\ m_2 \\ m_3 \end{pmatrix} = \mathbf{0},$$

where we are looking for a nontrivial solution. Such a solution is given by

$$\begin{pmatrix} m_0 \\ m_1 \\ m_2 \\ m_3 \end{pmatrix} = \begin{pmatrix} -12 \\ 16 \\ -7 \\ 1 \end{pmatrix},$$

and indeed the minimal polynomial for A is
$m_A(z) = -12 + 16z - 7z^2 + z^3 = (z-3)(z-2)^2$. This works for most choices of $\mathbf{u}$ and $\mathbf{v}$. What one needs to avoid is that certain eigenspaces are overlooked by $\mathbf{u}$ and $\mathbf{v}$. For instance, if $\mathbf{v}$ is an eigenvector of A at eigenvalue 3 (and thus $\mathbf{v}$ does not have any component in the generalized eigenspace with eigenvalue 2), one would find $m(z) = z - 3$ (as $s_j = u_1 3^j v_1$, in this case). Finally, note that we set $d = 3$, thus making use of some advance knowledge that in general one cannot count on. In an algorithm due to Elwyn R. Berlekamp and James L. Massey, which is beyond the scope of this book, one does not need to know the degree in advance.

Remember that finding u and v so that $u^2 - v^2 = (u - v)(u + v)$ is a multiple of n does not automatically lead to the desired factorization $n = pq$. Thus one needs to repeat the above steps several times. Using the ideas presented here, the 232-digit number RSA-768 was factored in 2009. It took two years to do the factorization and involved more than 10^{20} operations (performed on several computer clusters located in different countries).

7.8   Quantum computation and positive maps

Quantum computing provides another threat to RSA encryption. In a quantum computer information is stored in characteristics of physical particles. The elementary objects in quantum computing are called *qubits* as opposed to bits in a classical computer. A bit can take value 0 or 1 (on or off) while a qubit is mathematically described by a unit vector in $\mathbb{C}^2$, thus

$\begin{pmatrix} x_1 \\ x_2 \end{pmatrix} \in \mathbb{C}^2$ with $|x_1|^2 + |x_2|^2 = 1$. The numbers $|x_1|^2$ and $|x_2|^2$ are thought of as probabilities, with $|x_1|^2$ the probability the vector is in state $|0\rangle = \begin{pmatrix} 1 \\ 0 \end{pmatrix}$ and $|x_2|^2$ the probability the vector is in state $|1\rangle = \begin{pmatrix} 0 \\ 1 \end{pmatrix}$. While in quantum computing vectors are often denoted in "ket" notation $|\mathbf{x}\rangle$, we will stick to the notation as used in the rest of this book.

What makes quantum computing powerful is that many computations can be done simultaneously, though the outcome can only be measured with a certain possibility. One of the results that gave the development of a quantum computer a strong impulse of energy is the algorithm that Peter Shor developed to factor an integer in polynomial time with the help of a quantum computer. Thus, if a powerful enough quantum computer exists, the RSA encryption method will need to be rethought as the code can now be broken. Currently, the quantum computers that exists are very small. There are still significant problems in managing a large number of physical particles in close proximity of one another without destroying the stored information. Time will tell whether these physical problems can be overcome.

The mathematical development of quantum computing leads to many interesting linear algebra questions. In this section we will discuss one of these, the so-called separability problem.

We have seen that if $A \in \mathbb{C}^{n \times n}$ and $B \in \mathbb{C}^{m \times m}$ are positive semidefinite, then so is $A \otimes B \in \mathbb{C}^{(n+m) \times (n+m)}$. If we take sums of such tensor products then the resulting matrix is also positive semidefinite. The separability question is the reverse question: given $M \in \mathbb{C}^{(n+m) \times (n+m)}$, how can I determine whether M can be written as $M = \sum_{i=1}^{k} A_i \otimes B_i$ with $A_1, \ldots, A_k \in \mathbb{C}^{n \times n}$ and $B_1, \ldots, B_k \in \mathbb{C}^{m \times m}$ all positive semidefinite? When M is of this form, we say that M is (n, m)-*separable*. Positive semidefinite matrices that are not separable are called *entangled*. These entangled matrices provide in some sense the avenue to perform quantum (thus, non-classical) computations and therefore represent the power of quantum computing. Before we discuss the separablility problem in more detail, let us develop some more terminology.

As before, we let H_k be the vector space over $\mathbb{R}$ of $k \times k$ self-adjoint complex matrices. Thus $\dim_\mathbb{R} H_k = k^2$. A subset $\mathcal{C}$ of a vector space V over $\mathbb{R}$ is called a *(convex) cone* if

(i) $\mathbf{v} \in \mathcal{C}$ and $c \geq 0$ implies $c\mathbf{v} \in \mathcal{C}$, and

(ii) $\mathbf{v}, \mathbf{w} \in \mathcal{C}$ implies $\mathbf{v} + \mathbf{w} \in \mathcal{C}$.

For example,

$$\text{PSD}_k := \{A \in H_k : A \text{ is positive semidefinite}\}$$

is a cone in H_k. If the vector space V has a norm $\| \cdot \|$, then we say that the cone $\mathcal{C}$ is *closed* if

$$A_n \in \mathcal{C}, n = 1, 2, \ldots, \quad A \in V, \text{ and } \lim_{n \to \infty} \|A_n - A\| = 0 \text{ imply that } A \in \mathcal{C}.$$

The vector space H_k has an inner product

$$\langle A, B \rangle = \text{tr}(AB),$$

and therefore also an induced norm.

Proposition 7.8.1 *The set* PSD_k *is a closed cone in* H_k.

Proof. Let $A_n = (a_{ij}^{(n)})_{i,j=1}^k \in \text{PSD}_k$, $n = 1, 2, \ldots$, and $A = (a_{ij})_{i,j=1}^n \in H_k$ be so that $\lim_{n \to \infty} \|A_n - A\| = 0$. Then $\lim_{n \to \infty} a_{ij}^{(n)} = a_{ij}$ for all $1 \le i, j \le n$. If we let $\mathbf{x} \in \mathbb{C}^k$, then $\langle A_n \mathbf{x}, \mathbf{x} \rangle \ge 0$. Also $\lim_{n \to \infty} \langle A_n \mathbf{x}, \mathbf{x} \rangle = \langle A\mathbf{x}, \mathbf{x} \rangle$, and thus $\langle A\mathbf{x}, \mathbf{x} \rangle \ge 0$. As this is true for every $\mathbf{x} \in \mathbb{C}^k$, we obtain that $A \in \text{PSD}_k$. □

In H_{nm} we define

$$\text{SEP}_{n,m} = \{M \in H_{nm} : \text{ there exist } k \in \mathbb{N}, A_i \in \text{PSD}_n, B_i \in \text{PSD}_m$$

$$\text{so that } M = \sum_{i=1}^k A_i \otimes B_i\}.$$

It is easy to see that $\text{SEP}_{n,m}$ is a cone. It is actually a closed cone, but we will not provide the proof as it requires more analysis results than we are covering here. We next provide a first way of seeing how some elements of PSD_{nm} do not lie in $\text{SEP}_{n,m}$.

Proposition 7.8.2 *Let* $M \in \text{SEP}_{n,m}$, *and let us write* $M = (M_{ij})_{i,j=1}^n$ *where* $M_{ij} \in \mathbb{C}^{m \times m}$. *Then* $M^\Gamma := (M_{ij}^T) \in \text{SEP}_{n,m} \subseteq \text{PSD}_{nm}$.

Proof. As $M \in \text{SEP}_{n,m}$ we have that there exist $k \in \mathbb{N}$, and $A_i \in \text{PSD}_n, B_i \in \text{PSD}_m$, $i = 1, \ldots, k$, so that $M = (M_{ij})_{i,j=1}^n = \sum_{r=1}^k A_r \otimes B_r$. Notice that $M_{ij} = \sum_{r=1}^k a_{ij}^{(r)} B_r$, where $a_{ij}^{(r)}$ is the (i,j)th entry of A_r. But then $M_{ij}^T = \sum_{r=1}^k a_{ij}^{(r)} B_r^T$, and thus $M^\Gamma = \sum_{r=1}^k A_r \otimes B_r^T \in \text{SEP}_{n,m} \subseteq \text{PSD}_{nm}$, as the transpose of a positive semidefinite matrix B_i is also positive semidefinite. □

When $M^\Gamma \in \mathrm{PSD}_{nm}$, we say that M "passes the Peres test." Asher Peres discovered Proposition 7.8.2 in 1996. In addition, the map $M \mapsto M^\Gamma$ is referred to as taking the *partial transpose*.

Example 7.8.3 The matrix

$$M = \left(\begin{array}{cc|cc} 1 & 0 & 0 & 1 \\ 0 & 0 & 0 & 0 \\ \hline 0 & 0 & 0 & 0 \\ 1 & 0 & 0 & 1 \end{array}\right)$$

is not $(2,2)$-separable, as

$$M^\Gamma = \left(\begin{array}{cc|cc} 1 & 0 & 0 & 0 \\ 0 & 0 & 1 & 0 \\ \hline 0 & 1 & 0 & 0 \\ 0 & 0 & 0 & 1 \end{array}\right)$$

is not positive semidefinite. Thus M does not pass the Peres test.

Proposition 7.8.2 relies on the fact that taking the transpose maps PSD_k into PSD_k. For other maps that have this property the same test can be applied as well. We call a linear map $\Phi : \mathbb{C}^{m \times m} \to \mathbb{C}^{l \times l}$ *positive* if $\Phi(\mathrm{PSD}_m) \subseteq \mathrm{PSD}_l$. Thus, Φ is positive if it maps positive semidefinite matrices to positive semidefinite matrices.

Example 7.8.4 Let $S_i \in \mathbb{C}^{l \times m}$, $i = 1, \ldots, k$. Then

$$\Phi(X) = \sum_{i=1}^{k} S_i X S_i^* \tag{7.20}$$

is an example of a positive map. Indeed, if $X \in \mathrm{PSD}_m$, we have that $S_i X S_i^* \in \mathrm{PSD}_l$, $i = 1, \ldots, k$, and thus $\Phi(X) \in \mathrm{PSD}_l$. As a special case, we can take $S_i = \mathbf{e}_i^* \in \mathbb{C}^{1 \times m}$, $i = 1, \ldots, m$. Then

$$\Phi(X) = \sum_{i=1}^{m} \mathbf{e}_i^* X \mathbf{e}_i = \mathrm{tr} X.$$

Thus taking the trace is a positive map. If in addition, we let $T_i \in \mathbb{C}^{l \times m}$, $i = 1, \ldots, r$, then

$$\Phi(X) = \sum_{i=1}^{k} S_i X S_i^* + \sum_{i=1}^{r} T_i X^T T_i^* \tag{7.21}$$

defines a positive map.

We can now state a more general version of Proposition 7.8.2.

Proposition 7.8.5 *Let* $M \in \mathrm{SEP}_{n,m}$, *and let us write* $M = (M_{ij})_{i,j=1}^{n}$ *where* $M_{ij} \in \mathbb{C}^{m \times m}$. *Let* $\Phi : \mathbb{C}^{m \times m} \to \mathbb{C}^{l \times l}$ *be a positive map. Then*

$$(id_{\mathbb{C}^{n \times n}} \otimes \Phi)(M) = (\Phi(M_{ij}))_{i,j=1}^{n} \in \mathrm{SEP}_{n,l} \subseteq \mathrm{PSD}_{nl}.$$

Proof. As $M \in \mathrm{SEP}_{n,m}$ we have that there exist $k \in \mathbb{N}$, and $A_i \in \mathrm{PSD}_n, B_i \in \mathrm{PSD}_m, i = 1, \ldots, k$, so that $M = (M_{ij})_{i,j=1}^{n} = \sum_{r=1}^{k} A_r \otimes B_r$. Notice that $M_{ij} = \sum_{r=1}^{k} a_{ij}^{(r)} B_r$, where $a_{ij}^{(r)}$ is the (i,j)th entry of A_r. But then $\Phi(M_{ij}) = \sum_{r=1}^{k} a_{ij}^{(r)} \Phi(B_r)$, and thus

$$(id_{\mathbb{C}^{n \times n}} \otimes \Phi)(M) = \sum_{r=1}^{k} A_r \otimes \Phi(B_r) \in \mathrm{SEP}_{n,l} \subseteq \mathrm{PSD}_{nl}.$$

$\square$

There are positive maps Φ for which $id_{\mathbb{C}^{k \times k}} \otimes \Phi$ is positive for every $k \in \mathbb{N}$. We call such maps *completely positive*. These completely positive maps are useful in several contexts, however they are unable to identify $M \in \mathrm{PSD}_{mn}$ that are not separable, as $(id_{\mathbb{C}^{n \times n}} \otimes \Phi)(M)$ will in this case always be positive semidefinite. The completely positive maps are characterized in the following result, due to Man-Duen Choi.

Theorem 7.8.6 *Let* $\Phi : \mathbb{C}^{m \times m} \to \mathbb{C}^{l \times l}$ *be a linear map. Let* $\{E_{ij} : 1 \le i, j \le m\}$ *be the standard basis of* $\mathbb{C}^{m \times m}$. *Then the following are equivalent.*

(i) Φ *is completely positive.*

(ii) $id_{\mathbb{C}^{m \times m}} \otimes \Phi$ *is positive.*

(iii) *The matrix* $(\Phi(E_{ij}))_{i,j=1}^{m}$ *is positive semidefinite.*

(iv) *There exist* $S_r \in \mathbb{C}^{l \times m}, r = 1, \ldots, s$, *so that* $\Phi(X) = \sum_{r=1}^{s} S_r X S_r^*$.

When one (and thus all) of (i)–(iv) hold, then s *in (iv) can be chosen to be at most* ml.

Proof. (i) $\to$ (ii) is trivial, as when $id_{\mathbb{C}^{k \times k}} \otimes \Phi$ is positive for all $k \in \mathbb{N}$, then it is certainly positive for $k = m$.

(ii) $\to$ (iii): The matrix $H = (E_{ij})_{i,j=1}^{m} \in \mathbb{C}^{m^2 \times m^2}$ is easily seen to be positive semidefinite. As $id_{\mathbb{C}^{m\times m}} \otimes \Phi$ is positive, we thus get that $(id_{\mathbb{C}^{m\times m}} \otimes \Phi)(H) = (\Phi(E_{ij}))_{i,j=1}^m$ is positive semidefinite.

(iii) $\to$ (iv): Since $M = (\Phi(E_{ij}))_{i,j=1}^m \in \mathbb{C}^{ml\times ml}$ is positive semidefinite, we can find vectors $\mathbf{v}_r \in \mathbb{C}^{ml}$, $r = 1, \ldots, ml$, so that $M = \sum_{r=1}^{lm} \mathbf{v}_r\mathbf{v}_r^*$. Write the vectors $\mathbf{v}_r$ as

$$\mathbf{v}_r = \begin{pmatrix} \mathbf{v}_{r1} \\ \vdots \\ \mathbf{v}_{rm} \end{pmatrix}, \text{ where } \mathbf{v}_{r1}, \ldots, \mathbf{v}_{rm} \in \mathbb{C}^l.$$

We now have that

$$M_{ij} = \Phi(E_{ij}) = \sum_{r=1}^{ml} \mathbf{v}_{ri}\mathbf{v}_{rj}^*.$$

If we introduce $S_r = \begin{pmatrix} \mathbf{v}_{r1} & \cdots & \mathbf{v}_{rm} \end{pmatrix}$, we have that $\mathbf{v}_{ri}\mathbf{v}_{rj}^* = S_r E_{ij} S_r^*$. Thus we find that $M_{ij} = \Phi(E_{ij}) = \sum_{r=1}^{lm} \mathbf{v}_{ri}\mathbf{v}_{rj}^* = \sum_{r=1}^{ml} S_r E_{ij} S_r^*$. As any X is a linear combination of the basis elements E_{ij}, we thus find that $\Phi(X) = \sum_{i=1}^{ml} S_i X S_i^*$.

(iv) $\to$ (i): When $\Phi(X) = \sum_{r=1}^s S_r X S_r^*$, we may write for $M = (M_{ij})_{i,j=1}^k$ with $M_{ij} \in \mathbb{C}^{m\times m}$,

$$(id_{\mathbb{C}^{k\times k}} \otimes \Phi)(M) = (\sum_{r=1}^s S_r M_{ij} S_r^*)_{i,j=1}^m = \sum_{r=1}^n (I_k \otimes S_r) M (I_k \otimes S_r^*).$$

When M is positive semidefinite, then so is $(I_k \otimes S_r) M (I_k \otimes S_r)^*$, and thus also $\Phi(M)$ is positive semidefinite. □

Thus completely positive maps are well-characterized in the sense that there is a simple way to check that a linear map is completely positive, as well as that it is easy to generate all completely positive maps. The set of positive maps (which actually forms a cone) is not that well understood. First of all, it is typically not so easy to check whether a map is positive, and secondly there is not a way to generate all positive maps. Let us end this section with a positive map that is not completely positive, also due to Man-Duen Choi.

Example 7.8.7 Let $\Phi : \mathbb{C}^{3\times 3} \to \mathbb{C}^{3\times 3}$ be defined by

$$\Phi((a_{ij})_{i,j=1}^3) = \begin{pmatrix} a_{11} + 2a_{22} & -a_{12} & -a_{13} \\ -a_{21} & a_{22} + 2a_{33} & -a_{23} \\ -a_{31} & -a_{32} & a_{33} + 2a_{11} \end{pmatrix}.$$

Then

$$(\Phi(E_{ij})_{i,j=1}^3) = \begin{pmatrix} 1 & 0 & 0 & 0 & -1 & 0 & 0 & 0 & -1 \\ 0 & 0 & 0 & 0 & 0 & 0 & 0 & 0 & 0 \\ 0 & 0 & 2 & 0 & 0 & 0 & 0 & 0 & 0 \\ 0 & 0 & 0 & 2 & 0 & 0 & 0 & 0 & 0 \\ -1 & 0 & 0 & 0 & 1 & 0 & 0 & 0 & -1 \\ 0 & 0 & 0 & 0 & 0 & 0 & 0 & 0 & 0 \\ 0 & 0 & 0 & 0 & 0 & 0 & 0 & 0 & 0 \\ 0 & 0 & 0 & 0 & 0 & 0 & 0 & 2 & 0 \\ -1 & 0 & 0 & 0 & -1 & 0 & 0 & 0 & 1 \end{pmatrix},$$

which is not positive semidefinite. Thus, by Proposition 7.8.6, Φ is not completely positive. To show that Φ is positive, it suffices to show that $\Phi(\mathbf{xx}^*)$ is positive semidefinite for all $\mathbf{x} = (x_i)_{i=1}^3 \in \mathbb{C}^3$ as every positive semidefinite is a sum of positive semidefinite rank 1 matrices $\mathbf{xx}^*$. To show that $\Phi(\mathbf{xx}^*)$ is positive semidefinite, we need to show that $\mathbf{y}^*\Phi(\mathbf{xx}^*)\mathbf{y} \geq 0$ for all $\mathbf{y} = (y_i)_{i=1}^3 \in \mathbb{C}^3$. We show the proof in case $|x_3| \leq |x_2|$. In this case we observe that

$$\mathbf{y}^*\Phi(\mathbf{xx}^*)\mathbf{y} = |x_1\bar{y}_1 - x_2\bar{y}_2 + x_3\bar{y}_3|^2 + 2|x_3|^2|y_2|^2+$$

$$2(|x_2| - |x_3|^2)|y_1|^2 + 2|x_1y_3 - x_3y_1|^2 \geq 0.$$

The proof for the case $|x_2| \leq |x_3|$ is similar.

7.9 Exercises

Exercise 7.9.1 Let $p(n)$ be a polynomial in n of degree k, and let $\lambda \in \mathbb{C}$ be of modulus greater than one. Show that $\lim_{n\to\infty} \frac{p(n)}{\lambda^n} = 0$. (Hint: write $|\lambda| = 1 + \epsilon$, $\epsilon > 0$, and use the binomial formula to give that $|\lambda^n| = \sum_{j=0}^n \binom{n}{j}\epsilon^j$, which for n large enough can be bounded below by a polynomial of degree greater than k.)

Exercise 7.9.2 Let $A = (a_{ij})_{i,j=1}^n \in \mathbb{R}^{n\times n}$. Let A be *column-stochastic*, which means that $a_{ij} \geq 0$ for all $i, j = 1, \ldots, n$, and $\sum_{i=1}^n a_{ij} = 1$, $j = 1, \ldots, n$.

(i) Show that 1 is an eigenvalue of A.

(ii) Show that A^m is column-stochastic for all $m \in \mathbb{N}$. (Hint: use that $\mathbf{e}A = \mathbf{e}$.)

(iii) Show that for every $\mathbf{x}, \mathbf{y} \in \mathbb{R}^n$ we have that $|\mathbf{y}^T A^m \mathbf{x}| \leq (\sum_{j=1}^n |x_j|)(\sum_{j=1}^n |y_j|)$ for all $m \in \mathbb{N}$. In particular, the sequence $\{\mathbf{y}^T A^m \mathbf{x}\}_{m \in \mathbb{N}}$ is bounded.

(iv) Show that A cannot have Jordan blocks at 1 of size greater than 1. (Hint: use that when $k > 1$ some of the entries of $J_k(1)^m$ do not stay bounded as $m \to \infty$. With this observation, find a contradiction with the previous part.)

(v) Show that if $\mathbf{x}A = \lambda \mathbf{x}$, for some $\mathbf{x} \neq \mathbf{0}$, then $|\lambda| \leq 1$.

(vi) For a vector $\mathbf{v} = (v_i)_{i=1}^n$ we define $|\mathbf{v}| = (|v_i|)_{i=1}^n$. Show that if λ is an eigenvalue of A with $|\lambda| = 1$, and $\mathbf{x}A = \lambda \mathbf{x}$, then $\mathbf{y} := |\mathbf{x}|A - |\mathbf{x}|$ has all nonnegative entries.

For the remainder of this exercise, assume that A only has positive entries; thus $a_{ij} > 0$ for all $i, j = 1, \ldots, n$.

(vii) Show that $\mathbf{y} = \mathbf{0}$. (Hint: put $\mathbf{z} = |\mathbf{x}|A$, and show that $\mathbf{y} \neq \mathbf{0}$ implies that $\mathbf{z}A - \mathbf{z}$ has all positive entries. The latter can be shown to contradict $\sum_{i=1}^n a_{ij} = 1$, $j = 1, \ldots, n$.)

(viii) Show that if $\mathbf{x}A = \lambda \mathbf{x}$ with $|\lambda| = 1$, then $\mathbf{x}$ is a multiple of $\mathbf{e}$ and $\lambda = 1$. (Hint: first show that all entries of $\mathbf{x}$ have the same modulus.)

(ix) Conclude that we can apply the power method. Starting with a vector $\mathbf{v}_0$ with positive entries, show that there is a vector $\mathbf{w}$ with positive entries so that $A\mathbf{w} = \mathbf{w}$. In addition, show that $\mathbf{w}$ is unique when we require in addition that $\mathbf{e}^T\mathbf{w} = 1$.

Exercise 7.9.3 Let $\|\cdot\|$ be a norm on $\mathbb{C}^{n \times n}$, and let $A \in \mathbb{C}^{n \times n}$. Show that

$$\rho(A) = \lim_{k \to \infty} \|A^k\|^{\frac{1}{k}}, \tag{7.22}$$

where $\rho(\cdot)$ is the spectral radius. (Hint: use that for any $\epsilon > 0$ the spectral radius of $\frac{1}{\rho(A)+\epsilon}A$ is less than one, and apply Corollary 7.2.4.)

Exercise 7.9.4 Let $A = (a_{ij})_{i,j=1}^n, B = (b_{ij})_{i,j=1}^n \in \mathbb{C}^{n \times n}$ so that $|a_{ij}| \leq b_{ij}$ for $i, j = 1, \ldots, n$. Show that $\rho(A) \leq \rho(B)$. (Hint: use (7.22) with the *Frobenius norm* $\|M\| = \sqrt{\sum_{i,j=1}^n |m_{ij}|^2}$.)

Exercise 7.9.5 Show that if $\{\mathbf{u}_1, \ldots, \mathbf{u}_m\}$ and $\{\mathbf{v}_1, \ldots, \mathbf{v}_m\}$ are orthonormal sets, then the *coherence* $\mu := \max_{i,j} |\langle \mathbf{u}_i, \mathbf{v}_j \rangle|$, satisfies $\frac{1}{\sqrt{m}} \leq \mu \leq 1$.

Exercise 7.9.6 Show that if A has the property that every $2s$ columns are linearly independent, then the equation $Ax = b$ can have at most one solution x with at most s nonzero entries.

Exercise 7.9.7 Let $A = (a_{ij})_{i,j=1}^n$. Show that for all permutations σ on $\{1, \ldots, , n\}$ we have $a_{1,\sigma(1)} a_{2,\sigma(2)} \cdots a_{n,\sigma(n)} = 0$ if and only if there exist r $(1 \leq r \leq n)$ rows and $n + 1 - r$ columns in A so that the entries they have in common are all 0.

Exercise 7.9.8 We say that $A = (a_{ij})_{i,j=1}^n \in \mathbb{R}^{n \times n}$ is *row-stochastic* if A^T is columns-stochastic. We call A *doubly stochastic* if A is both column- and row-stochastic. The matrix $P = (p_{ij})_{i,j=1}^n$ is called a *permutation matrix* if every row and column of P has exactly one entry equal to 1 and all the others equal to zero.

(i) Show that a permutation matrix is doubly stochastic.

(ii) Show that if A is a doubly stochastic matrix, then there exists a permutation σ on $\{1, \ldots, , n\}$, so that $a_{1,\sigma(1)} a_{2,\sigma(2)} \cdots a_{n,\sigma(n)} \neq 0$.

(iii) Let σ be as in the previous part, and put $\alpha = \min_{j=1,\ldots,n} a_{j,\sigma(j)} (> 0)$, and let P_σ be the permutation matrix with a 1 in positions $(1, \sigma(1)), \ldots, (n, \sigma(n))$ and zeros elsewhere. Show that either A is a permutation matrix, or $\frac{1}{1-\alpha}(A - \alpha P_\sigma)$ is a doubly stochastic matrix with fewer nonzero entries than A.

(iv) Prove

Theorem 7.9.9 *(Birkhoff) Let A be doubly stochastic. Then there exists a $k \in \mathbb{N}$, permutation matrices $P_1, \ldots, P_k$ and positive numbers $\alpha_1, \ldots, \alpha_k$ so that*

$$A = \alpha_1 P_1 + \cdots + \alpha_k P_k, \quad \sum_{j=1}^k \alpha_j = 1.$$

In other words, every doubly stochastic matrix is a convex combination of permutation matrices.

(Hint: Use induction on the number of nonzero entries of A.)

Exercise 7.9.10 Write the matrix $\begin{pmatrix} 1/6 & 1/2 & 1/3 \\ 7/12 & 0 & 5/12 \\ 1/4 & 1/2 & 1/4 \end{pmatrix}$ as a convex combination of permutation matrices.

Exercise 7.9.11 (a) Show that

$$\min \operatorname{rank} \begin{pmatrix} A & ? \\ B & C \end{pmatrix} = \operatorname{rank} \begin{pmatrix} A \\ B \end{pmatrix} + \operatorname{rank} \begin{pmatrix} B & C \end{pmatrix}.$$

(b) Show that the lower triangular partial matrix

$$\mathcal{A} = \begin{pmatrix} A_{11} & & ? \\ \vdots & \ddots & \\ A_{n1} & \cdots & A_{nn} \end{pmatrix}$$

has minimal rank $\min \operatorname{rank} \mathcal{A}$ equal to

$$\sum_{i=1}^{n} \operatorname{rank} \begin{pmatrix} A_{i1} & \cdots & A_{ii} \\ \vdots & & \vdots \\ A_{n1} & \cdots & A_{ni} \end{pmatrix} - \sum_{i=1}^{n-1} \operatorname{rank} \begin{pmatrix} A_{i+1,1} & \cdots & A_{i+1,i} \\ \vdots & & \vdots \\ A_{n1} & \cdots & A_{ni} \end{pmatrix}. \quad (7.23)$$

Exercise 7.9.12 Show that all minimal rank completions of

$$\begin{pmatrix} ? & ? & ? \\ 1 & 0 & ? \\ 0 & 1 & 1 \end{pmatrix}$$

are

$$\begin{pmatrix} x_1 & x_2 & x_1 x_3 + x_2 \\ 1 & 0 & x_3 \\ 0 & 1 & 1 \end{pmatrix}.$$

Exercise 7.9.13 Consider the partial matrix

$$A = \begin{pmatrix} 1 & ? & ? \\ ? & 1 & ? \\ -1 & ? & 1 \end{pmatrix}.$$

Show that there exists a completion of A that is a Toeplitz matrix of rank 1, but that such a completion cannot be chosen to be real.

Exercise 7.9.14 Consider the $n \times n$ tri-diagonal Toeplitz matrix

$$A_n = \begin{pmatrix} 2 & -1 & 0 & \cdots & 0 \\ -1 & 2 & -1 & \cdots & 0 \\ \vdots & \ddots & \ddots & \ddots & \vdots \\ 0 & \cdots & -1 & 2 & -1 \\ 0 & \cdots & 0 & -1 & 2 \end{pmatrix}.$$

Show that $\lambda_j = 2 - 2\cos(j\theta)$, $j = 1, \ldots, n$, where $\theta = \frac{\pi}{n+1}$, are the eigenvalues. In addition, an eigenvector associated with λ_j is

$$\mathbf{v}_j = \begin{pmatrix} \sin(j\theta) \\ \sin(2j\theta) \\ \vdots \\ \sin(nj\theta) \end{pmatrix}.$$

Exercise 7.9.15 Let $A = (a_{ij})_{i,j=1}^n \in \mathbb{C}^{n \times n}$ be given.

(a) Let $U = \begin{pmatrix} 1 & 0 \\ 0 & U_1 \end{pmatrix} \in \mathbb{C}^{n \times n}$, with $U_1 \in \mathbb{C}^{(n-1) \times (n-1)}$ a unitary matrix chosen so that

$$U_1 \begin{pmatrix} a_{21} \\ a_{31} \\ \vdots \\ a_{n1} \end{pmatrix} = \begin{pmatrix} \sigma \\ 0 \\ \vdots \\ 0 \end{pmatrix}, \quad \sigma = \sqrt{\sum_{j=2}^n |a_{j1}|^2}.$$

Show that UAU^* has the form

$$UAU^* = \begin{pmatrix} a_{11} & * & * & \cdots & * \\ \sigma & * & * & \cdots & * \\ 0 & * & * & \cdots & * \\ \vdots & \vdots & \vdots & & \vdots \\ 0 & * & * & \cdots & * \end{pmatrix} = \begin{pmatrix} a_{11} & * \\ \sigma \mathbf{e}_1 & A_1 \end{pmatrix}.$$

(b) Show that there exists a unitary V so that VAV^* is upper Hessenberg. (Hint: after part (a), find a unitary $U_2 = \begin{pmatrix} 1 & 0 \\ 0 & * \end{pmatrix}$ so that $U_2 A_1 U_2^*$ has the form $\begin{pmatrix} * & * \\ \sigma_2 \mathbf{e}_1 & A_2 \end{pmatrix}$, and observe that

$$\hat{A} = \begin{pmatrix} 1 & 0 \\ 0 & U_2 \end{pmatrix} \begin{pmatrix} 1 & 0 \\ 0 & U_1 \end{pmatrix} A \begin{pmatrix} 1 & 0 \\ 0 & U_1^* \end{pmatrix} \begin{pmatrix} 1 & 0 \\ 0 & U_2^* \end{pmatrix}$$

has now zeros in positions $(2,1), \ldots, (n,1), (3,2), \ldots, (n,2)$. Continue the process.)

Remark. If one puts a matrix in upper Hessenberg form before starting the QR algorithm, it (in general) speeds up the convergence of the QR algorithm, so this is standard practice when numerically finding eigenvalues.

Exercise 7.9.16 The *adjacency matrix* A_G of a graph $G = (V, E)$ is an $n \times n$ matrix, where $n = |V|$ is the number of vertices of the graph, and the entry (i, j) equals 1 when $\{i, j\}$ is an edge, and 0 otherwise. For instance, the graph in Figure 7.6 has adjacency matrix

$$\begin{pmatrix} 0 & 1 & 0 & 0 & 1 & 0 \\ 1 & 0 & 1 & 0 & 1 & 0 \\ 0 & 1 & 0 & 1 & 0 & 0 \\ 0 & 0 & 1 & 0 & 1 & 1 \\ 1 & 1 & 0 & 1 & 0 & 0 \\ 0 & 0 & 0 & 1 & 0 & 0 \end{pmatrix}.$$

The adjacency matrix is a symmetric real matrix. Some properties of graphs can be studied by studying associated matrices. In this exercise we show this for the so-called *chromatic number* $\chi(G)$ of a graph G. It is defined as follows. A k-*coloring* of a graph is a function $c : V \to \{1, \dots, k\}$ so that $c(i) \neq c(j)$ whenever $\{i, j\} \in E$. Thus, there are k colors and adjacent vertices should not be given the same color. The smallest number k so that G has a k-coloring is defined to be the chromatic number $\chi(G)$ of the graph G.

(a) Find the chromatic number of the graph in Figure 7.6.

(b) The *degree* d_i of a vertex i is the number of vertices it is adjacent to. For instance, for the graph in Figure 7.6 we have that the degree of vertex 1 is 2, and the degree of vertex 6 is 1. Let $\mathbf{e} = \begin{pmatrix} 1 & \cdots & 1 \end{pmatrix}^T \in \mathbb{R}^n$. Show that $\mathbf{e}^T A_G \mathbf{e} = \sum_{i \in V} d_i$.

(c) For a real number x let $\lfloor x \rfloor$ denote the largest integer $\leq x$. For instance, $\lfloor \pi \rfloor = 3$, $\lfloor -\pi \rfloor = -4$, $\lfloor 5 \rfloor = 5$. Let $\alpha = \lambda_{\max}(A_G)$ be the largest eigenvalue of the adjacency matrix of G. Show that G must have a vertex of degree at most $\lfloor \alpha \rfloor$. (Hint: use Exercise 5.7.21(b).)

(d) Show that

$$\chi(G) \leq \lfloor \lambda_{\max}(A_G) \rfloor + 1, \tag{7.24}$$

which is a result due to Herbert S. Wilf. (Hint: use induction and Exercise 5.7.21(c).)

Exercise 7.9.17 Let

$$\rho_\alpha = \frac{1}{7} \begin{pmatrix} \frac{2}{3} & 0 & 0 & 0 & \frac{2}{3} & 0 & 0 & 0 & \frac{2}{3} \\ 0 & \frac{\alpha}{3} & 0 & 0 & 0 & 0 & 0 & 0 & 0 \\ 0 & 0 & \frac{5-\alpha}{3} & 0 & 0 & 0 & 0 & 0 & 0 \\ 0 & 0 & 0 & \frac{5-\alpha}{3} & 0 & 0 & 0 & 0 & 0 \\ \frac{2}{3} & 0 & 0 & 0 & \frac{2}{3} & 0 & 0 & 0 & \frac{2}{3} \\ 0 & 0 & 0 & 0 & 0 & \frac{\alpha}{3} & 0 & 0 & 0 \\ 0 & 0 & 0 & 0 & 0 & 0 & \frac{\alpha}{3} & 0 & 0 \\ 0 & 0 & 0 & 0 & 0 & 0 & 0 & \frac{5-\alpha}{3} & 0 \\ \frac{2}{3} & 0 & 0 & 0 & \frac{2}{3} & 0 & 0 & 0 & \frac{2}{3} \end{pmatrix},$$

where $0 \leq \alpha \leq 5$. We want to investigate when ρ_α is 3×3 separable.

(a) Show that ρ_α passes the Peres test if and only if $1 \leq \alpha \leq 4$.

(b) Let

$$Z = \begin{pmatrix} 1 & 0 & 0 & 0 & -1 & 0 & 0 & 0 & -1 \\ 0 & 0 & 0 & 0 & 0 & 0 & 0 & 0 & 0 \\ 0 & 0 & 2 & 0 & 0 & 0 & 0 & 0 & 0 \\ 0 & 0 & 0 & 2 & 0 & 0 & 0 & 0 & 0 \\ -1 & 0 & 0 & 0 & 1 & 0 & 0 & 0 & -1 \\ 0 & 0 & 0 & 0 & 0 & 0 & 0 & 0 & 0 \\ 0 & 0 & 0 & 0 & 0 & 0 & 0 & 0 & 0 \\ 0 & 0 & 0 & 0 & 0 & 0 & 0 & 2 & 0 \\ -1 & 0 & 0 & 0 & -1 & 0 & 0 & 0 & 1 \end{pmatrix}.$$

Show that for $x, y \in \mathbb{C}^3$ we have that $(x \otimes y)^* Z (x \otimes y) \geq 0$.

(c) Show that $\operatorname{tr}(\rho_\alpha Z) = \frac{1}{7}(3 - \alpha)$, and conclude that ρ_α is not 3×3 separable for $3 < \alpha \leq 5$.

(d) *(Honors)* Show that ρ_α is not 3×3 separable for $0 \leq \alpha < 2$.

(e) *(Honors)* Show that ρ_α is 3×3 separable for $2 \leq \alpha \leq 3$.

Exercise 7.9.18 *(Honors)* A matrix is $2 \times 2 \times 2$ separable if it lies in the cone generated by matrices of the form $A \otimes B \otimes C$ with $A, B, C \in \mathrm{PSD}_2$. Put

$$R = I - x_1 x_1^* - x_2 x_2^* - x_3 x_3^* - x_4 x_4^*,$$

where

$$x_1 = \begin{pmatrix} 1 \\ 0 \end{pmatrix} \otimes \begin{pmatrix} 0 \\ 1 \end{pmatrix} \otimes \begin{pmatrix} \frac{1}{2}\sqrt{2} \\ \frac{1}{2}\sqrt{2} \end{pmatrix}, \quad x_2 = \begin{pmatrix} 0 \\ 1 \end{pmatrix} \otimes \begin{pmatrix} \frac{1}{2}\sqrt{2} \\ \frac{1}{2}\sqrt{2} \end{pmatrix} \otimes \begin{pmatrix} 1 \\ 0 \end{pmatrix},$$

$$x_3 = \begin{pmatrix} \frac{1}{2}\sqrt{2} \\ \frac{1}{2}\sqrt{2} \end{pmatrix} \otimes \begin{pmatrix} 1 \\ 0 \end{pmatrix} \otimes \begin{pmatrix} 0 \\ 1 \end{pmatrix}, \quad x_4 = \begin{pmatrix} \frac{1}{2}\sqrt{2} \\ -\frac{1}{2}\sqrt{2} \end{pmatrix} \otimes \begin{pmatrix} \frac{1}{2}\sqrt{2} \\ -\frac{1}{2}\sqrt{2} \end{pmatrix} \otimes \begin{pmatrix} \frac{1}{2}\sqrt{2} \\ -\frac{1}{2}\sqrt{2} \end{pmatrix}.$$

Show that R is not $2 \times 2 \times 2$ separable.

Hint: Let

$$Z = \begin{pmatrix} 1 & -1 & -1 & 1 & -1 & 1 & 1 & -1 \\ -1 & 4 & 1 & 0 & 1 & 3 & -1 & 1 \\ -1 & 1 & 4 & 3 & 1 & -1 & 0 & 1 \\ 1 & 0 & 3 & 4 & -1 & 1 & 1 & -1 \\ -1 & 1 & 1 & -1 & 4 & 0 & 3 & 1 \\ 1 & 3 & -1 & 1 & 0 & 4 & 1 & -1 \\ 1 & -1 & 0 & 1 & 3 & 1 & 4 & -1 \\ -1 & 1 & 1 & -1 & 1 & -1 & -1 & 1 \end{pmatrix},$$

and show that trace$(RZ) = -\frac{3}{8}$ but

$$(v \otimes w \otimes z)^* Z(v \otimes w \otimes z) \geq 0,$$

for all $v, w, z \in \mathbb{C}^{2 \times 2}$.

Bibliography for Chapter 7

It is beyond the scope of this book to provide complete references for the topics discussed in this chapter. Rather, we provide just a few references, which can be a starting point for further reading on these topics. With the references in the papers and books below as well as the sources that refer to them (see the chapter "How to start your own research project" on how to look for these), we hope that you will be able to familiarize yourself in more depth with the topics of your interest.

- M. Bakonyi, H. J. Woerdeman, *Matrix completions, moments, and sums of Hermitian squares.* Princeton University Press, Princeton, NJ, 2011.

- W. W. Barrett, R. W. Forcade, A. D. Pollington, On the spectral radius of a (0,1) matrix related to Mertens' function. *Linear Algebra Appl.* 107 (1988), 151–159.

- M. Bellare, O. Goldreich, M. Sudan, Free bits, PCPs, and nonapproximability–towards tight results. *SIAM J. Comput.* 27 (1998), no. 3, 804–915.

- T. Blumensath, M. E. Davies, Iterative thresholding for sparse approximations. *J. Fourier Anal. Appl.* 14 (2008), no. 5–6, 629–654.

- R. P. Boyer and D. T. Parry, On the zeros of plane partition polynomials. *Electron. J. Combin.* 18 (2011), no. 2, Paper 30, 26 pp.

- K. Bryan and T. Leise, The $25,000,000,000 eigenvector. The linear algebra behind Google. *SIAM Rev.* 48 (2006), 569–581.

- S. Chandrasekaran, M. Gu, J. Xia and J. Zhu, A fast QR algorithm for companion matrices. *Operator Theory: Adv. Appl.*, 179 (2007), 111–143.

- M. D. Choi, Positive semidefinite biquadratic forms. *Linear Algebra and Appl.* 12 (1975), no. 2, 95–100.

- S. Foucart and H. Rauhut, *A mathematical introduction to compressive sensing*. Applied and Numerical Harmonic Analysis. Birkhäuser/ Springer, New York, 2013.

- M.X. Goemans and D.P. Williamson, Improved approximation algorithms for maximum cut and satisfiability problems using semidefinite programming, *J. ACM* 42 (1995) 1115–1145.

- K. Kaplan, Cognitech thinks it's got a better forensic tool: The firm uses complex math in video image-enhancing technology that helps in finding suspects, *Los Angeles Times*, September 5, 1994; http://articles.latimes.com/1994-09-05/business/fi-35101_1_image-enhancement.

- A. K. Lenstra and M. S. Manasse, Factoring with two large primes, *Math. Comp.* 63 (1994), no. 208, 785–798.

- P. J. Olver, Orthogonal bases and the QR algorithm, University of Minnesota, http://www.math.umn.edu/~olver/aims_/qr.pdf.

- L. Page, S. Brin, R. Motwani, T. Winograd, The PageRank citation ranking: Bringing order to the web (1999), http://ilpubs.stanford.edu:8090/422/1/1999-66.pdf.

- R. Redheffer, Eine explizit lösbare Optimierungsaufgabe. (German) Numerische Methoden bei Optimierungsaufgaben, Band 3 (Tagung, Math. Forschungsinst., Oberwolfach, (1976), pp. 213–216. *Internat. Ser. Numer. Math.*, Vol. 36, Birkhäuser, Basel, 1977.

- P. W. Shor, Polynomial-time algorithms for prime factorization and discrete logarithms on a quantum computer, *SIAM J. Comput.* 26 (1977), no. 5, 1484–1509.

- H. J. Woerdeman, Minimal rank completions for block matrices. *Linear Algebra Appl.* 121 (1989), 105–122.

- T. Zhang, J. M. Pauly, S. S. Vasanawala and M. Lustig, Coil compression for accelerated imaging with Cartesian sampling. *Magnetic Resonance in Medicine*, 69 (2013), 571–582.

How to Start Your Own Research Project

For a research problem you need

- A problem nobody solved,

- One that you will be able to make some headway on, and

- One that people are interested in.

So how do you go about finding such a problem?

In MathSciNet (a database of reviews of mathematical journal articles and books maintained by the American Mathematical Society) you can do a search. For instance, with search term "Anywhere" you can put a topic such as "Normal matrix," "QR algorithm," etc., and see what comes up. If you click on the review of a paper you can see in a box "Citations" what other papers or reviews in the database refer to this paper. Of course, very recent papers will have no or few citations, but earlier ones typically have some. The number of citations is a measure of the influence of the paper.

Of course, you can also search terms in any search engine. Some search engines, when you give them titles of papers, will indicate what other papers cite this paper. I find this a very useful feature. Again, it gives a sense of how that particular line of research is developing and how much interest there is for it.

If you want to get a sense of how hot a topic is, you can see if government agencies or private industry give grants for this line of research. For instance, in the United States the National Science Foundation (NSF) gives grants for basic research. On the NSF web page (www.nsf.gov) you can go to "Search Awards," and type terms like *eigenvalue, singular value decomposition*, etc., and see which funded grants have that term in the title or abstract. Again, it gives you an idea of what types of questions people are interested in, enough to put US tax dollars toward the research. Of course, many countries have government agencies that support research, for instance in the Netherlands it is the Nederlandse Organisatie voor Wetenschappelijk Onderzoek (NWO, www.nwo.nl). If you are searching in the Dutch language it is useful to know that "wiskunde" is the Dutch word for mathematics.

Another source for hot topics is to see what areas of mathematics receive the major prizes. The Fields medal and the Abel Prize are two well-known prestigious prizes for mathematical research, but there are many others. In addition, some of the prize winners and other well-known mathematicians started their own blogs, which are also a source for exciting ideas.

There is some time lag between finishing a paper and it appearing in a journal, as professional journals have a review process that in fact could take quite a while. But there are also so-called preprint servers where researchers can post their finished paper as soon as it is ready. One such example is the ArXiv (arxiv.org), which many mathematicians (and other scientists) use. So this is a source where you can find results of some very fresh research. ArXiv also has an option to get regular email updates on new articles in areas of your choice.

Of course, you should also leverage all your contacts. Your professor would be a good person to talk about this, or other professors at your school. In addition, don't be afraid to contact a person you do not know. It has been my experience that when you put some thought in an email message to a mathematician, a good number of them will take the effort to write back. For instance, if you would write to me and say something along the lines "I looked at your paper X, and I thought of changing the problem to Y. Would that be of interest? Has anyone looked at this?", you would probably get an answer from me. And if you don't within a few weeks, maybe just send the message again as it may just have ended up in a SPAM filter or it somehow fell off my radar screen.

Finally, let me mention that in my research I found it often useful to try out ideas numerically using MATLAB®, Maple™ or Mathematica®. In some cases I discovered patterns this way that turned out to be essential. In addition, try to write things up along the way as it will help you document what you have done, and it will lower the bar to eventually write a paper. Typically mathematical texts (such as this book) are written up using the program LaTeX (or TeX), so it is definitely useful to getting used to this freely available program. For instance, you can write up your homework using LaTeX, which will surely score some points with your professor.

It would be great if you picked up a research project. One thing about mathematical research: we will never run out of questions. In fact, when you answer a question, it often generates new ones. So, good luck, and maybe I will see you at a conference sometime when you present your result!

Answers to Exercises

Chapter 1

Exercise 1.5.1 The set of integers $\mathbb{Z}$ with the usual addition and multiplication is **not** a field. Which of the field axioms does $\mathbb{Z}$ satisfy, and which one(s) are not satisfied?

Answer: The only axiom that is not satisfied is number 10, involving the existence of a multiplicative inverse. For instance, 2 does not have a multiplicative inverse in $\mathbb{Z}$.

Exercise 1.5.2 Write down the addition and multiplication tables for $\mathbb{Z}_2$ and $\mathbb{Z}_5$. How is commutativity reflected in the tables?

Answer: Here are the tables for $\mathbb{Z}_2$ and $\mathbb{Z}_5$:

+	0	1
0	0	1
1	1	0

.	0	1
0	0	0
1	0	1

+	0	1	2	3	4
0	0	1	2	3	4
1	1	2	3	4	0
2	2	3	4	0	1
3	3	4	0	1	2
4	4	0	1	2	3

.	0	1	2	3	4
0	0	0	0	0	0
1	0	1	2	3	4
2	0	2	4	1	3
3	0	3	1	4	2
4	0	4	3	2	1

The symmetry in the tables is due to commutativity.

Exercise 1.5.3 The addition and multiplication defined in (1.4) also works when p is not prime. Write down the addition and multiplication tables for $\mathbb{Z}_4$. How can you tell from the tables that $\mathbb{Z}_4$ is **not** a field?

Answer: The tables for $\mathbb{Z}_4$ are:

+	0	1	2	3
0	0	1	2	3
1	1	2	3	0
2	2	3	0	1
3	3	0	1	2

.	0	1	2	3
0	0	0	0	0
1	0	1	2	3
2	0	2	0	2
3	0	3	2	1

In the multiplication table there is no 1 in the row involving 2. Indeed, 2 does not have a multiplicative inverse in $\mathbb{Z}_4$, so therefore it is not a field.

Exercise 1.5.4 Solve Bezout's identity for the following choices of a and b:

(i) $a = 25$ and $b = 7$;

Answer: $25 - 3 \cdot 7 = 4$, $7 - 1 \cdot 4 = 3$, $4 - 1 \cdot 3 = 1$, thus $1 = \gcd(25, 7)$, and we get

$$1 = 4 - 1 \cdot 3 = 4 - (7 - 1 \cdot 4) = -7 + 2 \cdot 4 = -7 + 2(25 - 3 \cdot 7) = 2 \cdot 25 - 7 \cdot 7.$$

Thus $m = 2$ and $n = -7$ is a solution to (1.5).

(ii) $a = -50$ and $b = 3$.

Answer: $-50 + 17 \cdot 3 = 1$, thus $1 = \gcd(-50, 3)$ and $m = 1$ and $n = 17$ is a solution to (1.5).

Exercise 1.5.5 In this exercise we are working in the field $\mathbb{Z}_3$.

(i) $2 + 2 + 2 =$

Answer: 0

(ii) $2(2 + 2)^{-1} =$

Answer: 2

(iii) Solve for x in $2x + 1 = 2$.

Answer: 2

(iv) Find $\det \begin{pmatrix} 1 & 2 \\ 1 & 0 \end{pmatrix}$.

Answer: 1

(v) Compute $\begin{pmatrix} 1 & 2 \\ 0 & 2 \end{pmatrix} \begin{pmatrix} 1 & 1 \\ 2 & 1 \end{pmatrix}$.

Answer: $\begin{pmatrix} 2 & 0 \\ 1 & 2 \end{pmatrix}$

(vi) Find $\begin{pmatrix} 2 & 0 \\ 1 & 1 \end{pmatrix}^{-1}$.

Answer: $\begin{pmatrix} 2 & 0 \\ 1 & 1 \end{pmatrix}$

Exercise 1.5.6 In this exercise we are working in the field $\mathbb{Z}_5$.

(i) $4 + 3 + 2 =$

Answer: 4

(ii) $4(1 + 2)^{-1} =$

Answer: 3

(iii) Solve for x in $3x + 1 = 3$.

Answer: 4

(iv) Find $\det \begin{pmatrix} 4 & 2 \\ 1 & 0 \end{pmatrix}$.

Answer: 3

(v) Compute $\begin{pmatrix} 1 & 2 \\ 3 & 4 \end{pmatrix} \begin{pmatrix} 0 & 1 \\ 2 & 1 \end{pmatrix}$.

Answer: $\begin{pmatrix} 4 & 3 \\ 3 & 2 \end{pmatrix}$.

(vi) Find $\begin{pmatrix} 2 & 2 \\ 4 & 3 \end{pmatrix}^{-1}$.

Answer: $\begin{pmatrix} 1 & 1 \\ 2 & 4 \end{pmatrix}$.

Exercise 1.5.7 In this exercise we are working in the field $\mathbb{C}$. Make sure you write the final answers in the form $a + bi$, with $a, b \in \mathbb{R}$. For instance, $\frac{1+i}{2-i}$ should not be left as a final answer, but be reworked as

$$\frac{1+i}{2-i} = \left(\frac{1+i}{2-i}\right)\left(\frac{2+i}{2+i}\right) = \frac{2+i+2i+i^2}{2^2+1^2} = \frac{1+3i}{5} = \frac{1}{5} + \frac{3i}{5}.$$

Notice that in order to get rid of i in the denominator, we decided to multiply both numerator and denominator with the complex conjugate of the denominator.

(i) $(1 + 2i)(3 - 4i) - (7 + 8i) =$
Answer: $4 - 6i$.

(ii) $\frac{1+i}{3+4i} =$ Answer: $\frac{7}{25} - \frac{i}{25}$.

(iii) Solve for x in $(3 + i)x + 6 - 5i = -3 + 2i$. Answer: $-2 + 3i$.

(iv) Find $\det \begin{pmatrix} 4+i & 2-2i \\ 1+i & -i \end{pmatrix}$. Answer: $-3 - 4i$.

(v) Compute $\begin{pmatrix} -1+i & 2+2i \\ -3i & -6+i \end{pmatrix} \begin{pmatrix} 0 & 1-i \\ -5+4i & 1-2i \end{pmatrix}$. Answer: $\begin{pmatrix} -18-2i & 6 \\ 26-29i & -7+10i \end{pmatrix}$.

(vi) Find $\begin{pmatrix} 2+i & 2-i \\ 4 & 4 \end{pmatrix}^{-1}$. Answer: $\begin{pmatrix} -\frac{i}{2} & \frac{1}{8}+\frac{i}{4} \\ \frac{i}{2} & \frac{1}{8}-\frac{i}{4} \end{pmatrix}$

Exercise 1.5.8 Here the field is $\mathbb{R}(t)$. Find the inverse of the matrix $\begin{pmatrix} 2+3t & \frac{1}{t^2+2t+1} \\ t+1 & \frac{3t-4}{1+t} \end{pmatrix}$, if it exists. Answer: $\frac{1}{9t^2-6t-9}\begin{pmatrix} 3t-4 & \frac{-1}{t+1} \\ -t^2-2t-1 & 3t^2+5t+2 \end{pmatrix}$.

Exercise 1.5.9 Let $\mathbb{F} = \mathbb{Z}_3$. Compute the product $\begin{pmatrix} 1 & 1 & 0 \\ 2 & 1 & 1 \end{pmatrix} \begin{pmatrix} 1 & 0 & 2 \\ 1 & 2 & 1 \\ 2 & 0 & 1 \end{pmatrix}$. Answer: $\begin{pmatrix} 2 & 2 & 0 \\ 2 & 2 & 0 \end{pmatrix}$.

Exercise 1.5.10 Let $\mathbb{F} = \mathbb{C}$. Compute the product $\begin{pmatrix} 2-i & 2+i \\ 2-i & -10 \end{pmatrix} \begin{pmatrix} 5+i & 6-i \\ 1-i & 2+i \end{pmatrix}$.

Answer: $\begin{pmatrix} 14-4i & 14-4i \\ 1+7i & -9-18i \end{pmatrix}$.

Exercise 1.5.11 Let $\mathbb{F} = \mathbb{Z}_5$. Put the matrix

$$\begin{pmatrix} 3 & 1 & 4 \\ 2 & 1 & 0 \\ 2 & 2 & 1 \end{pmatrix}$$

in row echelon form, and compute its determinant.

Answer: Multiply the first row with $3^{-1} = 2$ and row reduce:

$$\begin{pmatrix} 1 & 2 & 3 \\ 0 & 2 & 4 \\ 0 & 3 & 0 \end{pmatrix} \rightarrow \begin{pmatrix} 1 & 2 & 3 \\ 0 & 1 & 0 \\ 0 & 0 & 4 \end{pmatrix},$$

where we subsequently switched rows 2 and 3, and multiplied (the new) row 2 with 3^{-1}. Then

$$\det \begin{pmatrix} 3 & 1 & 4 \\ 2 & 1 & 0 \\ 2 & 2 & 1 \end{pmatrix} = -3 \cdot 3 \det \begin{pmatrix} 1 & 2 & 3 \\ 0 & 1 & 0 \\ 0 & 0 & 4 \end{pmatrix} = 4.$$

Exercise 1.5.12 Let $\mathbb{F} = \mathbb{Z}_3$. Find the set of all solutions to the system of linear equations

$$\begin{cases} 2x_1 + x_2 & = 1 \\ 2x_1 + 2x_2 + x_3 = 0 \end{cases}.$$

Answer: $\left(\begin{array}{ccc|c} 2 & 1 & 0 & 1 \\ 2 & 2 & 1 & 0 \end{array} \right) \rightarrow \left(\begin{array}{ccc|c} 1 & 2 & 0 & 2 \\ 0 & 1 & 1 & 2 \end{array} \right) \rightarrow \left(\begin{array}{ccc|c} 1 & 0 & 1 & 1 \\ 0 & 1 & 1 & 2 \end{array} \right)$, so all solutions are

$$\begin{pmatrix} x_1 \\ x_2 \\ x_3 \end{pmatrix} = \begin{pmatrix} 1 \\ 2 \\ 0 \end{pmatrix} + x_3 \begin{pmatrix} 2 \\ 2 \\ 1 \end{pmatrix}, x_3 \in \mathbb{Z}_3,$$

or, equivalently,

$$\left\{ \begin{pmatrix} 1 \\ 2 \\ 0 \end{pmatrix}, \begin{pmatrix} 0 \\ 1 \\ 1 \end{pmatrix}, \begin{pmatrix} 2 \\ 0 \\ 2 \end{pmatrix} \right\}.$$

Exercise 1.5.13 Let $\mathbb{F} = \mathbb{C}$. Determine whether $\mathbf{b}$ is a linear combination of $\mathbf{a}_1, \mathbf{a}_2, \mathbf{a}_3$, where

$$\mathbf{a}_1 = \begin{pmatrix} i \\ 1-i \\ 2-i \\ 1 \end{pmatrix}, \mathbf{a}_2 = \begin{pmatrix} 0 \\ 3+i \\ -1+i \\ -3 \end{pmatrix}, \mathbf{a}_3 = \begin{pmatrix} -i \\ 2+2i \\ -3+2i \\ 3 \end{pmatrix}, \mathbf{b} = \begin{pmatrix} 0 \\ 0 \\ 0 \\ 1 \end{pmatrix}.$$

Answer: Row reducing the augmented matrix yields the row echelon form

$$\left(\begin{array}{ccc|c} 1 & 0 & -1 & 0 \\ 0 & 1 & 1 & 0 \\ 0 & 0 & 7 & 1 \\ 0 & 0 & 0 & 0 \end{array} \right).$$

No pivot in the augmented column, thus $\mathbf{b}$ is a linear combination of $\mathbf{a}_1, \mathbf{a}_2, \mathbf{a}_3$; in fact

$$\mathbf{b} = \frac{1}{7}\mathbf{a}_1 - \frac{1}{7}\mathbf{a}_2 + \frac{1}{7}\mathbf{a}_3.$$

Exercise 1.5.14 Let $\mathbb{F} = \mathbb{Z}_5$. Compute the inverse of $\begin{pmatrix} 2 & 3 & 1 \\ 1 & 4 & 1 \\ 1 & 1 & 2 \end{pmatrix}$ in two different ways (row reduction and by applying (1.11)).

Answer: $\begin{pmatrix} 4 & 0 & 3 \\ 3 & 1 & 3 \\ 4 & 2 & 0 \end{pmatrix}.$

Exercise 1.5.15 Let $\mathbb{F} = \mathbb{C}$. Find bases of the column space, row space and null space of the matrix

$$A = \begin{pmatrix} 1 & 1+i & 2 \\ 1+i & 2i & 3+i \\ 1-i & 2 & 3+5i \end{pmatrix}.$$

Answer: Basis for $\mathrm{Col}A$ is $\left\{ \begin{pmatrix} 1 \\ 1+i \\ 1-i \end{pmatrix}, \begin{pmatrix} 2 \\ 3+i \\ 3+5i \end{pmatrix} \right\}.$

Basis for $\mathrm{Row}A$ is $\{ (1 \quad 1+i \quad 2), (0 \quad 0 \quad 1) \}.$

Basis for $\mathrm{Nul}A$ is $\left\{ \begin{pmatrix} 1+i \\ -1 \\ 0 \end{pmatrix} \right\}.$

Exercise 1.5.16 Let $\mathbb{F} = \mathbb{Z}_7$. Find a basis for the eigenspace of $A = \begin{pmatrix} 3 & 5 & 0 \\ 4 & 6 & 5 \\ 2 & 2 & 4 \end{pmatrix}$ corresponding to the eigenvalue $\lambda = 1$.

Answer: $\left\{ \begin{pmatrix} 1 \\ 1 \\ 1 \end{pmatrix} \right\}.$

Exercise 1.5.17 Let $\mathbb{F} = \mathbb{Z}_3$. Use Cramer's rule to find the solution to the system of linear equations

$$\begin{cases} 2x_1 + 2x_2 = 1 \\ x_1 + 2x_2 = 1 \end{cases}.$$

Answer: $x_1 = 0, x_2 = 2.$

Exercise 1.5.18 Let $\mathbb{F} = \mathbb{C}$. Consider the matrix vector equation $A\mathbf{x} = \mathbf{b}$ given by

$$\begin{pmatrix} i & 1-i & 2 \\ 1+i & \alpha & 0 \\ 1-i & 1+2i & 3+5i \end{pmatrix} \begin{pmatrix} x_1 \\ x_2 \\ x_3 \end{pmatrix} = \begin{pmatrix} 2 \\ 0 \\ 5i \end{pmatrix}.$$

Determine $\alpha \in \mathbb{C}$ so that A is invertible and $x_1 = x_2$.

Answer: $\alpha = -1 - i.$

Exercise 1.5.19 Let $\mathbb{F} = \mathbb{R}(t)$. Compute the adjugate of

$$A = \begin{pmatrix} \frac{1}{t} & 2+t^2 & 2-t \\ \frac{2}{1+t} & 3t & 1-t \\ 1 & 4+t^2 & 0 \end{pmatrix}.$$

Answer:

$$\mathrm{adj}(A) = \begin{pmatrix} -(1-t)(4+t^2) & (2-t)(4+t^2) & 2-8t+4t^2-t^3 \\ 1-t & t-2 & -\frac{1}{t}+1+\frac{4-2t}{1+t} \\ \frac{8+2t^2}{1+t}-3t & -\frac{4}{t}-t+2+t^2 & 3-\frac{4+2t^2}{1+t} \end{pmatrix}.$$

Exercise 1.5.20 Recall that the *trace* of a square matrix is defined to be the sum of its diagonal entries. Thus $\operatorname{tr}[(a_{ij})_{i,j=1}^n] = a_{11} + \cdots + a_{nn} = \sum_{j=1}^n a_{jj}$.

(a) Show that if $A \in \mathbb{F}^{n \times m}$ and $B \in \mathbb{F}^{m \times n}$, then $\operatorname{tr}(AB) = \operatorname{tr}(BA)$.

Answer: Write $A = (a_{ij})$ and $B = (b_{ij})$. Then

$$\operatorname{tr}(AB) = \sum_{k=1}^n (AB)_{kk} = \sum_{k=1}^n (\sum_{j=1}^m a_{kj} b_{jk}).$$

Similarly,

$$\operatorname{tr}(BA) = \sum_{j=1}^m (BA)_{jj} = \sum_{j=1}^m (\sum_{k=1}^n b_{jk} a_{kj}).$$

As $a_{kj} b_{jk} = b_{jk} a_{kj}$ for all j and k, the equality $\operatorname{tr}(AB) = \operatorname{tr}(BA)$ follows.

(b) Show that if $A \in \mathbb{F}^{n \times m}$, $B \in \mathbb{F}^{m \times k}$, and $C \in \mathbb{F}^{k \times n}$, then $\operatorname{tr}(ABC) = \operatorname{tr}(CAB) = \operatorname{tr}(BCA)$.

Answer: By the previous part, we have that $\operatorname{tr}((AB)C) = \operatorname{tr}(C(AB))$ and also $\operatorname{tr}(A(BC)) = \operatorname{tr}((BC)A)$. Thus $\operatorname{tr}(BCA) = \operatorname{tr}(ABC) = \operatorname{tr}(CAB)$ follows.

(c) Give an example of matrices $A, B, C \in \mathbb{F}^{n \times n}$ so that $\operatorname{tr}(ABC) \neq \operatorname{tr}(BAC)$.

Answer: For instance $A = \begin{pmatrix} 0 & 1 \\ 0 & 0 \end{pmatrix}$, $B = \begin{pmatrix} 0 & 0 \\ 1 & 0 \end{pmatrix}$, and $C = \begin{pmatrix} 0 & 0 \\ 0 & 1 \end{pmatrix}$. Then $\operatorname{tr}(ABC) = 0 \neq 1 = \operatorname{tr}(BAC)$.

Exercise 1.5.21 Let $A, B \in \mathbb{F}^{n \times n}$. The *commutator* $[A, B]$ of A and B is defined by $[A, B] := AB - BA$.

(a) Show that $\operatorname{tr}([A, B]) = 0$.

Answer: By the previous exercise $\operatorname{tr}(AB) = \operatorname{tr}(BA)$, and thus $\operatorname{tr}(AB - BA) = \operatorname{tr}(AB) - \operatorname{tr}(BA) = 0$.

(b) Show that when $n = 2$, we have that $[A, B]^2 = -\det([A, B])I_2$.

Answer: Write $A = (a_{ij})$ and $B = (b_{ij})$. Then $AB - BA$ equals

$$\begin{pmatrix} a_{12}b_{21} - b_{12}a_{21} & a_{11}b_{12} + a_{12}b_{22} - b_{11}a_{12} - b_{12}a_{22} \\ a_{21}b_{11} + a_{22}b_{21} - b_{21}a_{11} - b_{22}a_{21} & a_{21}b_{12} - b_{21}a_{12} \end{pmatrix},$$

which is of the form $\begin{pmatrix} x & y \\ z & -x \end{pmatrix}$. Then

$$[A, B]^2 = \begin{pmatrix} x & y \\ z & -x \end{pmatrix}^2 = \begin{pmatrix} x^2 + yz & 0 \\ 0 & x^2 + yz \end{pmatrix} = -\det([A, B])I_2,$$

since $\det([A, B]) = -x^2 - yz$.

(c) Show that if $C \in \mathbb{F}^{n \times n}$ as well, then $\operatorname{tr}(C[A, B]) = \operatorname{tr}([B, C]A)$.

Answer: Using the previous exercise

$$\operatorname{tr}(C(AB - BA)) = \operatorname{tr}(CAB) - \operatorname{tr}(CBA) =$$
$$\operatorname{tr}(BCA) - \operatorname{tr}(CBA) = \operatorname{tr}((BC - CB)A).$$

Exercise 1.5.22 *Answer:*

$$1 \cdot 3 + 2 \cdot 0 + 3 \cdot 3 + 4 \cdot 4 + 5 \cdot 8 + 6 \cdot 0 + 7 \cdot 6 + 8 \cdot 3 + 9 \cdot 8 + 10 \cdot 8 = \operatorname{rem}(286|11) = 0.$$

Exercise 1.5.23 *Answer:* AWESOME

Chapter 2

Exercise 2.6.1 For the proof of Lemma 2.1.1 provide a reason why each equality holds. For instance, the equality $\mathbf{0} = 0u + v$ is due to Axiom 5 in the definition of a vector space and $\mathbf{v}$ being the additive inverse of $0u$.

Answer:

$$\mathbf{0} = \text{(Axiom 5)} = 0\mathbf{u} + \mathbf{v} = \text{(Field Axiom 4)} = (0+0)\mathbf{u} + \mathbf{v} =$$
$$= \text{(Axiom 8)} = (0\mathbf{u} + 0\mathbf{u}) + \mathbf{v} = \text{(Axiom 2)} = 0\mathbf{u} + (0\mathbf{u} + \mathbf{v}) =$$
$$= \text{(Axiom 5)} = 0\mathbf{u} + \mathbf{0} = \text{(Axiom 4)} = 0\mathbf{u}.$$

Exercise 2.6.2 Consider $p(X), q(X) \in \mathbb{F}[X]$ with $\mathbb{F} = \mathbb{R}$ or $\mathbb{F} = \mathbb{C}$. Show that if $p(X) = q(X)$ if and only if $p(x) = q(x)$ for all $x \in \mathbb{F}$. (One way to do it is by using derivatives. Indeed, using calculus one can observe that if two polynomials are equal, then so are all their derivatives. Next observe that $p_j = \frac{1}{j!} \frac{d^j p}{dx^j}(0)$.) Where do you use in your proof that $\mathbb{F} = \mathbb{R}$ or $\mathbb{F} = \mathbb{C}$?

Answer: When $f(x) = g(x)$ for all $x \in \mathbb{R}$, then $\frac{f(x+h)-f(x)}{h} = \frac{g(x+h)-g(x)}{h}$ for all $x \in \mathbb{F}$ and all $h \in \mathbb{F} \setminus \{0\}$. And thus, after taking limits, we get $f'(x) = g'(x)$, assuming f (and thus g) is differentiable at x. Thus when two differentiable functions are equal, then so are their derivatives.

As $p(x) = q(x)$ for all $x \in \mathbb{F}$, we get that $p^{(j)}(x) = q^{(j)}(x)$ for all j. In particular, $p^{(j)}(0) = q^{(j)}(0)$ for all j. When $p(X) = \sum_{j=0}^{n} p_j X^j$ and $q(X) = \sum_{j=0}^{n} q_j X^j$, then $p_j = \frac{1}{j!} p^{(j)}(0) = \frac{1}{j!} q^{(j)}(0) = q_j$, for all j. This proves that $p(X) = q(X)$.

When we took derivatives we used that we are working over $\mathbb{F} = \mathbb{R}$ or $\mathbb{F} = \mathbb{C}$. For the other fields $\mathbb{F}$ we are considering in this chapter, derivatives of functions are not defined.

Exercise 2.6.3 When the underlying field is $\mathbb{Z}_p$, why does closure under addition automatically imply closure under scalar multiplication?

Answer: To show that $c\mathbf{x}$ lies in the subspace, one simply needs to observe that $c\mathbf{x} = \mathbf{x} + \cdots + \mathbf{x}$, where in the right-hand side there are c terms. When the subspace is closed under addition, $\mathbf{x} + \cdots + \mathbf{x}$ will be in the subspace, and thus $c\mathbf{x}$ lies in the subspace.

Exercise 2.6.4 Let $V = \mathbb{R}^{\mathbb{R}}$. For $W \subset V$, show that W is a subspace of V.

(a) $W = \{f : \mathbb{R} \to \mathbb{R} \ : \ f \text{ is continuous}\}$.

(b) $W = \{f : \mathbb{R} \to \mathbb{R} \ : \ f \text{ is differentiable}\}$.

Answer: (a). The constant zero function is continuous. As was shown in calculus, when f and g are continuous, then so are $f + g$ and cf. This gives that W is a subspace.

(b). The constant zero function is differentiable. As was shown in calculus, when f and g are differentiable, then so are $f + g$ and cf. This gives that W is a subspace.

Exercise 2.6.5 For the following choices of $\mathbb{F}$, V and W, determine whether W is a subspace of V over $\mathbb{F}$. In case the answer is yes, provide a basis for W.

(a) Let $\mathbb{F} = \mathbb{R}$ and $V = \mathbb{R}^3$,

$$W = \{ \begin{pmatrix} x_1 \\ x_2 \\ x_3 \end{pmatrix} \; : \; x_1, x_2, x_3 \in \mathbb{R}, x_1 - 2x_2 + x_3^2 = 0 \}.$$

Answer: Not closed under scalar multiplication. For example, $\mathbf{x} = \begin{pmatrix} -1 \\ 0 \\ 1 \end{pmatrix} \in W$, but

$(-1)\mathbf{x} \notin W$.

(b) $\mathbb{F} = \mathbb{C}$ and $V = \mathbb{C}^{3\times 3}$,

$$W = \{ \begin{pmatrix} a & b & c \\ 0 & a & b \\ 0 & 0 & a \end{pmatrix} \; : \; a, b, c \in \mathbb{C} \}.$$

Answer: This is a subspace: the zero matrix lies in W (choose $a = b = c = 0$), the sum of two matrices in W is again of the same type, and a scalar multiple of a matrix in W is again of the same type. In fact,

$$W = \text{Span} \left\{ \begin{pmatrix} 1 & 0 & 0 \\ 0 & 1 & 0 \\ 0 & 0 & 1 \end{pmatrix}, \begin{pmatrix} 0 & 1 & 0 \\ 0 & 0 & 1 \\ 0 & 0 & 0 \end{pmatrix}, \begin{pmatrix} 0 & 0 & 1 \\ 0 & 0 & 0 \\ 0 & 0 & 0 \end{pmatrix} \right\}.$$

(c) $\mathbb{F} = \mathbb{C}$ and $V = \mathbb{C}^{2\times 2}$,

$$W = \{ \begin{pmatrix} a & \bar{b} \\ b & c \end{pmatrix} \; : \; a, b, c \in \mathbb{C} \}.$$

Answer: Not closed under scalar multiplication. For example, $\begin{pmatrix} 0 & -i \\ i & 0 \end{pmatrix} \in W$, but

$i \begin{pmatrix} 0 & -i \\ i & 0 \end{pmatrix} = \begin{pmatrix} 0 & 1 \\ -1 & 0 \end{pmatrix} \notin W$.

(d) $\mathbb{F} = \mathbb{R}$, $V = \mathbb{R}_2[X]$ and

$$W = \{ p(x) \in V \; : \; \int_0^1 p(x) \cos x dx = 0 \}.$$

Answer: This is a subspace. If $p(x) \equiv 0$, then $\int_0^1 p(x) \cos x dx = 0$, so W contains $\mathbf{0}$. If $\int_0^1 p(x) \cos x dx = 0$ and $\int_0^1 q(x) \cos x dx = 0$, then $\int_0^1 (p + q)(x) \cos x dx = 0$ and $\int_0^1 (cp)(x) \cos x dx = 0$, thus W is closed under addition and scalar multiplication.

(e) $\mathbb{F} = \mathbb{R}$, $V = \mathbb{R}_2[X]$ and

$$W = \{ p(x) \in V \; : p(1) = p(2)p(3) \}.$$

Answer: Not closed under scalar multiplication (or addition). For example, $p(X) \equiv 1$ is in W, but $(2p)(X) \equiv 2$ is not in W.

(f) $\mathbb{F} = \mathbb{C}$, $V = \mathbb{C}^3$, and

$$W = \{ \begin{pmatrix} x_1 \\ x_2 \\ x_3 \end{pmatrix} \in \mathbb{C}^3 \; : \; x_1 - x_2 = x_3 - x_2 \}.$$

Answer: This is a subspace; it is in fact the kernel of the matrix $\begin{pmatrix} 1 & 0 & -1 \end{pmatrix}$.

Exercise 2.6.6 For the following vector spaces (V over $\mathbb{F}$) and vectors, determine whether the vectors are linearly independent or linearly independent.

(a) Let $\mathbb{F} = \mathbb{Z}_5$, $V = \mathbb{Z}_5^4$ and consider the vectors

$$\begin{pmatrix} 3 \\ 0 \\ 2 \\ 1 \end{pmatrix}, \begin{pmatrix} 2 \\ 1 \\ 0 \\ 3 \end{pmatrix}, \begin{pmatrix} 1 \\ 2 \\ 1 \\ 0 \end{pmatrix}.$$

Answer: Making these vectors the columns of a matrix, and performing row reduction yields that all columns have a pivot. Thus linearly independent.

(b) Let $\mathbb{F} = \mathbb{R}$, $V = \{f \mid f : (0, \infty) \to \mathbb{R}$ is a continuous function$\}$, and consider the vectors

$$t, t^2, \frac{1}{t}.$$

Answer: Suppose $at + bt^2 + c\frac{1}{t} \equiv 0$. As this equality holds for all t, we can choose for instance $t = 1$, $t = 2$ and $t = \frac{1}{2}$, giving the system

$$\begin{pmatrix} 1 & 1 & 1 \\ 1 & 4 & \frac{1}{2} \\ 1 & \frac{1}{4} & 2 \end{pmatrix} \begin{pmatrix} a \\ b \\ c \end{pmatrix} = \begin{pmatrix} 0 \\ 0 \\ 0 \end{pmatrix}.$$

Row reducing this matrix gives pivots in all columns, thus $a = b = c = 0$ is the only solution. Thus the vectors $t, t^2, \frac{1}{t}$ are linearly independent.

(c) Let $\mathbb{F} = \mathbb{Z}_5$, $V = \mathbb{Z}_5^4$ and consider the vectors

$$\begin{pmatrix} 4 \\ 0 \\ 2 \\ 3 \end{pmatrix}, \begin{pmatrix} 2 \\ 1 \\ 0 \\ 3 \end{pmatrix}, \begin{pmatrix} 1 \\ 2 \\ 1 \\ 0 \end{pmatrix}.$$

Answer: Making these vectors the columns of a matrix, and performing row reduction yields that not all columns have a pivot. Thus linearly dependent. In fact,

$$2\begin{pmatrix} 4 \\ 0 \\ 2 \\ 3 \end{pmatrix} + 3\begin{pmatrix} 2 \\ 1 \\ 0 \\ 3 \end{pmatrix} + 1\begin{pmatrix} 1 \\ 2 \\ 1 \\ 0 \end{pmatrix} = \begin{pmatrix} 0 \\ 0 \\ 0 \\ 0 \end{pmatrix}.$$

(d) Let $\mathbb{F} = \mathbb{R}$, $V = \{f \mid f : \mathbb{R} \to \mathbb{R}$ is a continuous function$\}$, and consider the vectors

$$\cos 2x, \sin 2x, \cos^2 x, \sin^2 x.$$

Answer: The equality $\cos 2x = \cos^2 x - \sin^2 x$ holds for all $x \in \mathbb{R}$. Thus $\cos 2x + 0(\sin 2x) - \cos^2 x + \sin^2 x = \mathbf{0}(x)$ for all $x \in \mathbb{R}$, thus the vectors are linearly dependent.

(e) Let $\mathbb{F} = \mathbb{C}$, $V = \mathbb{C}^{2 \times 2}$, and consider the vectors

$$\begin{pmatrix} i & 1 \\ -1 & -i \end{pmatrix}, \begin{pmatrix} 1 & 1 \\ i & -i \end{pmatrix}, \begin{pmatrix} -1 & i \\ -i & 1 \end{pmatrix}.$$

Answer: Suppose

$$a\begin{pmatrix} i & 1 \\ -1 & -i \end{pmatrix} + b\begin{pmatrix} 1 & 1 \\ i & -i \end{pmatrix} + c\begin{pmatrix} -1 & i \\ -i & 1 \end{pmatrix} = \begin{pmatrix} 0 & 0 \\ 0 & 0 \end{pmatrix}.$$

Rewriting we get

$$\begin{pmatrix} i & 1 & -1 \\ 1 & 1 & i \\ -1 & i & -i \\ -i & -i & 1 \end{pmatrix} \begin{pmatrix} a \\ b \\ c \end{pmatrix} = \begin{pmatrix} 0 \\ 0 \\ 0 \end{pmatrix}.$$

Row reducing this matrix gives no pivot in column three. We find that

$$\begin{pmatrix} a \\ b \\ c \end{pmatrix} = c \begin{pmatrix} -i \\ 0 \\ 1 \end{pmatrix}$$

is the general solution. Indeed,

$$-i \begin{pmatrix} i & 1 \\ -1 & -i \end{pmatrix} + 0 \begin{pmatrix} 1 & 1 \\ i & -i \end{pmatrix} + \begin{pmatrix} -1 & i \\ -i & 1 \end{pmatrix} = \begin{pmatrix} 0 & 0 \\ 0 & 0 \end{pmatrix},$$

and thus these vectors are linearly dependent.

(f) Let $\mathbb{F} = \mathbb{R}$, $V = \mathbb{C}^{2 \times 2}$, and consider the vectors

$$\begin{pmatrix} i & 1 \\ -1 & -i \end{pmatrix}, \begin{pmatrix} 1 & 1 \\ i & -i \end{pmatrix}, \begin{pmatrix} -1 & i \\ -i & 1 \end{pmatrix}.$$

Answer: Suppose

$$a \begin{pmatrix} i & 1 \\ -1 & -i \end{pmatrix} + b \begin{pmatrix} 1 & 1 \\ i & -i \end{pmatrix} + c \begin{pmatrix} -1 & i \\ -i & 1 \end{pmatrix} = \begin{pmatrix} 0 & 0 \\ 0 & 0 \end{pmatrix},$$

with now $a, b, c \in \mathbb{R}$. As before we find that this implies that $a = -ic$ and $b = 0$. As $a, b, c \in \mathbb{R}$, this implies that $a = b = c = 0$, and thus these vectors are linearly independent over $\mathbb{R}$.

(g) Let $\mathbb{F} = \mathbb{Z}_5$, $V = \mathbb{F}^{3 \times 2}$, and consider the vectors

$$\begin{pmatrix} 3 & 4 \\ 1 & 0 \\ 1 & 0 \end{pmatrix}, \begin{pmatrix} 1 & 1 \\ 4 & 2 \\ 1 & 2 \end{pmatrix}, \begin{pmatrix} 1 & 2 \\ 3 & 1 \\ 1 & 2 \end{pmatrix}.$$

Answer: Suppose

$$a \begin{pmatrix} 3 & 4 \\ 1 & 0 \\ 1 & 0 \end{pmatrix} + b \begin{pmatrix} 1 & 1 \\ 4 & 2 \\ 1 & 2 \end{pmatrix} + c \begin{pmatrix} 1 & 2 \\ 3 & 1 \\ 1 & 2 \end{pmatrix} = \begin{pmatrix} 0 & 0 \\ 0 & 0 \\ 0 & 0 \end{pmatrix}.$$

Rewriting, we get

$$\begin{pmatrix} 3 & 1 & 1 \\ 4 & 1 & 2 \\ 1 & 4 & 3 \\ 0 & 2 & 1 \\ 1 & 1 & 1 \\ 0 & 2 & 2 \end{pmatrix} \begin{pmatrix} a \\ b \\ c \end{pmatrix} = \begin{pmatrix} 0 \\ 0 \\ 0 \\ 0 \\ 0 \\ 0 \end{pmatrix}.$$

Row reducing this matrix gives pivots in all columns, thus $a = b = c = 0$ is the only solution. Thus the vectors are linearly independent.

(h) Let $\mathbb{F} = \mathbb{R}$, $V = \{f \mid f : \mathbb{R} \to \mathbb{R}$ is a continuous function$\}$, and consider the vectors

$$1, e^t, e^{2t}.$$

Answer: Suppose $a + be^t + ce^{2t} \equiv 0$. As this equality holds for all t, we can choose for instance $t = 0$, $t = \ln 2$ and $t = \ln 3$, giving the system

$$\begin{pmatrix} 1 & 1 & 1 \\ 1 & 2 & 4 \\ 1 & 3 & 9 \end{pmatrix} \begin{pmatrix} a \\ b \\ c \end{pmatrix} = \begin{pmatrix} 0 \\ 0 \\ 0 \end{pmatrix}.$$

Row reducing this matrix gives pivots in all columns, thus $a = b = c = 0$ is the only solution. Thus the vectors $1, e^t, e^{2t}$ are linearly independent.

Exercise 2.6.7 Let $\mathbf{v}_1, \mathbf{v}_2, \mathbf{v}_3$ be linearly independent vectors in a vector space V.

(a) For which k are $k\mathbf{v}_1 + \mathbf{v}_2, k\mathbf{v}_2 - \mathbf{v}_3, \mathbf{v}_3 + \mathbf{v}_1$ linearly independent?

(b) Show that if $\mathbf{v}$ is in the span of $\mathbf{v}_1, \mathbf{v}_2$ and in the span of $\mathbf{v}_2 + \mathbf{v}_3, \mathbf{v}_2 - \mathbf{v}_3$, then $\mathbf{v}$ is a multiple of $\mathbf{v}_2$.

Answer: (a) Suppose $a(k\mathbf{v}_1 + \mathbf{v}_2) + b(k\mathbf{v}_2 - \mathbf{v}_3) + c(\mathbf{v}_3 + \mathbf{v}_1) = \mathbf{0}$. Then

$$(ak + c)\mathbf{v}_1 + (a + bk)\mathbf{v}_2 + (-bk + c)\mathbf{v}_3 = \mathbf{0}.$$

As $\mathbf{v}_1, \mathbf{v}_2, \mathbf{v}_3$ are linearly independent, we get $ak + c = 0$, $a + bk = 0$, and $-b + c = 0$. Thus

$$\begin{pmatrix} k & 0 & 1 \\ 1 & k & 0 \\ 0 & -1 & 1 \end{pmatrix} \begin{pmatrix} a \\ b \\ c \end{pmatrix} = \begin{pmatrix} 0 \\ 0 \\ 0 \end{pmatrix}.$$

For this system to have a nontrivial solution, we need that the determinant of the matrix equals 0. This yields the equation $k^2 - 1 = 0$. Thus for $k = 1$ and $k = -1$ we get linearly dependent vectors.

(b) $v = a\mathbf{v}_1 + b\mathbf{v}_2$ and $v = c(\mathbf{v}_2 + \mathbf{v}_3) + d(\mathbf{v}_2 - \mathbf{v}_3)$, gives
$a\mathbf{v}_1 + b\mathbf{v}_2 = c(\mathbf{v}_2 + \mathbf{v}_3) + d(\mathbf{v}_2 - \mathbf{v}_3)$. Then $a\mathbf{v}_1 + (b - c - d)\mathbf{v}_2 + (-c + d)\mathbf{v}_3 = \mathbf{0}$. As $\mathbf{v}_1, \mathbf{v}_2, \mathbf{v}_3$ are linearly independent, we get $a = 0$, $b - c - d = 0$, and $-c + d = 0$. Since $a = 0$, we have $v = b\mathbf{v}_2$, and thus is $\mathbf{v}$ a multiple of $\mathbf{v}_2$.

Exercise 2.6.8 (a) Show that if the set $\{\mathbf{v}_1, \ldots, \mathbf{v}_k\}$ is linearly independent, and $\mathbf{v}_{k+1}$ is not in $\mathrm{Span}\{\mathbf{v}_1, \ldots, \mathbf{v}_k\}$, then the set $\{\mathbf{v}_1, \ldots, \mathbf{v}_k, \mathbf{v}_{k+1}\}$ is linearly independent.

(b) Let W be a subspace of an n-dimensional vector space V, and let $\{\mathbf{v}_1, \ldots, \mathbf{v}_p\}$ be a basis for W. Show that there exist vectors $\mathbf{v}_{p+1}, \ldots, \mathbf{v}_n \in V$ so that $\{\mathbf{v}_1, \ldots, \mathbf{v}_p, \mathbf{v}_{p+1}, \ldots, \mathbf{v}_n\}$ is a basis for V.

(Hint: once $\mathbf{v}_1, \ldots, \mathbf{v}_k$ are found and $k < n$, observe that one can choose $\mathbf{v}_{k+1} \in V \setminus (\mathrm{Span}\{\mathbf{v}_1, \ldots, \mathbf{v}_k\})$. Argue that this process stops when $k = n$, and that at that point a basis for V is found.)

Answer: (a) Let $c_1, \ldots, c_k, c_{k+1}$ be so that $c_1\mathbf{v}_1 + \cdots + c_k\mathbf{v}_k + c_{k+1}\mathbf{v}_{k+1} = \mathbf{0}$. Suppose that $c_{k+1} \neq 0$. Then $\mathbf{v}_{k+1} = -\frac{c_1}{c_{k+1}}\mathbf{v}_1 - \cdots - \frac{c_k}{c_{k+1}}\mathbf{v}_k \in \mathrm{Span}\{\mathbf{v}_1, \ldots, \mathbf{v}_k\}$. Contradiction. Thus we must have $c_{k+1} = 0$. Then we get that $c_1\mathbf{v}_1 + \cdots + c_k\mathbf{v}_k = \mathbf{0}$. As $\{\mathbf{v}_1, \ldots, \mathbf{v}_k\}$ is linearly independent, we now must have $c_1 = \cdots = c_k = 0$. Thus $c_1 = \cdots = c_k = c_{k+1} = 0$, and linear independence of $\{\mathbf{v}_1, \ldots, \mathbf{v}_k, \mathbf{v}_{k+1}\}$ follows.

(b) Suppose that $\mathbf{v}_1, \ldots, \mathbf{v}_k$ are found and $k < n$. Then $\mathrm{Span}\{\mathbf{v}_1, \ldots, \mathbf{v}_k\}$ is a k-dimensional subspace of V. As $\dim V = n > k$, there must exist a $\mathbf{v}_{k+1} \in V \setminus (\mathrm{Span}\{\mathbf{v}_1, \ldots, \mathbf{v}_k\})$. By (a) we have that the set $\{\mathbf{v}_1, \ldots, \mathbf{v}_k, \mathbf{v}_{k+1}\}$ is linearly independent. If $k + 1 < n$, one continues this process. Ultimately one finds a linearly independent set $\{\mathbf{v}_1, \ldots, \mathbf{v}_p, \mathbf{v}_{p+1}, \ldots, \mathbf{v}_n\}$. This set must span V. Indeed, if we take $v \in V$, then by Remark 2.4.5 $\{\mathbf{v}_1, \ldots, \mathbf{v}_n, v\}$ is a linear dependent set. Due to linear independence of $\{\mathbf{v}_1, \ldots, \mathbf{v}_n\}$ this implies that $\mathbf{v} \in \mathrm{Span}\{\mathbf{v}_1, \ldots, \mathbf{v}_n\}$. Thus $V = \mathrm{Span}\{\mathbf{v}_1, \ldots, \mathbf{v}_n\}$ and $\{\mathbf{v}_1, \ldots, \mathbf{v}_n\}$ is linearly independent, thus $\{\mathbf{v}_1, \ldots, \mathbf{v}_n\}$ is a basis for V.

Exercise 2.6.9 Let $V = \mathbb{R}_2[X]$ and

$$W = \{p \in V \ : \ p(2) = 0\}.$$

(a) Show that W is a subspace of V.

(b) Find a basis for W.

Answer: (a) We have $\mathbf{0}(2) = 0$, so $\mathbf{0} \in W$. Also, when $p, q \in W$ and $c \in \mathbb{R}$, we have $(p+q)(2) = p(2) + q(2) = 0 + 0 = 0$ and $(cp)(2) = cp(2) = c0 = 0$, so $p + q \in W$ and $cp \in W$. Thus W is a subspace.

(b) A general element in V is of the form $p_0 + p_1 X + p_2 X^2$. For this element to be in W we have the condition $p(2) = 0$, yielding $p_0 + 2p_1 + 4p_2 = 0$. Thus

$$\begin{pmatrix} 1 & 2 & 4 \end{pmatrix} \begin{pmatrix} p_0 \\ p_1 \\ p_2 \end{pmatrix} = 0.$$

With p_1 and p_2 as free variables, we find $p_0 = -2p_1 - 4p_2$, thus we get

$$p_0 + p_1 X + p_2 X^2 = -2p_1 - 4p_2 + p_1 X + p_2 X^2 = p_1(-2 + X) + p_2(-4 + X^2).$$

Thus $\{-2 + X, -4 + X^2\}$ is a basis for W.

Exercise 2.6.10 For the following choices of subspaces U and W in V, find bases for $U + W$ and $U \cap W$.

(a) $V = \mathbb{R}_5[X]$, $U = \mathrm{Span}\{X + 1, X^2 - 1\}$, $W = \{p(X) \ : \ p(2) = 0\}$.

(b) $V = \mathbb{Z}_5^4$,

$$U = \mathrm{Span}\left\{ \begin{pmatrix} 3 \\ 0 \\ 2 \\ 1 \end{pmatrix}, \begin{pmatrix} 2 \\ 1 \\ 0 \\ 0 \end{pmatrix} \right\}, \quad W = \mathrm{Span}\left\{ \begin{pmatrix} 1 \\ 2 \\ 1 \\ 0 \end{pmatrix}, \begin{pmatrix} 4 \\ 4 \\ 1 \\ 1 \end{pmatrix} \right\}.$$

Answer: (a) A general element in U is of the form $a(X + 1) + b(X^2 - 1)$. For this to be in W, we need $a(2 + 1) + b(4 - 1) = 0$. Thus $3a + 3b = 0$, yielding $a = -b$. Thus a general element in $U \cap W$ is of the form $a(X + 1 - (X^2 - 1)) = a(2 + X - X^2)$. A basis for $U \cap W$ is $\{2 + X - X^2\}$.

A basis for W is $\{-2 + X, -4 + X^2, -8 + X^3, -16 + X^4, -32 + X^5\}$, thus $U + W$ is spanned by $\{X + 1, X^2 - 1, -2 + X, -4 + X^2, -8 + X^3, -16 + X^4, -32 + X^5\}$. This is a linear dependent set. Removing $-4 + X^2$, makes it a basis for $U + W$, so we get that $\{X + 1, X^2 - 1, -2 + X, -8 + X^3, -16 + X^4, -32 + X^5\}$ is a basis for $U + W$. In fact, $U + W = \mathbb{R}_5[X]$, so we can also take $\{1, X, X^2, X^3, X^4, X^5\}$ as a basis for $U + W$.

(b) If $\mathbf{v} \in U \cap W$, then there exist a, b, c, d so that

$$\mathbf{v} = a \begin{pmatrix} 3 \\ 0 \\ 2 \\ 1 \end{pmatrix} + b \begin{pmatrix} 2 \\ 1 \\ 0 \\ 0 \end{pmatrix} = c \begin{pmatrix} 1 \\ 2 \\ 1 \\ 0 \end{pmatrix} + d \begin{pmatrix} 4 \\ 4 \\ 1 \\ 1 \end{pmatrix}.$$

This gives

$$\begin{pmatrix} 3 & 2 & 1 & 4 \\ 0 & 1 & 2 & 4 \\ 2 & 0 & 1 & 1 \\ 1 & 0 & 0 & 1 \end{pmatrix} \begin{pmatrix} a \\ b \\ -c \\ -d \end{pmatrix} = \begin{pmatrix} 0 \\ 0 \\ 0 \\ 0 \end{pmatrix}.$$

Row reduction yields the echelon form

$$\begin{pmatrix} 1 & 4 & 2 & 3 \\ 0 & 1 & 2 & 4 \\ 0 & 0 & 3 & 2 \\ 0 & 0 & 0 & 0 \end{pmatrix},$$

making d a free variable, and $c = d$. Thus

$$c\begin{pmatrix}1\\2\\1\\0\end{pmatrix} + c\begin{pmatrix}4\\4\\1\\1\end{pmatrix} = c\begin{pmatrix}0\\1\\2\\1\end{pmatrix}$$

is a general element of $U \cap W$. Thus

$$\{\begin{pmatrix}0\\1\\2\\1\end{pmatrix}\}$$

is a basis for $U \cap W$.

For a basis for $U + W$, we find a basis for the column space of

$$\begin{pmatrix}3 & 2 & 1 & 4\\0 & 1 & 2 & 4\\2 & 0 & 1 & 1\\1 & 0 & 0 & 1\end{pmatrix}.$$

From the calculations above, we see that the first three columns are pivot columns. Thus

$$\{\begin{pmatrix}3\\0\\2\\1\end{pmatrix}, \begin{pmatrix}2\\1\\0\\0\end{pmatrix}, \begin{pmatrix}1\\2\\1\\0\end{pmatrix}\}$$

is a basis for $U + W$.

Exercise 2.6.11 Let $\{v_1, v_2, v_3, v_4, v_5\}$ be linearly independent vectors in a vector space V. Determine whether the following sets are linearly dependent or linearly independent.

(a) $\{v_1 + v_2 + v_3 + v_4, v_1 - v_2 + v_3 - v_4, v_1 - v_2 - v_3 - v_4\}$

(b) $\{v_1 + v_2, v_2 + v_3, v_3 + v_4, v_4 + v_5, v_5 + v_2\}$

(c) $\{v_1 + v_3, v_4 - v_2, v_5 + v_1, v_4 - v_2, v_5 + v_3, v_1 + v_2\}$.

When you did this exercise, did you make any assumptions on the underlying field?

Answer: (a) Let a, b, c be so that

$$a(v_1 + v_2 + v_3 + v_4) + b(v_1 - v_2 + v_3 - v_4) + c(v_1 - v_2 - v_3 - v_4) = 0.$$

Rewriting, we get

$$(a + b + c)v_1 + (a - b - c)v_2 + (a + b - c)v_3 + (a - b - c)v_4 = 0.$$

As $\{v_1, v_2, v_3, v_4, v_5\}$ is linearly independent, we get

$$\begin{pmatrix}1 & 1 & 1\\1 & -1 & -1\\1 & 1 & -1\\1 & -1 & -1\end{pmatrix}\begin{pmatrix}a\\b\\c\end{pmatrix} = \begin{pmatrix}0\\0\\0\\0\end{pmatrix}.$$

Row reduction gives the echelon form

$$\begin{pmatrix}1 & 1 & 1\\0 & -2 & -2\\0 & 0 & -2\\0 & 0 & 0\end{pmatrix},$$

where we assumed that $-2 \neq 0$. As there is a pivot in every column, we get that $\{v_1 + v_2 + v_3 + v_4, v_1 - v_2 + v_3 - v_4, v_1 - v_2 - v_3 - v_4\}$ is linearly independent. We assumed that $\mathbb{F} \neq \mathbb{Z}_2$.

If $\mathbb{F} = \mathbb{Z}_2$, then $\{v_1 + v_2 + v_3 + v_4, v_1 - v_2 + v_3 - v_4, v_1 - v_2 - v_3 - v_4\}$ is linearly dependent.

(b) Here we obtain the matrix

$$\begin{pmatrix} 1 & 0 & 0 & 0 & 0 \\ 1 & 1 & 0 & 0 & 1 \\ 0 & 1 & 1 & 0 & 0 \\ 0 & 0 & 1 & 1 & 0 \\ 0 & 0 & 0 & 1 & 1 \end{pmatrix}.$$

The echelon form is

$$\begin{pmatrix} 1 & 0 & 0 & 0 & 0 \\ 0 & 1 & 0 & 0 & 1 \\ 0 & 0 & 1 & 0 & -1 \\ 0 & 0 & 0 & 1 & 1 \\ 0 & 0 & 0 & 0 & 0 \end{pmatrix}.$$

No pivot in the last column, so $\{v_1 + v_2, v_2 + v_3, v_3 + v_4, v_4 + v_5, v_5 + v_2\}$ is linearly dependent. This works for all fields.

(c) Here we have six vectors in the five-dimensional space $\mathrm{Span}\{v_1, v_2, v_3, v_4, v_5\}$. Thus these vectors are linearly dependent. This works for all fields.

Exercise 2.6.12

Let $\{v_1, v_2, v_3, v_4\}$ be a basis for a vector space V over $\mathbb{Z}_3$. Determine whether the following are also bases for V.

(a) $\{v_1 + v_2 + v_3 + v_4, v_1 - v_2 + v_3 - v_4, v_1 - v_2 - v_3 - v_4.\}$
(b) $\{v_1, v_2 + v_3 + v_4, v_1 - v_2 + v_3 - v_4, v_1 - v_2 - v_3 - v_4.\}$
(c) $\{v_1 + v_2 + v_3 + v_4, v_1 - v_2 + v_3 - v_4, v_1 - v_2 - v_3 - v_4, v_2 + v_4, v_1 + v_3.\}$

Answer: (a) These three vectors can never span the four-dimensional space V, so this is not a basis.

(b) Here we obtain the matrix

$$\begin{pmatrix} 1 & 0 & 1 & 1 \\ 0 & 1 & -1 & -1 \\ 0 & 1 & 1 & -1 \\ 0 & 1 & -1 & -1 \end{pmatrix}.$$

The echelon form is

$$\begin{pmatrix} 1 & 0 & 1 & 1 \\ 0 & 1 & -1 & -1 \\ 0 & 0 & -2 & 0 \\ 0 & 0 & 0 & 0 \end{pmatrix}.$$

No pivot in the last column, so linearly dependent. Thus not a basis.

(c) Here we have five vectors in a four-dimensional vector space, thus not a basis.

Exercise 2.6.13

For the following choices of vector spaces V over the field $\mathbb{F}$, bases $\mathcal{B}$ and vectors $\mathbf{v}$, determine $[\mathbf{v}]_\mathcal{B}$.

(a) Let $\mathbb{F} = \mathbb{Z}_5$, $V = \mathbb{Z}_5^4$,

$$\mathcal{B} = \{\begin{pmatrix} 3 \\ 0 \\ 2 \\ 1 \end{pmatrix}, \begin{pmatrix} 2 \\ 1 \\ 0 \\ 0 \end{pmatrix}, \begin{pmatrix} 1 \\ 2 \\ 1 \\ 0 \end{pmatrix}, \begin{pmatrix} 0 \\ 2 \\ 1 \\ 0 \end{pmatrix}\}, \mathbf{v} = \begin{pmatrix} 1 \\ 3 \\ 2 \\ 2 \end{pmatrix}.$$

(b) Let $\mathbb{F} = \mathbb{R}$, $\mathcal{B} = \{t, t^2, \frac{1}{t}\}$, $V = \text{Span}\mathcal{B}$ and $\mathbf{v} = \frac{t^3 + 3t^2 + 5}{t}$.

(c) Let $\mathbb{F} = \mathbb{C}$, $V = \mathbb{C}^{2 \times 2}$,

$$\mathcal{B} = \{\begin{pmatrix} 0 & 1 \\ -1 & -i \end{pmatrix}, \begin{pmatrix} 1 & 1 \\ i & -i \end{pmatrix}, \begin{pmatrix} i & 0 \\ -1 & -i \end{pmatrix}, \begin{pmatrix} i & 1 \\ -1 & -i \end{pmatrix}\}, \mathbf{v} = \begin{pmatrix} -2+i & 3-2i \\ -5-i & 10 \end{pmatrix}.$$

(d) Let $\mathbb{F} = \mathbb{R}$, $V = \mathbb{C}^{2 \times 2}$, and consider the vectors

$$\mathcal{B} = \{E_{11}, E_{12}, E_{21}, E_{22}, iE_{11}, iE_{12}, iE_{21}, iE_{22}\}, \mathbf{v} = \begin{pmatrix} -1 & i \\ -i & 1 \end{pmatrix}.$$

(e) Let $\mathbb{F} = \mathbb{Z}_5$, $V = \text{Span}\mathcal{B}$,

$$\mathcal{B} = \left\{ \begin{pmatrix} 3 & 4 \\ 1 & 0 \\ 1 & 0 \end{pmatrix}, \begin{pmatrix} 1 & 1 \\ 4 & 2 \\ 1 & 2 \end{pmatrix}, \begin{pmatrix} 1 & 2 \\ 3 & 3 \\ 3 & 0 \end{pmatrix} \right\}, \mathbf{v} = \begin{pmatrix} 0 & 2 \\ 3 & 0 \\ 0 & 2 \end{pmatrix}.$$

Answer: (a) $[\mathbf{v}]_\mathcal{B} = \begin{pmatrix} 2 \\ 2 \\ 1 \\ 2 \end{pmatrix}$.

(b) $[\mathbf{v}]_\mathcal{B} = \begin{pmatrix} 3 \\ 1 \\ 5 \end{pmatrix}$.

(c) $[\mathbf{v}]_\mathcal{B} = \begin{pmatrix} 4 - i \\ 2 + 7i \\ -3 + 12i \\ -3 - 8i \end{pmatrix}$.

(d) $[\mathbf{v}]_\mathcal{B} = \begin{pmatrix} -1 \\ 0 \\ 0 \\ 1 \\ 0 \\ 1 \\ -1 \\ 0 \end{pmatrix}$.

(e) $[\mathbf{v}]_\mathcal{B} = \begin{pmatrix} 1 \\ 1 \\ 1 \end{pmatrix}$.

Exercise 2.6.14 Given a matrix $A = (a_{jk})_{j=1,k=1}^{n\ \ \ \ m} \in \mathbb{C}^{n \times m}$, we define
$A^* = (\overline{a_{kj}})_{j=1,k=1}^{m\ \ \ \ n} \in \mathbb{C}^{m \times n}$. For instance,

$$\begin{pmatrix} 1+2i & 3+4i & 5+6i \\ 7+8i & 9+10i & 11+12i \end{pmatrix}^* = \begin{pmatrix} 1-2i & 7-8i \\ 3-4i & 9-10i \\ 5-6i & 11-12i \end{pmatrix}.$$

We call a matrix $A \in \mathbb{C}^{n \times n}$ *Hermitian* if $A^* = A$. For instance, $\begin{pmatrix} 2 & 1-3i \\ 1+3i & 5 \end{pmatrix}$ is
Hermitian. Let $H_n \subseteq \mathbb{C}^{n \times n}$ be the set of all $n \times n$ Hermitian matrices.

(a) Show that H_n is **not** a vector space over $\mathbb{C}$.

(b) Show that H_n is a vector space over $\mathbb{R}$. Determine $\dim_{\mathbb{R}} H_n$.

(Hint: Do it first for 2×2 matrices.)

Answer: (a) $\begin{pmatrix} 0 & i \\ -i & 0 \end{pmatrix} \in H_2$, but $i \begin{pmatrix} 0 & i \\ -i & 0 \end{pmatrix} = \begin{pmatrix} 0 & -1 \\ 1 & 0 \end{pmatrix} \notin H_2$.

(b) We observe that $(A+B)^* = A^* + B^*$ and $(cA)^* = cA^*$, when $c \in \mathbb{R}$. Observe that the
zero matrix is in H_n. Next, if $A, B \in H_n$, then $(A+B)^* = A^* + B^* = A + B$, thus
$A + B \in H_n$. Finally, if $c \in \mathbb{R}$ and $A \in H_n$, then $(cA)^* = cA^* = cA$, thus $cA \in H_n$. This
shows that H_n is a subspace over $\mathbb{R}$.

As a basis for H_n we can choose

$$\{E_{jj} : 1 \le j \le n\} \cup \{E_{jk} + E_{kj} : 1 \le j < k \le n\} \cup \{iE_{jk} - iE_{kj} : 1 \le j < k \le n\}.$$

There are $n + 2\sum_{j=1}^{n-1} j = n^2$ elements in this basis, thus $\dim_{\mathbb{R}} H_n = n^2$.

Exercise 2.6.15 (a) Show that for finite-dimensional subspaces U and W of V we have
that $\dim(U + W) = \dim U + \dim W - \dim(U \cap W)$.
(Hint: Start with a basis $\{\mathbf{v}_1, \ldots, \mathbf{v}_p\}$ for $U \cap W$. Next, find $\mathbf{u}_1, \ldots, \mathbf{u}_k$ so that
$\{\mathbf{v}_1, \ldots, \mathbf{v}_p, \mathbf{u}_1, \ldots, \mathbf{u}_k\}$ is a basis for U. Similarly, find $\mathbf{w}_1, \ldots, \mathbf{w}_l$ so that
$\{\mathbf{v}_1, \ldots, \mathbf{v}_p, \mathbf{w}_1, \ldots, \mathbf{w}_l\}$ is a basis for W. Finally, argue that
$\{\mathbf{v}_1, \ldots, \mathbf{v}_p, \mathbf{u}_1, \ldots, \mathbf{u}_k, \mathbf{w}_1, \ldots, \mathbf{w}_l\}$ is a basis for $U + W$.)

(b) Show that for a direct sum $U_1 \dotplus \cdots \dotplus U_k$ of finite-dimensional subspaces $U_1, \ldots, U_k$,
we have that
$$\dim(U_1 \dotplus \cdots \dotplus U_k) = \dim U_1 + \cdots + \dim U_k.$$

Answer: (a) Following the hint we need to show that $\{\mathbf{v}_1, \ldots, \mathbf{v}_p, \mathbf{u}_1, \ldots, \mathbf{u}_k, \mathbf{w}_1, \ldots, \mathbf{w}_l\}$
is a basis for $U + W$. First, let $\mathbf{v}$ in $U + W$. Then there exists a $\mathbf{u} \in U$ and a $\mathbf{w} \in W$ so
that $\mathbf{v} = \mathbf{u} + \mathbf{w}$. As $\mathbf{u} \in U$, there exists a_i and b_i so that

$$\mathbf{u} = \sum_{i=1}^{p} a_i \mathbf{v}_i + \sum_{i=1}^{k} b_i \mathbf{u}_i.$$

As $\mathbf{w} \in W$, there exists c_i and d_i so that

$$\mathbf{w} = \sum_{i=1}^{p} c_i \mathbf{v}_i + \sum_{i=1}^{l} d_i \mathbf{w}_i.$$

Then $\mathbf{v} = \mathbf{u} + \mathbf{w} = \sum_{i=1}^{p}(a_i + c_i)\mathbf{v}_i + \sum_{i=1}^{k} b_i \mathbf{u}_i + \sum_{i=1}^{l} d_i \mathbf{w}_i$, thus
$\{\mathbf{v}_1, \ldots, \mathbf{v}_p, \mathbf{u}_1, \ldots, \mathbf{u}_k, \mathbf{w}_1, \ldots, \mathbf{w}_l\}$ span $U + W$. Next, to show linear independence,

suppose that

$$\sum_{i=1}^{p} a_i \mathbf{v}_i + \sum_{i=1}^{k} b_i \mathbf{u}_i + \sum_{i=1}^{l} c_i \mathbf{w}_i = \mathbf{0}.$$

Then

$$\sum_{i=1}^{p} a_i \mathbf{v}_i + \sum_{i=1}^{k} b_i \mathbf{u}_i = -\sum_{i=1}^{l} c_i \mathbf{w}_i \in U \cap W.$$

As $\{\mathbf{v}_1, \ldots, \mathbf{v}_p\}$ is a basis for $U \cap W$, there exist d_i so that

$$-\sum_{i=1}^{l} c_i \mathbf{w}_i = \sum_{i=1}^{p} d_i \mathbf{v}_i.$$

Then $\sum_{i=1}^{p} d_i \mathbf{v}_i + \sum_{i=1}^{l} c_i \mathbf{w}_i = \mathbf{0}$. As $\{\mathbf{v}_1, \ldots, \mathbf{v}_p, \mathbf{w}_1, \ldots, \mathbf{w}_l\}$ is linearly independent, we get that $d_1 = \cdots = d_p = c_1 = \cdots = c_l = 0$. But then we get that $\sum_{i=1}^{p} a_i \mathbf{v}_i + \sum_{i=1}^{k} b_i \mathbf{u}_i = \mathbf{0}$. Using now that $\{\mathbf{v}_1, \ldots, \mathbf{v}_p, \mathbf{u}_1, \ldots, \mathbf{u}_k\}$ is linearly independent, we get $a_1 = \cdots = a_p = b_1 = \cdots = b_k = 0$. This shows that $\{\mathbf{v}_1, \ldots, \mathbf{v}_p, \mathbf{u}_1, \ldots, \mathbf{u}_k, \mathbf{w}_1, \ldots, \mathbf{w}_l\}$ is linearly independent, proving that it is a basis for $U + W$.

Thus $\dim U + W = p + k + l = (p + k) + (p + l) - p = \dim U + \dim W - \dim(U \cap W)$.

(b) We show this by induction. It is trivial for $k = 1$. Suppose we have proven the statement for $k - 1$, giving $\dim(U_1 \dotplus \cdots \dotplus U_{k-1}) = \dim U_1 + \cdots + \dim U_{k-1}$. Then, using (a) we get

$$\dim[(U_1 \dotplus \cdots \dotplus U_{k-1}) \dotplus U_k] = \dim(U_1 \dotplus \cdots \dotplus U_{k-1}) + \dim U_k -$$

$$\dim[(U_1 \dotplus \cdots \dotplus U_{k-1}) \cap U_k] = \dim(U_1 \dotplus \cdots \dotplus U_{k-1}) + \dim U_k - 0,$$

where we used that $(U_1 \dotplus \cdots \dotplus U_{k-1}) \cap U_k = \{0\}$. Now using the induction assumption, we get

$$\dim[(U_1 \dotplus \cdots \dotplus U_{k-1}) \dotplus U_k] = \dim(U_1 \dotplus \cdots \dotplus U_{k-1}) + \dim U_k =$$

$$\dim U_1 + \cdots + \dim U_{k-1} + \dim U_k.$$

This proves the statement.

Chapter 3

Exercise 3.4.1 Let $T : V \to W$ and $S : W \to X$ be linear maps. Show that the composition $S \circ T : V \to X$ is also linear.

Answer: $(S \circ T)(\mathbf{v} + \mathbf{w}) = S(T(\mathbf{v} + \mathbf{w})) = S(T(\mathbf{v}) + T(\mathbf{w})) = S(T(\mathbf{v})) + S(T(\mathbf{w})) = (S \circ T)(\mathbf{v}) + (S \circ T)(\mathbf{w})$, and
$(S \circ T)(c\mathbf{v}) = S(T(c\mathbf{v})) = S(cT(\mathbf{v})) = cS(T(\mathbf{v})) = c(S \circ T)(\mathbf{v})$, proving linearity.

Exercise 3.4.2 For the following choices of V, W and $T : V \to W$, determine whether T is linear or not.

(a) $V = \mathbb{R}^3$, $W = \mathbb{R}^4$,
$$T \begin{pmatrix} x_1 \\ x_2 \\ x_3 \end{pmatrix} = \begin{pmatrix} x_1 - 5x_3 \\ 7x_2 + 5 \\ 3x_1 - 6x_2 \\ 8x_3 \end{pmatrix}.$$

(b) $V = \mathbb{Z}_5^3$, $W = \mathbb{Z}_5^2$,
$$T \begin{pmatrix} x_1 \\ x_2 \\ x_3 \end{pmatrix} = \begin{pmatrix} x_1 - 2x_3 \\ 3x_2 x_3 \end{pmatrix}.$$

(c) $V = W = \mathbb{C}^{2 \times 2}$ (over $\mathbb{F} = \mathbb{C}$), $T(A) = A - A^T$.

(d) $V = W = \mathbb{C}^{2 \times 2}$ (over $\mathbb{F} = \mathbb{C}$), $T(A) = A - A^*$.

(e) $V = W = \mathbb{C}^{2 \times 2}$ (over $\mathbb{F} = \mathbb{R}$), $T(A) = A - A^*$.

(f) $V = \{f : \mathbb{R} \to \mathbb{R} \ : \ f \text{ is differentiable}\}$, $W = \mathbb{R}^{\mathbb{R}}$,
$$(T(f))(x) = f'(x)(x^2 + 5).$$

(g) $V = \{f : \mathbb{R} \to \mathbb{R} \ : \ f \text{ is continuous}\}$, $W = \mathbb{R}$,
$$T(f) = \int_{-5}^{10} f(x)\,dx.$$

Answer: (a) $T \begin{pmatrix} 0 \\ 0 \\ 0 \end{pmatrix} = \begin{pmatrix} 0 \\ 5 \\ 0 \\ 0 \end{pmatrix} \neq \mathbf{0}$, thus T is not linear.

(b) $2T \begin{pmatrix} 0 \\ 1 \\ 1 \end{pmatrix} = 2 \begin{pmatrix} -2 \\ 3 \end{pmatrix} \neq \begin{pmatrix} -4 \\ 12 \end{pmatrix} = T \begin{pmatrix} 0 \\ 2 \\ 2 \end{pmatrix}$, so T is not linear.

(c) $T(A + B) = A + B - (A + B)^T = A + B - A^T - B^T = T(A) + T(B)$ and $T(cA) = (cA) - (cA)^T = cA - cA^T = cT(A)$, thus T is linear.

(d) $T(i \begin{pmatrix} 0 & 1 \\ 0 & 0 \end{pmatrix}) = \begin{pmatrix} 0 & i \\ 0 & 0 \end{pmatrix} - \begin{pmatrix} 0 & 0 \\ -i & 0 \end{pmatrix}$, however $iT \begin{pmatrix} 0 & 1 \\ 0 & 0 \end{pmatrix} = i(\begin{pmatrix} 0 & 1 \\ 0 & 0 \end{pmatrix} - \begin{pmatrix} 0 & 0 \\ 1 & 0 \end{pmatrix})$, thus T does not satisfy the rule $T(cA) = cT(A)$.

(e) $T(A+B) = A+B-(A+B)^* = A+B-A^*-B^* = T(A)+T(B)$ and $T(cA) = (cA)-(cA)^* = cA-\bar{c}A^* = cT(A)$, where in the last step we used that $\bar{c} = c$ as c is real. Thus T is linear.

(f) $(T(f+g))(x) = (f+g)'(x)(x^2+5) = (f'(x)+g'(x))(x^2+5) = T(f)(x)+T(g)(x) = (T(f)+T(g))(x)$, and $T(cf)(x) = (cf)'(x)(x^2+5) = cf'(x)(x^2+5) = c(T(f))(x)$, thus T is linear.

(g) $T(f+g) = \int_{-5}^{10} f(x)+g(x)dx = \int_{-5}^{10} f(x)dx + \int_{-5}^{10} g(x)dx = T(f)+T(g)$, and $T(cf) = \int_{-5}^{10} cf(x)dx = c\int_{-5}^{10} f(x)dx = cT(f)$, and thus T is linear.

Exercise 3.4.3 Show that if $T : V \to W$ is linear and the set $\{T(\mathbf{v}_1),\ldots,T(\mathbf{v}_k)\}$ is linearly independent, then the set $\{\mathbf{v}_1,\ldots,\mathbf{v}_k\}$ is linearly independent.

Answer: Let $c_1,\ldots,c_k$ be so that $c_1\mathbf{v}_1 + \cdots + c_k\mathbf{v}_k = \mathbf{0}$. We need to show that $c_1 = \cdots = c_k = 0$. We have $T(c_1\mathbf{v}_1 + \cdots + c_k\mathbf{v}_k) = T(\mathbf{0})$, which gives that $c_1 T(\mathbf{v}_1) + \cdots + c_k T(\mathbf{v}_k) = \mathbf{0}$. As $\{T(\mathbf{v}_1),\ldots,T(\mathbf{v}_k)\}$ is linearly independent, we get $c_1 = \cdots = c_k = 0$.

Exercise 3.4.4 Show that if $T : V \to W$ is linear and onto, and $\{\mathbf{v}_1 \ldots,\mathbf{v}_k\}$ is a basis for V, then the set $\{T(\mathbf{v}_1),\ldots,T(\mathbf{v}_k)\}$ spans W. When is $\{T(\mathbf{v}_1),\ldots,T(\mathbf{v}_k)\}$ a basis for W?

Answer: We need to show that every $\mathbf{w} \in W$ is a linear combination of $T(\mathbf{v}_1),\ldots,T(\mathbf{v}_k)$. So, let $\mathbf{w} \in W$. As T is onto, there exists a $\mathbf{v} \in V$ so that $T(\mathbf{v}) = \mathbf{w}$. As $\{\mathbf{v}_1 \ldots,\mathbf{v}_k\}$ is a basis for V, there exist scalars $c_1,\ldots,c_k$ so that $\mathbf{v} = c_1\mathbf{v}_1 + \cdots + c_k\mathbf{v}_k$. Then

$$\mathbf{w} = T(\mathbf{v}) = T(c_1\mathbf{v}_1 + \cdots + c_k\mathbf{v}_k) = c_1 T(\mathbf{v}_1) + \cdots + c_k T(\mathbf{v}_k),$$

where in the last equality we use the linearity of T. Thus $\mathbf{w}$ is a linear combination of $T(\mathbf{v}_1),\ldots,T(\mathbf{v}_k)$.

When T is not one-to-one, then $\{T(\mathbf{v}_1),\ldots,T(\mathbf{v}_k)\}$ is linearly independent, and therefore a basis. Indeed, suppose that $c_1 T(\mathbf{v}_1) + \cdots + c_k T(\mathbf{v}_k) = \mathbf{0}$. Then $T(c_1\mathbf{v}_1 + \cdots + c_k\mathbf{v}_k) = T(\mathbf{0})$. When T is one-to-one, this implies $c_1\mathbf{v}_1 + \cdots + c_k\mathbf{v}_k = \mathbf{0}$. As $\{\mathbf{v}_1 \ldots,\mathbf{v}_k\}$ is linearly independent, this yields $c_1 = \cdots = c_k = 0$.

Exercise 3.4.5 Let $T : V \to W$ be linear, and let $U \subseteq V$ be a subspace of V. Define

$$T[U] := \{\mathbf{w} \in W \text{ ; there exists } \mathbf{u} \in U \text{ so that } \mathbf{w} = T(\mathbf{u})\}. \tag{3.25}$$

Observe that $T[V] = \operatorname{Ran} T$.

(a) Show that $T[U]$ is a subspace of W.

(b) Assuming $\dim U < \infty$, show that $\dim T[U] \leq \dim U$.

(c) If $\hat{U}$ is another subspace of V, is it always true that $T[U + \hat{U}] = T[U] + T[\hat{U}]$? If so, provide a proof. If not, provide a counterexample.

(d) If $\hat{U}$ is another subspace of V, is it always true that $T[U \cap \hat{U}] = T[U] \cap T[\hat{U}]$? If so, provide a proof. If not, provide a counterexample.

Answer: (a) First observe that $\mathbf{0} \in U$ and $T(\mathbf{0}) = \mathbf{0}$ gives that $\mathbf{0} \in T[U]$. Next, let $\mathbf{w}$, $\hat{\mathbf{w}} \in T[U]$ and $c \in \mathbb{F}$. Then there exist $\mathbf{u}$, $\hat{\mathbf{u}} \in U$ so that $T(\mathbf{u}) = \mathbf{w}$ and $T(\hat{\mathbf{u}}) = \hat{\mathbf{w}}$. Then $\mathbf{w} + \hat{\mathbf{w}} = T(\mathbf{u} + \hat{\mathbf{u}}) \in T[U]$ and $c\mathbf{w} = T(c\mathbf{u}) \in T[U]$. Thus, by Proposition 2.3.1, $T[U]$ is a subspace of W.

(b) Let $\{\mathbf{v}_1, \ldots, \mathbf{v}_p\}$ be a basis for U. We claim that $T[U] = \text{Span}\{T(\mathbf{v}_1), \ldots, T(\mathbf{v}_p)\}$, from which it then follows that $\dim T[U] \leq \dim U$.

Clearly, $T(\mathbf{v}_1), \ldots, T(\mathbf{v}_p) \in T[U]$, and since $T[U]$ is a subspace we have that $\text{Span}\{T(\mathbf{v}_1), \ldots, T(\mathbf{v}_p)\} \subseteq T[U]$. For the converse inclusion, let $\mathbf{w} \in T[U]$. Then there exists a $\mathbf{v} \in U$ so that $T(\mathbf{v}) = \mathbf{w}$. As $\{\mathbf{v}_1, \ldots, \mathbf{v}_p\}$ is a basis for U, there exist $c_1, \ldots, c_p \in \mathbb{F}$ so that $\mathbf{v} = c_1 \mathbf{v}_1 + \cdots + c_p \mathbf{v}_p$. Then

$$\mathbf{w} = T(\mathbf{v}) = T\left(\sum_{j=1}^{p} c_j \mathbf{v}_j\right) = \sum_{j=1}^{p} c_j T(\mathbf{v}_j) \in \text{Span}\{T(\mathbf{v}_1), \ldots, T(\mathbf{v}_p)\}.$$

Thus $T[U] \subseteq \text{Span}\{T(\mathbf{v}_1), \ldots, T(\mathbf{v}_p)\}$. We have shown both inclusions, and consequently $T[U] = \text{Span}\{T(\mathbf{v}_1), \ldots, T(\mathbf{v}_p)\}$ follows.

(c) Let $w \in T[U + \hat{U}]$. Then there exists a $\mathbf{v} \in U + \hat{U}$ so that $\mathbf{w} = T(\mathbf{v})$. As $\mathbf{v} \in U + \hat{U}$ there exists $\mathbf{u} \in U$ and $\hat{\mathbf{u}} \in \hat{U}$ so that $\mathbf{v} = \mathbf{u} + \hat{\mathbf{u}}$. Then $\mathbf{w} = T(\mathbf{v}) = T(\mathbf{u} + \hat{\mathbf{u}}) = T(\mathbf{u}) + T(\hat{\mathbf{u}}) \in T[U] + T[\hat{U}]$. This proves $T[U + \hat{U}] \subseteq T[U] + T[\hat{U}]$.

For the converse inclusion, let $\mathbf{w} \in T[U] + T[\hat{U}]$. Then there is an $\mathbf{x} \in T[U]$ and a $\hat{\mathbf{x}} \in T[\hat{U}]$, so that $\mathbf{w} = \mathbf{x} + \hat{\mathbf{x}}$. As $\mathbf{x} \in T[U]$, there exists a $\mathbf{u} \in U$ so that $\mathbf{x} = T(\mathbf{u})$. As $\hat{\mathbf{x}} \in T[\hat{U}]$, there exists a $\hat{\mathbf{u}} \in \hat{U}$ so that $\hat{\mathbf{x}} = T(\hat{\mathbf{u}})$. Then $\mathbf{w} = \mathbf{x} + \hat{\mathbf{x}} = T(\mathbf{u}) + T(\hat{\mathbf{u}}) = T(\mathbf{u} + \hat{\mathbf{u}}) \in T[U + \hat{U}]$. This proves $T[U] + T[\hat{U}] \subseteq T[U + \hat{U}]$, and we are done.

(d) Let $T : \mathbb{R}^2 \to \mathbb{R}^2$ be given via $T\begin{pmatrix} x_1 \\ x_2 \end{pmatrix} = \begin{pmatrix} x_1 + x_2 \\ 0 \end{pmatrix}$, and let $U = \text{Span}\{\mathbf{e}_1\}$ and $\hat{U} = \text{Span}\{\mathbf{e}_2\}$. Then $T[U \cap \hat{U}] = T[\{\mathbf{0}\}] = \{\mathbf{0}\}$, while $T[U] \cap T[\hat{U}] = \text{Span}\{\mathbf{e}_1\} \cap \text{Span}\{\mathbf{e}_1\} = \text{Span}\{\mathbf{e}_1\}$. So $T[U \cap \hat{U}] \neq T[U] \cap T[\hat{U}]$ in this case.

Exercise 3.4.6 Let $\mathbf{v}_1, \mathbf{v}_2, \mathbf{v}_3, \mathbf{v}_4$ be a basis for a vector space V.

(a) Let $T : V \to V$ be given by $T(\mathbf{v}_i) = \mathbf{v}_{i+1}, i = 1, 2, 3$, and $T(\mathbf{v}_4) = \mathbf{v}_1$. Determine the matrix representation of T with respect to the basis $\{\mathbf{v}_1, \mathbf{v}_2, \mathbf{v}_3, \mathbf{v}_4\}$.

(b) If the matrix representation of a linear map $S : V \to V$ with respect to the $\{\mathbf{v}_1, \mathbf{v}_2, \mathbf{v}_3, \mathbf{v}_4\}$ is given by

$$\begin{pmatrix} 1 & 0 & 1 & 1 \\ 0 & 2 & 0 & 2 \\ 1 & 2 & 1 & 3 \\ -1 & 0 & -1 & -1 \end{pmatrix},$$

determine $S(\mathbf{v}_1 - \mathbf{v}_4)$.

(c) Determine bases for Ran S and Ker S.

Answer: (a) $\begin{pmatrix} 0 & 0 & 0 & 1 \\ 1 & 0 & 0 & 0 \\ 0 & 1 & 0 & 0 \\ 0 & 0 & 1 & 0 \end{pmatrix}$.

(b) $S(\mathbf{v}_1 - \mathbf{v}_4) = S(\mathbf{v}_1) - S(\mathbf{v}_4) = \mathbf{v}_1 + \mathbf{v}_3 - \mathbf{v}_4 - (\mathbf{v}_1 + 2\mathbf{v}_2 + 3\mathbf{v}_3 - \mathbf{v}_4) = -2\mathbf{v}_2 - 2\mathbf{v}_3$.

(c) The reduced echelon form of the matrix representation in (b) is $\begin{pmatrix} 1 & 0 & 1 & 1 \\ 0 & 1 & 0 & 1 \\ 0 & 0 & 0 & 0 \\ 0 & 0 & 0 & 0 \end{pmatrix}$.

From this we deduce that with respect to the basis $\{v_1, v_2, v_3, v_4\}$ we have that

$\{\begin{pmatrix} -1 \\ 0 \\ 1 \\ 0 \end{pmatrix}, \begin{pmatrix} -1 \\ -1 \\ 0 \\ 1 \end{pmatrix}\}$ is a basis for Ker T, $\{\begin{pmatrix} 1 \\ 0 \\ 1 \\ -1 \end{pmatrix}, \begin{pmatrix} 0 \\ 2 \\ 2 \\ 0 \end{pmatrix}\}$ is a basis for Ran T. In other

words, $\{-v_1 + v_3, -v_1 - v_2 + v_4\}$ is a basis for Ker T, $\{v_1 + v_3 - v_4, 2v_2 + 2v_3\}$ is a basis for Ran T.

Exercise 3.4.7 Consider the linear map $T : \mathbb{R}_2[X] \to \mathbb{R}^2$ given by $T(p(X)) = \begin{pmatrix} p(1) \\ p(3) \end{pmatrix}$.

(a) Find a basis for the kernel of T.

(b) Find a basis for the range of T.

Answer: (a) $\{(X-1)(X-3)\} = \{X^2 - 4X + 3\}$.

(b) Ran $T = \mathbb{R}^2$, so a possible basis is $\{e_1, e_2\}$.

Exercise 3.4.8 Let $T : V \to W$ with $V = \mathbb{Z}_5^4$ and $W = \mathbb{Z}_5^{2\times 2}$ be defined by

$$T(\begin{pmatrix} a \\ b \\ c \\ d \end{pmatrix}) = \begin{pmatrix} a+b & b+c \\ c+d & d+a \end{pmatrix}.$$

(a) Find a basis for the kernel of T.

(b) Find a basis for the range of T.

Answer: (a) $\{\begin{pmatrix} 4 \\ 1 \\ 4 \\ 1 \end{pmatrix}\}$.

(b) $\{\begin{pmatrix} 1 & 0 \\ 0 & 1 \end{pmatrix}, \begin{pmatrix} 1 & 1 \\ 0 & 0 \end{pmatrix}, \begin{pmatrix} 0 & 1 \\ 1 & 0 \end{pmatrix}\}$.

Exercise 3.4.9 For the following $T : V \to W$ with bases $\mathcal{B}$ and $\mathcal{C}$, respectively, determine the matrix representation for T with respect to the bases $\mathcal{B}$ and $\mathcal{C}$. In addition, find bases for the range and kernel of T.

(a) $\mathcal{B} = \mathcal{C} = \{\sin t, \cos t, \sin 2t, \cos 2t\}$, $V = W = \text{Span } \mathcal{B}$, and $T = \frac{d^2}{dt^2} + \frac{d}{dt}$.

(b) $\mathcal{B} = \{1, t, t^2, t^3\}, \mathcal{C} = \{\begin{pmatrix} 1 \\ 0 \end{pmatrix}, \begin{pmatrix} 1 \\ -1 \end{pmatrix}\}$, $V = \mathbb{C}_3[X]$, and $W = \mathbb{C}^2$, and $T(p) = \begin{pmatrix} p(3) \\ p(5) \end{pmatrix}$.

(c) $\mathcal{B} = \mathcal{C} = \{e^t \cos t, e^t \sin t, e^{3t}, te^{3t}\}$, $V = W = \text{Span } \mathcal{B}$, and $T = \frac{d}{dt}$.

(d) $\mathcal{B} = \{1, t, t^2\}, \mathcal{C} = \{\begin{pmatrix} 1 \\ 1 \end{pmatrix}, \begin{pmatrix} 1 \\ 0 \end{pmatrix}\}$, $V = \mathbb{C}_2[X]$, and $W = \mathbb{C}^2$, and $T(p) = \begin{pmatrix} \int_0^1 p(t)dt \\ p(1) \end{pmatrix}$.

Answer: (a) $\begin{pmatrix} -1 & -1 & 0 & 0 \\ 1 & -1 & 0 & 0 \\ 0 & 0 & -4 & -2 \\ 0 & 0 & 2 & -4 \end{pmatrix}$, Ker $T = \{0\}$, $\mathcal{B}$ is a basis for Ran T.

(b) $\begin{pmatrix} 2 & 8 & 34 & 152 \\ -1 & -5 & -25 & -125 \end{pmatrix}$, $\{ \begin{pmatrix} 15 \\ -8 \\ 1 \\ 0 \end{pmatrix}, \begin{pmatrix} 120 \\ -49 \\ 0 \\ 1 \end{pmatrix} \}$ is a basis for Ker T, $\{e_1, e_2\}$ is a

basis for Ran T.

(c) $\begin{pmatrix} 1 & 1 & 0 & 0 \\ -1 & 1 & 0 & 0 \\ 0 & 0 & 3 & 1 \\ 0 & 0 & 0 & 3 \end{pmatrix}$, Ker $T = \{0\}$, $\{e_1, e_2, e_3, e_4\}$ is a basis for Ran T.

(d) $\begin{pmatrix} 1 & 1 & 1 \\ 0 & -\frac{1}{2} & -\frac{2}{3} \end{pmatrix}$, $\{\frac{1}{3} - \frac{4}{3}X + X^2\}$ is a basis for Ker T, $\{e_1, e_2\}$ is a basis for Ran T.

Exercise 3.4.10 Let $V = \mathbb{C}^{n \times n}$. Define $L : V \to V$ via $L(A) = \frac{1}{2}(A + A^T)$.

(a) Let

$$B = \{ \begin{pmatrix} 1 & 0 \\ 0 & 0 \end{pmatrix}, \begin{pmatrix} 0 & 1 \\ 0 & 0 \end{pmatrix}, \begin{pmatrix} 0 & 0 \\ 1 & 0 \end{pmatrix}, \begin{pmatrix} 0 & 0 \\ 0 & 1 \end{pmatrix} \}.$$

Determine the matrix representation of L with respect to the basis B.

(b) Determine the dimensions of the subspaces

$$W = \{A \in V \ : \ L(A) = A\} \text{ and Ker } L = \{A \in V \ : \ L(A) = 0\}.$$

(c) Determine the eigenvalues of L.

Answer: (a) $C := [L]_{B \leftarrow B} = \begin{pmatrix} 1 & 0 & 0 & 0 \\ 0 & \frac{1}{2} & \frac{1}{2} & 0 \\ 0 & \frac{1}{2} & \frac{1}{2} & 0 \\ 0 & 0 & 0 & 1 \end{pmatrix}$.

(b) Row reduce $C - I = \begin{pmatrix} 0 & 0 & 0 & 0 \\ 0 & -\frac{1}{2} & \frac{1}{2} & 0 \\ 0 & \frac{1}{2} & -\frac{1}{2} & 0 \\ 0 & 0 & 0 & 0 \end{pmatrix} \to \begin{pmatrix} 0 & -\frac{1}{2} & \frac{1}{2} & 0 \\ 0 & 0 & 0 & 0 \\ 0 & 0 & 0 & 0 \\ 0 & 0 & 0 & 0 \end{pmatrix}$, so dim $W = 3$.

Row reduce $C = \begin{pmatrix} 1 & 0 & 0 & 0 \\ 0 & \frac{1}{2} & \frac{1}{2} & 0 \\ 0 & \frac{1}{2} & \frac{1}{2} & 0 \\ 0 & 0 & 0 & 1 \end{pmatrix} \to \begin{pmatrix} 1 & 0 & 0 & 0 \\ 0 & \frac{1}{2} & \frac{1}{2} & 0 \\ 0 & 0 & 0 & 1 \\ 0 & 0 & 0 & 0 \end{pmatrix}$, so dim Ker $L = 1$.

(c) 0 and 1 are the only eigenvalues of L.

Exercise 3.4.11 Let $B = \{1, t, \ldots, t^n\}$, $C = \{1, t, \ldots, t^{n+1}\}$, $V = $ Span B and $W = $ Span C. Define $A : V \to W$ via

$$Af(t) := (2t^2 - 3t + 4)f'(t),$$

where f' is the derivative of f.

(a) Find the matrix representation of A with respect to the bases B and C.

(b) Find bases for Ran A and Ker A.

Answer: (a)
$$\begin{pmatrix} 0 & 4 & 0 & \cdots & & 0 \\ 0 & -3 & 8 & \ddots & & \vdots \\ & 2 & -6 & \ddots & & 0 \\ & & 4 & \ddots & & 4n \\ & & & \ddots & & -3n \\ 0 & & & & & 2n \end{pmatrix}.$$

(b) $\{1\}$ is a basis for Ker A.

$\{2t^2 - 3t + 4, 2t^3 - 3t^2 + 4t, 2t^4 - 3t^3 + 4t^2, \cdots, 2t^{n+1} - 3t^n + 4t^{n-1}\}$ is a basis for Ran A.

Chapter 4

Exercise 4.10.1 Let $\mathbb{F} = \mathbb{Z}_3$. Check the Cayley–Hamilton Theorem on the matrix

$$A = \begin{pmatrix} 1 & 0 & 2 \\ 2 & 1 & 0 \\ 2 & 2 & 2 \end{pmatrix}.$$

Answer: We have $p_A(\lambda) = \lambda^3 + 2\lambda^2 + \lambda$. Now

$$\begin{pmatrix} 1 & 1 & 1 \\ 2 & 0 & 1 \\ 2 & 1 & 0 \end{pmatrix} + 2 \begin{pmatrix} 2 & 1 & 0 \\ 1 & 1 & 1 \\ 1 & 0 & 2 \end{pmatrix} + \begin{pmatrix} 1 & 0 & 2 \\ 2 & 1 & 0 \\ 2 & 2 & 2 \end{pmatrix} = 0.$$

Exercise 4.10.2 For the following matrices A (and B) determine its Jordan canonical form J and a similarity matrix P, so that $P^{-1}AP = J$.

(a)

$$A = \begin{pmatrix} -1 & 1 & 0 & 0 \\ -1 & 0 & 1 & 0 \\ -1 & 0 & 0 & 1 \\ -1 & 0 & 0 & 1 \end{pmatrix}.$$

This matrix is nilpotent.

Answer:

$$P = \begin{pmatrix} 1 & 0 & 0 & 0 \\ 1 & 1 & 0 & 0 \\ 1 & 1 & 1 & 0 \\ 1 & 1 & 1 & 1 \end{pmatrix}, J = \begin{pmatrix} 0 & 1 & 0 & 0 \\ 0 & 0 & 1 & 0 \\ 0 & 0 & 0 & 1 \\ 0 & 0 & 0 & 0 \end{pmatrix}.$$

(b)

$$A = \begin{pmatrix} 10 & -1 & 1 & -4 & -6 \\ 9 & -1 & 1 & -3 & -6 \\ 4 & -1 & 1 & -3 & -1 \\ 9 & -1 & 1 & -4 & -5 \\ 10 & -1 & 1 & -4 & -6 \end{pmatrix}.$$

This matrix is nilpotent.

Answer:

$$P = \begin{pmatrix} 1 & 3 & 1 & -1 & 2 \\ 1 & 2 & 2 & -1 & 0 \\ 1 & 3 & 2 & 4 & 5 \\ 1 & 3 & 1 & 0 & 2 \\ 1 & 3 & 1 & -1 & -3 \end{pmatrix}, J = \begin{pmatrix} 0 & 1 & 0 & 0 & 0 \\ 0 & 0 & 0 & 0 & 0 \\ 0 & 0 & 0 & 1 & 0 \\ 0 & 0 & 0 & 0 & 1 \\ 0 & 0 & 0 & 0 & 0 \end{pmatrix}.$$

(c)

$$A = \begin{pmatrix} 0 & 1 & 0 \\ -1 & 0 & 0 \\ 1 & 1 & 1 \end{pmatrix}.$$

Answer: Eigenvalues are $1, i - i$.

$$P = \begin{pmatrix} 0 & -1 & -1 \\ 0 & -i & i \\ 1 & i & -i \end{pmatrix}, J = \begin{pmatrix} 1 & 0 & 0 \\ 0 & i & 0 \\ 0 & 0 & -i \end{pmatrix}$$

(d)

$$A = \begin{pmatrix} 2 & 0 & -1 & 1 \\ 0 & 1 & 0 & 0 \\ 1 & 0 & 0 & 0 \\ 0 & 0 & 0 & 1 \end{pmatrix}.$$

Answer: The only eigenvalue is 1. We have

$$(A - I)^2 = \begin{pmatrix} 0 & 0 & 0 & 1 \\ 0 & 0 & 0 & 0 \\ 0 & 0 & 0 & 1 \\ 0 & 0 & 0 & 0 \end{pmatrix}, \text{ and } (A - I)^3 = 0.$$

So we get

$$J = \begin{pmatrix} 1 & & & \\ & 1 & 1 & 0 \\ & 0 & 1 & 1 \\ & 0 & 0 & 1 \end{pmatrix}.$$

For P we can choose

$$\begin{pmatrix} 0 & 1 & 1 & 0 \\ 1 & 0 & 0 & 0 \\ 0 & 1 & 0 & 0 \\ 0 & 0 & 0 & 1 \end{pmatrix}.$$

(e)

$$B = \begin{pmatrix} 1 & -5 & 0 & -3 \\ 1 & 1 & -1 & 0 \\ 0 & -3 & 1 & -2 \\ -2 & 0 & 2 & 1 \end{pmatrix}.$$

(Hint: 1 is an eigenvalue)

Answer: We find that 1 is the only eigenvalue.

$$(B - I)^2 = \begin{pmatrix} 1 & 0 & -1 & 0 \\ 0 & -2 & 0 & -1 \\ 1 & 0 & -1 & 0 \\ 0 & 4 & 0 & 2 \end{pmatrix}, \text{ and } (A - I)^3 = \begin{pmatrix} 0 & -2 & 0 & -1 \\ 0 & 0 & 0 & 0 \\ 0 & -2 & 0 & 1 \\ 0 & 0 & 0 & 0 \end{pmatrix}.$$

So we get

$$J = \begin{pmatrix} 1 & 1 & 0 & 0 \\ 0 & 1 & 1 & 0 \\ 0 & 0 & 1 & 1 \\ 0 & 0 & 0 & 1 \end{pmatrix}.$$

For P we can choose

$$P = \begin{pmatrix} -1 & 0 & -3 & 0 \\ 0 & -1 & 0 & 0 \\ -1 & 0 & -2 & 0 \\ 0 & 2 & 0 & 1 \end{pmatrix}.$$

(f) For the matrix B, compute B^{100}, by using the decomposition $B = PJP^{-1}$.

Answer: As

$$P^{-1} = \begin{pmatrix} 2 & 0 & -3 & 0 \\ 0 & -1 & 0 & 0 \\ -1 & 0 & 1 & 0 \\ 0 & 2 & 0 & 1 \end{pmatrix}, J^{100} = \begin{pmatrix} 1 & 100 & 4950 & 161700 \\ 0 & 1 & 100 & 49500 \\ & 0 & 1 & 100 \\ 0 & 2 & 0 & 1 \end{pmatrix},$$

we find that

$$B^{100} = PJ^{100}P^{-1} = \begin{pmatrix} 4951 & -323900 & -4950 & -162000 \\ 100 & -9899 & -100 & -4950 \\ 4950 & -323700 & -4949 & -161900 \\ -200 & 19800 & 200 & 9901 \end{pmatrix}.$$

Exercise 4.10.3 Let

$$A \begin{pmatrix} 3 & 1 & 0 & 0 & 0 & 0 & 0 \\ 0 & 3 & 1 & 0 & 0 & 0 & 0 \\ 0 & 0 & 3 & 0 & 0 & 0 & 0 \\ 0 & 0 & 0 & 3 & 1 & 0 & 0 \\ 0 & 0 & 0 & 0 & 3 & 0 & 0 \\ 0 & 0 & 0 & 0 & 0 & 3 & 1 \\ 0 & 0 & 0 & 0 & 0 & 0 & 3 \end{pmatrix}.$$

Determine bases for the following spaces:

(a) $\text{Ker}(3I - A)$.

(b) $\text{Ker}(3I - A)^2$.

(c) $\text{Ker}(3I - A)^3$.

Answer: (a) $\{e_1, e_4, e_6\}$.

(b) $\{e_1, e_2, e_4, e_5, e_6, e_7\}$.

(c) $\{e_1, e_2, e_3, e_4, e_5, e_6, e_7\}$.

Exercise 4.10.4 Let M and N be 6×6 matrices over $\mathbb{C}$, both having minimal polynomial x^3.

(a) Prove that M and N are similar if and only if they have the same rank.

(b) Give a counterexample to show that the statement is false if 6 is replaced by 7.

(c) Compute the minimal and characteristic polynomials of the following matrix. Is it diagonalizable?

$$\begin{pmatrix} 5 & -2 & 0 & 0 \\ 6 & -2 & 0 & 0 \\ 0 & 0 & 0 & 6 \\ 0 & 0 & 1 & -1 \end{pmatrix}$$

Answer: (a) both M and N have only 0 as the eigenvalue, and at least one Jordan block at 0 is of size 3×3. So the possible Jordan forms are

$$\begin{pmatrix} 0 & 1 & 0 & 0 & 0 & 0 \\ 0 & 0 & 1 & 0 & 0 & 0 \\ 0 & 0 & 0 & 0 & 0 & 0 \\ 0 & 0 & 0 & 0 & 1 & 0 \\ 0 & 0 & 0 & 0 & 0 & 1 \\ 0 & 0 & 0 & 0 & 0 & 0 \end{pmatrix}, \begin{pmatrix} 0 & 1 & 0 & 0 & 0 & 0 \\ 0 & 0 & 1 & 0 & 0 & 0 \\ 0 & 0 & 0 & 0 & 0 & 0 \\ 0 & 0 & 0 & 0 & 1 & 0 \\ 0 & 0 & 0 & 0 & 0 & 0 \\ 0 & 0 & 0 & 0 & 0 & 0 \end{pmatrix}, \begin{pmatrix} 0 & 1 & 0 & 0 & 0 & 0 \\ 0 & 0 & 1 & 0 & 0 & 0 \\ 0 & 0 & 0 & 0 & 0 & 0 \\ 0 & 0 & 0 & 0 & 0 & 0 \\ 0 & 0 & 0 & 0 & 0 & 0 \\ 0 & 0 & 0 & 0 & 0 & 0 \end{pmatrix}.$$

Knowing the rank uniquely identifies the Jordan canonical form.

(b) $M = \begin{pmatrix} 0 & 1 & 0 & 0 & 0 & 0 & 0 \\ 0 & 0 & 1 & 0 & 0 & 0 & 0 \\ 0 & 0 & 0 & 0 & 0 & 0 & 0 \\ 0 & 0 & 0 & 0 & 1 & 0 & 0 \\ 0 & 0 & 0 & 0 & 0 & 1 & 0 \\ 0 & 0 & 0 & 0 & 0 & 0 & 0 \\ 0 & 0 & 0 & 0 & 0 & 0 & 0 \end{pmatrix}$ and $N = \begin{pmatrix} 0 & 1 & 0 & 0 & 0 & 0 & 0 \\ 0 & 0 & 1 & 0 & 0 & 0 & 0 \\ 0 & 0 & 0 & 0 & 0 & 0 & 0 \\ 0 & 0 & 0 & 0 & 1 & 0 & 0 \\ 0 & 0 & 0 & 0 & 0 & 0 & 0 \\ 0 & 0 & 0 & 0 & 0 & 0 & 1 \\ 0 & 0 & 0 & 0 & 0 & 0 & 0 \end{pmatrix}$ have the

same rank and same minimal polynomial x^3, but are not similar.

(c) $p_A(x) = (x-2)^2(x-1)(x+3)$ and $m_A(x) = (x-2)(x-1)(x+3)$. As all roots of $m_A(x)$ have multiplicity 1, the matrix A is diagonalizable.

Exercise 4.10.5 (a) Let A be a 7×7 matrix of rank 4 and with minimal polynomial equal to $q_A(\lambda) = \lambda^2(\lambda + 1)$. Give all possible Jordan canonical forms of A.

(b) Let $A \in \mathbb{C}^n$. Show that if there exists a vector $\mathbf{v}$ so that $\mathbf{v}, A\mathbf{v}, \ldots, A^{n-1}\mathbf{v}$ are linearly independent, then the characteristic polynomial of A equals the minimal polynomial of A. (Hint: use the basis $\mathcal{B} = \{\mathbf{v}, A\mathbf{v}, \ldots, A^{n-1}\mathbf{v}\}$.)

Answer: (a)

$$
\begin{pmatrix}
0 & 1 & 0 & 0 & 0 & 0 & 0 \\
0 & 0 & 0 & 0 & 0 & 0 & 0 \\
0 & 0 & 0 & 1 & 0 & 0 & 0 \\
0 & 0 & 0 & 0 & 0 & 0 & 0 \\
0 & 0 & 0 & 0 & 0 & 1 & 0 \\
0 & 0 & 0 & 0 & 0 & 0 & 0 \\
0 & 0 & 0 & 0 & 0 & 0 & -1
\end{pmatrix},
\begin{pmatrix}
0 & 1 & 0 & 0 & 0 & 0 & 0 \\
0 & 0 & 0 & 0 & 0 & 0 & 0 \\
0 & 0 & 0 & 1 & 0 & 0 & 0 \\
0 & 0 & 0 & 0 & 0 & 0 & 0 \\
0 & 0 & 0 & 0 & 0 & 0 & 0 \\
0 & 0 & 0 & 0 & 0 & -1 & 0 \\
0 & 0 & 0 & 0 & 0 & 0 & -1
\end{pmatrix},
$$

$$
\begin{pmatrix}
0 & 1 & 0 & 0 & 0 & 0 & 0 \\
0 & 0 & 0 & 0 & 0 & 0 & 0 \\
0 & 0 & 0 & 0 & 0 & 0 & 0 \\
0 & 0 & 0 & 0 & 0 & 0 & 0 \\
0 & 0 & 0 & 0 & -1 & 0 & 0 \\
0 & 0 & 0 & 0 & 0 & -1 & 0 \\
0 & 0 & 0 & 0 & 0 & 0 & -1
\end{pmatrix}.
$$

(b) As $\mathcal{B} = \{\mathbf{v}, A\mathbf{v}, \ldots, A^{n-1}\mathbf{v}\}$ is a linearly independent set with n elements in $\mathbb{C}^n$, it is a basis for $\mathbb{C}^n$. Now $[A]_{\mathcal{B}\leftarrow\mathcal{B}}$ has the form

$$
\hat{A} = \begin{pmatrix}
0 & 0 & \cdots & 0 & * \\
1 & 0 & \cdots & 0 & * \\
\vdots & \ddots & \ddots & \vdots & \\
0 & \cdots & 1 & 0 & * \\
0 & \cdots & 0 & 1 & *
\end{pmatrix}.
$$

Then

$$
\hat{A} - \lambda I_n = \begin{pmatrix}
-\lambda & 0 & \cdots & 0 & * \\
1 & -\lambda & \cdots & 0 & * \\
\vdots & \ddots & \ddots & \vdots & \\
0 & \cdots & 1 & -\lambda & * \\
0 & \cdots & 0 & 1 & *
\end{pmatrix},
$$

which has rank $\geq n - 1$. Thus $\dim \mathrm{Ker}\,(A - \lambda I_n) = \dim \mathrm{Ker}\,(\hat{A} - \lambda I_n) \leq 1$, and thus $w_1(A, \lambda) = 1$ for every eigenvalue of A. This shows that A is nonderogatory, and thus $p_A(t) = m_A(t)$.

Exercise 4.10.6 Let $A \in \mathbb{F}^{n\times n}$ and A^T denote its transpose. Show that $w_k(A, \lambda) = w_k(A^T, \lambda)$, for all $\lambda \in \mathbb{F}$ and $k \in \mathbb{N}$. Conclude that A and A^T have the same Jordan canonical form, and are therefore similar.

Answer: In general we have that for any matrix rank $B = \mathrm{rank} B^T$. If B is square of size $n \times n$, we therefore have that

$$\dim \mathrm{Ker}\,B = n - \mathrm{rank}\,B = n - \mathrm{rank} B^T = \dim \mathrm{Ker}\,B^T.$$

Applying this to $B = (A - \lambda I)^k$, we have $B^T = [(A - \lambda I)^k]^T = (A^T - \lambda I)^k$, we get

$\dim \operatorname{Ker}(A - \lambda I)^k = \dim \operatorname{Ker}(A^T - \lambda I)^k$. Then it follows that $w_k(A, \lambda) = w_k(A^T, \lambda)$, for all $\lambda \in \mathbb{F}$ and $k \in \mathbb{N}$. Thus A and A^T have the same Jordan canonical form, and are therefore similar.

Exercise 4.10.7 Let $A \in \mathbb{C}^{4 \times 4}$ matrix satisfying $A^2 = -I$.

(a) Determine the possible eigenvalues of A.

(b) Determine the possible Jordan structures of A.

Answer: (a) Let $m(t) = t^2 + 1 = (t - i)(t + i)$. Then $m(A) = 0$, so the minimal polynomial of A divides $m(A)$. Thus the only possible eigenvalues of A are i and $-i$.

(b) As the minimal polynomial of A only has roots of multiplicity 1, the Jordan canonical form will only have 1×1 Jordan blocks. Thus the Jordan canonical form of A is a diagonal matrix with i and/or $-i$ appearing on the diagonal.

Exercise 4.10.8 Let $p(x) = (x - 2)^2(x - 3)^2$. Determine a matrix A for which $p(A) = 0$ and for which $q(A) \neq 0$ for all nonzero polynomials q of degree ≤ 3. Explain why $q(A) \neq 0$ for such q.

Answer: Let $A = \begin{pmatrix} 2 & 1 & 0 & 0 \\ 0 & 2 & 0 & 0 \\ 0 & 0 & 3 & 1 \\ 0 & 0 & 0 & 3 \end{pmatrix}$. Then $p(x)$ is the minimal polynomial for A, and thus $p(A) = 0$, and for any nonzero polynomial $q(x)$ with degree less than $\deg p = 4$ we have $q(A) \neq 0$.

Exercise 4.10.9 Let $m_A(t) = (t - 1)^2(t - 2)(t - 3)$ be the minimal polynomial of $A \in M_6$.

(a) What possible Jordan forms can A have?

(b) If it is known that $\operatorname{rank}(A - I) = 3$, what possible Jordan forms can A have?

Answer: (a)

$$\begin{pmatrix} 1 & 1 & 0 & 0 & 0 & 0 \\ 0 & 1 & 0 & 0 & 0 & 0 \\ 0 & 0 & 2 & 0 & 0 & 0 \\ 0 & 0 & 0 & 3 & 0 & 0 \\ 0 & 0 & 0 & 0 & 1 & 1 \\ 0 & 0 & 0 & 0 & 0 & 1 \end{pmatrix} \text{ or } \begin{pmatrix} 1 & 1 & 0 & 0 & 0 & 0 \\ 0 & 1 & 0 & 0 & 0 & 0 \\ 0 & 0 & 2 & 0 & 0 & 0 \\ 0 & 0 & 0 & 3 & 0 & 0 \\ 0 & 0 & 0 & 0 & a & 0 \\ 0 & 0 & 0 & 0 & 0 & b \end{pmatrix}, \text{ where } a, b \in \{1, 2, 3\}.$$

(b)

$$\begin{pmatrix} 1 & 1 & 0 & 0 & 0 & 0 \\ 0 & 1 & 0 & 0 & 0 & 0 \\ 0 & 0 & 2 & 0 & 0 & 0 \\ 0 & 0 & 0 & 3 & 0 & 0 \\ 0 & 0 & 0 & 0 & 1 & 0 \\ 0 & 0 & 0 & 0 & 0 & 1 \end{pmatrix}.$$

Exercise 4.10.10 Let A be a 4×4 matrix satisfying $A^2 = -A$.

(a) Determine the possible eigenvalues of A.

(b) Determine the possible Jordan structures of A (Hint: notice that $(A + I)A = 0$.)

Answer: Let $m(t) = t^2 + t = t(t+1)$. Then $m(A) = 0$, and thus the minimal polynomial $m_P(t)$ of P divides $m(t)$. Thus there are three possibilities $m_A(t) = t$, $m_A(t) = t + 1$ or $m_A(t) = t(t+1)$. The only possible roots of A are therefore 0 or -1. Next, since the minimal polynomial has roots of multiplicity 1 only, the Jordan blocks are all of size 1×1. Thus the Jordan canonical of A is a diagonal matrix with 0 and/or -1 on the diagonal.

Exercise 4.10.11 Let $A \in \mathbb{C}^{n \times n}$. For the following answer True or False. Provide an explanation.

(a) If $\det(A) = 0$, then 0 is an eigenvalue of A.

Answer: True, if $\det(A) = 0$, then $p_A(0) = 0$.

(b) If $A^2 = 0$, then the rank of A is at most $\frac{n}{2}$.

Answer: True, when $m(t) = t^2$, then $m(A) = 0$. Thus $m_A(t)$ divides $m(t)$, and thus $m_A(t) = t$ or $m_A(t) = t^2$. When $m_A(t) = t$, then $A = 0$, thus rank $A = 0$. If $m_A(t) = t^2$, then the Jordan canonical form has 1×1 and 2×2 Jordan blocks at 0. The rank of A equals the number of 2×2 Jordan blocks at 0, which is at most $\frac{n}{2}$.

(c) There exists a matrix A with minimal polynomial $m_A(t) = (t-1)(t-2)$ and characteristic polynomial $p_A(t) = t^{n-2}(t-1)(t-2)$ (here $n > 2$).

Answer: False, since 0 is a root of $p_A(t)$ it must also be a root of $m_A(t)$.

(d) If all eigenvalues of A are 1, then $A = I_n$ (=the $n \times n$ identity matrix).

Answer: False, A can have a Jordan block at 1 of size 2×2 or larger. For example, $A = \begin{pmatrix} 1 & 1 \\ 0 & 1 \end{pmatrix}$ has only 1 as eigenvalue, but $A \neq I_2$.

Exercise 4.10.12 Show that if A is similar to B, then tr $A =$ tr B.

Answer: Let P be so that $A = PBP^{-1}$. Put now $C = PB$, $D = P^{-1}$. Then since $\mathrm{tr}(CD) = \mathrm{tr}(DC)$ we obtain tr $A = \mathrm{tr}(CD) = \mathrm{tr}(DC) =$ tr B.

Exercise 4.10.13 Let P be a matrix so that $P^2 = P$.

(a) Show that P only has eigenvalues 0 or 1.

(b) Show that rank $P =$ trace P. (Hint: determine the possible Jordan canonical form of P.)

Answer: (a) and (b). Let $m(t) = t^2 - t = t(t-1)$. Then $m(P) = 0$, and thus the minimal polynomial $m_P(t)$ of P divides $m(t)$. Thus there are three possibilities $m_P(t) = t$, $m_P(t) = t - 1$ or $m_P(t) = t(t-1)$. The only possible roots of P are therefore 0 or 1. Next, since the minimal polynomial has roots of multiplicity 1 only, the Jordan blocks are all of size 1×1. Thus the Jordan canonical J of P is a diagonal matrix with zeros and/or ones on the diagonal. The rank of J is equal to the sum of its diagonal entries, and as rank and trace do not change when applying a similarity, we get rank $P =$ trace P.

Exercise 4.10.14 Let $A = PJP^{-1}$. Show that Ran $A = P[\text{Ran } J]$ and Ker $A = P[\text{Ker } J]$. In addition, dim Ran $A =$ dim Ran J, and dim Ker $A =$ dim Ker J.

Answer: Let $\mathbf{v} \in \operatorname{Ran} A$. Then there exists an $\mathbf{x}$ so that $\mathbf{v} = A\mathbf{x} = P(JP^{-1}\mathbf{x})$. Then $\mathbf{v} \in P[\operatorname{Ran} J]$ follows. Conversely, let $\mathbf{v} \in P[\operatorname{Ran} J]$. Then there exists an $\mathbf{x}$ so that $\mathbf{v} = P(J\mathbf{x})$. Then $\mathbf{v} = PJ\mathbf{x} = A(P^{-1}\mathbf{x})$. Thus $\mathbf{v} \in \operatorname{Ran} A$.

Let $\mathbf{v} \in \operatorname{Ker} A$. Then $A\mathbf{v} = \mathbf{0}$. Thus $PJP^{-1}\mathbf{v} = \mathbf{0}$. Let $\mathbf{x} = P^{-1}\mathbf{v}$. Then $\mathbf{x} \in \operatorname{Ker} J$ and $\mathbf{v} = P\mathbf{x}$. Thus $\mathbf{v} \in P[\operatorname{Ker} J]$. Conversely, let $\mathbf{v} \in P[\operatorname{Ker} J]$. Then there exists a $\mathbf{x} \in \operatorname{Ker} J$ so that $\mathbf{v} = P\mathbf{x}$. Thus $J\mathbf{x} = \mathbf{0}$, and $A\mathbf{v} = PJP^{-1}\mathbf{v} = PJ\mathbf{x} = \mathbf{0}$. Thus $\mathbf{v} \in \operatorname{Ker} A$.

By Exercise 3.4.5, it follows that $\dim \operatorname{Ran} A = \dim P[\operatorname{Ran} J] \leq \dim \operatorname{Ran} J$. As $\operatorname{Ran} J = P^{-1}[\operatorname{Ran} J]$, it also follows $\dim \operatorname{Ran} J = \dim P^{-1}[\operatorname{Ran} A] \leq \dim \operatorname{Ran} A$. Thus $\dim \operatorname{Ran} A = \dim \operatorname{Ran} J$. Similarly, $\dim \operatorname{Ker} A = \dim \operatorname{Ker} J$.

Exercise 4.10.15 Show that matrices A and B are similar if and only if they have the same Jordan canonical form.

Answer: Suppose A and B have the same Jordan canonical form J. Then there exist invertible P and S so that $A = PJP^{-1}$ and $B = SJS^{-1}$. But then $A = P(S^{-1}JS)P^{-1} = (PS^{-1})J(PS^{-1})^{-1}$, and thus A and B are similar.

Next suppose that A and B are similar. Thus there exists an invertible P so that $A = PBP^{-1}$. Then $A - \lambda I_n = PBP^{-1} - \lambda PP^{-1} = P(B - \lambda I_n)P^{-1}$, and thus $A - \lambda I_n$ and $B - \lambda I_n$ are similar for all $\lambda \in \mathbb{F}$. Also,

$$(A - \lambda I_n)^k = (P(B - \lambda I_n)P^{-1})^k = P(B - \lambda I_n)^k P^{-1},$$

and thus $(A - \lambda I_n)^k$ and $(B - \lambda I_n)^k$ are similar for all $\lambda \in \mathbb{F}$ and $k \in \mathbb{N}$. By Exercise 4.10.14 it follows that $\dim \operatorname{Ker}(A - \lambda I_n)^k = \dim \operatorname{Ker}(B - \lambda I_n)^k$ for all $\lambda \in \mathbb{F}$ and $k \in \mathbb{N}$. Thus $w_k(A, \lambda) = w_k(B, \lambda)$ for all $\lambda \in \mathbb{F}$. Consequently, A and B have the same Jordan canonical form.

Exercise 4.10.16 Show that if A and B are square matrices of the same size, with A invertible, then AB and BA have the same Jordan canonical form.

Answer: $A^{-1}(AB)A = BA$, so AB and BA are similar, and thus have the same Jordan canonical form.

Exercise 4.10.17 Let $A \in \mathbb{F}^{n \times m}$ and $B \in \mathbb{F}^{m \times n}$. Observe that

$$\begin{pmatrix} I_n & -A \\ 0 & I_m \end{pmatrix} \begin{pmatrix} AB & 0 \\ B & 0_m \end{pmatrix} \begin{pmatrix} I_n & A \\ 0 & I_m \end{pmatrix} = \begin{pmatrix} 0_n & 0 \\ B & BA \end{pmatrix}. \tag{4.26}$$

(a) Show that the Weyr characteristics at $\lambda \neq 0$ of AB and BA satisfy

$$w_k(AB, \lambda) = w_k(BA, \lambda), \quad k \in \mathbb{N}.$$

(b) Show that $\lambda \neq 0$ is an eigenvalue of AB if and only if it is an eigenvalue of BA, and that AB and BA have the same Jordan structure at λ.

(c) Provide an example of matrices A and B so that AB and BA have different Jordan structures at 0.

Answer: (a) and (b). From (4.26) it follows that

$$\begin{pmatrix} I_n & -A \\ 0 & I_m \end{pmatrix} \begin{pmatrix} AB - \lambda I & 0 \\ B & -\lambda I \end{pmatrix}^k \begin{pmatrix} I_n & A \\ 0 & I_m \end{pmatrix} = \begin{pmatrix} -\lambda I & 0 \\ B & BA - \lambda I \end{pmatrix}^k, \quad k \in \mathbb{N}.$$

Thus we get that

$$\dim\operatorname{Ker}\begin{pmatrix} AB-\lambda I & 0 \\ B & -\lambda I \end{pmatrix}^k = \dim\operatorname{Ker}\begin{pmatrix} -\lambda I & 0 \\ B & BA-\lambda I \end{pmatrix}^k, \ k\in\mathbb{N}.$$

Next we observe that when $\lambda \neq 0$, we have that

$$\dim\operatorname{Ker}\begin{pmatrix} AB-\lambda I & 0 \\ B & -\lambda I \end{pmatrix}^k = \dim\operatorname{Ker}\begin{pmatrix} (AB-\lambda I)^k & 0 \\ * & (-\lambda)^k I \end{pmatrix} =$$

$$\dim\operatorname{Ker}(AB-\lambda I)^k,$$

and

$$\dim\operatorname{Ker}\begin{pmatrix} -\lambda I & 0 \\ B & BA-\lambda I \end{pmatrix}^k = \dim\operatorname{Ker}\begin{pmatrix} (-\lambda)^k I & 0 \\ * & (BA-\lambda I)^k \end{pmatrix} =$$

$$\dim\operatorname{Ker}(BA-\lambda I)^k.$$

Combining the above equalities, we get that

$$\dim\operatorname{Ker}(AB-\lambda I)^k = \dim\operatorname{Ker}(BA-\lambda I)^k, \ \lambda\in\mathbb{F}\setminus\{0\}, k\in\mathbb{N}.$$

Thus

$$w_k(AB,\lambda) = w_k(BA,\lambda), \ \lambda\in\mathbb{F}\setminus\{0\}, k\in\mathbb{N}.$$

From this it follows that AB and BA have the same Jordan structure at $\lambda \neq 0$.

(c) Let $A = \begin{pmatrix} 1 & 0 \end{pmatrix}, B = A^*$. Then $BA = \begin{pmatrix} 1 & 0 \\ 0 & 0 \end{pmatrix}$ and thus has a 1×1 Jordan block at 0. On the other hand, $AB = \begin{pmatrix} 1 \end{pmatrix}$ does not have 0 as an eigenvalue.

Exercise 4.10.18 Let $A, B \in \mathbb{C}^{n\times n}$ be such that $(AB)^n = 0$. Prove that $(BA)^n = 0$.

Answer: As $(AB)^n = 0$, we have that 0 is the only eigenvalue of AB. By Exercise 4.10.17 this means that 0 is also the only eigenvalue of BA. Thus BA is nilpotent, and thus $(BA)^n = 0$.

Exercise 4.10.19 (a) Let $A \in \mathbb{R}^{8\times 8}$ with characteristic polynomial
$p(x) = (x+3)^4(x^2+1)^2$ and minimal polynomial $m(x) = (x+3)^2(x^2+1)$. What are the possible Jordan canonical form(s) for A (up to permutation of Jordan blocks)?

(b) Suppose that $A \in \mathbb{C}^{n\times n}$ satisfies $A^k \neq 0$ and $A^{k+1} = 0$. Prove that there exists $\mathbf{x} \in \mathbb{C}^n$ such that $\{\mathbf{x}, A\mathbf{x}, \ldots, A^k\mathbf{x}\}$ is linearly independent.

(c) Let $A, B \in \mathbb{C}^{n\times n}$ be such that $A^2 - 2AB + B^2 = 0$. Prove that every eigenvalue of B is an eigenvalue of A, and conversely that every eigenvalue of A is an eigenvalue of B.

Answer: (a)
$$\left(\begin{array}{cccccccc} 3 & 1 & & & & & & \\ 0 & 3 & & & & & & \\ & & 3 & 1 & & & & \\ & & 0 & 3 & & & & \\ & & & & i & & & \\ & & & & & i & & \\ & & & & & & -i & \\ & & & & & & & -i \end{array}\right), \quad \left(\begin{array}{cccccccc} 3 & 1 & & & & & & \\ 0 & 3 & & & & & & \\ & & 3 & & & & & \\ & & & 3 & & & & \\ & & & & i & & & \\ & & & & & i & & \\ & & & & & & -i & \\ & & & & & & & -i \end{array}\right).$$

(b) Choose $\mathbf{x}$ so that $A^k\mathbf{x} \neq \mathbf{0}$. We claim that $\{\mathbf{x}, A\mathbf{x}, \ldots, A^k\mathbf{x}\}$ is linearly independent. Let $c_0, c_1, \ldots, c_k$ be so that

$$c_0\mathbf{x} + c_1 A\mathbf{x} + \cdots + c_k A^k\mathbf{x} = \mathbf{0}. \tag{4.27}$$

Multiply (4.27) by A^k. Using that $A^{k+1} = 0$, we now get that $c_0 A^k \mathbf{x} = \mathbf{0}$. As $A^k \mathbf{x} \neq \mathbf{0}$, we must have $c_0 = 0$. Next, multiply (4.27) with A^{k-1}. Then we get that $c_1 A^k \mathbf{x} = \mathbf{0}$. As $A^k \mathbf{x} \neq \mathbf{0}$, we must have $c_1 = 0$. Continuing this way, we also get $c_2 = 0$, $c_3 = 0, \ldots, c_k = 0$, thus showing the linear independence.

(c) Suppose $B\mathbf{v} = \lambda \mathbf{v}$, $\mathbf{v} \neq \mathbf{0}$. Then $A^2 \mathbf{v} - 2\lambda A \mathbf{v} + \lambda^2 \mathbf{v} = \mathbf{0}$, which gives that $(A - \lambda)^2 \mathbf{v} = \mathbf{0}$. Thus $(A - \lambda I_n)^2$ has a nontrivial kernel, and thus is not invertible. But then it follows that $A - \lambda I_n$ is not invertible, thus λ is an eigenvalue for A. In a similar way one proves that every eigenvalue of A is an eigenvalue of B.

Exercise 4.10.20 (a) Prove Proposition 4.6.1.

(b) Let $A = \begin{pmatrix} 0 & 0 & 0 & \cdots & & -a_0 \\ 1 & 0 & 0 & \cdots & & -a_1 \\ 0 & 1 & 0 & \cdots & & -a_2 \\ \vdots & \vdots & \ddots & \ddots & & \vdots \\ 0 & \cdots & 1 & 0 & & -a_{n-2} \\ 0 & \cdots & 0 & 1 & & -a_{n-1} \end{pmatrix}$. Show that

$$ p_A(t) = t^n + a_{n-1} t^{n-1} + \cdots + a_1 t + a_0 = m_A(t). $$

This matrix is called the *companion matrix* of the polynomial $p(t) = p_A(t)$. Thus a companion matrix is nonderogatory.

Answer: (a) The number of Jordan blocks of A at λ equals

$$ (w_1(A, \lambda) - w_2(A, \lambda)) + (w_2(A, \lambda) - w_3(A, \lambda)) + \cdots + (w_n(A, \lambda) - w_{n+1}(A, \lambda)), $$

which in turn equals $w_1(A, \lambda)$. Thus at λ there is one Jordan block if and only if $w_1(A, \lambda) = 1$. This gives the equivalence of (i) and (ii) in Proposition 4.6.1. Next the multiplicity of λ as a root in $p_A(t)$ equals the sum of the sizes of the Jordan blocks at λ, while the multiplicity in $m_A(t)$ corresponds to the size of the largest Jordan block at λ. The two multiplicities are the same if and only if there one Jordan block at λ. This shows the equivalence of (i) and (iii).

(b) Notice that

$$ A - \lambda I_n = \begin{pmatrix} -\lambda & 0 & 0 & \cdots & & -a_0 \\ 1 & -\lambda & 0 & \cdots & & -a_1 \\ 0 & 1 & -\lambda & \cdots & & -a_2 \\ \vdots & \vdots & \ddots & \ddots & & \vdots \\ 0 & \cdots & 1 & -\lambda & & -a_{n-2} \\ 0 & \cdots & 0 & 1 & & -a_{n-1} - \lambda \end{pmatrix}. $$

Leaving out the first column and the last row, one obtains an invertible $(n-1) \times (n-1)$ submatrix. Thus $\dim \mathrm{Ker}\, (A - \lambda I_n) \leq 1$, and thus $w_1(A, \lambda) = 1$ for every eigenvalue of A. This shows that A is nonderogatory. It is straightforward to check that $p_A(t)$ is as described.

Exercise 4.10.21 For the following pairs of matrices A and B, find a polynomial $p(t)$ so that $p(A) = B$, or show that it is impossible.

(a) $A = \begin{pmatrix} 1 & 1 & 0 \\ 0 & 1 & 1 \\ 0 & 0 & 1 \end{pmatrix}, B = \begin{pmatrix} 1 & 2 & 3 \\ 0 & 2 & 3 \\ 0 & 0 & 3 \end{pmatrix}.$

Answer: $AB \neq BA$ and A is nonderogatory, so no polynomial p exists.

(b) $A = \begin{pmatrix} 1 & 1 & 0 \\ 0 & 1 & 1 \\ 0 & 0 & 1 \end{pmatrix}, B = \begin{pmatrix} 1 & 2 & 3 \\ 0 & 1 & 2 \\ 0 & 0 & 1 \end{pmatrix}.$

Answer: $B = I + 2(A - I) + 3(A - I)^2$, thus $p(t) = 2 - 4t + 3t^2$ works.

Exercise 4.10.22 Solve the system of differential equations

$$x'(t) = Ax(t), \quad x(0) = \begin{pmatrix} 1 \\ -1 \\ 0 \end{pmatrix},$$

where

$$A = \begin{pmatrix} 1 & -1 & 1 \\ 0 & 1 & -1 \\ 0 & 1 & 0 \end{pmatrix} \begin{pmatrix} 2 & 1 & 0 \\ 0 & 2 & 1 \\ 0 & 0 & 2 \end{pmatrix} \begin{pmatrix} 1 & -1 & 1 \\ 0 & 1 & -1 \\ 0 & 1 & 0 \end{pmatrix}^{-1}.$$

Answer: $x(t) = \begin{pmatrix} 1 & -1 & 1 \\ 0 & 1 & -1 \\ 0 & 1 & 0 \end{pmatrix} \begin{pmatrix} e^{2t} & te^{2t} & \frac{1}{2}t^2 e^{2t} \\ 0 & e^{2t} & te^{2t} \\ 0 & 0 & e^{2t} \end{pmatrix} \begin{pmatrix} 1 & -1 & 1 \\ 0 & 1 & -1 \\ 0 & 1 & 0 \end{pmatrix}^{-1} \begin{pmatrix} 1 \\ -1 \\ 0 \end{pmatrix} =$
$\begin{pmatrix} (\frac{1}{2}t^2 - t + 1)e^{2t} \\ (t - 1)e^{2t} \\ te^{2t} \end{pmatrix}.$

Exercise 4.10.23 Solve the following systems of linear differential equations:

(a)
$$\begin{cases} x_1'(t) = 3x_1(t) - x_2(t) & x_1(0) = 1 \\ x_2'(t) = x_1(t) + x_2(t) \end{cases}, \quad x_2(0) = 2$$
Answer: $x_1(t) = e^{2t} - te^{2t}, x_2(t) = 2e^{2t} - te^{2t}.$

(b)
$$\begin{cases} x_1'(t) = 3x_1(t) + x_2(t) + x_3(t) & x_1(0) = 1 \\ x_2'(t) = 2x_1(t) + 4x_2(t) + 2x_3(t), & x_2(0) = -1 \\ x_3'(t) = -x_1(t) - x_2(t) + x_3(t) & x_3(0) = 1 \end{cases}$$
Answer: $x_1(t) = \frac{1}{2}e^{4t} + \frac{1}{2}e^{2t}, x_2(t) = e^{4t} - 2e^{2t}, x_3(t) = -\frac{1}{2}e^{4t} + \frac{3}{2}e^{2t}.$

(c)
$$\begin{cases} x_1'(t) = -x_2(t) & x_1(0) = 1 \\ x_2'(t) = x_1(t) \end{cases}, \quad x_2(0) = 2$$
Answer: $x_1(t) = \cos t - 2\sin t, x_2(t) = \sin t + 2\cos t.$

(d)
$$x''(t) - 6x'(t) + 9x(t) = 0, \quad x(0) = 2, x'(0) = 1.$$
Answer: $2e^{3t} - 5te^{3t}.$

(e)
$$x''(t) - 4x'(t) + 4x(t) = 0, \quad x(0) = 6, x'(0) = -1.$$
Answer: $6e^{2t} - 13te^{2t}.$

Exercise 4.10.24 For the following matrices we determined their Jordan canonical form in Exercise 4.10.2.

(a) Compute $\cos A$ for

$$A = \begin{pmatrix} -1 & 1 & 0 & 0 \\ -1 & 0 & 1 & 0 \\ -1 & 0 & 0 & 1 \\ -1 & 0 & 0 & 1 \end{pmatrix}.$$

Answer: We have $A = PJP^{-1}$, where

$$P = \begin{pmatrix} 1 & 0 & 0 & 0 \\ 1 & 1 & 0 & 0 \\ 1 & 1 & 1 & 0 \\ 1 & 1 & 1 & 1 \end{pmatrix}, J = \begin{pmatrix} 0 & 1 & 0 & 0 \\ 0 & 0 & 1 & 0 \\ 0 & 0 & 0 & 1 \\ 0 & 0 & 0 & 0 \end{pmatrix}.$$

Thus $\cos A = P(\cos J)P^{-1}$, with

$$\cos J = \begin{pmatrix} \cos 0 & -\sin 0 & -\frac{1}{2}\cos 0 & \frac{1}{6}\sin 0 \\ 0 & \cos 0 & -\sin 0 & -\frac{1}{2}\cos 0 \\ 0 & 0 & \cos 0 & -\sin 0 \\ 0 & 0 & 0 & \cos 0 \end{pmatrix} = \begin{pmatrix} 1 & 0 & -\frac{1}{2} & 0 \\ 0 & 1 & 0 & -\frac{1}{2} \\ 0 & 0 & 1 & 0 \\ 0 & 0 & 0 & 1 \end{pmatrix}.$$

Thus

$$\cos A = \begin{pmatrix} 1 & \frac{1}{2} & -\frac{1}{2} & 0 \\ 0 & \frac{3}{2} & 0 & -\frac{1}{2} \\ 0 & \frac{1}{2} & 1 & -\frac{1}{2} \\ 0 & \frac{1}{2} & 0 & \frac{1}{2} \end{pmatrix}.$$

(b) Compute A^{24} for

$$A = \begin{pmatrix} 0 & 1 & 0 \\ -1 & 0 & 0 \\ 1 & 1 & 1 \end{pmatrix}.$$

Answer: We have $A = PJP^{-1}$, where

$$P = \begin{pmatrix} 0 & -1 & -1 \\ 0 & -i & i \\ 1 & i & -i \end{pmatrix}, J = \begin{pmatrix} 1 & 0 & 0 \\ 0 & i & 0 \\ 0 & 0 & -i \end{pmatrix}.$$

As $J^{24} = I$, we find $A^{24} = I$.

(c) Compute e^A for

$$A = \begin{pmatrix} 2 & 0 & -1 & 1 \\ 0 & 1 & 0 & 0 \\ 1 & 0 & 0 & 0 \\ 0 & 0 & 0 & 1 \end{pmatrix}.$$

Answer: We have $A = PJP^{-1}$, where

$$P = \begin{pmatrix} 0 & 1 & 1 & 0 \\ 1 & 0 & 0 & 0 \\ 0 & 1 & 0 & 0 \\ 0 & 0 & 0 & 1 \end{pmatrix}, J = \begin{pmatrix} 1 & 0 & 0 & 0 \\ 0 & 1 & 1 & 0 \\ 0 & 0 & 1 & 1 \\ 0 & 0 & 0 & 1 \end{pmatrix}.$$

Thus

$$e^A = P \begin{pmatrix} e & 0 & 0 & 0 \\ 0 & e & e & \frac{e}{2} \\ 0 & 0 & e & e \\ 0 & 0 & 0 & e \end{pmatrix} P^{-1} = \begin{pmatrix} 2e & 0 & -e & \frac{3e}{2} \\ 0 & e & 0 & 0 \\ e & 0 & 0 & \frac{e}{2} \\ 0 & 0 & 0 & e \end{pmatrix}.$$

Exercise 4.10.25 (a) Find matrices $A, B \in \mathbb{C}^{n \times n}$ so that $e^A e^B \neq e^{A+B}$.

Answer: Let $A = \begin{pmatrix} 0 & 1 \\ 0 & 0 \end{pmatrix}, B = \begin{pmatrix} 0 & 0 \\ 1 & 0 \end{pmatrix}$. Then $e^A = \begin{pmatrix} 1 & 1 \\ 0 & 1 \end{pmatrix}, e^B = \begin{pmatrix} 1 & 0 \\ 1 & 1 \end{pmatrix}$, while

$$e^{A+B} = \frac{1}{2} \begin{pmatrix} e + \frac{1}{e} & e - \frac{1}{e} \\ e - \frac{1}{e} & e + \frac{1}{e} \end{pmatrix} \neq e^A e^B.$$

(b) When $AB = BA$, then $e^A e^B = e^{A+B}$. Prove this statement when A is nonderogatory.

Answer: When A is nonderogatory, and $AB = BA$, we have by Theorem 4.6.2 that $B = p(A)$ for some polynomial. We can now introduce the functions

$$f(t) = e^t, g(t) = e^{p(t)}, h(t) = e^{t+p(t)} = f(t)g(t).$$

It follows from Theorem 4.8.3 that $h(A) = f(A)g(A)$. But then, we obtain that $e^{A+p(A)} = e^A e^{p(A)}$, and thus $e^{A+B} = e^A e^B$.

Exercise 4.10.26 Compute the matrices P_{20}, P_{21}, P_{22} from Example 4.8.5.

Answer:

$$P_{20} = \begin{pmatrix} 1 & -\frac{1}{2} & \frac{1}{2} & 0 & 1 & \frac{1}{2} \\ 0 & 0 & 1 & 0 & 1 & 0 \\ 0 & -\frac{1}{2} & \frac{3}{2} & 0 & 1 & \frac{1}{2} \\ 0 & -1 & 2 & 0 & 1 & 1 \\ 0 & \frac{1}{2} & -\frac{1}{2} & 0 & 0 & -\frac{1}{2} \\ 0 & -\frac{1}{2} & \frac{1}{2} & 0 & 0 & \frac{1}{2} \end{pmatrix},$$

$$P_{21} = \begin{pmatrix} 0 & \frac{1}{2} & -\frac{1}{2} & 0 & 0 & -\frac{1}{2} \\ -1 & \frac{1}{2} & -\frac{3}{2} & 0 & -2 & -\frac{1}{2} \\ 0 & -\frac{1}{2} & \frac{1}{2} & 0 & 0 & \frac{1}{2} \\ 1 & -\frac{3}{2} & \frac{5}{2} & 0 & 2 & \frac{3}{2} \\ -1 & 1 & -2 & 0 & -2 & -1 \\ 1 & -1 & 2 & 0 & 2 & 1 \end{pmatrix},$$

$$P_{22} = \frac{1}{2} \begin{pmatrix} -1 & 1 & -2 & 0 & -2 & -1 \\ 1 & -1 & 2 & 0 & 2 & 1 \\ 1 & -1 & 2 & 0 & 2 & 1 \\ 1 & -1 & 2 & 0 & 2 & 1 \\ 0 & 0 & 0 & 0 & 0 & 0 \\ 0 & 0 & 0 & 0 & 0 & 0 \end{pmatrix}.$$

Exercise 4.10.27 (a) Show that if $A = A^*$, then e^A is positive definite.

Answer: Let $A = U\Lambda U^*$ be a spectral decomposition of A. Then $e^A = U e^\Lambda U^*$. Since Λ is diagonal, we have $e^\Lambda = \mathrm{diag}\left(e^{\lambda_i}\right), \lambda_i \in \mathbb{R}, i = 1, \ldots, n$. Since the exponential function maps $\mathbb{R} \mapsto \mathbb{R}^+$, we have that $e^A = U e^\Lambda U^* = \left(U e^\Lambda U^*\right)^*$ with all positive eigenvalues. Thus e^A is positive definite.

(b) If e^A is positive definite, is then A necessarily Hermitian?

Answer: No. For instance, let $A = (2\pi i)$. Then $e^A = (1)$ is positive definite, but A is not Hermitian.

(c) What can you say about e^A when A is skew-Hermitian?

Answer: Once again, let $A = U\Lambda U^*$ be a spectral decomposition of A. Because A is skew-Hermitian, we know that Λ has pure imaginary entries. Then we may write $e^A = U e^\Lambda U^*$ where $e^\Lambda = \mathrm{diag}\left(e^{i\lambda_i}\right), \lambda_i \in \mathbb{R}, i = 1, \ldots, n$. Because Λ commutes with itself, we have $e^A e^{A^*} = U e^\Lambda e^{-\Lambda} U^* = U I U^* = I$. Thus e^A must be unitary.

Exercise 4.10.28 Let $A = \begin{pmatrix} \frac{\pi}{2} & 1 & -1 \\ 0 & \frac{\pi}{2} & -\frac{\pi}{4} \\ 0 & 0 & \frac{\pi}{4} \end{pmatrix}$.

(a) Compute $\cos A$ and $\sin A$.

Answer: Computing the Jordan canonical form decomposition of A, we have

$$A = SJS^{-1} = \begin{pmatrix} 1 & 0 & 0 \\ 0 & 1 & -1 \\ 0 & 0 & -1 \end{pmatrix} \begin{pmatrix} \frac{\pi}{2} & 1 & 0 \\ 0 & \frac{\pi}{2} & 0 \\ 0 & 0 & \frac{\pi}{4} \end{pmatrix} \begin{pmatrix} 1 & 0 & 0 \\ 0 & 1 & -1 \\ 0 & 0 & -1 \end{pmatrix}.$$

Then $\cos A = \cos\left(SJS^{-1}\right) = S\cos\left(J\right)S^{-1}$, and because we have a 2×2 Jordan block,

$$\cos J = \begin{pmatrix} 0 & -1 & 0 \\ 0 & 0 & 0 \\ 0 & 0 & \frac{\sqrt{2}}{2} \end{pmatrix}$$

and finally we may compute

$$\cos A = \begin{pmatrix} 1 & 0 & 0 \\ 0 & 1 & -1 \\ 0 & 0 & -1 \end{pmatrix} \begin{pmatrix} 0 & -1 & 0 \\ 0 & 0 & 0 \\ 0 & 0 & \frac{\sqrt{2}}{2} \end{pmatrix} \begin{pmatrix} 1 & 0 & 0 \\ 0 & 1 & -1 \\ 0 & 0 & -1 \end{pmatrix} = \begin{pmatrix} 0 & -1 & 1 \\ 0 & 0 & \frac{\sqrt{2}}{2} \\ 0 & 0 & \frac{\sqrt{2}}{2} \end{pmatrix}.$$

Similarly, we can compute $\sin A$ as

$$\sin A = \begin{pmatrix} 1 & 0 & 0 \\ 0 & 1 & -1 \\ 0 & 0 & -1 \end{pmatrix} \begin{pmatrix} 1 & 0 & 0 \\ 0 & 1 & -1 \\ 0 & 0 & \frac{\sqrt{2}}{2} \end{pmatrix} \begin{pmatrix} 1 & 0 & 0 \\ 0 & 1 & -1 \\ 0 & 0 & -1 \end{pmatrix} = \begin{pmatrix} 1 & 0 & 0 \\ 0 & 1 & \frac{\sqrt{2}}{2} - 1 \\ 0 & 0 & \frac{\sqrt{2}}{2} \end{pmatrix}.$$

(b) Check that $(\cos A)^2 + (\sin A)^2 = I$.

Answer:

$$\begin{pmatrix} 0 & -1 & 1 \\ 0 & 0 & \frac{\sqrt{2}}{2} \\ 0 & 0 & \frac{\sqrt{2}}{2} \end{pmatrix}^2 + \begin{pmatrix} 1 & 0 & 0 \\ 0 & 1 & \frac{\sqrt{2}}{2} - 1 \\ 0 & 0 & \frac{\sqrt{2}}{2} \end{pmatrix}^2 = \begin{pmatrix} 0 & 0 & 0 \\ 0 & 0 & 1/2 \\ 0 & 0 & 1/2 \end{pmatrix} + \begin{pmatrix} 1 & 0 & 0 \\ 0 & 1 & -1/2 \\ 0 & 0 & 1/2 \end{pmatrix} = I.$$

Exercise 4.10.29 Show that for $A \in \mathbb{C}^{4 \times 4}$, one has that

$$\sin 2A = 2 \sin A \cos A.$$

Answer: Let $A = SJ_4(\lambda)S^{-1}$. Then

$$\sin A = S \begin{pmatrix} \sin\lambda & \cos\lambda & -\frac{1}{2}\sin\lambda & -\frac{1}{6}\cos\lambda \\ & \sin\lambda & \cos\lambda & -\frac{1}{2}\sin\lambda \\ & & \sin\lambda & \cos\lambda \\ & & & \sin\lambda \end{pmatrix} S^{-1},$$

$$\cos A = S \begin{pmatrix} \cos\lambda & -\sin\lambda & -\frac{1}{2}\cos\lambda & \frac{1}{6}\sin\lambda \\ & \cos\lambda & -\sin\lambda & -\frac{1}{2}\cos\lambda \\ & & \cos\lambda & -\sin\lambda \\ & & & \cos\lambda \end{pmatrix} S^{-1},$$

and

$$\sin 2A = S \begin{pmatrix} \sin 2\lambda & 2\cos 2\lambda & -4\sin 2\lambda & -\frac{8}{6}\cos 2\lambda \\ & \sin 2\lambda & 2\cos 2\lambda & -2\sin 2\lambda \\ & & \sin 2\lambda & 2\cos 2\lambda \\ & & & \sin 2\lambda \end{pmatrix} S^{-1}.$$

Using now double angle formulas such as

$$\sin 2\lambda = 2\sin\lambda\cos\lambda, \cos 2\lambda = \cos^2\lambda - \sin^2\lambda,$$

one checks that $\sin 2A = 2\sin A \cos A$ in this case. The more general case where $A = SJS^{-1}$, with J a direct sum of single Jordan blocks, now also follows easily.

Exercise 4.10.30 Solve the inhomogeneous system of differential equations

$$\begin{cases} x_1'(t) = x_1(t) + 2x_2(t) + e^{-2t}, \\ x_2'(t) = 4x_1(t) - x_2(t). \end{cases}$$

Answer:

$$\begin{pmatrix} x_1(t) \\ x_2(t) \end{pmatrix} = \begin{pmatrix} c_1 e^{3t} + c_2 e^{-3t} + \frac{1}{5} e^{-2t} \\ c_1 e^{3t} - 2c_2 e^{-3t} - \frac{7}{10} e^{-2t} \end{pmatrix}.$$

Exercise 4.10.31 With the notation of Section 4.9 show that

$$I = \frac{1}{2\pi i} \int_\gamma R(\lambda) d\lambda, \ A = \frac{1}{2\pi i} \int_\gamma \lambda R(\lambda) d\lambda.$$

Answer: First note that by Cauchy's integral formula (4.34) with $f(z) \equiv 1$, we have that

$$1 = \frac{1}{2\pi i} \int_\gamma \frac{1}{z - \lambda_j} dz, \ 0 = \frac{1}{2\pi i} \int_\gamma \frac{1}{(z - \lambda_j)^{k+1}} dz, k \geq 1.$$

Using now (4.33) we get that

$$\frac{1}{2\pi i} \int_\gamma R(z) dz = \sum_{l=1}^{m} \sum_{k=0}^{n_j - 1} \frac{k!}{2\pi i} \int_\gamma \frac{1}{(z - \lambda_j)^{k+1}} dz P_{lk} = \sum_{j=0}^{m} P_{j0} = I.$$

Next, by Cauchy's integral formula (4.34) with $f(z) = z$, we have that

$$\lambda_j = \frac{1}{2\pi i} \int_\gamma \frac{z}{z - \lambda_j} dz, 1 = \frac{1}{2\pi i} \int_\gamma \frac{z}{(z - \lambda_j)^2} dz,$$

$$0 = \frac{1}{2\pi i} \int_\gamma \frac{z}{(z - \lambda_j)^{k+1}} dz, k \geq 2.$$

Using (4.33) we get that

$$\frac{1}{2\pi i} \int_\gamma z R(z) dz = \sum_{l=1}^{m} \sum_{k=0}^{n_j - 1} k! [\frac{1}{2\pi i} \int_\gamma \frac{z}{(z - \lambda_j)^{k+1}} dz P_{lk} = \sum_{j=1}^{m} (\lambda_j P_{j0} + P_{j1}).$$

Using the definitions of P_{jk} as in Theorem 4.8.4, one sees that this equals A, as desired.

Exercise 4.10.32 Show that the resolvent satisfies

(a) $\frac{R(\lambda) - R(\mu)}{\lambda - \mu} = -R(\lambda) R(\mu)$.

(b) $\frac{dR(\lambda)}{d\lambda} = -R(\lambda)^2$.

(c) $\frac{d^j R(\lambda)}{d\lambda^j} = (-1)^j j! R(\lambda)^{j+1}, j = 1, 2, \ldots$.

Answer: (a) We observe that

$$(\lambda - A)^{-1} - (\mu - A)^{-1} = (\lambda - A)^{-1} [(\mu - A) - (\lambda - A)](\mu - A)^{-1} =$$
$$(\lambda - A)^{-1} [\mu - \lambda](\mu - A)^{-1}.$$

Divide now both sides by $\lambda - \mu$ and $\frac{R(\lambda) - R(\mu)}{\lambda - \mu} = -R(\lambda) R(\mu)$ follows.

(b) Using (a) we get that

$$\lim_{h \to 0} \frac{R(\lambda + h) - R(\lambda)}{h} = - \lim_{h \to 0} R(\lambda + h) R(\lambda) = -R(\lambda)^2.$$

(c) Similar to part (b), we have

$$\lim_{h \to 0} \frac{R(\lambda+h)^{k+1} R(\lambda)^l - R(\lambda+h)^k R(\lambda)^{l+1}}{h} =$$

$$\lim_{h \to 0} R(\lambda+h)^k \frac{R(\lambda+h) - R(\lambda)}{h} R(\lambda)^l =$$

$$- \lim_{h \to 0} R(\lambda+h)^k R(\lambda+h) R(\lambda) R(\lambda)^l = -R(\lambda)^{k+l+2}. \tag{4.28}$$

Let us now prove $\frac{d^j R(\lambda)}{d\lambda^j} = (-1)^j j! R(\lambda)^{j+1}$ by induction on j. The $j = 1$ case was covered in part (b). Assume now that it has been proven that

$$\frac{d^{j-1} R(\lambda)}{d\lambda^{j-1}} = (-1)^{j-1}(j-1)! R(\lambda)^j.$$

Then

$$\frac{d^j R(\lambda)}{d\lambda^j} = \frac{d}{d\lambda}(-1)^{j-1}(j-1)! R(\lambda)^j = (-1)^{j-1}(j-1)! \lim_{h \to 0} \frac{R(\lambda+h)^j - R(\lambda)^j}{h}.$$

Write now

$$R(\lambda+h)^j - R(\lambda)^j = \sum_{k=0}^{j-1} (R(\lambda+h)^{j-k} R(\lambda)^k - R(\lambda+h)^{j-k-1} R(\lambda)^{k+1}).$$

Using observation (4.28), we have that

$$\lim_{h \to 0} \frac{R(\lambda+h)^{j-k} R(\lambda)^k - R(\lambda+h)^{j-k-1} R(\lambda)^{k+1}}{h} =$$

$$-R(\lambda)^{j-k+k+1} = -R(\lambda)^{j+1}.$$

And thus

$$\lim_{h \to 0} \frac{R(\lambda+h)^j - R(\lambda)^j}{h} = \sum_{k=0}^{j-1} -R(\lambda)^{j+1} = -jR(\lambda)^{j+1}.$$

Consequently,

$$\frac{d^j R(\lambda)}{d\lambda^j} = (-1)^{j-1}(j-1)!(-j) R(\lambda)^{j+1} = (-1)^j j! R(\lambda)^{j+1},$$

as desired.

Exercise 4.10.33 With the notation of Theorem 4.9.3 show that

$$\lambda_j P_{j0} + P_{j1} = AP_{j0} = \frac{1}{2\pi i} \int_{\gamma_j} \lambda R(\lambda) d\lambda.$$

Answer: First note that by Cauchy's integral formula (4.34) with $f(z) = z$, we have that

$$\lambda_j = \frac{1}{2\pi i} \int_{\gamma_j} \frac{z}{z - \lambda_j} dz, \quad 0 = \frac{1}{2\pi i} \int_{\gamma_l} \frac{z}{z - \lambda_j} dz \text{ when } j \neq l,$$

$$1 = \frac{1}{2\pi i} \int_{\gamma_j} \frac{z}{(z - \lambda_j)^2} dz,$$

$$0 = \frac{1}{2\pi i} \int_{\gamma_l} \frac{z}{(z - \lambda_j)^2} dz \text{ when } j \neq l, \quad 0 = \frac{1}{2\pi i} \int_{\gamma_l} \frac{z}{(z - \lambda_j)^{k+1}} dz, k \geq 2.$$

Using now (4.33) we get that

$$\frac{1}{2\pi i} \int_{\gamma_j} zR(z) dz = \sum_{l=1}^{m} \sum_{k=0}^{n_j-1} k! [\frac{1}{2\pi i} \int_{\gamma_j} \frac{z}{(z - \lambda_j)^{k+1}} dz] P_{lk} = \lambda_j P_{j0} + P_{j1}.$$

Using Theorem 4.8.4 ones sees that

$$AP_{j0} = S(\oplus_{l=1}^{m} J(\lambda_l)) S^{-1} P_{j0} = S(0 \oplus \cdots \oplus J(\lambda_j) I_{n_j} \oplus \cdots \oplus 0) S^{-1} = \lambda_j P_{j0} + P_{j1}.$$

Chapter 5

Exercise 5.7.1 For the following, check whether $\langle \cdot, \cdot \rangle$ is an inner product.

(a) $V = \mathbb{R}^2$, $\mathbb{F} = \mathbb{R}$,

$$\left\langle \begin{pmatrix} x_1 \\ x_2 \end{pmatrix}, \begin{pmatrix} y_1 \\ y_2 \end{pmatrix} \right\rangle = 3x_1y_1 + x_1y_2 + x_2y_1 + 2x_2y_2.$$

(b) $V = \mathbb{C}^2$, $\mathbb{F} = \mathbb{C}$,

$$\left\langle \begin{pmatrix} x_1 \\ x_2 \end{pmatrix}, \begin{pmatrix} y_1 \\ y_2 \end{pmatrix} \right\rangle = 3x_1y_1 + x_1y_2 + x_2y_1 + 2x_2y_2.$$

(c) Let $V = \{f : [0,1] \to \mathbb{R} \ : f \text{ is continuous}\}$, $\mathbb{F} = \mathbb{R}$,

$$\langle f, g \rangle = f(0)g(0) + f(1)g(1) + f(2)g(2).$$

(d) Let $V = \mathbb{R}_2[X]$, $\mathbb{F} = \mathbb{R}$,

$$\langle f, g \rangle = f(0)g(0) + f(1)g(1) + f(2)g(2).$$

(e) Let $V = \{f : [0,1] \to \mathbb{C} \ : f \text{ is continuous}\}$, $\mathbb{F} = \mathbb{C}$,

$$\langle f, g \rangle = \int_0^1 f(x)\overline{g(x)}(x^2 + 1)dx.$$

Answer: (a) Write $\left\langle \begin{pmatrix} x_1 \\ x_2 \end{pmatrix}, \begin{pmatrix} y_1 \\ y_2 \end{pmatrix} \right\rangle = (x_1 + x_2)(y_1 + y_2) + 2x_1y_1 + x_2y_2$ and realizing that everything is over $\mathbb{R}$, it is easy to see that this defines an inner product.

(b) $\langle ie_1, ie_1 \rangle = -3$, thus this is not an inner product.

(c) Let $f(t) = t(t-1)(t-2)$, then $f \neq \mathbf{0}$, but $\langle f, f \rangle = 0$. Thus this is not an inner product.

(d) Nonnegativity, linearity and symmetry are easy to check. Next suppose that $\langle f, f \rangle = 0$. Then we get that $f(0) = f(1) = f(2) = 0$. As $f \in \mathbb{R}_2[X]$, this implies that $f = \mathbf{0}$ (as a degree ≤ 2 polynomial with three roots is the zero polynomial).

(e) Nonnegativity, linearity and (complex conjugate) symmetry are easy to check. Next suppose that $\langle f, f \rangle = 0$. This implies that $int_0^1 |f(x)|^2(x^2 + 1)dx = 0$. Since the integrand is continuous and nonnegative, we must have that $|f(x)|^2(x^2 + 1) = 0$ for $x \in [0,1]$. Thus $f(x) = 0$ for $x \in [0,1]$. Thus $f = \mathbf{0}$. This shows that this is an inner product.

Exercise 5.7.2 For the following, check whether $\|\cdot\|$ is a norm.

(a) $V = \mathbb{C}^2$, $\mathbb{F} = \mathbb{C}$,

$$\left\| \begin{pmatrix} x_1 \\ x_2 \end{pmatrix} \right\| = x_1^2 + x_2^2.$$

(b) $V = \mathbb{C}^2$, $\mathbb{F} = \mathbb{C}$,

$$\left\| \begin{pmatrix} x_1 \\ x_2 \end{pmatrix} \right\| = |x_1| + 2|x_2|.$$

(c) Let $V = \{f : [0, 2] \to \mathbb{R} : f \text{ is continuous}\}$, $\mathbb{F} = \mathbb{R}$,

$$\|f\| = \int_0^2 |f(x)|(1 - x)dx.$$

(d) Let $V = \{f : [0, 1] \to \mathbb{R} : f \text{ is continuous}\}$, $\mathbb{F} = \mathbb{R}$,

$$\|f\| = \int_0^1 |f(x)|(1 - x)dx.$$

Answer: (a) Not a norm. For instance, $\left\| \begin{pmatrix} i \\ 0 \end{pmatrix} \right\| = -1 \not\geq 0$.

(b) Clearly this quantity is always nonnegative, and when it equals 0 we need that $|x_1| = 0 = |x_2|$, yielding that $\mathbf{x} = 0$. Thus the first property of a norm is satisfied.

Next, $\|c\mathbf{x}\| = |cx_1| + 2|cx_2| = |c|(|x_1| + 2|x_2|) = |c|\|\mathbf{x}\|$.

Finally,

$$\|\mathbf{x} + \mathbf{y}\| = |x_1 + y_1| + 2|x_2 + y_2| \leq |x_1| + |y_1| + 2(|x_2| + |y_2|) = \|\mathbf{x}\| + \|\mathbf{y}\|,$$

yielding that $\|\cdot\|$ is a norm.

(c) This is not a norm. For instance, if $f(x) = 1 + x$, then

$$\|f\| = \int_0^2 (1 + x)(1 - x)dx = (x - \frac{x^3}{3})|_{x=0}^{x=2} = 2 - \frac{8}{3} - 0 < 0.$$

(d) Notice that $1 - x \geq 0$ when $0 \leq x \leq 1$, thus $\|f\| \geq 0$ for all $f \in V$. Next, suppose that $\|f\| = 0$. As $|f(x)|(1 - x) \geq 0$ on $[0, 1]$, the only way the integral can be zero is when $|f(x)|(1 - x) = 0$ for $x \in [0, 1]$. Thus $f(x) = 0$ for $x \in (0, 1]$, and then, by continuity, it also follows that $f(0) = \lim_{x \to 0+} f(x) = 0$. Thus f is the zero function. This takes care of the first condition of a norm.

For properties (ii) and (iii), observe that $|cf(x)| = |c||f(x)|$ and $|(f + g)(x)| = |f(x) + g(x)| \leq |f(x)| + |g(x)|$. Using this, it is easy to see that $\|cf\| = |c|\|f\|$ and $\|f + g\| \leq \|f\| + \|g\|$, giving that $\|\cdot\|$ is a norm.

Exercise 5.7.3 Let $\mathbf{v}_1, \ldots, \mathbf{v}_n$ be nonzero orthogonal vectors in an inner product space V. Show that $\{\mathbf{v}_1, \ldots, \mathbf{v}_n\}$ is linearly independent.

Answer: Let $c_1, \ldots, c_n$ be so that $c_1\mathbf{v}_1 + \cdots + c_n\mathbf{v}_n = \mathbf{0}$. We need to show that $c_j = 0$, $j = 1, \ldots, n$. For this, observe that for $j = 1, \ldots, n$,

$$0 = \langle \mathbf{0}, \mathbf{v}_j \rangle = \langle c_1\mathbf{v}_1 + \cdots + c_n\mathbf{v}_n, \mathbf{v}_j \rangle = c_j \langle \mathbf{v}_j, \mathbf{v}_j \rangle.$$

As $\mathbf{v}_j \neq \mathbf{0}$, we have that $\langle \mathbf{v}_j, \mathbf{v}_j \rangle \neq 0$, and thus $c_j = 0$ follows.

Exercise 5.7.4 Let V be an inner product space.

(a) Determine $\{\mathbf{0}\}^\perp$ and $V^\perp$.

(b) Let $V = \mathbb{C}^4$ and $W = \left\{ \begin{pmatrix} 1 \\ i \\ 1 + i \\ 2 \end{pmatrix}, \begin{pmatrix} 0 \\ -i \\ 1 + 2i \\ 0 \end{pmatrix} \right\}$. Find a basis for $W^\perp$.

(c) In case V is finite-dimensional and W is a subspace, show that $\dim W^\perp = \dim V - \dim W$. (Hint: start with an orthonormal basis for W and add vectors to it to obtain an orthonormal basis for V).

Answer: (a) $\{\mathbf{0}\}^\perp = V$ and $V^\perp = \{\mathbf{0}\}$.

(b) We need to find a basis for the null space of $\begin{pmatrix} 1 & -i & 1-i & 2 \\ 0 & i & 1-2i & 0 \end{pmatrix}$, which in row-reduced echelon form is the matrix $\begin{pmatrix} 1 & 0 & -3i & 2 \\ 0 & 1 & -2+i & 0 \end{pmatrix}$. This gives the basis

$$\left\{ \begin{pmatrix} 3i \\ 2-i \\ 1 \\ 0 \end{pmatrix}, \begin{pmatrix} -2 \\ 0 \\ 0 \\ 1 \end{pmatrix} \right\}$$

for $W^\perp$.

(c) Let $k = \dim W$ and $n = \dim V$. Clearly $k \leq n$. Let $\{\mathbf{v}_1, \ldots, \mathbf{v}_k\}$ be an orthonormal basis for W, and extend this basis to an orthonormal basis $\{\mathbf{v}_1, \ldots, \mathbf{v}_n\}$ for V (which can be done, as one can extend to a basis of V and then make it orthonormal via the Gram–Schmidt process). We now claim that $\{\mathbf{v}_{k+1}, \ldots, \mathbf{v}_n\}$ is a basis for $W^\perp$. Let $\mathbf{x} \in W^\perp$. As $\mathbf{x} \in V$, we have that there exists $c_1, \ldots, c_n$ so that $\mathbf{x} = \sum_{i=1}^n c_i \mathbf{v}_i$. Since $\mathbf{x} \perp \mathbf{v}_i$, $i = 1, \ldots, k$, we get that $c_1 = \cdots = c_k = 0$. Thus $\mathbf{x} \in \mathrm{Span}\{\mathbf{v}_{k+1}, \ldots, \mathbf{v}_n\}$, yielding that $W^\perp \subseteq \mathrm{Span}\{\mathbf{v}_{k+1}, \ldots, \mathbf{v}_n\}$. Due to orthonormality of the basis for V, we have that $\mathbf{v}_{k+1}, \ldots, \mathbf{v}_n$ are orthogonal to the vectors $\mathbf{v}_1, \ldots, \mathbf{v}_k$, and thus orthogonal to any vector in W. Thus $\mathrm{Span}\{\mathbf{v}_{k+1}, \ldots, \mathbf{v}_n\} \subseteq W^\perp$, and we obtain equality. As we already have the linear independence of $\{\mathbf{v}_{k+1}, \ldots, \mathbf{v}_n\}$, we obtain that it is a basis for $W^\perp$. Thus $\dim W^\perp = n - k$.

Exercise 5.7.5 Let $\langle \cdot, \cdot \rangle$ be the Euclidean inner product on $\mathbb{F}^n$, and $\| \cdot \|$ the associated norm.

(a) Let $\mathbb{F} = \mathbb{C}$. Show that $A \in \mathbb{C}^{n \times n}$ is the zero matrix if and only if $\langle A\mathbf{x}, \mathbf{x} \rangle = 0$ for all $\mathbf{x} \in \mathbb{C}^n$. (Hint: for $\mathbf{x}, \mathbf{y} \in \mathbb{C}$, use that $\langle A(\mathbf{x} + \mathbf{y}), \mathbf{x} + \mathbf{y} \rangle = 0 = \langle A(\mathbf{x} + i\mathbf{y}), \mathbf{x} + i\mathbf{y} \rangle$.)

(b) Show that when $\mathbb{F} = \mathbb{R}$, there exists nonzero matrices $A \in \mathbb{R}^{n \times n}$, $n > 1$, so that $\langle A\mathbf{x}, \mathbf{x} \rangle = 0$ for all $\mathbf{x} \in \mathbb{R}^n$.

(c) For $A \in \mathbb{C}^{n \times n}$ define

$$w(A) = \max_{\mathbf{x} \in \mathbb{C}^n, \|\mathbf{x}\|=1} |\langle A\mathbf{x}, \mathbf{x} \rangle|. \tag{5.29}$$

Show that $w(\cdot)$ is a norm on $\mathbb{C}^{n \times n}$. This norm is called the *numerical radius* of A.

(d) Explain why $\max_{\mathbf{x} \in \mathbb{R}^n, \|\mathbf{x}\|=1} |\langle A\mathbf{x}, \mathbf{x} \rangle|$ does not define a norm.

Answer: (a) Clearly, if $A = 0$, then $\langle A\mathbf{x}, \mathbf{x} \rangle = 0$ for all $\mathbf{x} \in \mathbb{C}^n$.

For the converse, assume that $\langle A\mathbf{x}, \mathbf{x} \rangle = 0$ for all $\mathbf{x} \in \mathbb{C}^n$. Let now $\mathbf{x}, \mathbf{y} \in \mathbb{C}$. Then

$$0 = \langle A(\mathbf{x} + \mathbf{y}), \mathbf{x} + \mathbf{y} \rangle = \langle A\mathbf{x}, \mathbf{x} \rangle + \langle A\mathbf{y}, \mathbf{x} \rangle + \langle A\mathbf{x}, \mathbf{y} \rangle + \langle A\mathbf{y}, \mathbf{y} \rangle = \langle A\mathbf{y}, \mathbf{x} \rangle + \langle A\mathbf{x}, \mathbf{y} \rangle \tag{5.30}$$

and, similarly,

$$0 = \langle A(\mathbf{x} + i\mathbf{y}), \mathbf{x} + i\mathbf{y} \rangle = i\langle A\mathbf{y}, \mathbf{x} \rangle - i\langle A\mathbf{x}, \mathbf{y} \rangle. \tag{5.31}$$

Combining (5.30) and (5.31), we obtain that $\langle A\mathbf{x}, \mathbf{y} \rangle = 0$ for all $\mathbf{x}, \mathbf{y} \in \mathbb{C}$. Applying this with $\mathbf{x} = \mathbf{e}_j$ and $\mathbf{y} = \mathbf{e}_k$, we obtain that the (k, j)th entry of A equals zero. As this holds for all $k, j = 1, \ldots, n$, we obtain that $A = 0$.

(b) When $n = 2$ one may choose $A = \begin{pmatrix} 0 & -1 \\ 1 & 0 \end{pmatrix}$. For larger n one can add zero rows and columns to this matrix.

(c) Clearly, $w(A) \geq 0$. Next, suppose that $w(A) = 0$. Then for all $\|\mathbf{x}\| = 1$, we have that $\langle A\mathbf{x}, \mathbf{x} \rangle = 0$. This implies that for all $\mathbf{x} \in \mathbb{C}^n$ we have that $\langle A\mathbf{x}, \mathbf{x} \rangle = 0$. By (a), this implies that $A = 0$. Next, for $\|\mathbf{x}\| = 1$, we have

$$|\langle (A + B)\mathbf{x}, \mathbf{x} \rangle| \leq |\langle A\mathbf{x}, \mathbf{x} \rangle| + |\langle B\mathbf{x}, \mathbf{x} \rangle| \leq w(A) + w(B),$$

and thus $w(A + B) \leq w(A) + w(B)$. Finally, when $c \in \mathbb{C}$, one has that $|\langle (cA)\mathbf{x}, \mathbf{x} \rangle| = |c||\langle A\mathbf{x}, \mathbf{x} \rangle|$, and thus $w(cA) = |c|w(A)$ follows easily.

(d) With A as in part (b), we have that $\max_{\mathbf{x} \in \mathbb{R}^n, \|\mathbf{x}\|=1} |\langle A\mathbf{x}, \mathbf{x} \rangle| = 0$, and thus the first property in the definition of a norm fails.

Exercise 5.7.6 Find an orthonormal basis for the subspace in $\mathbb{R}^4$ spanned by

$$\begin{pmatrix} 1 \\ 1 \\ 1 \\ 1 \end{pmatrix}, \begin{pmatrix} 1 \\ 2 \\ 1 \\ 2 \end{pmatrix}, \begin{pmatrix} 3 \\ 1 \\ 3 \\ 1 \end{pmatrix}.$$

Answer: Applying Gram–Schmidt, we find

$$Q = \begin{pmatrix} \frac{1}{2} & \frac{1}{2} & \frac{1}{2} \\ \frac{1}{2} & -\frac{1}{2} & \frac{1}{2} \\ \frac{1}{2} & \frac{1}{2} & -\frac{1}{2} \\ \frac{1}{2} & -\frac{1}{2} & -\frac{1}{2} \end{pmatrix}, R = \begin{pmatrix} 2 & 3 & 4 \\ 0 & -1 & 2 \\ 0 & 0 & 0 \end{pmatrix}.$$

Thus $\left\{ \begin{pmatrix} \frac{1}{2} \\ \frac{1}{2} \\ \frac{1}{2} \\ \frac{1}{2} \end{pmatrix}, \begin{pmatrix} \frac{1}{2} \\ -\frac{1}{2} \\ \frac{1}{2} \\ -\frac{1}{2} \end{pmatrix} \right\}$ is the requested orthonormal basis.

Exercise 5.7.7

Let $V = \mathbb{R}[t]$ over the field $\mathbb{R}$. Define the inner product

$$\langle p, q \rangle := \int_{-1}^{1} p(t)q(t)dt.$$

For the following linear maps on V determine whether they are self-adjoint.

(a) $Lp(t) := (t^2 + 1)p(t)$.

(b) $Lp(t) := \frac{dp}{dt}(t)$.

(c) $Lp(t) = -p(-t)$.

Answer: (a) $\langle L(p), q \rangle = \int_{-1}^{1} (t^2 + 1)p(t)q(t)dt = \langle p, L(q) \rangle$. Thus L is self-adjoint.

(b) Let $p(t) \equiv 1$ and $q(t) = t$. Then $\langle L(p), q \rangle = 0$, and $\langle p, L(q) \rangle = \int_{-1}^{1} 1dt = 2$. Thus L is not self-adjoint.

(c) $\langle L(p), q \rangle = \int_{-1}^{1} -p(-t)q(t)dt = \int_{1}^{-1} -p(s)q(-s)(-ds) = \langle p, L(q) \rangle$. Thus L is self-adjoint.

Exercise 5.7.8 Let $V = \mathbb{R}[t]$ over the field $\mathbb{R}$. Define the inner product

$$\langle p, q \rangle := \int_0^2 p(t)q(t)dt.$$

For the following linear maps on V determine whether they are unitary.

(a) $Lp(t) := tp(t)$.

(b) $Lp(t) = -p(2 - t)$.

Answer: (a) Let $p(t) = q(t) = t$. Then $\langle p, q \rangle = 2$, while $\langle L(p), L(q) \rangle = \frac{32}{5}$. Thus L is not unitary.

(b) Doing a change of variables $s = 2 - t$, we get
$\langle L(p), L(q) \rangle = \int_0^2 (-p(2-t))(-q(2-t))dt = \int_2^0 p(s)q(s)(-ds) = \langle p, q \rangle$. Thus L is unitary.

Exercise 5.7.9 Let $U : V \to V$ be unitary, where the inner product on V is denoted by $\langle \cdot, \cdot \rangle$.

(a) Show that $|\langle \mathbf{x}, U\mathbf{x} \rangle| \leq \|\mathbf{x}\|^2$ for all $\mathbf{x}$ in V.

(b) Show that $|\langle \mathbf{x}, U\mathbf{x} \rangle| = \|\mathbf{x}\|^2$ for all $\mathbf{x}$ in V, implies that $U = \alpha I$ for some $|\alpha| = 1$.

Answer: (a) By (5.1) we have that $|\langle \mathbf{x}, U\mathbf{x} \rangle| \leq \|\mathbf{x}\|\|U\mathbf{x}\| = \|\mathbf{x}\|^2$, where in the last step we used that $\|\mathbf{x}\| = \|U\mathbf{x}\|$ as U is unitary.

(b) Let $\mathbf{x}$ be a unit vector. As $|\langle \mathbf{x}, U\mathbf{x} \rangle| = \|\mathbf{x}\|^2$, we have by (the last part of) Theorem 5.1.10 that $U\mathbf{x} = \alpha\mathbf{x}$ for some α. As $\|\mathbf{x}\| = \|U\mathbf{x}\|$ we must have $|\alpha| = 1$. If we are in a one-dimensional vector space, we are done. If not, let $\mathbf{v}$ be a unit vector orthogonal to $\mathbf{v}$. As above, we get $U\mathbf{v} = \beta\mathbf{v}$ for some $|\beta| = 1$. In addition, we get that $U(\mathbf{x} + \mathbf{v}) = \mu(\mathbf{x} + \mathbf{v})$ with $|\mu| = 1$. Now, we get that

$$\mu = \langle \mu\mathbf{x}, \mathbf{x} \rangle = \langle \mu(\mathbf{x} + \mathbf{v}), \mathbf{x} \rangle = \langle U(\mathbf{x} + \mathbf{v}), \mathbf{x} \rangle = \langle \alpha\mathbf{x} + \beta\mathbf{v}), \mathbf{x} \rangle = \alpha.$$

Similarly, we prove $\mu = \beta$. Thus $\alpha = \beta$. Thus, show that $U\mathbf{y} = \alpha\mathbf{y}$ for all $\mathbf{y} \perp \mathbf{x}$ and also for $\mathbf{y} = \mathbf{x}$. But then the same holds for linear combinations of $\mathbf{x}$ and $\mathbf{y}$, and we obtain that $U = \alpha I$.

Exercise 5.7.10 Let $V = \mathbb{C}^{n \times n}$, and define

$$\langle A, B \rangle = \mathrm{tr}(AB^*).$$

(a) Let $W = \mathrm{span}\{\begin{pmatrix} 1 & 2 \\ 0 & 1 \end{pmatrix}, \begin{pmatrix} 1 & 0 \\ 2 & 1 \end{pmatrix}\}$. Find an orthonormal basis for W.

(b) Find a basis for $W^\perp := \{B \in V : B \perp C \text{ for all } C \in W\}$.

Answer: (a) Performing Gram–Schmidt, we get

$$\begin{pmatrix} 1 & 0 \\ 2 & 1 \end{pmatrix} - \frac{2}{6}\begin{pmatrix} 1 & 2 \\ 0 & 1 \end{pmatrix} = \begin{pmatrix} \frac{2}{3} & -\frac{2}{3} \\ 2 & \frac{2}{3} \end{pmatrix}.$$

Thus $\{\frac{1}{\sqrt{6}}\begin{pmatrix} 1 & 2 \\ 0 & 1 \end{pmatrix}, \frac{\sqrt{3}}{\sqrt{10}}\begin{pmatrix} \frac{2}{3} & -\frac{2}{3} \\ 2 & \frac{2}{3} \end{pmatrix}\}$ is the requested orthonormal basis.

(b) Let $B = \begin{pmatrix} a & b \\ c & d \end{pmatrix} \in W^{\perp}$. Then we get $a + 2c + d = 0, 1a + 2b + d = 0$. With c and d as free variables, we get

$$\begin{pmatrix} a & b \\ c & d \end{pmatrix} = \begin{pmatrix} -2c - d & c \\ c & d \end{pmatrix} = c \begin{pmatrix} -2 & 1 \\ 1 & 0 \end{pmatrix} + d \begin{pmatrix} -1 & 0 \\ 0 & 1 \end{pmatrix}.$$

Performing Gram–Schmidt, we get

$$\begin{pmatrix} -1 & 0 \\ 0 & 1 \end{pmatrix} - \frac{2}{6} \begin{pmatrix} -2 & 1 \\ 1 & 0 \end{pmatrix} = \begin{pmatrix} -\frac{1}{3} & -\frac{1}{3} \\ -\frac{1}{3} & 1 \end{pmatrix}.$$

Thus $\{ \frac{1}{\sqrt{6}} \begin{pmatrix} -2 & 1 \\ 1 & 0 \end{pmatrix}, \frac{\sqrt{3}}{2} \begin{pmatrix} -\frac{1}{3} & -\frac{1}{3} \\ -\frac{1}{3} & 1 \end{pmatrix} \}$ is the requested orthonormal basis.

Exercise 5.7.11 Let $A \in \mathbb{C}^{n \times n}$. Show that if A is normal and $A^k = 0$ for some $k \in \mathbb{N}$, then $A = 0$.

Answer: As A is normal, $A = UDU^*$ for some uniray U and diagonal D. Then $0 = A^k = UD^kU^*$, thus $D^k = 0$. As D is diagonal, we get $D = 0$. Thus $A = 0$.

Exercise 5.7.12 Let $A \in \mathbb{C}^{n \times n}$ and $a \in \mathbb{C}$. Show that A is normal if and only if $A - aI$ is normal.

Answer: If $AA^* = A^*A$, then
$(A - aI)(A - aI)^* = AA^* - \bar{a}A - aA^* + |a|^2 I = A^*A - \bar{a}A - aA^* + |a|^2 I = (A - aI)^*(A - aI).$

Exercise 5.7.13 Show that the sum of two Hermitian matrices is Hermitian. How about the product?

Answer: If $A = A^*$ and $B = B^*$, then $(A + B)^* = A^* + B^* = A + B$. Thus the sum of two Hermitian matrices is Hermitian.

A product of two Hermitian matrices is not necessarily Hermitian. For example, $A = \begin{pmatrix} 2 & 1 \\ -i & 2 \end{pmatrix}, B = \begin{pmatrix} 0 & 1 \\ 1 & 0 \end{pmatrix}$ are Hermitian but AB is not.

Exercise 5.7.14 Show that the product of two unitary matrices is unitary. How about the sum?

Answer: Let U and V be unitary. Then $(UV)(UV)^* = UVV^*U^* = UU^* = I$ and $(UV)^*(UV) = V^*U^*UV = V^*V = I$, thus UV is unitary.

The sum of two unitary matrices is in general not unitary. For example, $U = I$ is unitary, but $U + U = 2I$ is not.

Exercise 5.7.15 Is the product of two normal matrices normal? How about the sum?

Answer: No, e.g., $A = \begin{pmatrix} 2 & i \\ -i & 2 \end{pmatrix}, B = \begin{pmatrix} 0 & 1 \\ i & 0 \end{pmatrix}$ are normal, but neither AB nor $A + B$ is normal.

Exercise 5.7.16 Show that the following matrices are unitary.

(a) $\frac{1}{\sqrt{2}} \begin{pmatrix} 1 & 1 \\ 1 & -1 \end{pmatrix}$.

Answer: $\frac{1}{\sqrt{2}} \begin{pmatrix} 1 & 1 \\ 1 & -1 \end{pmatrix} \frac{1}{\sqrt{2}} \begin{pmatrix} 1 & 1 \\ 1 & -1 \end{pmatrix} = \frac{1}{2} \begin{pmatrix} 2 & 0 \\ 0 & 2 \end{pmatrix} = I_2.$

(b) $\frac{1}{\sqrt{3}} \begin{pmatrix} 1 & 1 & 1 \\ 1 & e^{\frac{2i\pi}{3}} & e^{\frac{4i\pi}{3}} \\ 1 & e^{\frac{4i\pi}{3}} & e^{\frac{8i\pi}{3}} \end{pmatrix}$.

Answer: $\frac{1}{\sqrt{3}} \begin{pmatrix} 1 & 1 & 1 \\ 1 & e^{\frac{2i\pi}{3}} & e^{\frac{4i\pi}{3}} \\ 1 & e^{\frac{4i\pi}{3}} & e^{\frac{8i\pi}{3}} \end{pmatrix} \frac{1}{\sqrt{3}} \begin{pmatrix} 1 & 1 & 1 \\ 1 & e^{-\frac{2i\pi}{3}} & e^{-\frac{4i\pi}{3}} \\ 1 & e^{-\frac{4i\pi}{3}} & e^{-\frac{8i\pi}{3}} \end{pmatrix} = \frac{1}{3} \begin{pmatrix} 3 & 0 & 0 \\ 0 & 3 & 0 \\ 0 & 0 & 3 \end{pmatrix} = I_3.$

(c) $\frac{1}{2} \begin{pmatrix} 1 & 1 & 1 & 1 \\ 1 & i & -1 & -i \\ 1 & -1 & 1 & -1 \\ 1 & -i & -1 & i \end{pmatrix}$.

Answer: $\frac{1}{2} \begin{pmatrix} 1 & 1 & 1 & 1 \\ 1 & i & -1 & -i \\ 1 & -1 & 1 & -1 \\ 1 & -i & -1 & i \end{pmatrix} \frac{1}{2} \begin{pmatrix} 1 & 1 & 1 & 1 \\ 1 & -i & -1 & i \\ 1 & -1 & 1 & -1 \\ 1 & i & -1 & -i \end{pmatrix} = \frac{1}{4} \begin{pmatrix} 4 & 0 & 0 & 0 \\ 0 & 4 & 0 & 0 \\ 0 & 0 & 4 & 0 \\ 0 & 0 & 0 & 4 \end{pmatrix} = I_4.$

(d) Can you guess the general rule?

Answer: The matrices in the previous parts are all Fourier matrices. The general form of a Fourier matrix is given before Proposition 7.4.3.

Exercise 5.7.17 For the following matrices A, find the spectral decomposition UDU^* of A.

(a) $A = \begin{pmatrix} 2 & i \\ -i & 2 \end{pmatrix}$.

(b) $A = \begin{pmatrix} 2 & \sqrt{3} \\ \sqrt{3} & 4 \end{pmatrix}$.

(c) $A = \begin{pmatrix} 3 & 1 & 1 \\ 1 & 3 & 1 \\ 1 & 1 & 3 \end{pmatrix}$.

(d) $A = \begin{pmatrix} 0 & 1 & 0 \\ 0 & 0 & 1 \\ 1 & 0 & 0 \end{pmatrix}$.

Answer: (a) $U = \begin{pmatrix} \frac{1}{\sqrt{2}} & \frac{1}{\sqrt{2}} \\ \frac{i}{\sqrt{2}} & -\frac{i}{\sqrt{2}} \end{pmatrix}, D = \begin{pmatrix} 1 & 0 \\ 0 & 3 \end{pmatrix}$.

(b) $U = \begin{pmatrix} -\frac{\sqrt{3}}{2} & \frac{1}{2} \\ \frac{1}{2} & \frac{\sqrt{3}}{2} \end{pmatrix}, D = \begin{pmatrix} 1 & 0 \\ 0 & 5 \end{pmatrix}$.

(c) $U = \begin{pmatrix} \frac{1}{\sqrt{6}} & \frac{1}{2} & \frac{1}{\sqrt{3}} \\ \frac{1}{\sqrt{6}} & -\frac{1}{2} & \frac{1}{\sqrt{3}} \\ -\frac{\sqrt{6}}{3} & 0 & \frac{1}{\sqrt{3}} \end{pmatrix}, D = \begin{pmatrix} 2 & 0 & 0 \\ 0 & 2 & 0 \\ 0 & 0 & 5 \end{pmatrix}$.

(d) $U = \frac{1}{\sqrt{3}} \begin{pmatrix} 1 & 1 & 1 \\ 1 & e^{\frac{2\pi i}{3}} & e^{\frac{4\pi i}{3}} \\ 1 & e^{\frac{2\pi i}{3}} & e^{\frac{2\pi i}{3}} \end{pmatrix}, D = \begin{pmatrix} 1 & 0 & 0 \\ 0 & e^{\frac{2\pi i}{3}} & 0 \\ 0 & 0 & e^{\frac{4\pi i}{3}} \end{pmatrix}.$

Exercise 5.7.18 Let $A = \begin{pmatrix} 3 & 2i \\ -2i & 3 \end{pmatrix}$.

(a) Show that A is positive semidefinite.

(b) Find the positive *square root* of A; that is, find a positive semidefinite B so that $B^2 = A$.

Answer: (a) $p_A(\lambda) = (3 - \lambda) - |2i|^2 = \lambda^2 - 6\lambda + 5$, which has roots $1, 5$. As A is Hermitian with nonnegative eigenvalues, A is positive semidefinite.

(b) Let $U = \begin{pmatrix} \frac{1}{\sqrt{2}} & \frac{1}{\sqrt{2}} \\ \frac{i}{\sqrt{2}} & -\frac{i}{\sqrt{2}} \end{pmatrix}, D = \begin{pmatrix} 1 & 0 \\ 0 & 5 \end{pmatrix}$. Then $A = UDU^*$. Now let $B = U\sqrt{D}U^*$,

where $\sqrt{D} = \begin{pmatrix} 1 & 0 \\ 0 & \sqrt{5} \end{pmatrix}$. Then B is positive semidefinite, and $B^2 = A$.

Exercise 5.7.19 Let $A \in \mathbb{C}^{n \times n}$ be positive semidefinite, and let $k \in \mathbb{N}$. Show that there exists a unique positive semidefinite B so that $B^k = A$. We call B the *k*th root of A and denote $B = A^{\frac{1}{k}}$.

Answer: Since A is positive semidefinite, there exists a unitary U and a diagonal $D = \mathrm{diag}(d_i)_{i=1}^n$ so that $A = UDU^*$. Moreover, the diagonal entries d_i of D are nonnegative, and let us order them so that $d_1 \geq \cdots \geq d_n$. Thus we may define $D^{\frac{1}{k}} := \mathrm{diag}(d_i^{\frac{1}{k}})_{i=1}^n$. Let now $B = UD^{\frac{1}{k}}U^*$. Then B is positive semidefinite and $B^k = A$.

Next, suppose that C is positive semidefinite with $C^k = A$. For uniqueness of the *k*th root, we need to show that $C = B$. As C is positive semidefinite, we may write $C = V\Lambda V^*$ with V unitary, and $\Lambda = \mathrm{diag}(\lambda_i)_{i=1}^n$ with $\lambda_1 \geq \cdots \geq \lambda_n (\geq 0)$. Then $C^k = V\Lambda^k V^* = A$, and as the eigenvalues of C^k are $\lambda_1^k \geq \cdots \geq \lambda_n^k$ and the eigenvalues of A are $d_1 \geq \cdots \geq d_n$, we must have that $\lambda_i^k = d_i$, $i = 1, \ldots, n$. And thus, since $\lambda_i \geq 0$ for all i, we have $\lambda_i = d_i^{\frac{1}{k}}$, $i = 1, \ldots, n$. From the equalities $V\Lambda^k V^* = UDU^*$ and $\Lambda^k = D$, we obtain that $(U^*V)D = D(U^*V)$. Let $W = U^*V$ and write $W = (w_{ij})_{i,j=1}^n$. Then $WD = DW$ implies that $w_{ij}d_j = d_i w_{ij}$ for all $i, j = 1, \ldots, n$. When $d_j \neq d_i$ we thus get that $w_{ij} = 0$ (since $w_{ij}(d_j - d_i) = 0$). But then it follows that $w_{ij}d_j^{\frac{1}{k}} = d_i^{\frac{1}{k}} w_{ij}$ for all $i, j = 1, \ldots, n$ (indeed, when $w_{ij} = 0$ this is trivial, and when $d_i = d_j$ this also follows from $w_{ij}d_j = d_i w_{ij}$). Now we obtain that $U^*VD^{\frac{1}{k}} = WD^{\frac{1}{k}} = D^{\frac{1}{k}}W = D^{\frac{1}{k}}U^*V$, and thus $C = VD^{\frac{1}{k}}V^* = UD^{\frac{1}{k}}U^* = B$.

Exercise 5.7.20 Let $A \in \mathbb{C}^{n \times n}$ be positive semidefinite. Show that

$$\lim_{k \to \infty} \mathrm{tr} A^{\frac{1}{k}} = \mathrm{rank} A.$$

(Hint: use that for $\lambda > 0$ we have that $\lim_{k \to \infty} \lambda^{\frac{1}{k}} = 1$.)

Answer: We may write $A = U\Lambda U^*$ with U unitary and $\Lambda = \mathrm{diag}(\lambda_i)_{i=1}^n$ with

$\lambda_1 \geq \cdots \geq \lambda_r > \lambda_{r+1} = \cdots = \lambda_n = 0$ and $r = \operatorname{rank} A$. Then for $i = 1, \ldots, r$ we have that $\lim_{k\to\infty} \lambda_i^{\frac{1}{k}} = 1$, while for $i = r+1, \ldots, n$ we have that $\lim_{k\to\infty} \lambda_i^{\frac{1}{k}} = 0$. Thus

$$\lim_{k\to\infty} \operatorname{tr} A^{\frac{1}{k}} = \lim_{k\to\infty} \lambda_1^{\frac{1}{k}} + \cdots + \lambda_n^{\frac{1}{k}} = 1 + \cdots + 1 + 0 + \cdots + 0 = r = \operatorname{rank} A.$$

Exercise 5.7.21 Let $A = A^*$ be an $n \times n$ Hermitian matrix, with eigenvalues $\lambda_1 \geq \cdots \geq \lambda_n$.

(a) Show $tI - A$ is positive semidefinite if and only if $t \geq \lambda_1$.
   Answer: Write $A = UDU^*$, with U unitary and $D = \operatorname{diag}(\lambda_i)_{i=1}^n$. Then $tI - A = U(tI - D)U^*$. Thus $tI - A$ is positive semidefinite if and only if $t - \lambda_i \geq 0$, $i = 1, \ldots, n$, which holds if and only if $t - \lambda_1 \geq 0$.

(b) Show that $\lambda_{\max}(A) = \lambda_1 = \max_{\langle \mathbf{x},\mathbf{x}\rangle = 1} \langle A\mathbf{x}, \mathbf{x}\rangle$, where $\langle \cdot, \cdot \rangle$ is the Euclidean inner product.
   Answer: By part (a) $\lambda_1 I - A$ is positive semidefinite, and thus $\langle (\lambda_1 I - A)\mathbf{x}, \mathbf{x}\rangle \geq 0$. This gives that $\lambda_1 \langle \mathbf{x}, \mathbf{x}\rangle \geq \langle A\mathbf{x}, \mathbf{x}\rangle$ for all vectors $\mathbf{x}$. Choosing $\mathbf{x}$ to be a unit eigenvector of A at λ_1 we obtain equality. This proves the result.

(c) Let $\hat{A}$ be the matrix obtained from A by removing row and column i. Then $\lambda_{\max}(\hat{A}) \leq \lambda_{\max}(A)$.
   Answer: For $\mathbf{y} \in \mathbb{F}^n$ a vector with a 0 in the ith position, we let $\hat{\mathbf{y}}$ denote the vector obtained from $\mathbf{y}$ by removing the ith coordinate. Note that $\langle \hat{\mathbf{y}}, \hat{\mathbf{y}}\rangle = \langle \mathbf{y}, \mathbf{y}\rangle$ and $\langle \hat{A}\hat{\mathbf{y}}, \hat{\mathbf{y}}\rangle = \langle A\mathbf{y}, \mathbf{y}\rangle$. By part (b), we have that

$$\lambda_{\max}(\hat{A}) = \max_{\langle \hat{\mathbf{y}}, \hat{\mathbf{y}}\rangle = 1} \langle \hat{A}\hat{\mathbf{y}}, \hat{\mathbf{y}}\rangle = \max_{\langle \mathbf{y}, \mathbf{y}\rangle = 1, y_i = 0} \langle A\mathbf{y}, \mathbf{y}\rangle \leq \max_{\langle \mathbf{x}, \mathbf{x}\rangle = 1} \langle A\mathbf{x}, \mathbf{x}\rangle = \lambda_{\max}(A).$$

Exercise 5.7.22 (a) Show that a square matrix A is Hermitian iff $A^2 = A^*A$.

(b) Let H be positive semidefinite, and write $H = A + iB$ where A and B are real matrices. Show that if A is singular, then H is singular as well.

Answer: (a) Clearly, if A is Hermitian, then $A^*A = AA = A^2$.

Conversely, suppose that A is so that $A^2 = A^*A$. Apply Schur's triangularization theorem to obtain a unitary U and an upper triangular T so that $A = UTU^*$. But then $A^2 = A^*A$ implies $T^2 = T^*T$. Write $T = (t_{ij})_{i,j=1}^n$, with $t_{ij} = 0$ when $i > j$. Then the $(1,1)$ entry of the equation $T^2 = T^*T$ gives that $t_{11}^2 = |t_{11}|^2$, which shows that $t_{11} \in \mathbb{R}$. Next the $(2,2)$ entry of $T^2 = T^*T$ yields that $t_{22}^2 = |t_{12}|^2 + |t_{22}|^2$, which yields that $t_{22}^2 \geq 0$, and thus $t_{22}^2 = |t_{22}|^2$. But then it follows that $t_{12} = 0$ and $t_{22} \in \mathbb{R}$. In general, from the (k,k)th entry of $T^2 = T^*T$, one finds that $t_{ik} = 0$, $i < k$, and $t_{kk} \in \mathbb{R}$. Thus T is a real diagonal matrix, which gives that $A = UTU^*$ is Hermitian.

(b) Let $\mathbf{0} \neq \mathbf{v} \in \mathbb{R}^n$ be so that $A\mathbf{v} = \mathbf{0}$. Then $\mathbf{v}^* H\mathbf{v} \geq 0$, and thus $0 \leq \mathbf{v}^*(A + iB)\mathbf{v} = i\mathbf{v}^T B\mathbf{v}$, and thus we must have that $\mathbf{v}^T B\mathbf{v} = 0$. This gives that $\mathbf{v}^* H\mathbf{v} = 0$. Next write $H = UDU^*$, with $D = (d_{ij})_{i,j=1}^n$ a nonnegative diagonal matrix.

Let $\mathbf{v} = \begin{pmatrix} v_1 \\ \vdots \\ v_n \end{pmatrix}$. Then $d_{11}|v_1|^2 + \cdots + d_{nn}|v_n|^2 = 0$. As $\mathbf{v} \neq \mathbf{0}$, some v_i is nonzero. But

then $0 \leq d_{ii}|v_i|^2 \leq d_{11}|v_1|^2 + \cdots + d_{nn}|v_n|^2 = 0$, implies that $d_{ii} = 0$. Thus H is singular.

Exercise 5.7.23 (a) Let A be positive definite. Show that $A + A^{-1} - 2I$ is positive semidefinite.

(b) Show that A is normal if and only if $A^* = AU$ for some unitary matrix U.

Answer: (a) Clearly, since A is Hermitian, we have that $A + A^{-1} - 2I$ is Hermitian. Next, every eigenvalue of $A + A^{-1} - 2I$ is of the form $\lambda + \lambda^{-1} - 2 = (\lambda - 1)^2/\lambda$, where λ is an eigenvalue of A. As $\lambda > 0$, we have that $(\lambda - 1)^2/\lambda \geq 0$.

(b) If $A^* = AU$ for some unitary matrix U, then $A^*A = AU(U^*A^*) = AA^*$. Thus A is normal.

Conversely, let A be normal. Then there exists a diagonal D (with diagonal entries $|d_{jj}|$) and unitary V so that $A = VDV^*$. Let W be a unitary diagonal matrix so that $DW = D^*$ (by taking $w_{jj} = \frac{\overline{d_{jj}}}{d_{jj}}$, when $d_{jj} \neq 0$, and $w_{jj} = 1$ when $d_{jj} = 0$). Then $U = VWV^*$ is unitary and $AU = A(VWV^*) = VDV^*VWV^* = VDWV^* = VD^*V^* = A^*$.

Exercise 5.7.24 Find a QR factorization of $\begin{pmatrix} 1 & 1 & 0 \\ 1 & 0 & 1 \\ 0 & 1 & 1 \end{pmatrix}$.

Answer:

$$Q = \begin{pmatrix} \frac{\sqrt{2}}{2} & \frac{1}{\sqrt{6}} & -\frac{1}{\sqrt{3}} \\ \frac{\sqrt{2}}{2} & -\frac{1}{\sqrt{6}} & \frac{1}{\sqrt{3}} \\ 0 & \frac{\sqrt{6}}{3} & \frac{1}{\sqrt{3}} \end{pmatrix}, R = \begin{pmatrix} \sqrt{2} & -\frac{\sqrt{2}}{2} & -\frac{\sqrt{2}}{2} \\ 0 & \frac{\sqrt{6}}{2} & \frac{1}{\sqrt{6}} \\ 0 & 0 & \frac{2\sqrt{3}}{3} \end{pmatrix}.$$

Exercise 5.7.25 Find the Schur factorization $A = UTU^*$, with U unitary and T triangular, for the matrix

$$A = \begin{pmatrix} -1 & -2 & 3 \\ 2 & 4 & -2 \\ 1 & -2 & 1 \end{pmatrix}.$$

Note: 2 is an eigenvalue of A.

Answer:

$$T = \begin{pmatrix} 2 & 2\sqrt{2} & 2 \\ 0 & 4 & 2\sqrt{2} \\ 0 & 0 & -2 \end{pmatrix}, U = \begin{pmatrix} -\frac{\sqrt{2}}{2} & 0 & \frac{\sqrt{2}}{2} \\ 0 & 1 & 0 \\ -\frac{\sqrt{2}}{2} & 0 & -\frac{\sqrt{2}}{2} \end{pmatrix}.$$

Exercise 5.7.26 Let

$$T = \begin{pmatrix} A & B \\ C & D \end{pmatrix} \tag{5.32}$$

be a block matrix, and suppose that D is invertible. Define the *Schur complement* S of D in T by $S = A - BD^{-1}C$. Show that rank $T = \text{rank}(A - BD^{-1}C) + \text{rank } D$.

Answer: Observe that

$$\begin{pmatrix} I & -BD^{-1} \\ 0 & I \end{pmatrix} \begin{pmatrix} A & B \\ C & D \end{pmatrix} \begin{pmatrix} I & 0 \\ -D^{-1}C & I \end{pmatrix} = \begin{pmatrix} A - BD^{-1}C & 0 \\ 0 & D \end{pmatrix}.$$

But then we get that

$$\text{rank } T = \text{rank} \begin{pmatrix} A - BD^{-1}C & 0 \\ 0 & D \end{pmatrix} = \text{rank}(A - BD^{-1}C) + \text{rank } D.$$

Exercise 5.7.27 Using Sylvester's law of inertia, show that if

$$M = \begin{pmatrix} A & B \\ B^* & C \end{pmatrix} = M^* \in \mathbb{C}^{(n+m)\times(n+m)}$$

with C invertible, then

$$\text{In } M = \text{In } C + \text{In}(A - BC^{-1}B^*). \tag{5.33}$$

(Hint: Let $S = \begin{pmatrix} I & 0 \\ -B^*A^{-1} & I \end{pmatrix}$ and compute SMS^*.)

Answer: Let

$$S = \begin{pmatrix} I & 0 \\ -B^*A^{-1} & I \end{pmatrix},$$

and observe that

$$SMS^* = \begin{pmatrix} I & 0 \\ -B^*A^{-1} & I \end{pmatrix} \begin{pmatrix} A & B \\ B^* & C \end{pmatrix} \begin{pmatrix} I & -A^{-1}B \\ 0 & I \end{pmatrix} = \begin{pmatrix} A & 0 \\ 0 & C - B^*A^{-1}B \end{pmatrix}.$$

Theorem 5.5.5 now yields that

$$\text{In } M = \text{In } \begin{pmatrix} A & 0 \\ 0 & C - B^*A^{-1}B \end{pmatrix} = \text{In } A + \text{In}(C - B^*A^{-1}B). \tag{5.34}$$

Exercise 5.7.28 Determine the singular value decomposition of the following matrices.

(a) $A = \begin{pmatrix} 1 & 1 & 2\sqrt{2}i \\ -1 & -1 & 2\sqrt{2}i \\ \sqrt{2}i & -\sqrt{2}i & 0 \end{pmatrix}$.

(b) $A = \begin{pmatrix} -2 & 4 & 5 \\ 6 & 0 & -3 \\ 6 & 0 & -3 \\ -2 & 4 & 5 \end{pmatrix}$.

Answer: (a) $V = \begin{pmatrix} \frac{\sqrt{2}i}{2} & -\frac{1}{2} & \frac{1}{2} \\ \frac{\sqrt{2}i}{2} & \frac{1}{2} & -\frac{1}{2} \\ 0 & -\frac{\sqrt{2}i}{2} & -\frac{\sqrt{2}i}{2} \end{pmatrix}$, $\Sigma = \begin{pmatrix} 4 & 0 & 0 \\ 0 & 2 & 0 \\ 0 & 0 & 2 \end{pmatrix}$, $W = \begin{pmatrix} 0 & 0 & 1 \\ -1 & 0 & 0 \\ 0 & 1 & 0 \end{pmatrix}$.

(b) $V = \begin{pmatrix} -\frac{1}{2} & -\frac{1}{2} & -\frac{1}{2} & -\frac{1}{2} \\ \frac{1}{2} & -\frac{1}{2} & \frac{1}{2} & -\frac{1}{2} \\ \frac{1}{2} & -\frac{1}{2} & -\frac{1}{2} & \frac{1}{2} \\ -\frac{1}{2} & -\frac{1}{2} & \frac{1}{2} & \frac{1}{2} \end{pmatrix}$, $\Sigma = \begin{pmatrix} 12 & 0 & 0 \\ 0 & 6 & 0 \\ 0 & 0 & 0 \\ 0 & 0 & 0 \end{pmatrix}$, $W = \begin{pmatrix} \frac{2}{3} & -\frac{2}{3} & -\frac{1}{3} \\ -\frac{1}{3} & -\frac{2}{3} & \frac{2}{3} \\ -\frac{2}{3} & -\frac{1}{3} & -\frac{2}{3} \end{pmatrix}$.

Exercise 5.7.29 Let A be a 4×4 matrix with spectrum $\sigma(A) = \{-2i, 2i, 3+i, 3+4i\}$ and singular values $\sigma_1 \geq \sigma_2 \geq \sigma_3 \geq \sigma_4$.

(a) Determine the product $\sigma_1\sigma_2\sigma_3\sigma_4$.

(b) Show that $\sigma_1 \geq 5$.

(c) Assuming A is normal, determine $\text{tr}(A + AA^*)$.

Answer: (a) Write $A = V\Sigma W^*$, then $|\det A| = |\det V||\det \Sigma||\det W^*| = \sigma_1\sigma_2\sigma_3\sigma_4$ as the determinant of a unitary matrix has absolute value 1. Next observe that $|\det A| = |-2i||2i||3+i||3+4i| = 20\sqrt{10}$, which is the answer.

(b) Let $\mathbf{v}$ be a unit vector at the eigenvalue $3 + 4i$. Then

$$\sigma_1 = \max_{\|\mathbf{x}\|=1} \|A\mathbf{x}\| \geq \|A\mathbf{v}\| = \|(3+4i)\mathbf{v}\| = |3+4i| = 5.$$

(c) Since A is normal, we can write $A = UDU^*$ with U unitary and

$$D = \begin{pmatrix} -2i & 0 & 0 & 0 \\ 0 & 2i & 0 & 0 \\ 0 & 0 & 3+i & 0 \\ 0 & 0 & 0 & 3+4i \end{pmatrix}.$$

Thus

$$\mathrm{tr}(A + AA^*) = \mathrm{tr}(D + DD^*) = (-2i + 2i + 3 + i + 3 + 4i) + (4 + 4 + 10 + 25) = 49 + 5i.$$

Exercise 5.7.30 Let $A = \begin{pmatrix} P & Q \\ R & S \end{pmatrix} \in \mathbb{C}^{(k+l)\times(m+n)}$, where P is of size $k \times m$. Show that

$$\sigma_1(P) \leq \sigma_1(A).$$

Conclude that $\sigma_1(Q) \leq \sigma_1(A), \sigma_1(R) \leq \sigma_1(A), \sigma_1(S) \leq \sigma_1(A)$ as well.

Answer: By (5.17),

$$\sigma_1(P) = \max_{\|\mathbf{x}\|=1, \mathbf{x}\in\mathbb{C}^m} \|P\mathbf{x}\| \leq \max_{\|\mathbf{x}\|=1, \mathbf{x}\in\mathbb{C}^m} \left\|A\begin{pmatrix}\mathbf{x}\\\mathbf{0}\end{pmatrix}\right\| \leq$$

$$\max_{\|\mathbf{z}\|=1, \mathbf{z}\in\mathbb{C}^{m+n}} \|A\mathbf{z}\| = \sigma_1(A).$$

The same type of reasoning can be applied to obtain the other inequalities. Alternatively, one can use that permuting block rows and/or block columns does not change the singular values of a matrix. For instance, the singular values of $\begin{pmatrix} P & Q \\ R & S \end{pmatrix}$ and $\begin{pmatrix} Q & P \\ S & R \end{pmatrix}$ are the same, as multiplying on the left with the unitary matrix $J = \begin{pmatrix} 0 & I_m \\ I_n & 0 \end{pmatrix}$ does not change the singular values (it only changes the singular value decomposition from $V\Sigma W^*$ to $V\Sigma W^* J = V\Sigma(J^*W)^*$).

Exercise 5.7.31 This is an exercise that uses MATLAB®, and its purpose it to show what happens with an image if you take a low rank approximation of it.

Answer:

Figure 5.7: The original image (of size $672 \times 524 \times 3$).

(a) Using 10 singular values.   (b) Using 30 singular values.   (c) Using 50 singular values.

Exercise 5.7.32 The *condition number* $\kappa(A)$ of an invertible $n \times n$ matrix A is given by $\kappa(A) = \frac{\sigma_1(A)}{\sigma_n(A)}$, where $\sigma_1(A) \geq \cdots \geq \sigma_n(A)$ are the singular values of A. Show that for all invertible matrices A and B, we have that $\kappa(AB) \leq \kappa(A)\kappa(B)$. (Hint: use that $\sigma_1(A^{-1}) = (\sigma_n(A))^{-1}$ and (5.18).)

Answer: Notice that for any invertible matrix A, $\kappa(A) = \sigma_1(A)\sigma_1(A^{-1})$. So by (5.18),
$$\sigma_1(AB)\sigma_1(B^{-1}A^{-1}) \leq \sigma_1(A)\sigma_1(A^{-1})\sigma_1(B)\sigma_1(B^{-1}) = \frac{\sigma_1(A)}{\sigma_n(A)}\frac{\sigma_1(B)}{\sigma_n(B)} = \kappa(A)\kappa(B).$$

Exercise 5.7.33 Prove that if X and Y are positive definite $n \times n$ matrices such that $Y - X$ is positive semidefinite, then $\det X \leq \det Y$. Moreover, $\det X = \det Y$ if and only if $X = Y$.

Answer: Notice that $Y^{-\frac{1}{2}}(Y - X)Y^{-\frac{1}{2}} = I - Y^{-\frac{1}{2}}XY^{-\frac{1}{2}}$ is positive semidefinite. Thus the eigenvalues $\mu_1, \ldots, \mu_n$ of $Y^{-\frac{1}{2}}XY^{-\frac{1}{2}}$ satisfy $0 \leq \mu_j \leq 1$, $j = 1, \ldots, n$. But then $\frac{\det X}{\det Y} = \det(Y^{-\frac{1}{2}}XY^{-\frac{1}{2}}) = \prod_{j=1}^{n} \mu_j \leq 1$. Next, $\det X = \det Y$ if and only if

$\mu_1 = \cdots = \mu_n = 1$, which in turn holds if and only if $Y^{-\frac{1}{2}}XY^{-\frac{1}{2}} = I_n$. The latter holds if and only if $X = Y$.

Exercise 5.7.34 *(Least squares solution)* When the equation $Ax = b$ does not have a solution, one may be interested in finding an $\mathbf{x}$ so that $\|Ax - b\|$ is minimal. Such an $\mathbf{x}$ is called a *least squares solution* to $Ax = b$. In this exercise we will show that if $A = QR$, with R invertible, then the least squares solution is given by $\mathbf{x} = R^{-1}Q^*\mathbf{b}$. Let $A \in \mathbb{F}^{n \times m}$ with rank $A = m$.

(a) Let $A = QR$ be a QR-factorization of A. Show that Ran $A = \text{Ran} Q$.

(b) Observe that $QQ^*\mathbf{b} \in \text{Ran } Q$. Show that for all $\mathbf{v} \in \text{Ran } Q$ we have $\|\mathbf{v} - \mathbf{b}\| \geq \|QQ^*\mathbf{b} - \mathbf{b}\|$ and that the inequality is strict if $\mathbf{v} \neq QQ^*\mathbf{b}$.

(c) Show that $\mathbf{x} := R^{-1}Q^*\mathbf{b}$ is the least squares solution to $Ax = b$.

(d) Let $A = \begin{pmatrix} 1 & 1 \\ 2 & 1 \\ 3 & 1 \end{pmatrix}$ and $\mathbf{b} = \begin{pmatrix} 3 \\ 5 \\ 4 \end{pmatrix}$. Find the least squares solution to $Ax = b$.

(e) In trying to fit a line $y = cx + d$ through the points $(1, 3)$, $(2, 5)$, and $(3, 4)$, one sets up the equations
$$3 = c + d, 5 = 2c + d, 4 = 3c + d.$$
Writing this in matrix form we get
$$A \begin{pmatrix} c \\ d \end{pmatrix} = \mathbf{b},$$
where A and $\mathbf{b}$ as above. One way to get a "fitting line" $y = cx + d$, is to solve for c and d via least squares, as we did in the previous part. This is the most common way to find a so-called *regression line*. Plot the three points $(1, 3)$, $(2, 5)$, and $(3, 4)$ and the line $y = cx + d$, where c and d are found via least squares as in the previous part.

Answer: (a) Since rank $A = m$, the columns of A are linearly independent. This gives that the $m \times m$ matrix R is invertible. Thus Ran $A = $ Ran Q follows.

(b) Clearly, $QQ^*\mathbf{b} \in \text{Ran } Q$. Let $\mathbf{v} \in \text{Ran } Q$. Thus there exists a $\mathbf{w}$ so that $\mathbf{v} = Q\mathbf{w}$. Then, since $(I - QQ^*)\mathbf{b} \perp \mathbf{x}$ for every $\mathbf{x} \in \text{Ran } Q$ (use that $Q^*Q = I$),
$$\|\mathbf{v} - \mathbf{b}\|^2 = \|Q\mathbf{w} - QQ^*\mathbf{b} + QQ^*\mathbf{b} - \mathbf{b}\|^2 = \|Q\mathbf{w} - QQ^*\mathbf{b}\|^2 + \|QQ^*\mathbf{b} - \mathbf{b}\|^2.$$
Thus $\|\mathbf{v} - \mathbf{b}\| \geq \|QQ^*\mathbf{b} - \mathbf{b}\|$, and equality only holds when $\mathbf{v} = QQ^*\mathbf{b}$.

(c) For $\|Ax - b\|$ to be minimal, we need $Ax = QQ^*\mathbf{b}$, as $QQ^*\mathbf{b}$ is the element in Ran $A = $ RanQ closest to $\mathbf{b}$. Now, $Ax = QQ^*\mathbf{b}$, gives $QRx = QQ^*\mathbf{b}$. Putting $\mathbf{x} = R^{-1}Q^*\mathbf{b}$, we indeed get that $QRx = QQ^*\mathbf{b}$.

(d) $\mathbf{x} = \begin{pmatrix} \frac{1}{2} \\ 3 \end{pmatrix}$.   (e)

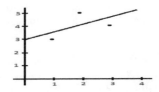

Exercise 5.7.35 Let A, X be $m \times m$ matrices such that $A = A^*$ is invertible and

$$H := A - X^* A X \tag{5.35}$$

is positive definite.

(a) Show that X has no eigenvalues on the unit circle $\mathbb{T} = \{z \in \mathbb{C} : |z| = 1\}$.

(b) Show that A is positive definite if and only if X has all eigenvalues in $\mathbb{D} = \{z \in \mathbb{C} : |z| < 1\}$. (Hint: When X has all eigenvalues in $\mathbb{D}$, we have that $X^n \to 0$ as $n \to \infty$. Use this to show that $A = H + \sum_{k=1}^{\infty} X^{*k} H X^k$.)

Answer: (a) Suppose that $\mathbf{x}$ is an eigenvector of X with eigenvalue λ. Then (5.35) yields that

$$0 < \mathbf{x}^* H \mathbf{x} = (1 - |\lambda|^2) \mathbf{x}^* A \mathbf{x},$$

and thus $|\lambda| \neq 1$.

(b) Assuming that A is positive definite we get that

$$A^{-\frac{1}{2}} H A^{-\frac{1}{2}} = I - (A^{-\frac{1}{2}} X^* A^{\frac{1}{2}})(A^{\frac{1}{2}} X A^{-\frac{1}{2}})$$

is positive definite, and thus $\sigma_1(A^{\frac{1}{2}} X A^{-\frac{1}{2}}) < 1$. Consequently, the eigenvalues of $A^{\frac{1}{2}} X A^{-\frac{1}{2}}$ lie in $\mathbb{D}$. But then, the same follows for X (as X is similar $A^{\frac{1}{2}} X A^{-\frac{1}{2}}$).

For the converse, suppose that the eigenvalues of S lie in $\mathbb{D}$. Then $X^n \to 0$ as $n \to \infty$. Rewriting (5.35) and reusing it over and over again we get that

$$A = H + X^* A X = H + X^* H X + X^{*2} A X^2 = \cdots =$$

$$\sum_{k=0}^{n-1} X^{*k} H X^k + X^{*n} A X^n \to \sum_{k=0}^{\infty} X^{*k} H X^k = H + \sum_{k=1}^{\infty} X^{*k} H X^k. \tag{5.36}$$

Thus $A = H + \sum_{k=1}^{\infty} X^{*k} H X^k$ is positive definite.

Chapter 6

Exercise 6.7.1 The purpose of this exercise is to show (the vector form of) *Minkowski's inequality*, which says that for complex numbers x_i, y_i, $i = 1, \ldots, n$, and $p \geq 1$, we have

$$\left(\sum_{i=1}^{n} |x_i + y_i| \right)^{\frac{1}{p}} \leq \left(\sum_{i=1}^{n} |y_i| \right)^{\frac{1}{p}} + \left(\sum_{i=1}^{n} |y_i| \right)^{\frac{1}{p}}. \tag{6.37}$$

Recall that a real-valued function f defined on an interval in $\mathbb{R}$ is called *convex* if for all c, d in the domain of f, we have that $f(tc + (1-t)d) \leq tf(c) + (1-t)f(d)$, $0 \leq t \leq 1$.

(a) Show that $f(x) = -\log x$ is a convex function on $(0, \infty)$. (One can do this by showing that $f''(x) \geq 0$.)

Answer: $f''(x) = \frac{1}{x^2} > 0$.

(b) Use (a) to show that for $a, b > 0$ and $p, q \geq 1$, with $\frac{1}{p} + \frac{1}{q} = 1$, we have $ab \leq \frac{a^p}{p} + \frac{b^q}{q}$. This inequality is called *Young's inequality*.

Answer: Taking $c = a^p$ and $d = b^q$, $t = \frac{1}{p}$ (and thus $1 - t = \frac{1}{q}$), we obtain from the convexity of $-\log$ that

$$-\log\left(\frac{1}{p}a^p + \frac{1}{q}b^q\right) \leq -\frac{1}{p}\log a^p - \frac{1}{q}\log b^q.$$

Multiplying by -1 and applying $s \mapsto e^s$ on both sides gives

$$\frac{1}{p}a^p + \frac{1}{q}b^q \geq (a^p)^{\frac{1}{p}}(b^q)^{\frac{1}{q}} = ab.$$

(c) Show *Hölder's inequality*: when $a_i, b_i \geq 0$, $i = 1, \ldots, n$, then

$$\sum_{i=1}^{n} a_i b_i \leq \left(\sum_{i=1}^{n} a_i^p \right)^{\frac{1}{p}} \left(\sum_{i=1}^{n} b_i^q \right)^{\frac{1}{q}}.$$

(Hint: Let $\lambda = \left(\sum_{i=1}^{n} a_i^p\right)^{\frac{1}{p}}$ and $\mu = \left(\sum_{i=1}^{n} b_i^q\right)^{\frac{1}{q}}$, and divide on both sides a_i by λ and b_i by μ. Use this to argue that it is enough to prove the inequality when $\lambda = \mu = 1$. Next use (b)).

Answer: If λ or μ equals 0, the inequality is trivial, so let us assume $\lambda, \mu > 0$. Put $\alpha_i = \frac{a_i}{\lambda}$ and $\beta_i = \frac{b_i}{\mu}$, $i = 1, \ldots, n$. Then $\left(\sum_{i=1}^{n} \alpha_i^p\right)^{\frac{1}{p}} = 1$, and thus $\sum_{i=1}^{n} \alpha_i^p = 1$. Similarly, $\sum_{i=1}^{n} \beta_i^q = 1$. We need to prove that $\sum_{i=1}^{n} \alpha_i \beta_i \leq 1$. By (b) we have that $\alpha_i \beta_i \leq \frac{1}{p}\alpha_i^p + \frac{1}{q}\beta_i^q$, $i = 1, \ldots, n$. Taking the sum, we obtain

$$\sum_{i=1}^{n} \alpha_i \beta_i \leq \frac{1}{p}\sum_{i=1}^{n} \alpha_i^p + \frac{1}{q}\sum_{i=1}^{n} \beta_i^q = \frac{1}{p} + \frac{1}{q} = 1,$$

and we are done.

(d) Use (c) to prove (6.37) in the case when $x_i, y_i \geq 0$. (Hint: Write $(x_i + y_i)^p = x_i(x_i + y_i)^{p-1} + y_i((x_i + y_i)^{p-1}$, take the sum on both sides, and now apply Hölder's inequality to each of the terms on the right-hand side. Rework the resulting inequality, and use that $p + q = pq$.)

Answer: Using (c) we have that

$$\sum_{i=1}^{n}(x_i + y_i)^p = \sum_{i=1}^{n} x_i(x_i + y_i)^{p-1} + \sum_{i=1}^{n} y_i(x_i + y_i)^{p-1} \leq$$

$$(\sum_{i=1}^{n} x_i^p)^{\frac{1}{p}} (\sum_{i=1}^{n} (x_i + y_i)^{(p-1)q})^{\frac{1}{q}} + (\sum_{i=1}^{n} y_i^p)^{\frac{1}{p}} (\sum_{i=1}^{n} (x_i + y_i)^{(p-1)q})^{\frac{1}{q}} =$$

$$[(\sum_{i=1}^{n} x_i^p)^{\frac{1}{p}} + (\sum_{i=1}^{n} y_i^p)^{\frac{1}{p}}] (\sum_{i=1}^{n} (x_i + y_i)^p)^{\frac{1}{q}},$$

where in the last step we used that $(p - 1)q = p$. Dividing both sides by $(\sum_{i=1}^{n} (x_i + y_i)^p)^{\frac{1}{q}}$, we obtain

$$(\sum_{i=1}^{n} (x_i + y_i)^p)^{1-\frac{1}{q}} \leq (\sum_{i=1}^{n} x_i^p)^{\frac{1}{p}} + (\sum_{i=1}^{n} y_i^p)^{\frac{1}{p}},$$

and using that $1 - \frac{1}{q} = \frac{1}{p}$, we are done.

(e) Prove Minkowski's inequality (6.37).

Answer: We just need to observe that for complex numbers x_i and y_i we have that $|x_i + y_i| \leq |x_i| + |y_i|$, and thus $\sum_{i=1}^{n} |x_i + y_i|^p \leq \sum_{i=1}^{n} (|x_i| + |y_i|)^p$. Using (d) we obtain

$$(\sum_{i=1}^{n} (|x_i| + |y_i|)^p)^{\frac{1}{p}} \leq (\sum_{i=1}^{n} |x_i|^p)^{\frac{1}{p}} + (\sum_{i=1}^{n} |y_i|^p)^{\frac{1}{p}},$$

and we are done.

(f) Show that when V_i has a norm $\| \cdot \|_i$, $i = 1, \ldots, k$, then for $p \geq 1$ we have that

$$\| \begin{pmatrix} \mathbf{v}_1 \\ \vdots \\ \mathbf{v}_k \end{pmatrix} \|_p := \left(\sum_{i=1}^{k} \|\mathbf{v}_i\|_i^p \right)^{\frac{1}{p}}$$

defines a norm on $V_1 \times \cdots \times V_k$.

Answer: The only part that is not trivial is the triangle inequality. For this we need to observe that $\|\mathbf{v}_i + \mathbf{w}_i\|_i \leq \|\mathbf{v}_i\|_i + \|\mathbf{w}_i\|_i$, and thus $\sum_{i=1}^{n} \|\mathbf{v}_i + \mathbf{w}_i\|_i^p \leq \sum_{i=1}^{n} (\|\mathbf{v}_i\|_i + \|\mathbf{w}_i\|_i)^p$. Now we can apply (d) with $x_i = \|\mathbf{v}_i\|_i$ and $y_i = \|\mathbf{w}_i\|_i$, and obtain

$$(\sum_{i=1}^{n} (\|\mathbf{v}_i + \mathbf{w}_i\|_i)^p)^{\frac{1}{p}} \leq (\sum_{i=1}^{n} (\|\mathbf{v}_i\|_i + \|\mathbf{w}_i\|_i)^p)^{\frac{1}{p}} \leq (\sum_{i=1}^{n} \|\mathbf{v}_i\|_i^p)^{\frac{1}{p}} + (\sum_{i=1}^{n} \|\mathbf{w}_i\|_i^p)^{\frac{1}{p}},$$

proving the triangle inequality.

Exercise 6.7.2 Let V and Z be vector spaces over $\mathbb{F}$ and $T : V \to Z$ be linear. Suppose $W \subseteq \text{Ker } T$. Show there exists a linear transformation $S : V/W \to \text{Ran } T$ such that $S(\mathbf{v} + W) = T\mathbf{v}$ for $\mathbf{v} \in V$. Show that S is surjective and that $\text{Ker } S$ is isomorphic to $(\text{Ker } T)/W$.

Answer: Define $S : V/W \to \text{Ran } T$ via $S(\mathbf{v} + W) = T\mathbf{v}$. We need to check that S is well-defined. For this, suppose that $\mathbf{v} + W = \mathbf{x} + W$. Then $\mathbf{v} - \mathbf{x} \in W$. As $W \subseteq \text{Ker } T$, we thus have that $\mathbf{v} - \mathbf{x} \in \text{Ker } T$, which implies that $T\mathbf{v} = T\mathbf{x}$. This shows that S is well-defined.

Next, to show surjectivity of S, let $\mathbf{y} \in \text{Ran} T$. Then there exists a $\mathbf{v} \in V$ so that $T\mathbf{v} = \mathbf{y}$. As $S(\mathbf{v} + W) = T\mathbf{v} = \mathbf{y}$, we obtain that $\mathbf{y} \in \text{Ran} S$, showing surjectivity.

Finally, let us define $\phi : (\text{Ker } T)/W \to \text{Ker } S$ via $\phi(\mathbf{v} + W) = \mathbf{v} + W$, where $\mathbf{v} \in \text{Ker} T$. We claim that ϕ is an isomorphism. First note that $S(\mathbf{v} + W) = T\mathbf{v} = \mathbf{0}$, as $\mathbf{v} \in \text{Ker} T$. Clearly, ϕ is linear and one-to-one, so it remains to check that ϕ is surjective. When $\mathbf{v} + W \in \text{Ker} S$, then we must have that $T\mathbf{v} = \mathbf{0}$. Thus $\mathbf{v} \in \text{Ker} T$, yielding that $\mathbf{v} + W \in (\text{Ker } T)/W$. Clearly, $\phi(\mathbf{v} + W) = \mathbf{v} + W$, and thus $\mathbf{v} + W \in \text{Ran} \phi$.

Exercise 6.7.3 Consider the vector space $\mathbb{F}^{n \times m}$, where $\mathbb{F} = \mathbb{R}$ or $\mathbb{F} = \mathbb{C}$, and let $\| \cdot \|$ be norm on $\mathbb{F}^{n \times m}$.

(a) Let $A = (a_{ij})_{i=1,j=1}^{n \quad m}$, $A_k = (a_{ij}^{(k)})_{i=1,j=1}^{n \quad m}$, $k = 1, 2, \ldots$, be matrices in $\mathbb{F}^{n \times m}$. Show that $\lim_{k \to \infty} \|A_k - A\| = 0$ if and only if $\lim_{k \to \infty} |a_{ij}^{(k)} - a_{ij}| = 0$ for every $i = 1, \ldots, n$ and $j = 1, \ldots, m$.

(b) Let $n = m$. Show that $\lim_{k \to \infty} \|A_k - A\| = 0$ and $\lim_{k \to \infty} \|B_k - B\| = 0$ imply that $\lim_{k \to \infty} \|A_k B_k - AB\| = 0$.

Answer: (a) Notice that if $c\|A_k - A\|_a \leq \|A_k - A\|_b \leq C\|A_k - A\|_a$, for some $c, C > 0$, and $\lim_{k \to \infty} \|A_k - A\|_a = 0$, then $\lim_{k \to \infty} \|A_k - A\|_n = 0$. Thus, by Theorem 5.1.25, when we have $\lim_{k \to \infty} \|A_k - A\| = 0$ in one norm on $\mathbb{F}^{n \times m}$, we automatically have it for every norm on $\mathbb{F}^{n \times m}$. Let us use the norm

$$\|M\|_\infty = \|(m_{ij})_{i=1,j=1}^{n \quad m}\| := \max_{i=1,\ldots,n; j=1\ldots,m} |m_{ij}|.$$

Notice that $|m_{ij}| \leq \|M\|_\infty$ for every i and j.

Suppose that $\lim_{k \to \infty} \|A_k - A\|_\infty = 0$. Then for every $i = 1, \ldots, n$ and $j = 1, \ldots, m$, we have $|a_{ij}^{(k)} - a_{ij}| \leq \|A_k - A\|_\infty$, and thus

$0 \leq \lim_{k \to \infty} |a_{ij}^{(k)} - a_{ij}| \leq \lim_{k \to \infty} \|A_k - A\| = 0$, giving $\lim_{k \to \infty} |a_{ij}^{(k)} - a_{ij}| = 0$.

Next, let $\lim_{k \to \infty} |a_{ij}^{(k)} - a_{ij}| = 0$ for every i and j. Let $\epsilon > 0$. Then for every i and j, there exists a $K_{ij} \in \mathbb{N}$ so that for $k > K_{ij}$ we have $|a_{ij}^{(k)} - a_{ij}| < \epsilon$. Let now $K = \max_{i=1,\ldots,n; j=1\ldots,m} K_{ij}$. Then for every $k > K$ we have that $\|A_k - A\|_\infty = \max_{i=1,\ldots,n; j=1\ldots,m} |a_{ij}^{(k)} - a_{ij}| < \epsilon$. Thus, by definition of a limit, we have $\lim_{k \to \infty} \|A_k - A\|_\infty = 0$.

(b) For scalars we have that $\lim_{k \to \infty} |a_k - a| = 0 = \lim_{k \to \infty} |b_k - b|$ implies $\lim_{k \to \infty} |(a_k + b_k) - (a + b)|$ and $\lim_{k \to \infty} |a_k b_k - ab| = 0$ (which you can prove by using inequalities like $|a_k b_k - ab| = |a_k b_k - a_k b + a_k b - ab| \leq |a_k||b_k - b| + |a_k - a||b|$). Equivalently, $\lim_{k \to \infty} a_k = a$ and $\lim_{k \to \infty} b_k = b$ implies $\lim_{k \to \infty} a_k b_k = ab$.

Suppose now that $\lim_{k \to \infty} \|A_k - A\| = 0 = \lim_{k \to \infty} \|B_k - B\| = 0$. Then, using (a), $\lim_{k \to \infty} a_{ij}^{(k)} = a_{ij}$ and $\lim_{k \to \infty} b_{ij}^{(k)} = b_{ij}$ for all $i, j = 1, \ldots, n$. Now, for the (r, s) element of the product $A_k B_k$ we obtain

$$\lim_{k \to \infty} (A_k B_k)_{rs} = \lim_{k \to \infty} \sum_{j=1}^{n} a_{rj}^{(k)} b_{js}^{(k)} = \sum_{j=1}^{n} (\lim_{k \to \infty} a_{rj}^{(k)})(\lim_{k \to \infty} b_{js}^{(k)}) =$$

$$\sum_{j=1}^{n} a_{rj} b_{js} = (AB)_{rs}, r, s = 1, \ldots, n.$$

Again using (a), we may conclude $\lim_{k \to \infty} \|A_k B_k - AB\| = 0$.

Exercise 6.7.4 Given $A \in \mathbb{C}^{n \times n}$, we define its *similarity orbit* to be the set of matrices

$$\mathcal{O}(A) = \{SAS^{-1} : S \in \mathbb{C}^{n \times n} \text{ is invertible}\}.$$

Thus the similarity orbit of a matrix A consists of all matrices that are similar to A.

(a) Show that if A is diagonalizable, then its similarity orbit $\mathcal{O}(A)$ is closed. (Hint: notice that due to A being diagonalizable, we have that $B \in \mathcal{O}(A)$ if and only if $m_A(B) = 0$.)

(b) Show that if A is not diagonalizable, then its similarity orbit is not closed.

Answer: (a) Suppose that $B_k \in \mathcal{O}(A)$, $k \in \mathbb{N}$, and that $\lim_{n\to\infty} \|B_k - B\| = 0$. We need to show that $B \in \mathcal{O}(A)$, or equivalently, $m_A(B) = 0$. Write $m_A(t) = a_n t^n + a_{n-1} t^{n-1} + \cdots + a_0$ (where $a_n = 1$). By exercise 6.7.3(b) we have that $\lim_{n\to\infty} \|B_k - B\| = 0$ imples that $\lim_{n\to\infty} \|B_k^j - B^j\| = 0$ for all $j \in \mathbb{N}$. But then

$$\lim_{k\to\infty} \|m_A(B_k) - m_A(B)\| \leq \lim_{k\to\infty} \sum_{j=0}^{n} |a_j| \|B_k^j - B^j\| = 0.$$

As $m_A(B_k) = 0$ for every k, we thus also have that $m_A(B) = 0$. Thus $B \in \mathcal{O}(A)$ follows.

(b) First let $A = J_k(\lambda)$, $k \geq 2$, be a Jordan block. For $\epsilon > 0$ put $D_\epsilon = \mathrm{diag}(\epsilon^j)_{j=0}^{k-1}$. Then

$$A_\epsilon := D_\epsilon^{-1} J_k(\lambda) D_\epsilon = \begin{pmatrix} \lambda & \epsilon & 0 & \cdots & 0 \\ 0 & \lambda & \epsilon & \cdots & 0 \\ \vdots & & \ddots & \ddots & \vdots \\ 0 & 0 & \cdots & \lambda & \epsilon \\ 0 & 0 & \cdots & 0 & \lambda \end{pmatrix} \in \mathcal{O}(A). \tag{6.38}$$

Notice that $\lim_{m\to\infty} A_{\frac{1}{m}} = \lambda I_k \notin \mathcal{O}(A)$, and thus $\mathcal{O}(A)$ is not closed.

Using the reasoning above, one can show that if $A = SJS^{-1}$ with $J = \oplus_{l=1}^{s} J_{n_l}(\lambda_l)$ and some $n_l > 1$, then $S(\oplus_{j=1}^{s} \lambda_l I_{n_l}) S^{-1} \notin \mathcal{O}(A)$ is the limit of elements in $\mathcal{O}(A)$. This gives that $\mathcal{O}(A)$ is not closed.

Exercise 6.7.5 Suppose that V is an infinite-dimensional vector space with basis $\{v_j\}_{j\in J}$. Let $f_j \in V'$, $j \in J$, be so that $f_j(v_j) = 1$ and $f_j(v_k) = 0$ for $k \neq j$. Show that $\{f_j\}_{j\in J}$ is a linearly independent set in V' but is not a basis of V'.

Answer: Consider a finite linear combination $f = \sum_{r=1}^{s} c_r f_{j_r}$ and set it equal to 0. Then $f(v_{j_k}) = 0$, and thus $0 = \sum_{i=1}^{r} c_r f_{j_r}(v_{j_k}) = c_k$. As this holds for all $k = 1, \ldots, s$, we get that $c_1 = \ldots = c_s = 0$, proving linear independence.

Next, let $f \in V'$ be defined by $f(v_j) = 1$ for all $j \in J$. In other words, if $v = \sum_{r=1}^{s} c_r v_{j_r}$ is a vector in V, then $f(v) = \sum_{r=1}^{s} c_r$. Clearly, f is a linear functional on V. In addition, f is not a finite linear combination of elements in $\{f_j : j \in J\}$. Indeed, suppose that $f = \sum_{r=1}^{s} c_r f_{j_r}$. Choose now a $j \in J \setminus \{j_1, \ldots, j_r\}$, which can always be done since J is infinite. Then $f(v_j) = 1$, while $f_{j_r}(v_j) = 0$ as $j \neq j_r$. Thus $f(v_j) \neq \sum_{r=1}^{s} c_r f_{j_r}(v_j)$, giving that $f \notin \mathrm{Span}\{f_j : j \in J\}$.

Exercise 6.7.6 Describe the linear functionals on $\mathbb{C}_n[X]$ that form the dual basis of $\{1, X, \ldots, X^n\}$.

Answer: If $\Phi_0, \ldots, \Phi_n$ are the dual basis elements, and $p(X) = p_0 + p_1 X + \cdots + p_n X^n$, then we need that

$$\Phi_j(p(X)) = p_j, j = 0, \ldots, n.$$

One way to find the number $j! p_j$ is to take the jth derivative of $p(X)$, and evaluate this jth derivative at 0. Thus we can describe Φ_j as

$$\Phi_j(p(X)) = \frac{1}{j!} \frac{d^j p}{dX^j}(0), j = 1, \ldots, n.$$

Exercise 6.7.7 Let $a_0, \ldots, a_n$ be different complex numbers, and define $E_j \in (\mathbb{C}_n[X])'$, $j = 0, \ldots, n$, via $E_j(p(X)) = p(a_j)$. Find a basis of $\mathbb{C}_n[X]$ for which $\{E_0, \ldots, E_n\}$ is the dual basis.

Answer: If we let $\{q_0(X), \ldots, q_n(X)\}$ be the basis of $\mathbb{C}_n[X]$ we are looking for, then we need that $E_j(q_k(X)) = 1$ if $j = k$, and $E_j(q_k(X)) = 0$ if $j \neq k$. Thus, we need to find a polynomial $q_k(X)$ so that $q_k(a_k) = 1$, while $a_0, \ldots, a_{k-1}, a_{k+1}, \ldots, a_n$ are roots of $q_k(X)$. Thus

$$q_k(X) = c(X - a_0) \cdots (X - a_{k-1})(X - a_{k+1}) \cdots (X - a_n),$$

with c chosen so that $q_k(a_k) = 1$. Thus we find

$$q_k(X) = \prod_{r=0,\ldots,n; r \neq k} \frac{X - a_r}{a_k - a_r},$$

which are called the *Lagrange interpolation polynomials*.

Exercise 6.7.8 Let $V = W \dot{+} X$.

(a) Show how given $f \in W'$ and $g \in X'$, one can define $h \in V'$ so that $h(\mathbf{w}) = f(\mathbf{w})$ for $\mathbf{w} \in W$ and $h(\mathbf{x}) = g(\mathbf{x})$ for $\mathbf{x} \in X$.

(b) Using the construction in part (a), show that $V' = W' \dot{+} X'$. Here it is understood that we view W' as a subspace of V', by letting $f \in W'$ be defined on all of V by putting $f(\mathbf{w} + \mathbf{x}) = f(\mathbf{w})$, when $\mathbf{w} \in W$ and $\mathbf{x} \in X$. Similarly, we view X' as a subspace of V', by letting $g \in W'$ be defined on all of V by putting $g(\mathbf{w} + \mathbf{x}) = g(\mathbf{x})$, when $\mathbf{w} \in W$ and $\mathbf{x} \in X$.

Answer: (a) Let $f \in W'$ and $g \in X'$, and $\mathbf{v} \in V$. As $V = W \dot{+} X$, there exist unique $\mathbf{w} \in W$ and $\mathbf{x} \in X$ so that $\mathbf{v} = \mathbf{w} + \mathbf{x}$. We now define $h(\mathbf{v}) = f(\mathbf{w}) + g(\mathbf{x})$. Then $h \in V'$ and satisfies the desired conditions.

(b) We first show that $W' \cap X' = \{0\}$. Indeed, let $f \in W' \cap X'$. By the way of viewing $f \in W'$ as a function on all of V, we have that $f(\mathbf{x}) = 0$ for all $\mathbf{x} \in X$. Similarly, by the way of viewing $f \in X'$ as a function on all of V, we have that $f(\mathbf{w}) = 0$ for all $\mathbf{w} \in W$. But then for a general $\mathbf{v} \in V$, which can always be written as $\mathbf{v} = \mathbf{w} + \mathbf{x}$ for some $\mathbf{w} \in W$ and $\mathbf{x} \in X$, we have that $f(\mathbf{v}) = f(\mathbf{w} + \mathbf{x}) = f\mathbf{w}) + f(\mathbf{x}) = 0 + 0 = 0$. Thus f is the zero functional, yielding $W' \cap X' = \{0\}$.

Next, when $h \in V'$, we can define $f \in W'$ and $g \in X'$ as by $f(\mathbf{w}) = h(\mathbf{w})$, $\mathbf{w} \in W$ and $g(\mathbf{x}) = h(\mathbf{x})$, $\mathbf{x} \in X$. Then, with the understanding as in (b), we have that $h = f + g$. This shows that $V' = W' + X'$. Together with $W' \cap X' = \{0\}$, we obtain $V' = W' \dot{+} X'$.

Exercise 6.7.9 Let W be a subspace of V. Define

$$W_{\mathrm{ann}} = \{f \in V' : f(\mathbf{w}) = 0 \text{ for all } \mathbf{w} \in W\},$$

the *annihilator* of W.

(a) Show that W_{ann} is a subspace of V'.

(b) Determine the annihilator of $\mathrm{Span}\{\begin{pmatrix} 1 \\ -1 \\ 2 \\ -2 \end{pmatrix}, \begin{pmatrix} 1 \\ 0 \\ 1 \\ 0 \end{pmatrix}\} \subseteq \mathbb{C}^4$.

(c) Determine the annihilator of $\mathrm{Span}\{1 + 2X, X + X^2\} \subseteq \mathbb{R}_3[X]$.

Answer: (a) Let $f, g \in W_{\text{ann}}$ and c be a scalar. Then for $\mathbf{w} \in W$ we have that $(f + g)(\mathbf{w} = f(\mathbf{w}) + g(\mathbf{w}) = 0 + 0 = 0$ and $(cf)(\mathbf{w} = cf(\mathbf{w}) = c0 = 0$. This shows that $f + g, cf \in W_{\text{ann}}$, and thus W_{ann} is a subspace.

(b) This amounts to finding the null space of $\begin{pmatrix} 1 & -1 & 2 & -2 \\ 1 & 0 & 1 & 0 \end{pmatrix}$, which in row-reduced

echelon form is $\begin{pmatrix} 1 & 0 & 1 & 0 \\ 0 & 1 & -1 & 2 \end{pmatrix}$. The null space is spanned by

$$\mathbf{v}_1 = \begin{pmatrix} -1 \\ 1 \\ 1 \\ 0 \end{pmatrix}, \mathbf{v}_2 = \begin{pmatrix} 0 \\ -2 \\ 0 \\ 1 \end{pmatrix}.$$ Thus $W_{\text{ann}} = \text{Span}\{f_1, f_2\}$, where (using the Euclidean inner

product) $f_i(\mathbf{v}) = \langle \mathbf{v}, \mathbf{v}_1 \rangle$, $i = 1, 2$.

(c) This amounts to finding the null space of $\begin{pmatrix} 1 & 2 & 0 & 0 \\ 0 & 1 & 1 & 0 \end{pmatrix}$, which is spanned by

$$\mathbf{v}_1 = \begin{pmatrix} 2 \\ -1 \\ 1 \\ 0 \end{pmatrix}, \mathbf{v}_2 = \begin{pmatrix} 0 \\ 0 \\ 0 \\ 1 \end{pmatrix}.$$ Now define $f_1(p_0 + p_1 X + p_2 X^2 + p_3 X^3) = 2p_0 - p_1 + p_2$ and

$f_2(p_0 + p_1 X + p_2 X^2 + p_3 X^3) = p_3$. Then $W_{\text{ann}} = \text{Span}\{f_1, f_2\}$.

Exercise 6.7.10 Let V be a finite-dimensional vector space over $\mathbb{R}$, and let $\{\mathbf{v}_1, \ldots, \mathbf{v}_k\}$ be linearly independent. We define

$$\mathcal{C} = \{\mathbf{v} \in V : \text{ there exist } c_1, \ldots, c_k \geq 0 \text{ so that } \mathbf{v} = \sum_{i=1}^{k} c_i \mathbf{v}_i\}.$$

Show that $\mathbf{v} \in \mathcal{C}$ if and only if for all $f \in V'$ with $f(\mathbf{v}_j) \geq 0$, $j = 1, \ldots, k$, we have that $f(\mathbf{v}) \geq 0$.

Remark. The statement is also true when $\{\mathbf{v}_1, \ldots, \mathbf{v}_k\}$ are not linearly independent, but in that case the proof is more involved. The corresponding result is the Farkas–Minkowski Theorem, which plays an important role in linear programming.

Answer: Clearly, when $\mathbf{v} = \sum_{i=1}^{k} c_i \mathbf{v}_i \in \mathcal{C}$ and $f(\mathbf{v}_j) \geq 0$, $j = 1, \ldots, k$, then $f(\mathbf{v}) = \sum_{i=1}^{k} c_i f(\mathbf{v}_i) \geq 0$, since $c_i \geq 0$ and $f(\mathbf{v}_i) \geq 0$, $i = 1, \ldots, n$.

Conversely, suppose that $\mathbf{v} \in V$ has the property that for all $f \in V'$ with $f(\mathbf{v}_j) \geq 0$, $j = 1, \ldots, k$, we have that $f(\mathbf{v}) \geq 0$. First, we show that $\mathbf{v} \in \text{Span}\{\mathbf{v}_1, \ldots, \mathbf{v}_k\}$. If not, we can find a linear functional so that on the $(k+1)$-dimensional space $\text{Span}\{\mathbf{v}, \mathbf{v}_1, \ldots, \mathbf{v}_k\}$ we have $f(\mathbf{v}) = -1$ and $f(\mathbf{v}_j) = 0$, $j = 1, \ldots, k$. But this contradicts that $\mathbf{v} \in V$ has the property that for all $f \in V'$ with $f(\mathbf{v}_j) \geq 0$, $j = 1, \ldots, k$, we have that $f(\mathbf{v}) \geq 0$.

As $\mathbf{v} \in \text{Span}\{\mathbf{v}_1, \ldots, \mathbf{v}_k\}$, we may write $\mathbf{v} = \sum_{i=1}^{k} c_i \mathbf{v}_i$, for some scalars $c_1, \ldots, c_k$. Fix a $j \in \{1, \ldots, k\}$. Let now $f \in V'$ be so that $f(\mathbf{v_j}) = 1$ and $f(\mathbf{v}_r) = 0$, $r \neq j$. Then $f \in V'$ with $f(\mathbf{v}_r) \geq 0$, $r = 1, \ldots, k$, and thus we must have that $f(\mathbf{v}) \geq 0$. As this number equals c_j, we obtain that $c_j \geq 0$. This holds for every $j = 1, \ldots k$, and thus we find that $\mathbf{v} \in \mathcal{C}$.

Exercise 6.7.11 Let V and W be finite-dimensional vector spaces and $A : V \to W$ a linear map. Show that $A\mathbf{v} = \mathbf{w}$ has a solution if and only if for all $f \in (\text{Ran} A)_{\text{ann}}$ we have that $f(\mathbf{w}) = 0$. Here the definition of the annihilator is used as defined in Exercise 6.7.9.

Answer: If $A\mathbf{v} = \mathbf{w}$ and $f \in (\text{Ran} A)_{\text{ann}}$, then $0 = f(A\mathbf{v}) = f(\mathbf{w})$, proving the only if statement. Next, suppose that for all $f \in (\mathrm{Ran}A)_{\mathrm{ann}}$ we have that $f(\mathbf{w}) = 0$. If $\mathbf{w} \notin \mathrm{Ran}A$, then letting $\{\mathbf{w}_1, \ldots, \mathbf{w}_k\}$ be a basis of $\mathrm{Ran}A$, we can find a linear functional so that $f(\mathbf{w}_j) = 0$, $j = 1, \ldots, k$, and $f(\mathbf{w}) = 1$. Then $f \in (\mathrm{Ran}A)_{\mathrm{ann}}$, but $f(\mathbf{w}) \neq 0$, giving a contradiction. Thus we must have that $\mathbf{w} \in \mathrm{Ran}A$, yielding the existence of a $\mathbf{v}$ so that $A\mathbf{v} = \mathbf{w}$.

Exercise 6.7.12 For $\mathbf{x}, \mathbf{y} \in \mathbb{R}^3$, let the cross product $\mathbf{x} \times \mathbf{y}$ be defined as in (6.17).

(a) Show that $\langle \mathbf{x}, \mathbf{x} \times \mathbf{y} \rangle = \langle \mathbf{y}, \mathbf{x} \times \mathbf{y} \rangle = 0$.

(b) Show that $\mathbf{x} \times \mathbf{y} = -\mathbf{y} \times \mathbf{x}$.

(c) Show that $\mathbf{x} \times \mathbf{y} = \mathbf{0}$ if and only if $\{\mathbf{x}, \mathbf{y}\}$ is linearly dependent.

Answer: (a) and (b) are direct computations. For (c), let $\mathbf{x} \times \mathbf{y} = \mathbf{0}$, and assume that one of the entries of $\mathbf{x}$ and $\mathbf{y}$ is nonzero (otherwise, we are done). Without loss of generalization, we assume that $x_1 \neq 0$. Then, reworking the equations one obtains from $\mathbf{x} \times \mathbf{y} = \mathbf{0}$, one sees that $\mathbf{y} = \frac{y_1}{x_1} \mathbf{x}$.

Exercise 6.7.13 Let

$$A = \begin{pmatrix} i & 1-i & 2-i \\ 1+i & -2 & -3+i \end{pmatrix}, B = \begin{pmatrix} -1 & 0 \\ -2 & 5 \\ 1 & 3 \end{pmatrix}.$$

Compute $A \otimes B$ and $B \otimes A$, and show that they are similar via a permutation matrix.

Answer:

$$A \otimes B = \begin{pmatrix} -i & 0 & -1+i & 0 & -2+i & 0 \\ -2i & 5i & -2+2i & 5-5i & -4+2i & 10-5i \\ i & 3i & 1-i & 3-3i & 2-i & 6-3i \\ -1-i & 0 & 2 & 0 & 3-i & 0 \\ -2-2i & 5+5i & 4 & -10 & 6-2i & -15+5i \\ 1+i & 3+3i & -2 & -6 & -3+i & -9+3i \end{pmatrix},$$

$$B \otimes A = \begin{pmatrix} -i & -1+i & -2+i & 0 & 0 & 0 \\ -1-i & 2 & 3-i & 0 & 0 & 0 \\ -2i & -2+2i & -4+2i & 5i & 5-5i & 10-5i \\ -2-2i & 4 & 6-2i & 5+5i & -10 & -15+5i \\ i & 1-i & 2-i & 3i & 3-3i & 6-3i \\ 1+i & -2 & -3+i & 3+3i & -6 & -9+3i \end{pmatrix}.$$

We have that $A \otimes B = P(B \otimes A)P^T$, where

$$P = \begin{pmatrix} 1 & 0 & 0 & 0 & 0 & 0 \\ 0 & 0 & 0 & 1 & 0 & 0 \\ 0 & 1 & 0 & 0 & 0 & 0 \\ 0 & 0 & 0 & 0 & 1 & 0 \\ 0 & 0 & 1 & 0 & 0 & 0 \\ 0 & 0 & 0 & 0 & 0 & 1 \end{pmatrix}.$$

Exercise 6.7.14 Let $A \in \mathbb{F}^{n \times n}$ and $B \in \mathbb{F}^{m \times m}$.

(a) Show that $\mathrm{tr}(A \otimes B) = (\mathrm{tr}\, A)(\mathrm{tr}\, B)$.

(b) Show that $\mathrm{rank}(A \otimes B) = (\mathrm{rank}\, A)(\mathrm{rank}\, B)$.

Answer: (a) The diagonal entries of $A \otimes B$ are $a_{ii}b_{jj}$, $i = 1, \ldots, n$, $j = 1, \ldots, m$. Adding them all up, we obtain

$$\text{tr}(A \otimes B) = \sum_{i=1}^{n} \sum_{j=1}^{m} a_{ii}b_{jj} = \sum_{i=1}^{n} [a_{ii}(\sum_{j=1}^{m} b_{jj})] = \sum_{i=1}^{n} [a_{ii}(\text{tr } B)] = (\text{tr } A)(\text{tr } B).$$

(b) From the Gaussian elimination algorithm we know that we can write $A = SET$ and $B = \hat{S}\hat{E}\hat{T}$, where S, T, $\hat{S}$ and $\hat{T}$ are invertible, and

$$E = \begin{pmatrix} I_k & 0 \\ 0 & 0 \end{pmatrix}, \hat{E} = \begin{pmatrix} I_l & 0 \\ 0 & 0 \end{pmatrix},$$

where $k = \text{rank} A$ and $l = \text{rank} B$. Then

$$A \otimes B = (S \otimes \hat{S})(E \otimes \hat{E})(T \otimes \hat{T}).$$

Notice that $E \otimes \hat{E}$ has rank kl (as there are exactly kl entries equal to 1 in different rows and in different columns, and all the other entries are 0). Since $(S \otimes \hat{S})$ and $(T \otimes \hat{T})$ are invertible, we get that

$$\text{rank}(A \otimes B) = \text{rank}(E \otimes \hat{E}) = kl = (\text{rank } A)(\text{rank } B).$$

Exercise 6.7.15 Given Schur triangularization decompositions for A and B, find a Schur triangularization decomposition for $A \otimes B$. Conclude that if $\lambda_1, \ldots, \lambda_n$ are the eigenvalues for A and $\mu_1, \ldots, \mu_m$ are the eigenvalues for B, then $\lambda_i \mu_j$, $i = 1, \ldots, n$, $j = 1, \ldots, m$, are the nm eigenvalues of $A \otimes B$.

Answer: If $A = UTU^*$ and $B = VSV^*$, with U, V unitary and T, S upper triangular, then

$$A \otimes B = (U \otimes V)(T \otimes S)(U \otimes V)^*. \tag{6.39}$$

We have that $U \otimes V$ is unitary, and it is easy to see that $T \otimes S$ is upper triangular with its diagonal entries equal to the products of diagonal entries of T and S. Thus $\lambda_i \mu_j$, $i = 1, \ldots, n$, $j = 1, \ldots, m$, are the nm diagonal entries (and thus eigenvalues) of $T \otimes S$, and (6.39) a Schur triangularization decomposition for $A \otimes B$.

Exercise 6.7.16 Given singular value decompositions for A and B, find a singular value decomposition for $A \otimes B$. Conclude that if $\sigma_1, \ldots, \sigma_k$ are the nonzero singular values for A and $\hat{\sigma}_1, \ldots, \hat{\sigma}_l$ are the nonzero singular values for B, then $\sigma_i \hat{\sigma}_j$, $i = 1, \ldots, k$, $j = 1, \ldots, l$, are the kl nonzero singular values of $A \otimes B$.

Answer: If $A = V\Sigma W^*$ and $B = \hat{V}\hat{\Sigma}\hat{W}^*$, are singular value decompositions, then

$$A \otimes B = (V \otimes \hat{V})(\Sigma \otimes \hat{\Sigma})(W \otimes \hat{W})^*. \tag{6.40}$$

We have that $V \otimes \hat{V}$ and $W \otimes \hat{W}$ are unitary, and $\Sigma \otimes \hat{\Sigma}$ is up to permutation of rows and columns of the form

$$\begin{pmatrix} R & 0 \\ 0 & 0 \end{pmatrix},$$

where R is a $kl \times kl$ diagonal matrix with diagonal entries $\sigma_i \hat{\sigma}_j$, $i = 1, \ldots, k$, $j = 1, \ldots, l$. Thus a singular value decomposition of $A \otimes B$ is given by

$$A \otimes B = [(V \otimes \hat{V})P^T]P(\Sigma \otimes \hat{\Sigma})P^T[P(W \otimes \hat{W})^*], \tag{6.41}$$

where the permutation matrix P is chosen to that $P(\Sigma \otimes \hat{\Sigma})P^T$ has the nonzero singular values $\sigma_i \hat{\sigma}_j$, $i = 1, \ldots, k$, $j = 1, \ldots, l$ in nonincreasing order in the entries $(1,1), (2,2), \ldots, (kl, kl)$ and zeros everywhere else.

Exercise 6.7.17 Show that $\det(I \otimes A + A \otimes I) = (-1)^n \det p_A(-A)$, where $A \in \mathbb{C}^{n \times n}$.

Answer: Let $A = UTU^*$ be a Schur triangularization decomposition, where the diagonal entries of T are $\lambda_1, \ldots, \lambda_n$. Then

$$I \otimes A + A \otimes I = (U \otimes U)(I \otimes T + T \otimes I)(U \otimes U)^*.$$

Notice that $I \otimes T + T \otimes I$ is upper triangular with diagonal entries $\lambda_i + \lambda_j$, $i, j = 1, \ldots, n$. Thus $\det(I \otimes A + A \otimes I) = \prod_{i,j=1}^n (\lambda_i + \lambda_j)$. On the other hand, $p_A(t) = \prod_{j=1}^n (t - \lambda_j)$, so $p_A(-A) = U p_A(-T) U^* = U(-T - \lambda_1 I) \cdots (-T - \lambda_n I) U^*$. This gives that

$$\det p_A(-A) = \det(-T - \lambda_1) \cdots \det(-T - \lambda_n) =$$

$$\prod_{j=1}^n [\prod_{i=1}^n (-\lambda_i - \lambda_j)] = (-1)^{n^2} \prod_{i,j=1}^n (\lambda_i + \lambda_j).$$

It remains to observe that $(-1)^{n^2} = (-1)^n$ since n^2 is even if and only if n is even.

Exercise 6.7.18 Show that if A is a matrix and f a function, so that $f(A)$ is well-defined, then $f(I_m \otimes A)$ is well-defined as well, and $f(I_m \otimes A) = I_m \otimes f(A)$.

Answer: Let $A = SJS^{-1}$ be a Jordan canonical decomposition of A. Then $I_m \otimes A = (I_m \otimes S)(I_m \otimes J)(I_m \otimes S)^{-1}$. Since $I_m \otimes J$ is a direct sum of m copies of J, we have that $I_m \otimes J$ gives the Jordan canonical form of $I_m \otimes A$. Thus $f(I_m \otimes A) = (I_m \otimes S)f(I_m \otimes J)(I_m \otimes S)^{-1}$. Moreover, as $I_m \otimes J$ is a direct sum of m copies of J, we obtain that $f(I_m \otimes J) = I_m \otimes f(J)$. Now

$$f(I_m \otimes A) = (I_m \otimes S)(I_m \otimes f(J))(I_m \otimes S)^{-1} = I_m \otimes (Sf(J)S^{-1}) = I_m \otimes f(A).$$

Exercise 6.7.19 For a diagonal matrix $A = \operatorname{diag}(\lambda_i)_{i=1}^n$, find matrix representations for $A \wedge A$ and $A \vee A$ using the canonical (lexicographically ordered) bases for $\mathbb{F}^n \wedge \mathbb{F}^n$ and $\mathbb{F}^n \vee \mathbb{F}^n$, respectively.

Answer: The diagonal elements of the diagonal matrix $A \wedge A$ are ordered as

$$\lambda_1 \lambda_2, \ldots, \lambda_1 \lambda_n, \lambda_2 \lambda_3, \ldots, \lambda_2 \lambda_n, \ldots, \lambda_{n-2} \lambda_{n-1}, \lambda_{n-2} \lambda_n, \lambda_{n-1} \lambda_n.$$

The diagonal elements of the diagonal matrix $A \vee A$ are ordered as

$$\lambda_1 \lambda_1, \ldots, \lambda_1 \lambda_n, \lambda_2 \lambda_2, \ldots, \lambda_2 \lambda_n, \ldots$$

$$\ldots, \lambda_{n-2} \lambda_{n-2}, \ldots, \lambda_{n-2} \lambda_n, \lambda_{n-1} \lambda_{n-1}, \lambda_{n-1} \lambda_n, \lambda_n \lambda_n.$$

Exercise 6.7.20 Show that $\langle \mathbf{v}_1 \wedge \cdots \wedge \mathbf{v}_k, \mathbf{w}_1 \wedge \cdots \wedge \mathbf{w}_k \rangle = k! \det(\langle \mathbf{v}_i, \mathbf{w}_j \rangle)_{i,j=1}^k$.

Answer: Applying the definition of the anti-symmetric wedge product and using the linearity of the inner product we have that

$$\langle \mathbf{v}_1 \wedge \cdots \wedge \mathbf{v}_k, \mathbf{w}_1 \wedge \cdots \wedge \mathbf{w}_k \rangle =$$

$$\sum_{\sigma \in S_k} \sum_{\tau \in S_k} (-1)^\sigma (-1)^\tau \langle \mathbf{v}_{\sigma(1)} \otimes \cdots \otimes \mathbf{v}_{\sigma(k)}, \mathbf{w}_{\tau(1)} \otimes \cdots \otimes \mathbf{w}_{\tau(k)} \rangle.$$

Using the definition of the inner product on the tensor space, we obtain that the above equals

$$\sum_{\sigma \in S_k} \sum_{\tau \in S_k} (-1)^\sigma (-1)^\tau \langle \mathbf{v}_{\sigma(1)}, \mathbf{w}_{\tau(1)} \rangle \cdots \langle \mathbf{v}_{\sigma(k)}, \mathbf{w}_{\tau(k)} \rangle.$$

Since $\prod_{i=1}^{k}\langle\mathbf{v}_{\sigma(i)},\mathbf{w}_{\tau(i)}\rangle = \prod_{i=1}^{k}\langle\mathbf{v}_{i},\mathbf{w}_{\tau\circ\sigma^{-1}(i)}\rangle$, we obtain that

$$\sum_{\sigma\in S_k}\sum_{\tau\in S_k}[(-1)^{\tau\circ\sigma^{-1}}\prod_{i=1}^{k}\langle\mathbf{v}_i,\mathbf{w}_{\tau\circ\sigma^{-1}(i)}\rangle] = \sum_{\sigma\in S_k}\det(\langle\mathbf{v}_i,\mathbf{w}_j\rangle)_{i,j=1}^{k} =$$

$$k!\det(\langle\mathbf{v}_i,\mathbf{w}_j\rangle)_{i,j=1}^{k}.$$

Exercise 6.7.21 Find an orthonormal basis for $\vee^2\mathbb{C}^3$.

Answer: $\{\frac{1}{2}\mathbf{e}_1\vee\mathbf{e}_1, \frac{1}{\sqrt{2}}\mathbf{e}_1\vee\mathbf{e}_2, \frac{1}{2}\mathbf{e}_2\vee\mathbf{e}_2, \frac{1}{\sqrt{2}}\mathbf{e}_1\vee\mathbf{e}_3, \frac{1}{\sqrt{2}}\mathbf{e}_2\vee\mathbf{e}_3, \frac{1}{2}\mathbf{e}_3\vee\mathbf{e}_3\}$.

Exercise 6.7.22 (a) Let $A = (a_{ij})_{i=1,j=1}^{2,m} \in \mathbb{F}^{2\times m}$ and $B = (b_{ij})_{i=1,j=1}^{m,2} \in \mathbb{F}^{m\times 2}$. Find the matrix representations for $A\wedge A$, $B\wedge B$ and $AB\wedge AB$ using the canonical (lexicographically ordered) bases for $\wedge^k\mathbb{F}^n$, $k=2$, $n=2,m,1$, respectively.

(b) Show that the equality $AB\wedge AB = (A\wedge A)(B\wedge B)$ implies that

$$(\sum_{j=1}^{m}a_{1j}b_{j1})(\sum_{j=1}^{m}a_{2j}b_{j2}) - (\sum_{j=1}^{m}a_{1j}b_{j2})(\sum_{j=1}^{m}a_{2j}b_{j1}) =$$

$$\sum_{1\le j<k\le m}(a_{1j}a_{2k}-a_{1k}a_{2j})(b_{1j}b_{2k}-b_{1k}b_{2j}). \tag{6.42}$$

(c) Let $M = \{1,\ldots,m\}$ and $P = \{1,\ldots,p\}$. For $A \in \mathbb{F}^{p\times m}$ and $B \in \mathbb{F}^{m\times p}$, show that

$$\det AB = \sum_{S\subseteq M, |S|=p}\det(A[P,S])\det(B[S,P]). \tag{6.43}$$

(Hint: Use that $(\wedge^p A)(\wedge^p B) = \wedge^p(AB) = \det AB$.)

Remark: Equation (6.43) is called the *Cauchy–Binet identity*. When $p=2$ it reduces to (6.42), which when $B = A^T$ (or $B = A^*$ when $\mathbb{F} = \mathbb{C}$) is called the *Lagrange identity*.

Answer: (a)

$$A\wedge A = (a_{11}a_{22}-a_{12}a_{21} \quad a_{11}a_{23}-a_{13}a_{21} \quad \cdots \quad a_{11}a_{2m}-a_{1m}a_{21} \quad a_{12}a_{23}-a_{13}a_{22} \quad \cdots$$

$$a_{12}a_{2m}-a_{1m}a_{22} \quad \cdots \quad \cdots \quad a_{1,m-1}a_{2m}-a_{2,m-1}a_{1m}).$$

$$B\wedge B = \begin{pmatrix} b_{11}b_{22}-b_{21}b_{12} \\ b_{11}b_{32}-b_{31}b_{12} \\ \vdots \\ b_{11}b_{m2}-b_{m1}b_{12} \\ b_{21}b_{32}-b_{31}b_{22} \\ b_{21}b_{m2}-b_{m1}b_{22} \\ \vdots \\ \vdots \\ b_{m-1,1}b_{m2}-b_{m-1,2}b_{m1} \end{pmatrix}.$$

As $AB = \begin{pmatrix} \sum_{j=1}^{m}a_{1j}b_{j1} & \sum_{j=1}^{m}a_{1j}b_{j2} \\ \sum_{j=1}^{m}a_{2j}b_{j1} & \sum_{j=1}^{m}a_{2j}b_{j2} \end{pmatrix}$, we get that $AB\wedge AB$ is the 1×1 matrix

$$AB\wedge AB = \left((\sum_{j=1}^{m}a_{1j}b_{j1})(\sum_{j=1}^{m}a_{2j}b_{j2}) - (\sum_{j=1}^{m}a_{1j}b_{j2})(\sum_{j=1}^{m}a_{2j}b_{j1})\right).$$

(b) This follows immediately from using (a) and multiplying $A \wedge A$ with $B \wedge B$.

(c) To show (6.43) one needs to use that $(\wedge^p A)(\wedge^p B) = \wedge^p(AB) = \det(AB)$, where in the last step we used that AB is of size $p \times p$. The $1 \times \binom{m}{p}$ matrix $\wedge^p A$ is given by

$$\wedge^p A = (A[P, S])_{S \subseteq M, |S| = p}.$$

Similarly, the $\binom{m}{p} \times 1$ matrix $\wedge^p B$ is given by

$$\wedge^p B = (B[S, P])_{S \subseteq M, |S| = p}.$$

Equation (6.43) now immediately follows from $(\wedge^p A)(\wedge^p B) = \wedge^p(AB)$.

Exercise 6.7.23 For $\mathbf{x}, \mathbf{y} \in \mathbb{R}^3$, let the cross product $\mathbf{x} \times \mathbf{y}$ be defined as in (6.17). Show, using (6.42) (with $B = A^T$), that

$$\|\mathbf{x} \times \mathbf{y}\|^2 = \|\mathbf{x}\|^2 \|\mathbf{y}\|^2 - (\langle \mathbf{x}, \mathbf{y} \rangle)^2. \tag{6.44}$$

Notice that this equality implies the Cauchy–Schwarz inequality.

Answer: Let $A = \begin{pmatrix} x_1 & x_2 & x_3 \\ y_1 & y_2 & y_3 \end{pmatrix} = B^T$. Then $AB = \begin{pmatrix} \|\mathbf{x}\|^2 & \langle \mathbf{x}, \mathbf{y} \rangle \\ \langle \mathbf{y}, \mathbf{x} \rangle & \|\mathbf{y}\|^2 \end{pmatrix}$, thus $\det AB = \|\mathbf{x}\|^2 \|\mathbf{y}\|^2 - (\langle \mathbf{x}, \mathbf{y} \rangle)^2$. Next $B \wedge B = \mathbf{x} \times \mathbf{y} = (A \wedge A)^T$. And thus $(A \wedge A)(B \wedge B) = \|\mathbf{x} \times \mathbf{y}\|^2$. As this equals $AB \wedge AB = \det AB$, we obtain (6.44). Since $\|\mathbf{x} \times \mathbf{y}\|^2 \geq 0$, equation (6.44) implies the Cauchy–Schwarz inequality for the Euclidean inner product on $\mathbb{R}^3$.

Chapter 7

Exercise 7.9.1 Let $p(n)$ be a polynomial in n of degree k, and let $\lambda \in \mathbb{C}$ be of modulus greater than one. Show that $\lim_{n \to \infty} \frac{p(n)}{\lambda^n} = 0$. (Hint: write $|\lambda| = 1 + \epsilon$, $\epsilon > 0$, and use the binomial formula to give that $|\lambda^n| = \sum_{j=0}^{n} \binom{n}{j} \epsilon^j$, which for n large enough can be bounded below by a polynomial of degree greater than k.)

Answer: Let $n > 3k$, then

$$\binom{n}{k-1} = \frac{n(n-1)\cdots(n-k)}{(k-1)(k-2)\cdots 1} \geq \frac{1}{(k-1)!}(\frac{2n}{3})^{k+1}.$$

Thus

$$|\frac{p(n)}{\lambda^n}| = \frac{|p_0 + \cdots + p_k n^k|}{\sum_{j=0}^{n} \binom{n}{j} \epsilon^j} \leq \frac{(k-1)! \sum_{j=0}^{k} |p_j|}{(2/3)^{k+1} \epsilon^{(k-1)}} \frac{n^k}{n^{k+1}} \to 0 \text{ as } n \to \infty.$$

Exercise 7.9.2 Let $A = (a_{ij})_{i,j=1}^{n} \in \mathbb{R}^{n \times n}$. Let A be *column-stochastic*, which means that $a_{ij} \geq 0$ for all $i, j = 1, \ldots, n$, and $\sum_{i=1}^{n} a_{ij} = 1$, $j = 1, \ldots, n$.

(i) Show that 1 is an eigenvalue of A.

Answer: If we let $\mathbf{e} = \begin{pmatrix} 1 & \cdots & 1 \end{pmatrix}$ be the row vector of all ones, then $\mathbf{e}A = \mathbf{e}$, and thus 1 is an eigenvalue of A (with left eigenvector $\mathbf{e}$).

(ii) Show that A^m is column-stochastic for all $m \in \mathbb{N}$. (Hint: use that $\mathbf{e}A = \mathbf{e}$.)

Answer: Clearly A^m has all nonnegative entries, and the equality $\mathbf{e}A^m = \mathbf{e}$ gives that the column sums of A^m are 1.

(iii) Show that for every $\mathbf{x}, \mathbf{y} \in \mathbb{R}^n$ we have that $|\mathbf{y}^T A^m \mathbf{x}| \leq (\sum_{j=1}^{n} |x_j|)(\sum_{j=1}^{n} |y_j|)$ for all $m \in \mathbb{N}$. In particular, the sequence $\{\mathbf{y}^T A^m \mathbf{x}\}_{m \in \mathbb{N}}$ is bounded.

Answer: As A^m is column-stochastic, we have that each entry $(A^m)_{ij}$ of A^m satisfies $0 \leq (A^m)_{ij} \leq 1$. Then $|\mathbf{y}^T A^m \mathbf{x}| = \sum_{i,j=1}^{n} |y_i (A^m)_{ij} x_j| \leq \sum_{i,j=1}^{n} |y_i||x_j| = (\sum_{i=1}^{n} |y_i|)(\sum_{j=1}^{n} |x_j|)$.

(iv) Show that A cannot have Jordan blocks at 1 of size greater than 1. (Hint: Use that when $k > 1$ some of the entries of $J_k(1)^m$ do not stay bounded as $m \to \infty$. With this observation, find a contradiction with the previous part.)

Answer: First notice that when $k > 1$ the $(1,2)$ entry of $J_k(1)^m$ equals m. Suppose now that the Jordan canonical decomposition $A = SJS^{-1}$ of A has $J_k(1)$ in the upper left corner of J for some $k > 1$. Put $\mathbf{x} = Se_2$ and $\mathbf{y} = (S^T)^{-1}e_1$. Then $\mathbf{y}^T A^m \mathbf{x} = m \to \infty$ as $m \to \infty$. This is in contradiction with the previous part.

(v) Show that if $\mathbf{x}A = \lambda\mathbf{x}$, for some $\mathbf{x} \neq \mathbf{0}$, then $|\lambda| \leq 1$.

Answer: If $\mathbf{x} = \begin{pmatrix} x_1 & \cdots & x_n \end{pmatrix}$, let k be so that $|x_k| = \max_{j=1,\ldots,n} |x_j|$. Note that $|x_k| > 0$. Then the kth component of $\mathbf{x}A$ satisfies

$$|\lambda x_k| = |(\mathbf{x}A)_k| = |\sum_{i=1}^{n} a_{ij}x_i| \leq \sum_{i=1}^{n} a_{ij}|x_k| = |x_k|,$$

and thus after dividing by $|x_k|$, we get $|\lambda| \leq 1$.

(vi) For a vector $\mathbf{v} = (v_i)_{i=1}^{n}$ we define $|\mathbf{v}| = (|v_i|)_{i=1}^{n}$. Show that if λ is an eigenvalue of A with $|\lambda| = 1$, and $\mathbf{x}A = \lambda\mathbf{x}$, then $\mathbf{y} := |\mathbf{x}|A - |\mathbf{x}|$ has all nonnegative entries.

Answer: We have

$$|x_j| = |\lambda x_j| = |(\mathbf{x}A)_j| = |\sum_{i=1}^{n} x_i a_{ij}| \le \sum_{i=1}^{n} |x_i| a_{ij} = (|\mathbf{x}|A)_j, \ j = 1, \dots, n.$$

For the remainder of this exercise, assume that A only has positive entries; thus $a_{ij} > 0$ for all $i, j = 1, \dots, n$.

(vii) Show that $\mathbf{y} = \mathbf{0}$. (Hint: Put $\mathbf{z} = |\mathbf{x}|A$, and show that $\mathbf{y} \ne \mathbf{0}$ implies that $\mathbf{z}A - \mathbf{z}$ has all positive entries. The latter can be shown to contradict $\sum_{i=1}^{n} a_{ij} = 1$, $j = 1, \dots, n$.)

Answer: Suppose that $\mathbf{y} \ne \mathbf{0}$. Then $\mathbf{y}A = |\mathbf{x}|A^2 - |\mathbf{x}|A$ has all positive entries (as at least one entry of $\mathbf{y}$ is positive and the others are nonnegative). Put $\mathbf{z} = |\mathbf{x}|A$. Then $\mathbf{z}A - \mathbf{z}$ has all positive entries. If we let $z_k = \max_{j=1,\dots,n} z_j$. then $(\mathbf{z}A)_k = \sum_{i=1}^{n} z_i a_{ik} \le z_k \sum_{i=1}^{n} a_{ik} = z_k$, which contradicts that $\mathbf{z}A - \mathbf{z}$ has all positive entries. Thus we must have $\mathbf{y} = \mathbf{0}$.

(viii) Show that if $\mathbf{x}A = \lambda\mathbf{x}$ with $|\lambda| = 1$, then $\mathbf{x}$ is a multiple of $\mathbf{e}$ and $\lambda = 1$. (Hint: first show that all entries of $\mathbf{x}$ have the same modulus.)

Answer: Let k be so that $|x_k| = \max_{j=1,\dots,n} |x_j|$. Suppose that $|x_r| < |x_k|$ for some $r = 1, \dots, n$. Then

$$|x_k| = |\lambda x_k| = (|\mathbf{x}A|)_k = |\sum_{i=1}^{n} x_i a_{ik}| \le \sum_{i=1}^{n} |x_i| a_{ik} < |x_k| \sum_{i=1}^{n} a_{ik} = |x_k|,$$

giving a contradiction. Thus $|x_k| = |x_j|$ for $j = 1, \dots, n$. Now

$$|x_k| = |\lambda x_k| = (|\mathbf{x}A|)_k = |\sum_{i=1}^{n} x_i a_{ik}| \le \sum_{i=1}^{n} |x_i| a_{ik} = |x_k| \sum_{i=1}^{n} a_{ik} = |x_k|,$$

implies that we have

$$|\sum_{i=1}^{n} x_i a_{ik}| = \sum_{i=1}^{n} |x_i| a_{ik}.$$

But then, using Corollary 5.1.21, we must have that $x_j = e^{i\theta}|x_j|$, $j = 1, \dots, n$, for some $\theta \in \mathbb{R}$. Thus it follows that $\mathbf{x} = e^{i\theta}|x_k|\mathbf{e}$. As $\mathbf{e}A = \mathbf{e}$, it follows that $\lambda = 1$.

(ix) Conclude that we can apply the power method. Starting with a vector $\mathbf{v}_0$ with positive entries, show that there is a vector $\mathbf{w}$ with positive entries so that $A\mathbf{w} = \mathbf{w}$. In addition, show that $\mathbf{w}$ is unique when we require in addition that $\mathbf{e}^T\mathbf{w} = 1$.

Answer: The previous parts show that $\lambda_1 = 1$ is the eigenvalue of A of largest modulus, and that $1 > \max_{j=2,\dots,n} |\lambda_j|$. The vectors $\{\mathbf{e}, \mathbf{e} + \mathbf{e}_1, \dots, \mathbf{e} + \mathbf{e}_{n-1}\}$ span $\mathbb{R}^n$, so at least one of the vectors does not lie in $\mathrm{Ker} \prod_{j=2}^{n}(A - \lambda_j)$. Choose such a vector as $\mathbf{v}_0$, and apply Theorem 7.2.1. All the vectors $\mathbf{v}_k$ have nonnegative entries, and thus so does $\mathbf{w}$. As $\mathbf{w} \ne \mathbf{0}$, we get that $A\mathbf{w}$ has all positive entries, and thus so does $\mathbf{w}$. Since $\dim \mathrm{Ker}(A - I) = 1$, the vector $\mathbf{w}$ is unique up to multiplying with a scalar. Thus if we require that $\mathbf{e}^T\mathbf{w} = 1$, we get that $\mathbf{w}$ is unique.

Exercise 7.9.3 Let $\| \cdot \|$ be a norm on $\mathbb{C}^{n \times n}$, and let $A \in \mathbb{C}^{n \times n}$. Show that

$$\rho(A) = \lim_{k \to \infty} \|A^k\|^{\frac{1}{k}}, \tag{7.45}$$

where $\rho(\cdot)$ is the spectral radius. (Hint: use that for any $\epsilon > 0$ the spectral radius of $\frac{1}{\rho(A)+\epsilon}A$ is less than one, and apply Corollary 7.2.4.)

Answer: As $\lim_{k\to\infty} C^{\frac{1}{k}} = 1$ for all $C > 0$, it follows from Theorem 5.1.25 that the limit in (7.45) is independent of the chosen norm. Let us choose $\| \cdot \| = \sigma_1(\cdot)$.

If λ is an eigenvalue and $\mathbf{x}$ a corresponding unit eigenvector, then

$$|\lambda|^k = \|\lambda^k \mathbf{x}\| = \|A^k \mathbf{x}\| \leq \max_{\|\mathbf{y}\|=1} \|A^k \mathbf{y}\| = \sigma_1(A^k),$$

and thus

$$|\lambda| \leq (\sigma_1(A^k))^{\frac{1}{k}}.$$

This also holds for the eigenvalue of maximal modulus, and thus

$$\rho(A) \leq (\sigma_1(A^k))^{\frac{1}{k}}. \tag{7.46}$$

Next, let $\epsilon > 0$. Then the spectral radius of $B = \frac{1}{\rho(A)+\epsilon} A$ is less than one. Thus, by Corollary 7.2.4, we have that $B^k \to 0$ as $k \to \infty$. In particular, there exists a K so that for $k > K$ we have that $\sigma_1(B^k) \leq 1$. Then $\sigma_1(A^k) \leq (\rho(A) + \epsilon)^k$, which gives that

$$(\sigma_1(A^k))^{\frac{1}{k}} \leq \rho(A) + \epsilon.$$

Together with (7.46), this now gives that $\lim_{k \to \infty} (\sigma_1(A))^{\frac{1}{k}} = \rho(A)$.

Exercise 7.9.4 Let $A = (a_{ij})_{i,j=1}^n, B = (b_{ij})_{i,j=1}^n \in \mathbb{C}^{n \times n}$ so that $|a_{ij}| \leq b_{ij}$ for $i, j = 1, \ldots, n$. Show that $\rho(A) \leq \rho(B)$. (Hint: use (7.45) with the *Frobenius norm* $\|M\| = \sqrt{\sum_{i,j=1}^n |m_{ij}|^2}$.)

Answer: If we denote $A^k = (a_{ij}^{(k)})_{i,j=1}^n, B^k = (b_{ij}^{(k)})_{i,j=1}^n$, then is is easy to check that $|a_{ij}^{(k)}| \leq b_{ij}^{(k)}$ for all i, j, k. Using the Frobenius norm this implies that $\|A^k\| \leq \|B^k\|$ for all $k \in \mathbb{N}$. But then

$$\rho(A) = \lim_{k \to \infty} \|A^k\|^{\frac{1}{k}} \leq \lim_{k \to \infty} \|B^k\|^{\frac{1}{k}} = \rho(B)$$

follows.

Exercise 7.9.5 Show that if $\{\mathbf{u}_1, \ldots, \mathbf{u}_m\}$ and $\{\mathbf{v}_1, \ldots, \mathbf{v}_m\}$ are orthonormal sets, then the *coherence* $\mu := \max_{i,j} |\langle \mathbf{u}_i, \mathbf{v}_j \rangle|$, satisfies $\frac{1}{\sqrt{m}} \leq \mu \leq 1$.

Answer: By Proposition 5.1.10 we have that $|\langle \mathbf{u}_i, \mathbf{v}_j \rangle| \leq \|\mathbf{u}_i\| \|\mathbf{v}_j\| = 1$. Thus $\mu \leq 1$ follows. Next, suppose that $\mu < \frac{1}{\sqrt{m}}$. As $\mathbf{v}_1 = \sum_{i=1}^m \langle \mathbf{u}_i, \mathbf{v}_1 \rangle \mathbf{u}_i$, we have $1 = \|\mathbf{v}_1\|^2 = \sum_{i=1}^m |\langle \mathbf{u}_i, \mathbf{v}_1 \rangle|^2 < \sum_{i=1}^m \frac{1}{m} = 1$, giving a contradiction. Thus we have $\mu \geq \frac{1}{\sqrt{m}}$.

Exercise 7.9.6 Show that if A has the property that every $2s$ columns are linearly independent, then the equation $A\mathbf{x} = \mathbf{b}$ can have at most one solution $\mathbf{x}$ with at most s nonzero entries.

Answer: Suppose that $A\mathbf{x}_1 = \mathbf{b} = A\mathbf{x}_2$, where both $\mathbf{x}_1$ and $\mathbf{x}_2$ have at most s nonzero entries. Then $A(\mathbf{x}_1 - \mathbf{x}_2) = \mathbf{0}$, and $\mathbf{x}_1 - \mathbf{x}_2$ has at most $2s$ nonzero entries. If $\mathbf{x}_1 - \mathbf{x}_2 \neq \mathbf{0}$ we obtain that the columns of A that hit a nonzero entry in $\mathbf{x}_1 - \mathbf{x}_2$ are linearly independent. This contradicts the assumption that every $2s$ columns in A are linearly independent. Thus $\mathbf{x}_1 = \mathbf{x}_2$.

Exercise 7.9.7 Let $A = (a_{ij})_{i,j=1}^n$. Show that for all permutation σ on $\{1, \ldots, , n\}$ we have $a_{1,\sigma(1)} a_{2,\sigma(2)} \cdots a_{n,\sigma(n)} = 0$ if and only if there exist r ($1 \leq r \leq n$) rows and $n + 1 - r$ columns in A so that the entries they have in common are all 0.

Answer: When there exist rows $j_1, \ldots j_r$, $1 \leq r \leq n-1$, and columns $k_1, \ldots, k_{n+r-1}$ in A so that the entries they have in common are all 0, then for all permutations σ we have that

$$\{\sigma(j_1), \ldots, \sigma(j_r)\} \cap \{k_1, \ldots, k_{n+r-1}\} \neq \emptyset.$$

If l lies in this intersection, we have that $a_{j_l, \sigma(j_l)} = 0$, and thus $\prod_{i=1}^{n} a_{i,\sigma(i)} = 0$. When $r \in \{1, n\}$, a full row or column is 0, and thus $\prod_{i=1}^{n} a_{i,\sigma(i)} = 0$ follows as well.

For the converse, we use induction on the size of the matrix n. When $n = 1$, the statement is trivial, so suppose that the result holds for matrices of size up to $n - 1$. Let now $A = (a_{ij})_{i,j=1}^{n}$ and suppose that $a_{1,\sigma(1)} a_{2,\sigma(2)} \cdots a_{n,\sigma(n)} = 0$ for all σ. If $A = 0$, we are done. Next, let A have a nonzero entry, say $a_{i_0, j_0} \neq 0$. Deleting the row and column of this nonzero entry, we must have that the resulting $(n-1) \times (n-1)$ submatrix has a zero in every of its generalized diagonals $\{(1, \tau(1)), \ldots, (n-1, \tau(n-1))\}$ with τ a permutation on $\{1, \ldots, n-1\}$. By the induction assumption, we can identify rows $j_1, \ldots, j_r \in \{1, \ldots, n\} \setminus \{i_0\}$ and columns $k_1, \ldots, k_{n-r} \in \{1, \ldots, n\} \setminus \{j_0\}$, so that the entries they have in common are all 0. By permuting rows and columns of A, we may assume $\{j_1, \ldots, j_r\} = \{1, \ldots, r\}$ and $\{k_1, \ldots, k_{n-r}\} = \{r+1, \ldots, n\}$. Thus we have that

$$A = \begin{pmatrix} A_{11} & 0 \\ A_{12} & A_{22} \end{pmatrix},$$

where A_{11} is $r \times r$ and A_{22} is $(n-r) \times (n-r)$. Due to the assumption on A, we must have that either A_{11} or A_{22} also has the property that each of its generalized diagonals has a zero element. By applying the induction assumption on A_{11} or A_{22}, we obtain that one of these matrices has (possibly after a permutation of rows and columns) an upper triangular zero block which includes a diagonal entry. But then A has an upper triangular zero block which includes a diagonal zero entry, and thus we obtain the desired s rows and $n - s + 1$ columns.

Exercise 7.9.8 We say that $A = (a_{ij})_{i,j=1}^{n} \in \mathbb{R}^{n \times n}$ is *row-stochastic* if A^T is columns-stochastic. We call A *doubly stochastic* if A is both column- and row-stochastic. The matrix $P = (p_{ij})_{i,j=1}^{n}$ is called a *permutation matrix* if every row and column of P has exactly one entry equal to 1 and all the others equal to zero.

(i) Show that a permutation matrix is doubly stochastic.

Answer: Every row and column has exactly one entry equal to 1 and all others equal to 0, so all row and column sums equal 1. In addition, all entries (being either 0 or 1) are nonnegative.

(ii) Show that if A is a doubly stochastic matrix, then there exists a permutation σ on $\{1, \ldots, n\}$, so that $a_{1,\sigma(1)} a_{2,\sigma(2)} \cdots a_{n,\sigma(n)} \neq 0$.

Answer: Suppose that $a_{1,\sigma(1)} a_{2,\sigma(2)} \cdots a_{n,\sigma(n)} = 0$ for all permutations σ. By Exercise 7.9.7 there exist permutation matrices P_1 and P_2 so that

$$P_1 A P_2 = \begin{pmatrix} B & C \\ 0 & D \end{pmatrix},$$

where the 0 has size $r \times (n+1-r)$. As B has $n+1-r$ columns, all its entries sum up to $n+1-r$. As $\begin{pmatrix} B & C \end{pmatrix}$ has $n-r$ rows, all its entries add up to $n-r$, leading to a contradiction as the entries of C are nonnegative.

(iii) Let σ be as in the previous part, and put $\alpha = \min_{j=1,\ldots,n} a_{j,\sigma(j)} (> 0)$, and let P_σ be the permutation matrix with a 1 in positions $(1, \sigma(1)), \ldots, (n, \sigma(n))$ and zeros elsewhere. Show that either A is a permutation matrix, or $\frac{1}{1-\alpha}(A - \alpha P_\sigma)$ is a doubly stochastic matrix with fewer nonzero entries than A.

Answer: If A is not a permutation matrix, then $\alpha < 1$. By the definition of α we have that $A - \alpha P_\sigma$ only has nonnegative entries. In addition, notice that each row and column sum of $A - \alpha P_\sigma$ is $1 - \alpha$. Thus $\frac{1}{1-\alpha}(A - \alpha P_\sigma)$ is doubly stochastic. Finally, the entry in A that corresponds to α is zero in $A - \alpha P_\sigma$, and all zero entries in A are still zero in $A - \alpha P_\sigma$. Thus $\frac{1}{1-\alpha}(A - \alpha P_\sigma)$ has fewer nonzero entries than A.

(iv) Prove

Theorem 7.9.9 *(Birkhoff) Let A be doubly stochastic. Then there exist a $k \in \mathbb{N}$, permutation matrices $P_1, \ldots, P_k$ and positive numbers $\alpha_1, \ldots, \alpha_k$ so that*

$$A = \alpha_1 P_1 + \cdots + \alpha_k P_k, \quad \sum_{j=1}^{k} \alpha_j = 1.$$

In other words, every doubly stochastic matrix is a convex combination of permutation matrices.

(Hint: Use induction on the number of nonzero entries of A.)

Answer: Since A is doubly stochastic, every column has a nonzero entry, thus A has at least n nonzero entries. If A has exactly n nonzero entries, then A is a permutation matrix, and we are done. Next, suppose as our induction hypothesis that Birkhoff's theorem holds when A has at most l nonzero entries, where $l \geq n$. Next, let A have $l + 1$ nonzero entries. Then by the previous part we can identify a permutation σ and an $0 < \alpha < 1$ so that $\hat{A} = \frac{1}{1-\alpha}(A - \alpha P_\sigma)$ is a doubly stochastic matrix so that $\hat{A}$ has at most l nonzero entries. By our induction assumption we have that $\hat{A} = \sum_{j=1}^{\hat{k}} \beta_j P_j$ with P_j permutation matrices and β_j nonnegative so that $\sum_{j=1}^{\hat{k}} \beta_j = 1$. But then

$$A = \alpha P_\sigma + (1 - \alpha)\hat{A} = \alpha P_\sigma + \sum_{j=1}^{\hat{k}} (1 - \alpha)\beta_j P_j$$

is of the desired form.

Exercise 7.9.10 Write the matrix $\begin{pmatrix} 1/6 & 1/2 & 1/3 \\ 7/12 & 0 & 5/12 \\ 1/4 & 1/2 & 1/4 \end{pmatrix}$ as a convex combination of permutation matrices.

Answer:

$$\frac{1}{6} \begin{pmatrix} 1 & 0 & 0 \\ 0 & 0 & 1 \\ 0 & 1 & 0 \end{pmatrix} + \frac{1}{4} \begin{pmatrix} 0 & 1 & 0 \\ 1 & 0 & 0 \\ 0 & 0 & 1 \end{pmatrix} + \frac{1}{4} \begin{pmatrix} 0 & 1 & 0 \\ 0 & 0 & 1 \\ 1 & 0 & 0 \end{pmatrix} + \frac{1}{3} \begin{pmatrix} 0 & 0 & 1 \\ 1 & 0 & 0 \\ 0 & 1 & 0 \end{pmatrix}.$$

Exercise 7.9.11 (a) Show that

$$\min \operatorname{rank} \begin{pmatrix} A & ? \\ B & C \end{pmatrix} = \operatorname{rank} \begin{pmatrix} A \\ B \end{pmatrix} + \operatorname{rank} \begin{pmatrix} B & C \end{pmatrix}.$$

(b) Show that the lower triangular partial matrix

$$\mathcal{A} = \begin{pmatrix} A_{11} & & ? \\ \vdots & \ddots & \\ A_{n1} & \cdots & A_{nn} \end{pmatrix}$$

has minimal $\min \operatorname{rank} \mathcal{A}$ equal to rank

$$\sum_{i=1}^{n} \operatorname{rank} \begin{pmatrix} A_{i1} & \cdots & A_{ii} \\ \vdots & & \vdots \\ A_{n1} & \cdots & A_{ni} \end{pmatrix} - \sum_{i=1}^{n-1} \operatorname{rank} \begin{pmatrix} A_{i+1,1} & \cdots & A_{i+1,i} \\ \vdots & & \vdots \\ A_{n1} & \cdots & A_{ni} \end{pmatrix}. \tag{7.47}$$

Answer: We prove (b), as it will imply (a). For a matrix M, we let $\operatorname{col}_i(M)$ denote the ith scalar column of the matrix M. For $p = 1, \ldots, n$ we let $J_p \subseteq \{1, \ldots, \mu_p\}$ be a smallest possible set such that the columns

$$\operatorname{col}_i \begin{pmatrix} A_{pp} \\ \vdots \\ A_{np} \end{pmatrix}, \ i \in J_p, \tag{7.48}$$

satisfy

$$\operatorname{Span} \left\{ \operatorname{col}_i \begin{pmatrix} A_{pp} \\ \vdots \\ A_{np} \end{pmatrix} : i \in J_p \right\} + \operatorname{Ran} \begin{pmatrix} A_{p1} & \cdots & A_{p,p-1} \\ \vdots & & \vdots \\ A_{n1} & \cdots & A_{n,p-1} \end{pmatrix}$$

$$= \operatorname{Ran} \begin{pmatrix} A_{p1} & \cdots & A_{pp} \\ \vdots & & \vdots \\ A_{n1} & \cdots & A_{np} \end{pmatrix}.$$

Note that the number of elements in J_p equals

$$\operatorname{rank} \begin{pmatrix} A_{p1} & \cdots & A_{pp} \\ \vdots & & \vdots \\ A_{n1} & \cdots & A_{np} \end{pmatrix} - \operatorname{rank} \begin{pmatrix} A_{p1} & \cdots & A_{p,p-1} \\ \vdots & & \vdots \\ A_{n1} & \cdots & A_{n,p-1} \end{pmatrix}.$$

Thus $\sum_{p=1}^{n} \operatorname{card} J_p$ equals the right-hand side of (7.47). It is clear that regardless of the choice for A_{ij}, $i < j$, the collection of columns

$$\operatorname{col}_i \begin{pmatrix} A_{1p} \\ \vdots \\ A_{np} \end{pmatrix}, \ i \in J_p, \ p = 1, \ldots, n, \tag{7.49}$$

will be linearly independent. This gives that the minimal rank is greater than or equal to the right-hand side of (7.47). On the other hand, when one has identified the columns (7.48) one can freely choose entries above these columns. Once such a choice is made, every other column of the matrix can be written as a linear combination of the columns (7.49), and thus a so constructed completion has rank equal to the right-hand side of (7.47). This yields (7.47).

Exercise 7.9.12 Show that all minimal rank completions of

$$\begin{pmatrix} ? & ? & ? \\ 1 & 0 & ? \\ 0 & 1 & 1 \end{pmatrix}$$

are

$$\begin{pmatrix} x_1 & x_2 & x_1 x_3 + x_2 \\ 1 & 0 & x_3 \\ 0 & 1 & 1 \end{pmatrix}.$$

Answer: Let $\begin{pmatrix} x_1 & x_2 & x_4 \\ 1 & 0 & x_3 \\ 0 & 1 & 1 \end{pmatrix}$ be a completion. As $\begin{pmatrix} 1 & 0 \\ 0 & 1 \end{pmatrix}$ is a submatrix, the ranks is at least 2. For the rank to equal 2, we need that the determinant is 0. This leads to $x_4 = x_1 x_3 + x_2$.

Exercise 7.9.13 Consider the partial matrix

$$A = \begin{pmatrix} 1 & ? & ? \\ ? & 1 & ? \\ -1 & ? & 1 \end{pmatrix}.$$

Show that there exists a completion of A that is a Toeplitz matrix of rank 1, but that such a completion cannot be chosen to be real.

Answer: Let $\begin{pmatrix} 1 & b & c \\ a & 1 & b \\ -1 & a & 1 \end{pmatrix}$ be a Toeplitz completion. For this to be of rank 1 we need all

2×2 submatrices to have determinant 0. Thus $0 = \det \begin{pmatrix} a & 1 \\ -1 & a \end{pmatrix} = a^2 + 1$, giving $a = \pm i$,

and thus $a \notin \mathbb{R}$. Next $0 = \det \begin{pmatrix} 1 & b \\ a & 1 \end{pmatrix} = 1 - ab$, thus $b = \frac{1}{a}$. Finally, $\det \begin{pmatrix} b & c \\ 1 & b \end{pmatrix} = b^2 - c$,

giving $c = b^2$. We find that $\begin{pmatrix} 1 & -i & -1 \\ i & 1 & -i \\ -1 & i & 1 \end{pmatrix}$ is a rank 1 Toeplitz completion.

Exercise 7.9.14 Consider the $n \times n$ tri-diagonal Toeplitz matrix

$$A_n = \begin{pmatrix} 2 & -1 & 0 & \cdots & 0 \\ -1 & 2 & -1 & \cdots & 0 \\ \vdots & \ddots & \ddots & \ddots & \vdots \\ 0 & \cdots & -1 & 2 & -1 \\ 0 & \cdots & 0 & -1 & 2 \end{pmatrix}.$$

Show that $\lambda_j = 2 - 2\cos(j\theta)$, $j = 1, \ldots, n$, where $\theta = \frac{\pi}{n+1}$, are the eigenvalues. In addition, an eigenvector associated with λ_j is

$$\mathbf{v}_j = \begin{pmatrix} \sin(j\theta) \\ \sin(2j\theta) \\ \vdots \\ \sin(nj\theta) \end{pmatrix}.$$

Answer: Let $k \in \{1, \ldots, n\}$, and compute $\sin(kj\theta - j\theta) + \sin(kj\theta + j\theta) =$
$$\sin(kj\theta)\cos(j\theta) - \cos(kj\theta)\sin(j\theta) + \sin(kj\theta)\cos(j\theta) + \cos(kj\theta)\sin(j\theta) =$$
$$2\sin(kj\theta)\cos(j\theta).$$
Thus
$$-\sin((k-1)j\theta) + 2\sin(kj\theta) - \sin((k+1)j\theta) = (2 - 2\cos(j\theta))\sin(kj\theta).$$
Using this, and the observation that for $k = 1$ we have $\sin((k-1)j\theta) = 0$, and for $k = n$ we have $\sin((k+1)j\theta) = 0$ (here is where the definition of θ is used), it follows that
$$A_n \mathbf{v}_j = (2 - 2\cos(j\theta))\mathbf{v}_j, j = 1, \ldots, n.$$

Exercise 7.9.15 Let $A = (a_{ij})_{i,j=1}^n \in \mathbb{C}^{n \times n}$ be given.

(a) Let $U = \begin{pmatrix} 1 & 0 \\ 0 & U_1 \end{pmatrix} \in \mathbb{C}^{n \times n}$, with $U_1 \in \mathbb{C}^{(n-1) \times (n-1)}$ a unitary matrix chosen so that

$$U_1 \begin{pmatrix} a_{21} \\ a_{31} \\ \vdots \\ a_{n1} \end{pmatrix} = \begin{pmatrix} \sigma \\ 0 \\ \vdots \\ 0 \end{pmatrix}, \quad \sigma = \sqrt{\sum_{j=2}^n |a_{j1}|^2}.$$

Show that UAU^* has the form

$$UAU^* = \begin{pmatrix} a_{11} & * & * & \cdots & * \\ \sigma & * & * & \cdots & * \\ 0 & * & * & \cdots & * \\ \vdots & \vdots & \vdots & & \vdots \\ 0 & * & * & \cdots & * \end{pmatrix} = \begin{pmatrix} a_{11} & * \\ \sigma \mathbf{e}_1 & A_1 \end{pmatrix}.$$

(b) Show that there exists a unitary V so that VAV^* is upper Hessenberg. (Hint: after part (a), find a unitary $U_2 = \begin{pmatrix} 1 & 0 \\ 0 & * \end{pmatrix}$ so that $U_2 A_1 U_2^*$ has the form $\begin{pmatrix} * & * \\ \sigma_2 \mathbf{e}_1 & A_2 \end{pmatrix}$, and observe that

$$\hat{A} = \begin{pmatrix} 1 & 0 \\ 0 & U_2 \end{pmatrix} \begin{pmatrix} 1 & 0 \\ 0 & U_1 \end{pmatrix} A \begin{pmatrix} 1 & 0 \\ 0 & U_1^* \end{pmatrix} \begin{pmatrix} 1 & 0 \\ 0 & U_2^* \end{pmatrix}$$

has now zeros in positions $(2,1),\ldots,(n,1),(3,2),\ldots,(n,2)$. Continue the process.)

Remark. If one puts a matrix in upper Hessenberg form before starting the QR algorithm, it (in general) speeds up the convergence of the QR algorithm, so this is standard practice when numerically finding eigenvalues.

Answer: (a) Writing $A = \begin{pmatrix} a_{11} & A_{12} \\ A_{21} & A_{22} \end{pmatrix}$, we have that

$$UAU^* = \begin{pmatrix} a_{11} & \sigma \mathbf{e}_1 \\ A_{21} U_1^* & U_1 A_{22} U_1^* \end{pmatrix},$$

and is thus of the required form.

(b) As U_2 has the special form, the first column of $\hat{A}$ coincides with the first column of UAU^*, and has therefore zeros in positions $(3,1),\ldots,(n,1)$. Next, the second column of $\hat{A}$ below the main diagonal corresponds to $\sigma_2 \mathbf{e}_1$. Thus $\hat{A}$ also has zeros in positions $(4,2),\ldots,(n,2)$. Continuing this way, one can find U_k, $k = 3,\ldots,n-2$, making new zeros in positions $(k+2,k),\ldots,(n,k)$, while keeping the previously obtained zeros. Letting V equal the product of the unitaries, we obtain the desired result.

Exercise 7.9.16 The *adjacency matrix* A_G of a graph $G = (V, E)$ is an $n \times n$ matrix, where $n = |V|$ is the number of vertices of the graph, and the entry (i, j) equals 1 when $\{i, j\}$ is an edge, and 0 otherwise. For instance, the graph in Figure 7.6 has adjacency matrix

$$\begin{pmatrix} 0 & 1 & 0 & 0 & 1 & 0 \\ 1 & 0 & 1 & 0 & 1 & 0 \\ 0 & 1 & 0 & 1 & 0 & 0 \\ 0 & 0 & 1 & 0 & 1 & 1 \\ 1 & 1 & 0 & 1 & 0 & 0 \\ 0 & 0 & 0 & 1 & 0 & 0 \end{pmatrix}.$$

The adjacency matrix is a symmetric real matrix. Some properties of graphs can be studied by studying associated matrices. In this exercise we show this for the so-called *chromatic number* $\chi(G)$ of a graph G. It is defined as follows. A *k-coloring* of a graph is a function $c : V \to \{1,\ldots,k\}$ so that $c(i) \neq c(j)$ whenever $\{i, j\} \in E$. Thus, there are k colors and adjacent vertices should not be given the same color. The smallest number k so that G has a k-coloring is defined to be the chromatic number $\chi(G)$ of the graph G.

(a) Find the chromatic number of the graph in Figure 7.6.

Answer: The answer is 3. Indeed, for the vertices 1, 2 and 5, which are all adjacent to one another, we need at least three colors. Giving then 3 and 4 the same color as 1, and 6 the same color as 2, yields a 3-coloring of the graph.

(b) The *degree* d_i of a vertex i is the number of vertices it is adjacent to. For instance, for the graph in Figure 7.6 we have that the degree of vertex 1 is 2, and the degree of vertex 6 is 1. Let $\mathbf{e} = \begin{pmatrix} 1 & \cdots & 1 \end{pmatrix}^T \in \mathbb{R}^n$. Show that $\mathbf{e}^T A_G \mathbf{e} = \sum_{i \in V} d_i$.

Answer: Notice that d_i is equal to the sum of the entries in the ith row of A_G. Next, $\mathbf{e}^T A_G \mathbf{e}$ is the sum of all the entries of A_G, which thus equals $\sum_{i \in V} d_i$.

(c) For a real number x let $\lfloor x \rfloor$ denote the largest integer $\leq x$. For instance, $\lfloor \pi \rfloor = 3$, $\lfloor -\pi \rfloor = -4$, $\lfloor 5 \rfloor = 5$. Let $\alpha = \lambda_{\max}(A_G)$ be the largest eigenvalue of the adjacency matrix of G. Show that G must have a vertex of degree at most $\lfloor \alpha \rfloor$. (Hint: use Exercise 5.7.21(b).)

Answer: If we take $\mathbf{y} = \frac{1}{\sqrt{n}}\mathbf{e}$, we get that by Exercise 5.7.21 and part (b) that

$$\alpha = \max_{\langle \mathbf{x}, \mathbf{x} \rangle} \mathbf{x}^T A \mathbf{x} \geq \mathbf{y}^T A \mathbf{y} = \frac{1}{n} \sum_{i \in V} d_i. \tag{7.50}$$

If every vertex i has the property that $d_i > \alpha$, then $\sum_{i \in V} d_i > n\alpha$, which contradicts (7.50). Thus, for some i we have $d_i \leq \alpha$. As d_i is an integer, this implies $d_i \leq \lfloor \alpha \rfloor$.

(d) Show that

$$\chi(G) \leq \lfloor \lambda_{\max}(A_G) \rfloor + 1, \tag{7.51}$$

which is a result due to Herbert S. Wilf. (Hint: use induction and Exercise 5.7.21(c).)

Answer: Denote $\alpha = \lambda_{\max}(A_G)$. We use induction. When the graph has one vertex, we have that $A_G = (0)$ and $\chi(G) = 1$ (there is only one vertex to color), and thus inequality (7.51) holds.

Let us assume that (7.51) holds for all graphs with at most $n - 1$ vertices, and let $G = (V, E)$ have n vertices. By part (c) there is a vertex i so that $d_i \leq \lfloor \alpha \rfloor$. Let us remove vertex i (and the edges with endpoint i) from the graph G, to give us a graph $\hat{G} = (\hat{V}, \hat{E})$. Notice that $A_{\hat{G}}$ is obtained from A_G by removing row and column i. By Exercise 5.7.21(c) we have that $\lambda_{\max}(A_{\hat{G}}) \leq \lambda_{\max}(A_G) = \alpha$. Using the induction assumption on $\hat{G}$ (which has $n - 1$ vertices), we obtain that

$$\chi(\hat{G}) \leq \lfloor \lambda_{\max}(A_{\hat{G}}) \rfloor + 1 \leq \lfloor \alpha \rfloor + 1.$$

Thus $\hat{G}$ has a $(\lfloor \alpha \rfloor + 1)$-coloring. As the vertex i in G has degree $\leq \lfloor \alpha \rfloor$, there is at least one color left for the vertex i, and thus we find that G also has a $(\lfloor \alpha \rfloor + 1)$-coloring.

Exercise 7.9.17 Let

$$\rho_\alpha = \frac{1}{7} \begin{pmatrix} \frac{2}{3} & 0 & 0 & 0 & \frac{2}{3} & 0 & 0 & 0 & \frac{2}{3} \\ 0 & \frac{\alpha}{3} & 0 & 0 & 0 & 0 & 0 & 0 & 0 \\ 0 & 0 & \frac{5-\alpha}{3} & 0 & 0 & 0 & 0 & 0 & 0 \\ 0 & 0 & 0 & \frac{5-\alpha}{3} & 0 & 0 & 0 & 0 & 0 \\ \frac{2}{3} & 0 & 0 & 0 & \frac{2}{3} & 0 & 0 & 0 & \frac{2}{3} \\ 0 & 0 & 0 & 0 & 0 & \frac{\alpha}{3} & 0 & 0 & 0 \\ 0 & 0 & 0 & 0 & 0 & 0 & \frac{\alpha}{3} & 0 & 0 \\ 0 & 0 & 0 & 0 & 0 & 0 & 0 & \frac{5-\alpha}{3} & 0 \\ \frac{2}{3} & 0 & 0 & 0 & \frac{2}{3} & 0 & 0 & 0 & \frac{2}{3} \end{pmatrix},$$

where $0 \leq \alpha \leq 5$. We want to investigate when ρ_α is 3×3 separable.

(a) Show that ρ_α passes the Peres test if and only if $1 \leq \alpha \leq 4$.

(b) Let

$$
Z = \begin{pmatrix}
1 & 0 & 0 & 0 & -1 & 0 & 0 & 0 & -1 \\
0 & 0 & 0 & 0 & 0 & 0 & 0 & 0 & 0 \\
0 & 0 & 2 & 0 & 0 & 0 & 0 & 0 & 0 \\
0 & 0 & 0 & 2 & 0 & 0 & 0 & 0 & 0 \\
-1 & 0 & 0 & 0 & 1 & 0 & 0 & 0 & -1 \\
0 & 0 & 0 & 0 & 0 & 0 & 0 & 0 & 0 \\
0 & 0 & 0 & 0 & 0 & 0 & 0 & 0 & 0 \\
0 & 0 & 0 & 0 & 0 & 0 & 0 & 2 & 0 \\
-1 & 0 & 0 & 0 & -1 & 0 & 0 & 0 & 1
\end{pmatrix}.
$$

Show that for $x, y \in \mathbb{C}^3$ we have that $(x \otimes y)^* Z(x \otimes y) \geq 0$.

(c) Show that $\mathrm{tr}(\rho_\alpha Z) = \frac{1}{7}(3 - \alpha)$, and conclude that ρ_α is not 3×3 separable for $3 < \alpha \leq 5$.

Answer: (a) Applying the Peres test, we need to check whether

$$
\rho_\alpha^\Gamma = \frac{1}{7}\begin{pmatrix}
\frac{2}{3} & 0 & 0 & 0 & 0 & 0 & 0 & 0 & 0 \\
0 & \frac{\alpha}{3} & 0 & \frac{2}{3} & 0 & 0 & 0 & 0 & 0 \\
0 & 0 & \frac{5-\alpha}{3} & 0 & 0 & 0 & \frac{2}{3} & 0 & 0 \\
0 & \frac{2}{3} & 0 & \frac{5-\alpha}{3} & 0 & 0 & 0 & 0 & 0 \\
0 & 0 & 0 & 0 & \frac{2}{3} & 0 & 0 & 0 & 0 \\
0 & 0 & 0 & 0 & 0 & \frac{\alpha}{3} & 0 & \frac{2}{3} & 0 \\
0 & 0 & \frac{2}{3} & 0 & 0 & 0 & \frac{\alpha}{3} & 0 & 0 \\
0 & 0 & 0 & 0 & 0 & \frac{2}{3} & 0 & \frac{5-\alpha}{3} & 0 \\
0 & 0 & 0 & 0 & 0 & 0 & 0 & 0 & \frac{2}{3}
\end{pmatrix}
$$

is positive semidefinite. This matrix is essentially a direct sum of diagonal elements and the 2×2 submatrix

$$
\begin{pmatrix}
\frac{\alpha}{3} & \frac{2}{3} \\
\frac{2}{3} & \frac{5-\alpha}{3}
\end{pmatrix}.
$$

Computing its determinant we obtain that $(4 - \alpha)(\alpha - 1) \geq 0$, which gives $1 \leq \alpha \leq 4$. It is easy to see that for these values ρ_α^Γ is indeed positive semidefinite.

(b) Note that $x \otimes y = \begin{pmatrix} x_1 y_1 & x_1 y_2 & x_1 y_3 & x_2 y_1 & x_2 y_2 & x_2 y_3 & x_3 y_1 & x_3 y_2 & x_3 y_3 \end{pmatrix}^T$. If we assume that $|y_2| \geq |y_1|$, we write $(x \otimes y)^* Z(x \otimes y)$ as

$$
|x_1 y_1 - x_2 y_2 + x_3 y_3|^2 + 2|x_1 \bar{y}_3 - x_3 \bar{y}_1|^2 + 2|x_2 y_1|^2 + 2|x_3|^2(|y_2|^2 - |y_1|^2),
$$

which is nonnegative. The case $|y_1| \geq |y_2|$ can be dealt with in a similar manner.

(c) It is straightforward to compute that $\mathrm{tr}(\rho_\alpha Z) = \frac{1}{7}(3 - \alpha)$, which is negative when $3 < \alpha \leq 5$. If ρ_α is separable, it can be written as $\sum_{i=1}^{k} A_i \otimes B_i$, with A_i and B_i positive semidefinite. As each positive semidefinite can be written as $\sum_{j=1}^{l} \mathbf{v}_j \mathbf{v}_j^*$, with $\mathbf{v}_j$ vectors, we can actually write the separable ρ_α as

$$
\rho_\alpha = \sum_{j=1}^{s} \mathbf{x}_j \mathbf{x}_j^* \otimes \mathbf{y}_j \mathbf{y}_j^*,
$$

where $\mathbf{x}_j, \mathbf{y}_j \in \mathbb{C}^3$, $j = 1, \ldots, s$. Observe now that (b) yields

$$
\mathrm{tr}((\mathbf{x}_j \mathbf{x}_j^* \otimes \mathbf{y}_j \mathbf{y}_j^*)Z) = (\mathbf{x}_j^* \otimes \mathbf{y}_j^*)Z(\mathbf{x}_j \otimes \mathbf{y}_j) \geq 0,
$$

which implies

$$
\mathrm{tr}(\rho_\alpha Z) = \mathrm{tr}(\sum_{j=1}^{s} \mathbf{x}_j \mathbf{x}_j^* \otimes \mathbf{y}_j \mathbf{y}_j^*)Z \geq 0.
$$

When $3 < \alpha \leq 5$, we have reached a contradiction, thus ρ_α is not separable for these values of α.

Bibliography

- S. Axler, *Linear algebra done right*. Second edition. Undergraduate Texts in Mathematics. Springer-Verlag, New York, 1997.

- R. Bhatia, *Matrix analysis*. Graduate Texts in Mathematics, 169. Springer-Verlag, New York, 1997.

- J. B. Carrell, *Fundamentals of linear algebra*, http://www.math.ubc.ca/ carrell/NB.pdf.

- K. Hoffman, R. Kunze, *Linear algebra*. Second edition Prentice-Hall, Inc., Englewood Cliffs, N.J. 1971.

- R. A. Horn, C. R. Johnson, *Matrix analysis*. Second edition. Cambridge University Press, Cambridge, 2013.

- R. A. Horn, C. R. Johnson, *Topics in matrix analysis*. Corrected reprint of the 1991 original. Cambridge University Press, Cambridge, 1994.

- S. H. Friedberg, A. J. Insel, L. E. Spence, *Linear algebra*. Second edition. Prentice Hall, Inc., Englewood Cliffs, NJ, 1989.

- P. Lancaster, M. Tismenetsky, *The theory of matrices*. Second edition. Computer Science and Applied Mathematics. Academic Press, Inc., Orlando, FL, 1985.

- P. D. Lax, *Linear algebra*. Pure and Applied Mathematics (New York). A Wiley-Interscience Publication. John Wiley & Sons, Inc., New York, 1997.

- D. C. Lay, *Linear algebra and its applications*, Third Edition. Addison-Wesley, 2003.

- M. Marcus, *Finite dimensional multilinear algebra. Part 1*. Pure and Applied Mathematics, Vol. 23. Marcel Dekker, Inc., New York, 1973.

- B. Noble, J. W. Daniel, *Applied linear algebra*. Second edition. Prentice-Hall, Inc., Englewood Cliffs, N.J., 1977.

- G. Strang, *Linear algebra and its applications*. Second edition. Academic Press [Harcourt Brace Jovanovich, Publishers], New York–London, 1980.

- S. Treil, *Linear algebra done wrong*,
  http://www.math.brown.edu/ treil/papers/LADW/LADW.html.

- F. Zhang, *Matrix theory: Basic results and techniques*, Second edition,
  Universitext. Springer, New York, 2011.

Index

kth root of a matrix, 141, 294

absolute value, 8
additive inverse, field, 4
additive inverse, vector space, 28
adjacency matrix, 243, 320
adjoint, 123
adjugate, 19
algebraic number, 26
algebraically closed, 78
analytic, 99
annihilator, 191, 306
anti-symmetric tensor product,
        179
argument, 8
associativity for scalar
        multiplication, vector
        space, 28
associativity of addition, field, 3
associativity of addition, vector
        space, 28
associativity of multiplication,
        field, 4

basis, 41
bijective, 60
bilinear, 166
Birkhoff theorem, 240, 317
bounded functional, 160

canonical isomorphism for tensor
        product, 173
Cartesian product, 147
Cauchy's integral formula, 99
Cauchy–Binet identity, 193, 311
Cayley–Hamilton Theorem, 70
characteristic polynomial, 69

chromatic number, 243, 320
closed, 152
closed cone, 234
closure of addition, field, 3
closure of addition, vector space,
        27
closure of multiplication, field, 4
closure of scalar multiplication,
        vector space, 28
cofactor, 19
coherence, 239, 315
coloring, 243, 320
column-stochastic, 238, 313
common divisor, 76
commutativity of addition, field, 4
commutativity of addition, vector
        space, 28
commutativity of multiplication,
        field, 4
commutator, 24, 254
commute, 85
companion matrix, 104
completely positive map, 236
completion, 218
complex conjugate, 8
complex plane, 8
compound matrix, 183
compressed sensing, 203
condition number, 144, 299
cone, 233
conjugate linear, 159
convex function, 189, 302
coordinates (of a vector), 46
coprime, 76
cross product, 167
cut, 225

degree (of a polynomial), 36
degree of a vertex, 243, 321
diagonalizable, 84
dimension, 41
direct sum, 33
distance matrix, 220
distributive law (first), vector
    space, 28
distributive law (second), vector
    space, 28
distributive law, field, 4
divides (polynomial), 75
double dual, 162
doubly stochastic, 240, 316
dual basis, 158
dual space, 157

edge, 225
eigenspace, 16
eigenspace, generalized, 71
eigenvalue, 16
eigenvector, 16
entangled, 233
equivalence class, 149
equivalent norms, 117
Euclidean inner product, 110
Euclidean norm, 116
evaluation map, 162

field, 3
Fourier matrix, 212
frequency, 210
Frobenius norm, 116, 239, 315
functional, 157

Galerkin method, 214
gradient, 163
Gram–Schmidt process, 120
graph, 225
greatest common divisor, 76

Hölder's inequality, 189, 302
Hermitian, 52, 129
Hermitian form, 109

identity mapping, 60

imaginary part, 7
induced operator norm, 163
inertia, 130
injective, 58
inner product, 110
inner product space, 110
invariant subspace, 80
isometrically isomorphic, 122
isometry, 122, 125
isomorphic, 60
isomorphism, 46, 60

Jordan block, 71
Jordan canonical form, 78

kernel, 58
Krylov spaces, 201

Lagrange interpolation
    polynomial, 306
Lagrange's identity, 193, 311
least squares solution, 145
Leibniz rule, 92
lexicographical order, 173
linear dependence relation, 38
linear map, 55
linear transformation, 55
linearly dependent, 38
linearly independent, 37
Lyapunov equation, 146

maximum cut, 226
minimal polynomial, 82
minimal rank, 218
minimal rank completion, 218
Minkowski's inequality, 189, 302
modulus, 8
monic, 70
multilinear, 166
multilinear functional, 166
multiplicative inverse, field, 4

neutral element for addition, field,
    4
neutral element for addition,
    vector space, 28

neutral element for multiplication,
　　field, 4
nilpotent, 71
nonderogatory, 84
nontrivial solution, 38
norm, 113
normal, 128
numerical radius, 138, 289

one-to-one, 58
onto, 57
orthogonal, 119
orthogonal basis, 121
orthonormal, 119
orthonormal basis, 121

partial matrix, 218
partial transpose, 235
pattern, 218
permanent, 187
permutation matrix, 240, 316
polynomial time algorithm, 226
positive definite, 129
positive map, 235
positive semidefinite, 129
Power method, 198
pre-inner product, 155

QR factorization, 125
qubit, 232
quotient space, 150

range, 57
real part, 7
Redheffer matrix, 224
regression line, 145
representative, 149
resolvent, 98
restriction, 80
Riemann hypothesis, 223
Riesz representation theorem, 158
row-stochastic, 240, 316

Schur complement, 142, 296
Schur triangularization, 127
secret sharing, 213

self-adjoint, 123
separable, 233
similar, 65
similarity orbit, 190, 304
simple tensor, 169
singular values, 134
skew-Hermitian, 129
Span, 39
sparse matrix, 198
spectral radius, 200
spectrum, 98
square root of a matrix, 141, 294
standard basis, 45
standard basis for $\mathbb{F}^n$, 44
standard inner product, 110
Stein equation, 146
subfield, 6
subspace, 32
sum of functions, 29
surjective, 57
Sylvester's Law of Inertia, 130
symmetric tensor product, 184

trace, 24, 254
transcendental, 26
triangle inequality, 113
trivial solution, 38

unbounded linear map, 165
unit multiplication rule, vector
　　space, 28
unitary, 125, 129
upper Hessenberg, 206

Vandermonde matrix, 211
vector space, 27

Weyr characteristic, 71

Young's inequality, 189, 302